Springer Series in Computational Neuroscience

Volume 8

For further volumes:
http://www.springer.com/series/8164

Alain Destexhe • Michelle Rudolph-Lilith

Neuronal Noise

Springer

Dr. Alain Destexhe
CNRS, UPR-2191
Unité de Neuroscience,
Information et Complexité
av. de la Terrasse 1
91198 Gif-sur-Yvette
Bat. 32-33
France
destexhe@iaf.cnrs-gif.fr

Dr. Michelle Rudolph-Lilith
CNRS, UPR-2191
Unité de Neuroscience,
Information et Complexité
av. de la Terrasse 1
91198 Gif-sur-Yvette
Bat. 32-33
France
rudolph@iaf.cnrs-gif.fr

ISBN 978-0-387-79019-0 e-ISBN 978-0-387-79020-6
DOI 10.1007/978-0-387-79020-6
Springer New York Dordrecht Heidelberg London

Library of Congress Control Number: 2011944121

Printed on acid-free paper

Springer is part of Springer Science+Business Media (www.springer.com)

To Our Families

Foreword

Any student of the human condition or of animal behavior will be struck by its unpredictability, its randomness. While crowds of people, schools of fish, flocks of geese or clouds of gnats behave in predictable ways, the actions of individuals are often highly idiosyncratic. The roots of this unpredictability is apparent at every level of the brain, the organ responsible for behavior. Whether patch-clamping individual ionic channels, recording the electrical potential from inside or from outside neurons or measuring the local field potential (LFP) via EEG electrodes on the skull, one is struck by the ceaseless commotion, the feverish traces that move up and down irregularly, riding on top of slower fluctuations. The former are consequences of various noise sources. By far the biggest is synaptic noise.

Single synapses are unreliable—the release of a puff of neurotransmitter molecules can occur as infrequently as one out of every ten times that an action potential (AP) invades the presynaptic terminal. This should be contrasted with the reliability of transistors in integrated silicon circuits, whose switching probability is very, very close to unity. Is this a bug—an unavoidable consequence of packing roughly one billion synapses into one cubic millimeter of cortical tissue—and/or a feature—something that is exploited for functional purposes?

This high variability, together with the nature of the post-synaptic process—causing an increase in the electrical membrane conductance—leads to what the authors call "high-conductance states" during active processing in an intact animal (rather than in brain slices, whose closest equivalent in the living animal is something approaching coma), in which the total conductance of the neuron is dominated by the large number of excitatory and inhibitory inputs impinging on the neuron. The two authors here bring impressive mathematical and computational tools to bear on this problem—including the development of an empirical study of the biophysics of single cortical neurons in slices and in anesthetized, sleeping or awake cats from their laboratory over the past decade. They develop a number of novel experimental (active electrode compensation) and numerical methods to analyze their recordings. This leads to a wealth of insights into the working of the cerebral cortex, this universal scalable computational medium responsible—in mammals—for most higher level perceptual and cognitive processing. In one of their memorable phrases "The measurement of synaptic noise . . . allows one to see the network activity through the intracellular measurement of a single cell".

Most important is the realization that cortical networks are primarily driven by stochastic, internal fluctuations of inhibition rather than by excitatory, feed-forward input—spikes are preceded by a momentary reduction in inhibition rather than by an increase in excitation. As the neuroscience community begins to confront the awe-inspiring complexity of cortex with its hundred or more distinct cell types, the detailed modeling of their interactions will be key to any understanding. Central to such modeling—and to this book—are the dynamics of the membrane potential in individual nerve cells, tightly embedded into a vast sheet of bustling companion cells. For it is out of this ever flickering activity that perception, memory, and that most elusive and prized of all phenomenon, consciousness, arise.

Christof Koch

Lois and Victor Troendle Professor of Cognitive
& Behavioral Biology
California Institute of Technology
Pasadena, CA

Chief Scientific Officer
Allen Institute for Brain Science
Seattle, WA

Preface

We are living today in very exciting times for neuroscience, times in which a growing number of theoretical and experimental researchers are working hand-in-hand. This interplay of theory and experiments now has become a primary necessity for achieving a deeper understanding of the inner workings of nature in general. Neuroscience, as a relatively young field in science, is no exception to this, but, in fact, must be cited as one of the foremost scientific arenas in which theory and experiments are intimately associated. Here, "Neuronal Noise," both the title and thematic goal of the pages ahead, constitutes a subject where this interplay of theoretical and experimental approaches has been, and still is, one of the most particular, spectacular, fruitful and, yet, demanding enterprises.

As the Reader will discover, for most of the topics explored in this book, experiment and theory complement each other, and for many of them this combination cannot be dissociated. This is exemplified by the *dynamic-clamp* experiments, where models and living neurons interact with each other in real time. This approach is very powerful to reveal the underlying biophysical principles which govern the most obvious building blocks of our nervous system. Another example is the *Active Electrode Compensation* (AEC) technique, in which real-time recording of neurons is achievable with unprecedented accuracy through the use of a mathematical model.

In the past decades, together with our theoreticians and experimentalist colleagues and friends, we have collectively realized an impressive amount of work and achieved tremendous progress in understanding the effect of "noise" on neurons. The picture which has started to emerge from this work shows that this "noise" does not at all live up to its original conception as an "unwanted" artifact accompanying so many other natural systems, but that this "noise" emerges as an integral, even "required" part of the internal dynamics of biological neuronal systems. Its effects, those which will be foremost covered in this book, are best understood in single cells. However, what today's picture can only vaguely reveal is how deep the rabbit hole goes. With modern experimental techniques, together with the advances in computational capabilities and theoretical understanding, we begin to see that neuronal "noise" is present at almost every level of the nervous system. Along with its beneficial effects, noise seems to be an integral, natural part of the computational principles which give rise to the unparalleled complexity and power of natural neuronal systems.

The work which will be reviewed in this book saw its beginning many years ago. The results and emerging picture are starting to become clear enough to be assembled into a coherent framework, comprising both theory and experiments. For this reason, we felt that it is a good time to write a monograph on this subject. Another answer to the "why" of writing a book on "Neuronal Noise" is to acknowledge the work of exceptional quality by the many researchers, postdocs, and students who have been involved in this research.

Importantly, the field of "Neuronal Noise" is extremely vast, reaching from the microscopic aspects of noise at the level of ion channels, the noise seen at the synaptic arborization on the level of single neurons, and the irregular activity states characterizing the dynamics at the network levels, to the probabilistic aspects of neuronal computations that derive from these properties. In this book, we cannot and, intentionally, will not cover all of these aspects, but instead will focus on the largest noise source in neurons, *synaptic noise*, and its consequences on neuronal integrative properties and neuronal computations. Although we tried to provide a comprehensive overview of a field which is still under active investigation today, the contents of this book necessarily represents only a selection and, thus, must be considered as a mere snapshot of the present state of knowledge of the subject, as envisioned by the Authors. Many relevant areas might not be discussed or cited, or done so in a manner which does not pay due to their relevance and importance, a shortcoming for which we sincerely apologize.

Gif-sur-Yvette

Alain Destexhe
Michelle Rudolph-Lilith

Acknowledgements

This book would not have been possible without the work and intellectual contribution of a vast number of colleagues and friends. We would like to first thank all of our experimentalist colleagues with whom we collaborated and whose work is reported here: Mathilde Badoual, Thierry Bal, Diego Contreras, Damien Debay, Jean-Marc Fellous, Julien Fournier, Yann Le Franc, Yves Frégnac, Hélène Gaudreau, Andrea Hasenstaub, Eric Lang, Gwendael Le Masson, Manuel Levy, Olivier Marre, David McCormick, Cyril Monier, Denis Paré, Joe-Guillaume Pelletier, Zuzanna Piwkowska, Noah Roy, Mavi Sanchez-Vives, Eric Shink, Yusheng Shu, Julia Sliwa, Mircea Steriade, Alex Thomson, Igor Timofeev, and Jacob Wolfart.

We are also deeply grateful to our colleagues, theoreticians, and postdocs for insightful discussions and exciting collaborations: Fabian Alvarez, Claude Bédard, Romain Brette, Andrew Davison, Jose Gomez-Gonzalez, Helmut Kröger and Terrence Sejnowski.

Finally, we would like to thank our students and postdocs, who made an immmense contribution to the material presented here: Sami El Boustani, Nicolas Hô, Martin Pospischil, and Quan Zou. We also thank Marco Brigham, Nima Dehghani, Lyle Muller, Sebastien Behuret, Pierre Yger, and all members of the UNIC for many stimulating discussions and continuous support.

Last but not least, we also acknowledge the support from several institutions and grants, *Centre National de la Recherche Scientifique* (CNRS), *Agence Nationale de la Recherche* (ANR, grant HR-Cortex), *National Institutes of Health* (NIH, grant R01-NS37711), *Medical Research Council of Canada* (MRC grant MT-13724), *Human Frontier Science Program* (HFSP grant RGP 25–2002), and the European Community *Future and Emerging Technologies* program (FET grants FACETS FP6-015879 and BrainScaleS FP7-269921).

Contents

Chapter 1
Introduction

One of the characteristics of brain activity is that its function is always associated with considerable amounts of "noise." Noise is present at all levels, from the gating of single ion channels up to large-scale brain activity as seen, for instance, in electroencephalogram (EEG) signals. In this book, we will explore the presence of noise and its physiological role. Across the different chapters, we will see that noise is not a signal representing a nuisance, but that it can have many advantageous consequences for the computations performed by single neurons and, perhaps, also by neuronal networks. By using experiments, theory and computer simulations, we will explore the possibility that "noise" might indeed be an integral component of brain computations.

1.1 The Different Sources of Noise in the Brain

The central nervous system is subject to many different forms of noise, which have fascinated researchers since the beginning of electrophysiological recordings. At microscopic scales, a considerable amount of noise is present due to molecular agitation and collisions. This *thermal noise* (Johnson-Nyquist noise; Johnson 1927; Nyquist 1928) has many consequences for the function of neuronal ion channels. An ion channel is a macromolecule which is embedded into a constantly fluctuating medium, either the membrane with phospholipids or the intracellular and extracellular solutions. The thermal fluctuations present in these media, and the resulting numerous collisions with the ion channel, will trigger spontaneous conformation changes, some of which can open or close the channel. The ion channel will therefore appear to open and close in a stochastic manner, a phenomenon called *channel noise*. These types of noise have been theoretically and experimentally studied in great detail and were covered in several excellent reviews (e.g., Verveen and De Felice 1974; De Felice 1981; Manwani and Koch 1999a; White et al. 2000).

A. Destexhe and M. Rudolph-Lilith, *Neuronal Noise*, Springer Series in Computational Neuroscience 8, DOI 10.1007/978-0-387-79020-6_1,

There are a number of other sources of noise in neuronal membranes which are directly or indirectly linked to thermal and channel noise, but studied separately. The flux of ions through open ion gates gives rise to *1/f noise* (also *flicker noise* or *excess noise*), as seen in the power spectral density (PSD) of the membrane potential (Derksen 1965; Verveen and Derksen 1968; Poussart 1971; Siebenga and Verveen 1972; Fishman 1973; Lundstrom and McQueen 1974; Clay and Shlesinger 1977; Neumcke 1978). This migration of ions through open and leak channels and pores is the source of another type of noise called *shot noise* (Rice 1944, 1945; Frehland and Faulhaber 1980; Frehland 1982). Its origin can be found in the quantum mechanical properties of ions and, more generally, the particles involved, specifically the discreteness of their electric charge. In contrast to the thermal (Johnson–Nyquist) noise mentioned above, the appearance of shot noise is inevitably linked to nonequilibrium states of the system in question and plays, therefore, a role in transitional and dynamical states.

Finally, noise due to the carrier-mediated ion transport in ionic pumps across bilayer membranes (Kolb and Frehland 1980), *burst noise* (also *popcorn noise* or *bi-stable noise*) caused by sudden step-like erratic transitions in systems with two or more discrete voltage or current levels, and *avalanche noise* contribute to the noise observed in cellular membranes as well. However, although all of the above mentioned and identified noise sources might significantly shape the biophysical dynamics of nerve cells along with their function, they will not be covered in this book.

In central neurons, and cerebral cortex in particular, the by far largest-amplitude noise source is *synaptic noise*, which was found to be dominant in intracellular recordings in vivo. Synaptic noise describes the continuous and noisy "bombardment" of central neurons by irregular synaptic inputs. In particular, the cerebral cortex in vivo is characterized by sustained and irregular neuronal activity, which combined with the very high cortical interconnectivity, is responsible for a considerable and noisy synaptic activity in any given cortical neuron, which crucially shapes its intrinsic dynamical properties and responses. The origin of this synaptic noise, its experimental characterization, its theoretical description and its effects on the neuronal dynamics do constitute the main focus of this book.

1.2 The Highly Irregular Nature of Neuronal Activity

One of the most striking characteristics of awake and attentive states is the highly complex nature of cortical activity. Global measurements, such as the EEG or local field potentials (LFPs), display low-amplitude and very irregular activity, so-called *desynchronized EEG* (Steriade 2003). This activity exhibits very low spatiotemporal coherence between multiple sites in cortex, which contrasts with the widespread synchronization in slow-wave sleep (SWS) (Destexhe et al. 1999). Local measurements, such as extracellular (unit activity) or intracellular recordings of single neurons, also demonstrate very irregular spike discharge and high levels of fluctuations similar to noise (Steriade et al. 2001), as shown in Fig. 1.1. Multiple unit activity (Fig. 1.1a) reveals that the firing is irregular and of low correlation between different cells, while intracellular recordings (Fig. 1.1b) show that the membrane potential is dominated by intense fluctuations. As alluded above, this *synaptic noise* is the dominant source of noise in neurons.

1.3 Integrative Properties in the Presence of Noise

Since the classical view of passive dendritic integration, proposed for motoneurones some 50 years ago (Fatt 1957), the introduction of new experimental techniques, such as intradendritic recordings (Llinás and Nicholson 1971; Wong et al. 1979) and visually guided patch-clamp recording (Stuart et al. 1993; Yuste and Tank 1996), have revolutionized this area. These new approaches revealed that the dendrites of pyramidal neurons are actively involved in the integration of excitatory postsynaptic potentials (EPSPs) and that the activation of only a few synapses has powerful effects at the soma in brain slices (Mason et al. 1991; Markram et al. 1997; Thomson and Deuchars 1997). Although remarkably precise data has been obtained in slices, little is known about the integrative properties of the same neurons in vivo.

The synaptic connectivity of the neocortex is very dense. Each pyramidal cell receives 5,000–60,000 synapses (Cragg 1967; DeFelipe and Fariñas 1992), 90% of which originate from other cortical neurons (Szentagothai 1965; Gruner et al. 1974; Binzegger et al. 2004). Given that neocortical neurons spontaneously fire at 5–20 Hz in awake animals (Hubel 1959; Evarts 1964; Steriade 1978), cortical cells must experience tremendous synaptic currents.

Following the first observations of "neuronal noise" (Fatt and Katz 1950; Brock et al. 1952; Verveen and Derksen 1968), it was realized that neurons constantly operate in noisy conditions. How neurons integrate synaptic inputs in such noisy conditions is a problem which was investigated in early work on motoneurons (Barrett and Crill 1974; Barrett 1975), which was followed by studies in Aplysia (Bryant and Segundo 1976) and cerebral cortex (Holmes and Woody 1989). This early work motivated further studies using compartmental models in cortex (Bernander et al. 1991) and cerebellum (Rapp et al. 1992; De Schutter and Bower 1994). These studies pointed out that the integrative properties of neurons can be drastically different in such noisy states. However, at that time, no precise experimental measurements were available to characterize the noise sources in neurons.

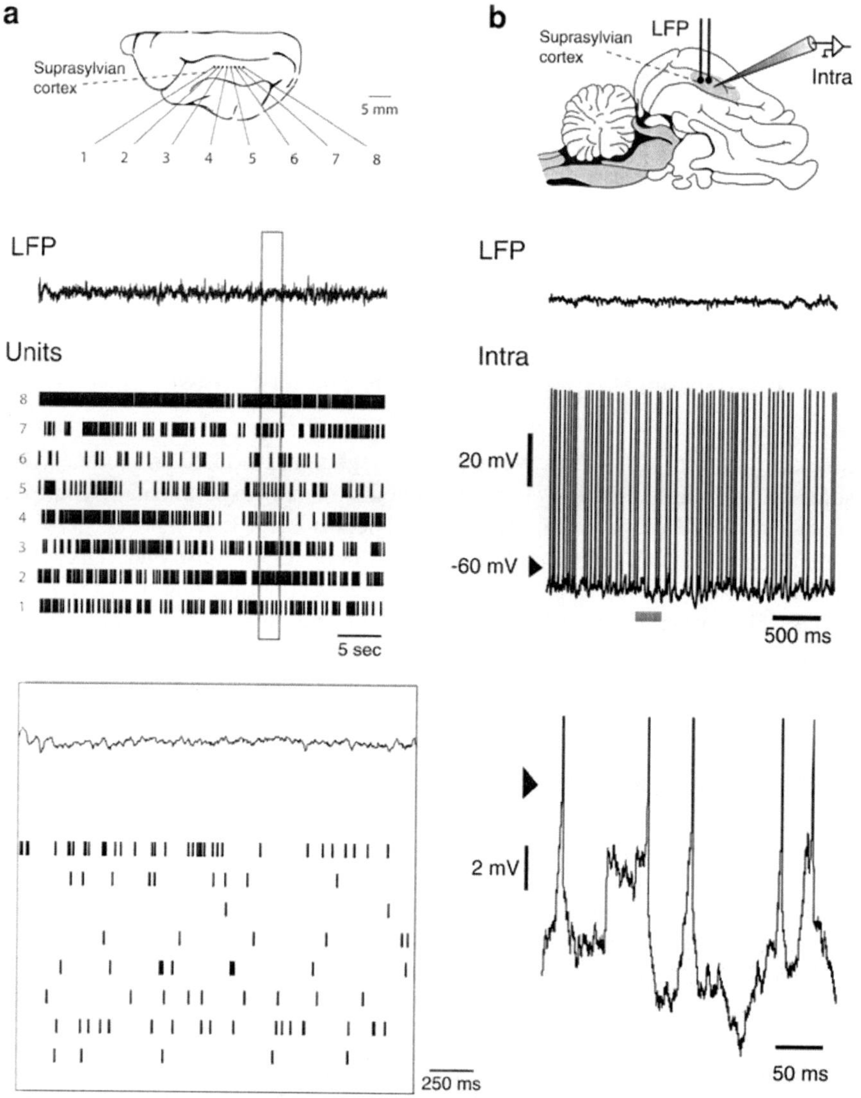

Fig. 1.1 Highly complex and "noisy" cortical activity during wakefulness. (**a**) Irregular firing activity of 8 multiunits shown at the same time as the LFP recorded in electrode 1 (scheme on top). During wakefulness, the LFP is of low amplitude and irregular activity ("desynchronized") and unit activity is sustained and irregular (see magnification below; 20 times higher temporal resolution). (**b**) Intracellular activity in the same brain region during wakefulness. Spiking activity was sustained and irregular, while the membrane potential displayed intense fluctuations around a relatively depolarized state (around −65 mV in this cell; see magnification below). Panel A modified from Destexhe et al. 1999; Panel B modified from Steriade et al. 2001

1.4 Structure of the Book

This is the point where the present book starts. After having reviewed basic concepts of neuronal biophysics (Chap. 2), we will review the measurement of irregular activity in cortex, both globally and at the level of synaptic noise in single neurons (Chap. 3). The chapter will emphasize the first quantitative measurement of synaptic noise in neurons.

Chapter 4 will be devoted to the development of computational models of synaptic noise. Models will be considered at two levels, both detailed biophysical models of neurons with complex dendritic morphologies, and highly simplified models. These models are based on the quantitative measurements and, thus, can simulate intracellular activity with unprecedented realism.

In Chap. 5, these models are used to investigate the important question of the consequences of synaptic noise on neurons. How neurons integrate their inputs in noisy states and, more generally, how neurons represent and process information in such noisy states will be reviewed here. It will be shown that many properties conferred by noise are beneficial to the cell's responsiveness, which will be compared to the well-known phenomenon of *stochastic resonance* studied by physicists.

Chapter 6 will cover a technique in which models are put in direct interaction with living neurons. This *dynamic-clamp* technique allows to inject computer-generated synaptic noise in neurons, and precisely control all of its parameters. Thus, this type of experiment is a very powerful tool to study the effect of noise on neurons. Not only can it be used to test the predictions of models concerning the consequences of noise but it can also be used to uncover qualitatively new consequences.

The formalization of synaptic noise in more mathematical terms is the subject of Chap. 7. This chapter will set the theoretical basis for Chap. 8, which is devoted to a new family of stochastic methods for analyzing synaptic noise. These methods were tested and their validity assessed using dynamic-clamp experiments.

Chapter 9 summarizes the concepts presented in this book into a few case studies, where both traditional and stochastic methods are applied to intracellular recordings of cortical neurons in vivo, in concert with computational models. The chapter illustrates that in a few cases, such as for example in awake animals, the amount of synaptic noise is dominant and sets the neuron into a radically different mode of input integration.

The concluding Chap. 10 will provide a final overview over the concepts presented in this book and transpose them at the network level. The goal is to understand what are the computations performed in stochastic network states. One of the main ideas supported by the results shown in this book is that the interplay of noise and nonlinear intrinsic properties do confer to networks particularly powerful computational capabilities, whose details, however, are still to be discovered.

Chapter 2
Basics

This chapter briefly covers the basic electrophysiological concepts that are used in the remainder of the book. Here, we overview models used to capture essential neuronal properties, such as ion channel dynamics and membrane excitability, the modeling of synaptic interactions, as well as the cable formalism that describes the spread of electrical activity along the dendritic tree of neurons.

2.1 Ion Channels and Membrane Excitability

2.1.1 Ion Channels and Passive Properties

Neuronal membranes at rest maintain an electric potential called the *resting membrane potential*, which is due to the selective permeability of several ionic species, in particular Na^+, K^+ and Cl^-. The active maintenance (via electrogenic pumps) of different ionic concentrations inside and outside of the cellular membrane, coupled with *ion channels* which have a selective permeability to one or several ions, establish a net difference of electric charges between the intracellular medium (globally charged negatively) and the exterior of the cell (globally charged positively). This charge difference is responsible for the membrane potential. Excellent reviews have been published where the details of this process have been exposed (e.g., Hille 2001).

Each type of ion u associated with ion channels is characterized by two important electric parameters. First, the *equilibrium potential* E_u, which represents the potential to which the membrane would asymptotically converge if this ion was the only one involved in the ionic permeability. The equilibrium potential can be calculated as a function of the ion concentrations using well-known relations, such as the Nernst equation. Secondly, the *conductance* g_u, which is the inverse of the resistance, and quantifies the permeability of the membrane to this ion in electrical terms. The presence of several ions can be represented using an equivalent

A. Destexhe and M. Rudolph-Lilith, *Neuronal Noise*, Springer Series in Computational Neuroscience 8, DOI 10.1007/978-0-387-79020-6_2,

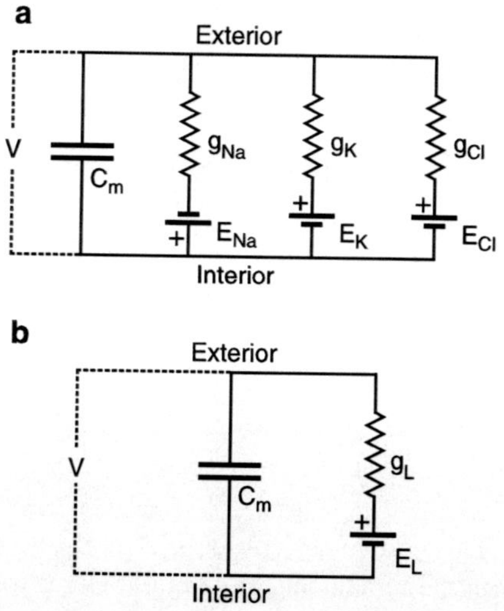

Fig. 2.1 Equivalent circuit of the membrane. (**a**) Equivalent circuit for a membrane with specific permeability for three types of ions: Na^+, K^+ and Cl^-. (**b**) Equivalent circuit where the three ionic conductances are merged into a single leak current. See text for description of symbols

electrical circuit composed of resistances (conductances of each ionic species) and electromotive forces (batteries). By using Ohm's law, the current flowing through the ion channel can be written as

$$I_u = g_u(V - E_u)\,, \tag{2.1}$$

where V denotes the membrane potential.

Besides ionic currents, the membrane itself (without ion channels) has an extremely low permeability to ions, so it acts like a capacitor characterized by a value of the *membrane capacitance C*. Together with ionic currents, these elements form the *equivalent electrical circuit of the membrane*, which is represented in Fig. 2.1. This circuit forms the basis of computational models of neuronal membranes and their passive properties. It can predict the value of the membrane potential, and the behavior of the membrane in response to stimuli such as current pulses or conductance changes. It can also predict the basic properties of membrane excitability if appropriate mechanisms are added, as we show in the next section.

2.1.2 *Membrane Excitability and Voltage-Dependent Ion Channels*

Not only do neuronal membranes maintain a stable membrane potential but they are also equipped with the machinery to modify this membrane potential through changes in the permeability of ions. In particular, ion channels can have their permeability dependent on the membrane potential; these are called *voltage-dependent ion channels*. Voltage-dependent channels can provide powerful mechanisms for nonlinear membrane behavior, since the membrane potential also depends on ionic conductances. This constitutes the electrophysiological basis for many properties of neurons, as exemplified by membrane excitability and the action potential (AP). In the first half of the twentieth century, a large number of physiologists have designed specific experiments to understand the mechanisms underlying APs. Among them, Alan Hodgkin and Andrew Huxley have shown that APs of the squid axon are generated by voltage-dependent Na^+ and K^+ conductances, and also provided a model to explain their results (Hodgkin et al. 1952; Hodgkin and Huxley 1952a–d).

Hodgkin and Huxley used a technique called *voltage clamp*, consisting of a specific recording configuration in which the membrane potential is forced to remain (clamped) at a fixed value, while a specific electrical circuit neutralizes the current produced by the membrane. This current is the inverse image of the current generated by ion channels, and, thus, this technique can be used to record ion currents at specific values of the voltage. If a given voltage-dependent conductance can be isolated (e.g., by chemically blocking all other conductances), then performing voltage-clamp recordings at different values of the clamped voltage allows the experimentalist to directly quantify the voltage dependence of this conductance. This type of protocol was realized by Hodgkin and Huxley separately for the Na^+ and K^+ currents, allowing them to characterize the kinetic properties and the voltage dependence of these currents (Hodgkin et al. 1952; Hodgkin and Huxley 1952a–c).

A mathematical model was necessary to establish that the identified kinetic properties of voltage dependence were sufficient to explain the genesis of APs. The model introduced by Hodgkin and Huxley (1952d) incorporated the results of their voltage-clamp experiments and successfully accounted for the main properties of APs, which represents a very convincing evidence that their postulated mechanism is plausible.

2.1.3 The Hodgkin–Huxley Model

The Hodgkin–Huxley model is based on a membrane equation describing three ionic currents in an isopotential compartment:

$$C_{\mathrm{m}} \frac{\mathrm{d}V}{\mathrm{d}t} = -g_{\mathrm{L}}(V - E_{\mathrm{L}}) - g_{\mathrm{Na}}(V)(V - E_{\mathrm{Na}}) - g_{\mathrm{K}}(V)(V - E_{\mathrm{K}}), \qquad (2.2)$$

where C_{m} is the membrane capacitance, V is the membrane potential, g_{L}, g_{Na} and g_{K} are the membrane conductances for leak currents, Na^+ and K^+ currents, respectively, and E_{L}, E_{Na} and E_{K} are their respective (equilibrium) reversal potentials.

The critical step in the Hodgkin–Huxley model is to specify how the conductances $g_{\mathrm{Na}}(V)$ and $g_{\mathrm{K}}(V)$ depend on the membrane potential V. Hodgkin and Huxley hypothesized that ionic currents result from the assembly of several independent *gating particles* which must occupy a given position in the membrane to allow the flow of Na^+ or K^+ ions (Hodgkin and Huxley 1952d). Each gating particle can be on either side of the membrane and bears a net electronic charge such that the membrane potential can switch its position from the inside to the outside or vice-versa. The transition from these two states is therefore voltage dependent, according to the diagram:

$$\text{(outside)} \underset{\beta_m(V)}{\overset{\alpha_m(V)}{\rightleftharpoons}} \text{(inside)}, \tag{2.3}$$

where α and β are, respectively, the forward and backward rate constants for the transitions from the outside to the inside position in the membrane. If m is defined as the fraction of particles in the inside position, and $(1-m)$ as the fraction outside the membrane, one obtains the first-order kinetic equation

$$\frac{\mathrm{d}m}{\mathrm{d}t} = \alpha_m(V)(1-m) - \beta_m(V)m. \tag{2.4}$$

Assuming that particles must occupy the inside position to conduct ions, then the conductance must be proportional to some function of m. In the case of the squid giant axon, Hodgkin and Huxley (1952a–c) found that the nonlinear behavior of the Na^+ and K^+ currents, their delayed activation and their sigmoidal rising phase was best fit by assuming that the conductance is proportional to the product of several of such variables (Hodgkin and Huxley 1952d)

$$\begin{aligned} g_{\mathrm{Na}} &= \bar{g}_{\mathrm{Na}} m^3 h \\ g_{\mathrm{K}} &= \bar{g}_{\mathrm{K}} n^4, \end{aligned} \tag{2.5}$$

where $\bar{g}_{\mathrm{Na}}$ and $\bar{g}_{\mathrm{K}}$ are the maximal values of the conductances, while m, h, n represent the fraction of three different types of gating particles in the inside of the cellular membrane. This equation allowed them to fit voltage-clamp data of the currents accurately. The interpretation is that the assembly of three gating particles of type m and one of type h is required for Na^+ ions to flow through the membrane, while the assembly of four gating particles of type n is necessary for the flow of K^+ ions. These particles operate independently of each other, leading to the m^3h and n^4 terms in the equation above.

Long after the work of Hodgkin and Huxley, it was established that ionic currents are mediated by the opening and closing of ion channels, the gating particles were reinterpreted as *gates* inside the pore of the channel. Thus, the reinterpretation of Hodgkin and Huxley's hypothesis is that the pore of the channel is controlled by four internal gates, that these gates operate independently of each other, and that all four gates must be open to enable the channel to conduct ions.

The rate constants $\alpha(V)$ and $\beta(V)$ of m and n are such that depolarization promotes opening the gate, a process which is called *activation*. On the other hand, the rate constants of h are such that depolarization promotes closing of the gate (and therefore closing of the entire channel because all gates must be open for the channel to conduct ions), which process is called *inactivation*. Thus, the experiments of Hodgkin and Huxley established that three identical activation gates (m^3) and a single inactivation gate (h) are sufficient to explain the Na^+ current's characteristics. The K^+ current does not have inactivation and can be well described by four identical activation gates (n^4).

Taking together all the steps above, one can write the following set of differential equations, called *Hodgkin–Huxley equations* (Hodgkin and Huxley 1952d):

$$\begin{aligned}
C_\mathrm{m}\frac{\mathrm{d}V}{\mathrm{d}t} &= -g_\mathrm{L}(V-E_\mathrm{L}) - \bar{g}_\mathrm{Na}m^3h(V-E_\mathrm{Na}) - \bar{g}_\mathrm{K}n^4(V-E_\mathrm{K}) \\
\frac{\mathrm{d}m}{\mathrm{d}t} &= \alpha_m(V)(1-m) - \beta_m(V)m \\
\frac{\mathrm{d}h}{\mathrm{d}t} &= \alpha_h(V)(1-h) - \beta_h(V)h \\
\frac{\mathrm{d}n}{\mathrm{d}t} &= \alpha_n(V)(1-n) - \beta_n(V)n\,.
\end{aligned} \tag{2.6}$$

The rate constants (α_i and β_i) were estimated by fitting empirical functions of voltage to the experimental data (Hodgkin and Huxley 1952d). These functions are:

$$\begin{aligned}
\alpha_m &= \frac{-0.1(V-V_r-25)}{\exp\left[-\frac{1}{4}(V-V_r-25)\right]-1} \\
\beta_m &= 4\exp\left[-\frac{1}{18}(V-V_r)\right] \\
\alpha_h &= 0.07\exp\left[-\frac{1}{20}(V-V_r)\right] \\
\beta_h &= \frac{1}{\exp\left[-\frac{1}{10}(V-V_r+30)\right]+1} \\
\alpha_n &= \frac{-0.01(V-V_r+10)}{\exp\left[-\frac{1}{10}(V-V_r+10)\right]-1} \\
\beta_n &= 0.125\exp\left[-\frac{1}{80}(V-V_r)\right].
\end{aligned} \tag{2.7}$$

These functions were estimated at a temperature of 6°C and the voltage axis was reversed in polarity, with voltage values given with respect to the resting membrane potential V_r.

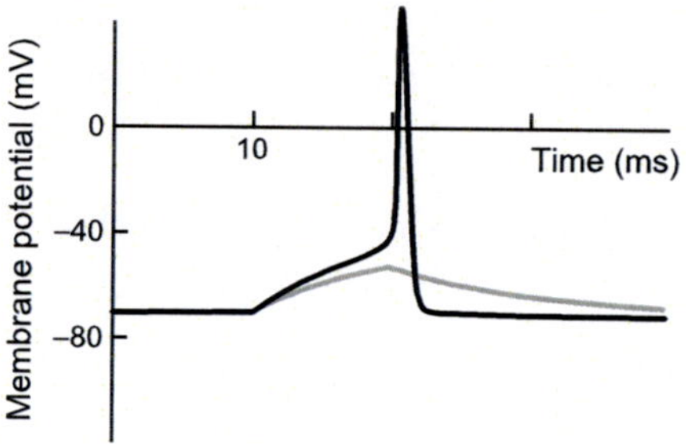

Fig. 2.2 Hodgkin–Huxley model and membrane excitability. Model neuron with Na^+ and K^+ described by a Hodgkin–Huxley type model, and submitted to a 10 ms depolarizing current pulse (amplitude of 0.8 nA and 1.1 nA). Below 1 nA, the current pulse was subthreshold (*gray*), but over 1 nA amplitude, the pulse evoked an action potential (*black*). For longer current pulses, this model can generate repetitive firing

The Hodgkin–Huxley model (2.4) is often written in an equivalent but more convenient form in order to fit experimental data:

$$\frac{\mathrm{d}m}{\mathrm{d}t} = \frac{1}{\tau_m(V)}(m_\infty(V) - m), \tag{2.8}$$

where

$$m_\infty(V) = \frac{\alpha(V)}{[\alpha(V) + \beta(V)]}$$
$$\tau_m(V) = \frac{1}{[\alpha(V) + \beta(V)]}. \tag{2.9}$$

Here, m_∞ is the *steady-state activation* and τ_m is the *activation time constant* of the Na^+ current (n_∞ and τ_n represent the same quantities for the K^+ current). In the case of h, h_∞ and τ_h are called *steady-state inactivation* and *inactivation time constant*, respectively. These quantities are important because they can easily be determined from voltage-clamp experiments (see below).

The properties of the Hodgkin–Huxley model were analyzed in detail, but the most prominent is the genesis of membrane excitability as illustrated in Fig. 2.2.

2.1.4 Markov Models

The formalism introduced by Hodgkin and Huxley (1952d) was remarkably forward looking, and closely reproduced the behavior of macroscopic currents. However, Hodgkin–Huxley models are not exact and, in fact, rest on several crucial approximations, while some of its features are inconsistent with experiments.

Measurements on Na^+ channels, for instance, have shown that activation and inactivation must necessarily be coupled (Armstrong 1981; Aldrich et al. 1983; Bezanilla 1985), which is in contrast with the independence of these processes in the Hodgkin–Huxley model. Na^+ channels may also show an inactivation which is not voltage dependent, as in the Hodgkin–Huxley model, but state-dependent (Aldrich 1981). Although the latter can be modeled with a modified Hodgkin–Huxley kinetics (Marom and Abbott 1994), these phenomena are best described using *Markov models*, a formalism more appropriate to describe single channels.

Markov models represent the gating of a channel as occurring through a series of conformational changes of the ion channel protein. They assume that the transition probability between conformational states depends only on the present state. The sequence of conformations involved in this process can be described by state diagrams of the form

$$S_1 \rightleftharpoons S_2 \rightleftharpoons \dots \rightleftharpoons S_n\,, \tag{2.10}$$

where S_1 ... S_n represent distinct conformational states of the ion channel. Defining $P(S_i,t)$ as the probability of being in a state S_i at time t and $P(S_i \to S_j)$ as the *transition probability* from state S_i to state S_j $(j = 1,...,n)$, according to

$$S_i \underset{P(S_j \to S_i)}{\overset{P(S_i \to S_j)}{\rightleftharpoons}} S_j\,, \tag{2.11}$$

then the following equation for the time evolution of $P(S_i,t)$ can be written down:

$$\frac{\mathrm{d}P(S_i,t)}{\mathrm{d}t} = \sum_{j=1}^{n} P(S_j,t)P(S_j \to S_i) - \sum_{j=1}^{n} P(S_i,t)P(S_i \to S_j)\,. \tag{2.12}$$

This equation is called the *master equation* (see, e.g., Stevens 1978; Colquhoun and Hawkes 1981). The left-hand term represents the "source" contribution of all transitions entering state S_i, while the right-hand term represents the "sink" contribution of all transitions leaving state S_i. In this equation, the time evolution depends only on the present state of the system, and is defined entirely by knowledge of the set of transition probabilities (Markovian system). In the limit of large numbers of identical channels, the quantities given in the master equation can be replaced by their macroscopic interpretation. The probability of being in a state S_i becomes the *fraction of channels* in state S_i, noted s_i, and the transition probabilities from state S_i to state S_j become the *rate constants* r_{ij} of the reactions

$$S_i \underset{r_{ji}}{\overset{r_{ij}}{\rightleftharpoons}} S_j\,. \tag{2.13}$$

In this case, one can rewrite the master equation as

$$\frac{\mathrm{d}s_i}{\mathrm{d}t} = \sum_{j=1}^{n} s_j r_{ji} - \sum_{j=1}^{n} s_i r_{ij}\,, \tag{2.14}$$

which is a conventional kinetic equation for the various states of the system. Here the rate constants can also be voltage dependent.

Stochastic Markov models (as in (2.12)) are adequate to describe the stochastic behavior of ion channels as recorded using single-channel recording techniques (see Sakmann and Neher 1995). In other cases, where a larger area of membrane is recorded and large numbers of ion channels are involved, the macroscopic currents are nearly continuous and more adequately described by conventional kinetic equations, as in (2.14) (see Johnston and Wu 1995). In the following, only systems of the latter type will be considered.

Finally, it must be noted that Markov models are more general than the Hodgkin–Huxley formalism, and include it as a subclass. Any Hodgkin–Huxley model can be written as a Markov scheme, while the opposite is not true. For example, the Markov model corresponding to the Hodgkin–Huxley sodium channel is (Fitzhugh 1965):

$$\begin{array}{ccccccc}
C_3 & \underset{\beta_m}{\overset{3\alpha_m}{\rightleftharpoons}} & C_2 & \underset{2\beta_m}{\overset{2\alpha_m}{\rightleftharpoons}} & C_1 & \underset{3\beta_m}{\overset{\alpha_m}{\rightleftharpoons}} & O \\
\alpha_h \downarrow\uparrow \beta_h & & \alpha_h \downarrow\uparrow \beta_h & & \alpha_h \downarrow\uparrow \beta_h & & \alpha_h \downarrow\uparrow \beta_h \\
I_3 & \underset{\beta_m}{\overset{3\alpha_m}{\rightleftharpoons}} & I_2 & \underset{2\beta_m}{\overset{2\alpha_m}{\rightleftharpoons}} & I_1 & \underset{3\beta_m}{\overset{\alpha_m}{\rightleftharpoons}} & I
\end{array}\;.$$

Here, the different states represent the channel with the inactivation gate in the open state (top) or closed state (bottom) and (from left to right) three, two, one or none of the activation gates closed. To be equivalent to the m^3 formulation, the rates must have the 3:2:1 ratio in the forward direction and the 1:2:3 ratio in the backward direction. Only the O state is conducting.

The squid delayed rectifier potassium current modeled by Hodgkin and Huxley (1952d) with four activation gates can be treated analogously (Fitzhugh 1965; Armstrong 1969), giving

$$C_4 \underset{\beta_m}{\overset{4\alpha_m}{\rightleftharpoons}} C_3 \underset{2\beta_m}{\overset{3\alpha_m}{\rightleftharpoons}} C_2 \underset{3\beta_m}{\overset{2\alpha_m}{\rightleftharpoons}} C_1 \underset{4\beta_m}{\overset{\alpha_m}{\rightleftharpoons}} O\,.$$

2.2 Models of Synaptic Interactions

Synaptic interactions are essential to neural network models at all levels of complexity, as well as for the representation of synaptic activity at the levels of single neurons. Synaptic currents are mediated by ion channels activated by neurotransmitter released from presynaptic terminals. Here, *kinetic models* are a powerful formalism for the description of channel behavior (as seen in the previous section for Markov models), and are also well suited to the description of synaptic interactions. Although a full representation of the molecular details of the synapse generally requires highly complex kinetic models, we focus here on simpler versions which are very efficient to compute. These models capture the time courses of several types of synaptic responses as well as the important phenomena of summation, saturation and desensitization.

For spiking neurons, a popular model of postsynaptic currents (PSCs) is the *alpha function*

$$r(t-t_0) = \frac{(t-t_0)}{\tau} \exp\left[-\frac{t-t_0}{\tau}\right], \tag{2.15}$$

where $r(t)$ resembles the time course of experimentally recorded postsynaptic potentials (PSPs) with a time constant τ (Rall 1967). The alpha function, and its double-exponential generalization, can be used to approximate most synaptic currents with a small number of parameters and, if implemented properly, at low computational and storage requirements (Srinivasan and Chiel 1993). Other types of template function were also proposed for spiking neurons (Traub and Miles 1991; Tsodyks et al. 1998). The disadvantages of the alpha function, or related heuristic approaches, include the lack of correspondence to a plausible biophysical mechanism and the absence of a natural method for handling the summation of successive PSCs from a train of presynaptic impulses. It must be noted that alpha functions were originally introduced to model the membrane potential or PSPs (Rall 1967), thus rendering the use of these functions for modelling postsynaptic conductances or PSCs erroneous, mainly because the slow rise time of alpha functions does not match the steep variations of most postsynaptic conductances (Destexhe et al. 1994b).

The most fundamental way to model synaptic currents is based on the kinetic properties of the underlying synaptic ion channels. The kinetic approach is closely related to the well-known model of Hodgkin and Huxley for voltage-dependent ion channels (see Sect. 2.1.3). Kinetic models are powerful enough to describe in great detail the properties of synaptic ion channels, and they can be integrated coherently with chemical kinetic models for enzymatic cascades underlying signal transduction and neuromodulation (Destexhe et al. 1994b). The drawback of kinetic models is that they are often complex, with several coupled differential equations, thus making them in many cases too costly to be used in simulations involving large populations of neurons.

In some cases, however, kinetic models can be simplified to become analytically solvable, yielding very fast algorithms to simulate synaptic currents. The rationale behind this simplification comes from voltage-clamp recordings of synaptic currents, which show that a square pulse of transmitter (about 1 ms duration and 1 mM concentration) reproduced PSCs that were similar to those recorded in the intact synapse (Hestrin 1992; Colquhoun et al. 1992; Standley et al. 1993). Models were then designed assuming that the transmitter, either glutamate or γ-aminobutyric acid (GABA), is released according to a pulse when an AP invades the presynaptic terminal (Destexhe et al. 1994a). Then, a two-state (open/closed) kinetic scheme, combined with such a pulse of transmitter, can be solved analytically (Destexhe et al. 1994a). The same approach also yields simplified algorithms for three-state and more complex schemes (Destexhe et al. 1994b). As a consequence, extremely fast algorithms can be used to simulate most types of synaptic receptors, as detailed below for four of the main receptor types encountered in the central nervous system.

2.2.1 Glutamate AMPA Receptors

AMPA receptors mediate the prototypical fast excitatory synaptic currents in the brain. In specialized auditory nuclei, AMPA receptor kinetics may be extremely rapid, with rise and decay time constants in the sub-millisecond range (Raman et al. 1994). In the cortex and hippocampus, responses are somewhat slower (e.g., see Hestrin et al. 1990). The 10–90% rise time of the fastest currents measured at the soma (representing those with least cable filtering) is 0.4–0.8 ms in cortical pyramidal neurons, while the decay time constant is about 5 ms (e.g., Hestrin 1993). It is worth noting that inhibitory interneurons express AMPA receptors with significantly different properties. They are about twice as fast in rise and decay time as those in pyramidal neurons (Hestrin 1993), and they also have a significant Ca^{2+} permeability (Koh et al. 1995).

The rapid time course of AMPA/kainate responses is thought to be due to a combination of rapid clearance of neurotransmitter and rapid channel closure (Hestrin 1992). Desensitization of these receptors does occur, but is somewhat slower than deactivation. The physiological significance of AMPA receptor desensitization, however, has not been well established yet. Although desensitization may contribute to the fast synaptic depression observed at neocortical synapses (Thomson and Deuchars 1994; Markram and Tsodyks 1996), a study of paired-pulse facilitation in the hippocampus suggested a minimal contribution of desensitization even at 7 ms intervals (Stevens and Wang 1995).

The simplest model that approximates the kinetics of the fast AMPA type of glutamate receptors can be represented by the two-state diagram:

$$C + T \underset{\beta}{\overset{\alpha}{\rightleftharpoons}} O, \tag{2.16}$$

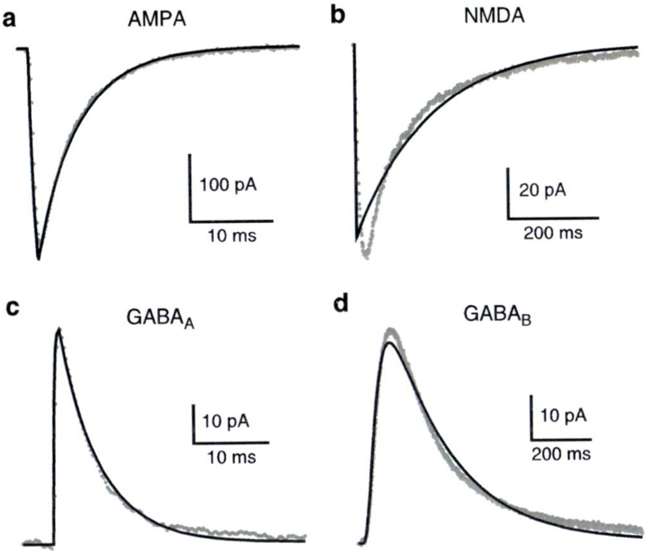

Fig. 2.3 Best fits of simplified kinetic models to averaged postsynaptic currents obtained from whole-cell recordings. (**a**) AMPA-mediated currents (recording from Xiang et al. 1992 31°C). (**b**) NMDA-mediated currents (recording from Hessler et al. 1993 22–25°C in Mg^{2+}-free solution). (**c**) $GABA_A$-mediated currents. (**d**) $GABA_B$-mediated currents (C-D recorded at 33–35°C by Otis et al. 1992, 1993). For all graphs, averaged whole-cell recordings of synaptic currents (*gray noisy traces*) are represented with the best fit obtained using the simplest kinetic models (*black solid*). Transmitter time course was a pulse of 1 mM and 1 ms duration in all cases (see text for parameters)

where α and β are voltage-independent forward and backward rate constants. If r is defined as the fraction of the receptors in the open state, the dynamics is then described by the following first-order kinetic equation

$$\frac{dr}{dt} = \alpha [T] (1 - r) - \beta r, \tag{2.17}$$

and the PSC I_{AMPA} is given by:

$$I_{AMPA} = \bar{g}_{AMPA}\, r (V - E_{AMPA}), \tag{2.18}$$

where $\bar{g}_{AMPA}$ is the maximal conductance, E_{AMPA} is the reversal potential and V is the postsynaptic membrane potential.

The best fit of this kinetic scheme to whole-cell recorded AMPA/kainate currents (Fig. 2.3a) gives $\alpha = 1.1 \times 10^6$ M^{-1}s^{-1} and $\beta = 190$ s^{-1} with $E_{AMPA} = 0$ mV. In neocortical and hippocampal pyramidal cells, measurements of miniature synaptic currents (10–30 pA amplitude; see McBain and Dingledine 1992; Burgard and Hablitz 1993) and quantal analysis (e.g., Stricker et al. 1996) lead to estimates of maximal conductance around 0.35–1.0 ns for AMPA-mediated currents in a single synapse.

2.2.2 *NMDA Receptors*

NMDA receptors mediate synaptic currents that are substantially slower than AMPA currents, with a rise time of about 20 ms and decay time constants of about 25–125 ms at 32°C (Hestrin et al. 1990). The slow kinetics of activation is due to the requirement that two agonist molecules must bind to open the receptor, as well as a relatively slow channel opening rate of bound receptors (Clements and Westbrook 1991). The slowness of decay is believed to be primarily due to the slow unbinding of glutamate from the receptor (Lester and Jahr 1992; Bartol and Sejnowski 1993).

An unique and important property of the NMDA receptor channel is its sensitivity to block by physiological concentrations of Mg^{2+} (Nowak et al. 1984; Jahr and Stevens 1990a,b). The Mg^{2+}-block is voltage dependent, allowing NMDA receptor channels to conduct ions only when depolarized. The necessity of both presynaptic and postsynaptic gating conditions (presynaptic neurotransmitter and postsynaptic depolarization) make the NMDA receptor a molecular coincidence detector. Furthermore, NMDA currents are carried partly by Ca^{2+} ions, which have a prominent role in triggering many intracellular biochemical cascades. Together, these properties are crucial to the NMDA receptor's role in synaptic plasticity (Bliss and Collingridge 1993) and activity-dependent development (Constantine-Paton et al. 1990).

The NMDA type of glutamate receptors can be represented with a two-state model similar to AMPA/kainate receptors, with a voltage-dependent term representing magnesium block. Using the same scheme as in (2.16) and (2.17), the PSC is given by

$$I_{\mathrm{NMDA}} = \bar{g}_{\mathrm{NMDA}} B(V)\, r\, (V - E_{\mathrm{NMDA}}), \tag{2.19}$$

where $\bar{g}_{\mathrm{NMDA}}$ is the maximal conductance, E_{NMDA} is the reversal potential and $B(V)$ represents the magnesium block given by Jahr and Stevens (1990b):

$$B(V) = \frac{1}{1 + \exp(-0.062V)[\mathrm{Mg}^{2+}]_o/3.57}. \tag{2.20}$$

Here, $[\mathrm{Mg}^{2+}]_o$ is the external magnesium concentration, which takes values between 1 mM and 2 mM in physiological conditions.

The best fit of this kinetic scheme to whole-cell recorded NMDA currents (Fig. 2.3b) gave $\alpha = 7.2 \times 10^4\ \mathrm{M}^{-1}\mathrm{s}^{-1}$ and $\beta = 6.6\ \mathrm{s}^{-1}$ with $E_{\mathrm{NMDA}} = 0$ mV. Miniature excitatory synaptic currents also have an NMDA-mediated component (McBain and Dingledine 1992; Burgard and Hablitz 1993) and the conductance of dendritic NMDA channels have been reported to be a fraction of AMPA channels, between 3% and 62% (Zhang and Trussell 1994; Spruston et al. 1995), leading to estimates of the maximal conductance of NMDA-mediated currents at a single synapse around $\bar{g}_{\mathrm{NMDA}} = 0.01$–$0.6$ ns.

2.2.3 GABA$_A$ Receptors

In the central nervous system, fast inhibitory postsynaptic potentials (IPSPs) are mostly mediated by GABA$_A$ receptors. GABA$_A$-mediated IPSPs are elicited following minimal stimulation, in contrast to GABA$_B$ responses (see next section), which require strong stimuli (Dutar and Nicoll 1988; Davies et al. 1990; Huguenard and Prince 1994). GABA$_A$ receptors have a high affinity for GABA and are believed to be saturated by release of a single vesicle of neurotransmitter (see Mody et al. 1994; Thompson 1994). They also have at least two binding sites for GABA and show a weak desensitization (Busch and Sakmann 1990; Celentano and Wong 1994). However, blocking uptake of GABA reveals prolonged GABA$_A$ currents that last for more than a second (Thompson and Gähwiler 1992; Isaacson et al. 1993), suggesting that, as with AMPA receptors, deactivation following transmitter removal is the main determinant of the decay time.

GABA$_A$ receptors can also be represented by the scheme in (2.16) and (2.17), with the postsynaptic current given by:

$$I_{\mathrm{GABA}_A} = \bar{g}_{\mathrm{GABA}_A}\, r\,(V - E_{\mathrm{GABA}_A}), \tag{2.21}$$

where $\bar{g}_{\mathrm{GABA}_A}$ is the maximal conductance and E_{GABA_A} is the reversal potential.

The best fit of this kinetic scheme to whole-cell recorded GABA$_A$ currents (Fig. 2.3c) gave $\alpha = 5 \times 10^6\ \mathrm{M^{-1}s^{-1}}$ and $\beta = 180\ \mathrm{s^{-1}}$ with $E_{\mathrm{GABA}_A} = -80$ mV. Estimation of the maximal conductance at a single GABAergic synapse from miniature GABA$_A$-mediated currents (Ropert et al. 1990; De Koninck and Mody 1994) leads to $\bar{g}_{\mathrm{GABA}_A} = 0.25$–$1.2$ ns.

2.2.4 GABA$_B$ Receptors and Neuromodulators

In the three types of synaptic receptors discussed so far, the receptor and ion channel are both part of the same protein complex. Besides these *ionotropic receptors*, other classes of synaptic response are mediated by an ion channel that is not directly coupled to a receptor, but rather is activated (or deactivated) by an intracellular *second messenger* that is produced when neurotransmitter binds to a separate receptor molecule. Such *metabotropic* receptors include glutamate metabotropic receptors, GABA (through GABA$_B$ receptors), acetylcholine (through muscarinic receptors), noradrenaline, serotonin, dopamine, histamine, opioids, and others. These receptors typically mediate slow intracellular responses. We mention here only models for GABA$_B$ receptors, whose response is mediated by K^+ channels that are activated by G-proteins (Dutar and Nicoll 1988).

Unlike GABA$_A$ receptors, which respond to weak stimuli, responses from GABA$_B$ responses require high levels of presynaptic activity (Dutar and Nicoll 1988; Davies et al. 1990; Huguenard and Prince 1994). This property might be due to extrasynaptic localization of $GABA_B$ receptors (Mody et al. 1994), but a detailed model of synaptic transmission on GABAergic receptors suggests that this effect could also be due to cooperativity in the activation kinetics of $GABA_B$ responses (Destexhe and Sejnowski 1995). The prediction that this nonlinearity arises from mechanisms intrinsic to the synapse was confirmed by dual recordings in thalamic slices (Kim et al. 1997) and in cortical slices (Thomson and Destexhe 1999). The typical properties of $GABA_B$-mediated responses in cortical, hippocampal and thalamic slices can be reproduced assuming that several G-proteins bind to the associated K^+ channels (Destexhe and Sejnowski 1995), leading to the following scheme:

$$\frac{dr}{dt} = K_1\,[T]\,(1-r) - K_2 r$$

$$\frac{ds}{dt} = K_3\,r - K_4 s$$

$$I_{GABA_B} = \bar{g}_{GABA_B}\frac{s^n}{s^n + K_d}(V - E_K)\,, \qquad (2.22)$$

where T is the transmitter (GABA), r is the fraction of receptor bound to GABA, s (in μM) is the concentration of activated G-protein, $\bar{g}_{GABA_B} = 1$ ns is the maximal conductance of K^+ channels, $E_K = -95$ mV is the potassium reversal potential, and K_d is the dissociation constant of the binding of G on the K^+ channels. Fitting of this model to whole-cell recorded $GABA_B$ currents (Fig. 2.3d) gave the following values: $K_d = 100\ \mu M^4$, $K_1 = 9 \times 10^4\ M^{-1}\ s^{-1}$, $K_2 = 1.2\ s^{-1}$, $K_3 = 180\ s^{-1}$ and $K_4 = 34\ s^{-1}$ with $n = 4$ binding sites.

As discussed above, $GABA_B$-mediated responses typically require high stimulus intensities to be evoked. Miniature GABAergic synaptic currents indeed never contain a $GABA_B$-mediated component (Otis and Mody 1992a,b; Thompson and Gähwiler 1992; Thompson 1994). As a consequence, $GABA_B$-mediated unitary IPSPs are difficult to obtain experimentally and the estimation of the maximal conductance of $GABA_B$ receptors in a single synapse is difficult. A peak $GABA_B$ conductance of around 0.06 ns was reported using release evoked by local application of sucrose (Otis et al. 1992).

Other neuromodulatory actions can also be modeled in a similar manner. Glutamate metabotropic receptors, muscarinic receptors, noradrenergic receptors, serotonergic receptors, and others, have been shown to also act through the intracellular activation of G proteins, which may affect ionic currents as well as the metabolism of the cell. As with GABA acting on $GABA_B$ receptors, the main electrophysiological target of many neuromodulators is to open or close K^+ channels (see Brown 1990; Brown and Birnbaumer 1990; McCormick 1992). The model of $GABA_B$ responses outlined here could, thus, be used to model these currents as well, with rate constants adjusted to fit the time courses reported for the particular responses (see Destexhe et al. 1994b).

2.3 Cable Formalism for Dendrites

Most neurons are endowed with beautiful and elaborated dendritic structures which provide the necessary sites for excitatory, inhibitory and neuromodulatory synaptic inputs. Besides their role as receivers of inputs from other neurons in the surrounding network, dendrites also serve as synaptic integrators where most of the arriving information is preprocessed before it reaches the soma. Despite the vast variety in dendritic shapes and functional architectures, the basic principles underlying the spread of electrical activity along dendrites or axons, formulated mathematically in terms of core conductors or electrical cables, is, however, in all cases the same.

In the early second half of the 19th century, William Thomson (Lord Kelvin) used the analogy to heat conduction in a wire to arrive at a mathematical formulation for the signal decay in submarine telegraphic cables (Smith and Wise 1989; Hunt 1997). Shortly after, Hermann applied the same formalism for describing the axonal electrotonus (Hermann 1874, 1879; Hoorweg 1898). He and, independently, Cremer extended this model to arrive at a theory of signal conduction in nerve fibers (Hermann 1899; Cremer 1900; Hermann 1905). Later, this theory was complemented by models based on cable theory as well as experimental studies in seminal papers by Cole and Curtis (1939), Cole and Hodgkin (1939), Hodgkin (1936, 1937a,b, 1939), Hodgkin and Rushton (1946), Offner et al. (1940), Rushton (1951) as well as Davis and Lorente de No (1947). Recently, the process of nerve conduction was re-examined in studies by Tasaki and Matsumoto (2002).

The idea behind *core conductors*, or *linear cable theory*, is the idealization of the electrical cables (or conducting structure) by a cylinder consisting of a conductive core surrounded by a membrane with specific electrical properties, such as transmembrane currents. The membrane may be passive or excitable due to the presence of ion channels (see Sect. 2.1), thus leading to different mathematical models which will be briefly outlined in the remainder of this section.

2.3.1 *Signal Conduction in Passive Cables*

The electrical equivalent circuit for a small passive membrane patch is shown if Fig. 2.1. Lining up such patches with their equivalent circuits in parallel, provides a spatially discretized model for a one-dimensional conducting cable (Fig. 2.4). For infinitesimal small membrane patches, this approach leads to the partial differential equation

$$\lambda^2 \frac{\partial^2 V(x,t)}{\partial x^2} = \tau_m \frac{\partial V(x,t)}{\partial t} + V(x,t) \tag{2.23}$$

for the membrane potential $V(x,t)$ at site x and time t.

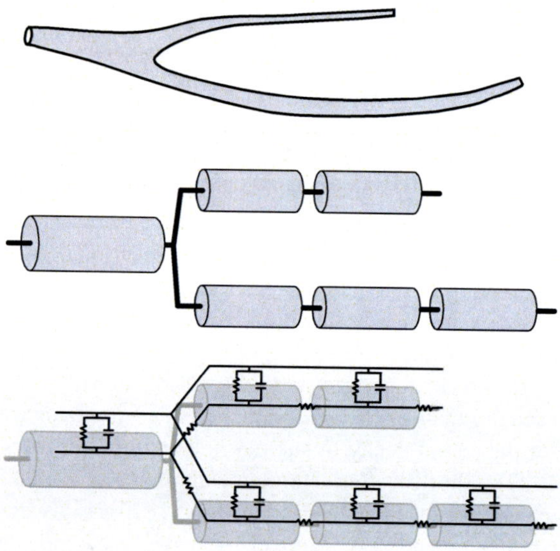

Fig. 2.4 In order to derive a biophysical model of neuronal cables, the membrane is split up in small patches. Placing an electrical equivalent circuit in each of these patches yields a (discretized) model for neuronal cables. For infinitesimal small patches, the continuous cable equation (2.23) is obtained

Equation (2.23) is called *passive cable equation* and describes the electrotonic spread of electrical signals, or *electrotonus*, in a passive cable such as passive dendrites. Here,

$$\lambda = \sqrt{\frac{aR_{\rm m}}{2R_i}}$$

denotes the *space* or *length constant*, with $R_{\rm m}$ and R_i as the specific membrane and intracellular resistivity, respectively, and a the radius of the cable. The term

$$\tau_m = R_{\rm m}C_{\rm m}\,,$$

with $C_{\rm m}$ being the specific membrane capacitance, denotes the *membrane time constant*. It is interesting to note that both τ_m and λ depend on the membrane resistivity $R_{\rm m}$ and increase proportional or proportional to the square root of $R_{\rm m}$, respectively. These relationships will play a crucial role in later chapters of the book when models of synaptic conductances and their impact on the cellular membrane will be considered.

Equation (2.23) can explicitly be solved in the limit of an infinite cable and injection of a constant current I_0, thus yielding a coarse model for passive signal conduction in long axons or thin unbranched dendrites. At steady state, i.e., for $t \to \infty$, the membrane potential follows an exponential decay from the site of the current injection (Fig. 2.5a):

$$V(x, t \to \infty) = \frac{R_i I_0 \lambda}{2\pi a^2} \mathrm{e}^{-x/\lambda}\,. \tag{2.24}$$

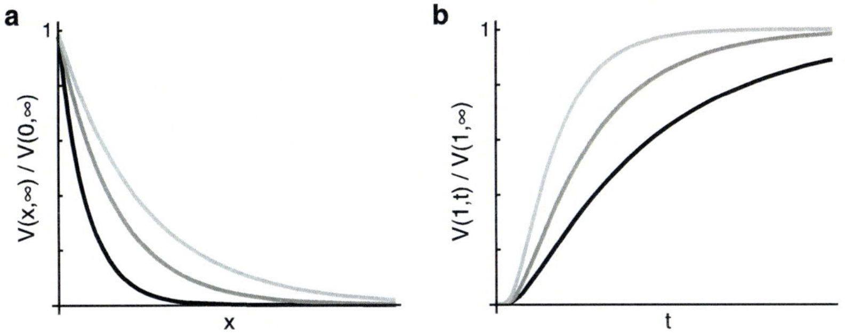

Fig. 2.5 (**a**) Attenuation of the membrane potential after constant current injection (2.24). The membrane potential follows an exponential decay and attenuates less for increasing space constant (*black*: $\lambda = \lambda_0$, *gray*: $\lambda = 2\lambda_0$, *light gray*: $\lambda = 3\lambda_0$). (**b**) Membrane potential at a fixed position ($x = 1$) during charging of the membrane through constant current injection (2.25). The smaller the membrane resistance (hence the membrane space and time constant), the faster a given membrane potential value is reached at a given position (*black*: $R_{\mathrm{m}} = R_{m0}$, *gray*: $R_{\mathrm{m}} = 0.5R_{m0}$, *light gray*: $R_{\mathrm{m}} = 0.25R_{m0}$)

For increasing membrane resistivity and, thus, space constant λ, the signal shows less *attenuation*, i.e., a larger portion of the signal will reach sites of the cable distal to the site of injection (Fig. 2.5a, *gray*). By doubling the membrane potential amplitude at $x = 0$, (2.24) describes the solution for a semi-infinite cable starting at the site of the current injection and extending infinitely to one side.

So far only the membrane potential along the cable after reaching its steady state was considered. However, before reaching the steady state, the membrane potential at a specific site x changes according to the transient solution of the membrane equation

$$V(x,t) = \frac{R_i I_0 \lambda}{4\pi a^2} \left\{ \mathrm{e}^{-x/\lambda} \mathrm{erfc}\left[\frac{x}{2\lambda\sqrt{\frac{t}{\tau_m}}} - \sqrt{\frac{t}{\tau_m}} \right] - \mathrm{e}^{x/\lambda} \mathrm{erfc}\left[\frac{x}{2\lambda\sqrt{\frac{t}{\tau_m}}} + \sqrt{\frac{t}{\tau_m}} \right] \right\}, \tag{2.25}$$

where $\mathrm{erfc}[x]$ denotes the error function. Charging of the membrane occurs faster for smaller membrane resistivity R_{m}, hence the smaller the membrane space and time constants (Fig. 2.5b). Furthermore, charging occurs later and is slower the further the recording site is away from the site of the current injection. As in the case of the steady-state solution described above, (2.25) holds for semi-infinite cables with twice the amplitude for the membrane potential.

Considering sites along the membrane with the same potential, we arrive at the notion of *propagation speed*, or *conductance velocity*, of a passive wave or signal. This speed of a passive wave is defined in terms of the time to conduct half of the maximum steady-state vale at site x, and is given by

$$\theta = \frac{2\lambda}{\tau_m} = \sqrt{\frac{2a}{R_{\mathrm{m}} R_i C_{\mathrm{m}}^2}}. \tag{2.26}$$

Above, only infinite cables were considered. In reality, however, this model provides a good approximation only for signal conduction along axons, for which the diameter is negligible compared to their length. A more realistic model for dendrites is given by the solution of (2.23) for a finite cable of length l, in which case boundary conditions at the ends of the dendritic cable have to be considered. Here we will focus only on the biophysically relevant case of an open circuit condition, in which the end of the cable is sealed, i.e., no current flows across the end of the cable. In this case, the steady-state solution is given by

$$V(x,t\to\infty) = \frac{R_i I_0 \lambda}{2\pi a^2} \frac{\cosh\left[\frac{l-x}{\lambda}\right]}{\cosh\left[\frac{l}{\lambda}\right]} . \tag{2.27}$$

In the last equation, l/λ defines the *electrotonic length* of the cable. In contrast to the (semi-) infinite cable, the decay, or *attenuation*, with distance is lower. However, as before, one observe a more pronounced attenuation the lower R_{m}, i.e., the more leaky the membrane is. The transient solution of (2.23) for a finite cable is given by an infinite sum:

$$V(x,t) = \frac{R_i I_0 \lambda}{4\pi a^2} \sum_{n=-\infty}^{\infty} \left\{ \mathrm{e}^{-\frac{|x-2nl|}{\lambda}} \operatorname{erfc}\left[\frac{\frac{|x-2nl|}{\lambda} - \frac{2t}{\tau_m}}{2\sqrt{\frac{t}{\tau_m}}}\right] - \mathrm{e}^{\frac{|x-2nl|}{\lambda}} \operatorname{erfc}\left[\frac{\frac{|x-2nl|}{\lambda} + \frac{2t}{\tau_m}}{2\sqrt{\frac{t}{\tau_m}}}\right]\right\}. \tag{2.28}$$

The above cases describe the voltage response of a passive cable to current injection. Real neurons, however, receive synaptic inputs which lead to a local transient change in the membrane conductance and, hence, transient current flow across the membrane. Although for such PSPs similar conclusions apply as in the case of constant current injection, e.g., PSPs are attenuated and distorted in shape along the dendritic cable, such models are far more difficult to solve and provide, in most cases, no explicit solution. For the simple case of an exponential-shaped synaptic conductance time course

$$G_s^{\exp}(t) = \begin{cases} 0 & \text{for } t < t_0 \\ \overline{G}\mathrm{e}^{-\frac{t-t_0}{\tau_s}} & \text{for } t \geq t_0 , \end{cases} \tag{2.29}$$

the PSP is given by (Rudolph and Destexhe 2006c)

$$\begin{aligned} V(t) &= \exp\left[-\frac{t}{\tau_m^L} + \frac{\tau_s}{\Delta\tau_m^s}\mathrm{e}^{-\frac{t}{\tau_s}}\right] \\ &\times\left\{E_{\mathrm{L}}\mathrm{e}^{-\frac{\tau_s}{\Delta\tau_m^s}} - E_s\left(\mathrm{e}^{-\frac{\tau_s}{\Delta\tau_m^s}} - \exp\left[\frac{t}{\tau_m^L} - \frac{\tau_s}{\Delta\tau_m^s}\mathrm{e}^{-\frac{t}{\tau_s}}\right]\right)\right. \\ &\left. -\Gamma\left[-\frac{\tau_s}{\tau_m^L}, \frac{\tau_s}{\Delta\tau_m^s}\mathrm{e}^{-\frac{t}{\tau_s}}, \frac{\tau_s}{\Delta\tau_m^s}\right]\left(\frac{\tau_s}{\Delta\tau_m^s}\right)^{\frac{\tau_s}{\tau_m^L}} \frac{\tau_s}{\tau_m^L}(E_s - E_{\mathrm{L}})\right\}, \end{aligned} \tag{2.30}$$

where $\Gamma[z,a,b] = \int_a^b \mathrm{d}t\, t^{z-1}\mathrm{e}^{-t}$ denotes the generalized incomplete gamma function. Here E_{L} and E_s denote the leak and synaptic reversal potentials, respectively. $\overline{G}$ denotes the maximal conductance which is linked to the update of the membrane time constant τ_m at time t_0 by $\Delta\tau_m^s = C/\overline{G}$. This equation holds for both EPSPs and IPSPs.

Numerical simulations of (2.30) show that distal synaptic signals are attenuated upon reaching the soma (decreasing peak amplitude with increasing distance from the synaptic input), that the rise time is progressively slowed and delayed for inputs at increasing distance from soma, but that the decay time is the same.

2.3.2 *Signal Conduction in Passive Dendritic Trees*

In the previous section, the signal conduction in a uniform passive cable was considered. However, most neurons form complex and elaborated dendrites by branching into beautiful tree-like structures. A theoretical framework for the spread of signals in dendritic trees is given by the Rall model (Rall 1962, 1964). In its simplest case, i.e., uniform semi-infinite cables comprised of passive membranes and a soma described by an isopotential sphere, the total input conductance at a branch point X, where the parent dendrite P branches into N daughter dendrites D_i, equals the total input conductance of the parent dendrite extended to infinity as long as the diameters of the dendrites obey the relation

$$d_P^{\frac{3}{2}} = \sum_{i=1}^{N} d_{D_i}^{\frac{3}{2}}. \tag{2.31}$$

This equation, also called the *3/2 power rule*, is a direct consequence of the total input conductance of a semi-infinite cable,

$$G_{\mathrm{in}} = \frac{\pi a^{\frac{3}{2}}\sqrt{2}}{\sqrt{R_{\mathrm{m}}R_i}} \propto d^{\frac{3}{2}}. \tag{2.32}$$

By applying this 3/2 power rule to all dendrites, the complicated tree structure can be mathematically reduced to a single equivalent semi-infinite cable of diameter d_0.

In the case of dendrites described as sealed finite-length cables all terminating at the same electrotonic length L, the dendritic tree can be reduced to a single equivalent cable of diameter d_0 and electrotonic length L if the condition

$$L = \sum_i \frac{l_i}{\lambda_l} \tag{2.33}$$

is fulfilled. In the last equation, l_i and λ_i denote the length and space constant of each dendritic segment, respectively.

The complexity of dendritic tree morphologies can also be reduced to 2- or 3-compartmental models using numerical fits of experimental data. For instance, the method introduced by Destexhe (2001) preserves the total membrane area, input resistance, time constant, and voltage attenuation of a detailed morphology in a 3-compartmental model. The underlying idea here is, while choosing for the dendritic cable of the simplified model the typical physical length of the dendritic tree, to adjust the diameters of the cables in a way which preserve the total membrane area. After that, the passive cellular properties must be adjusted by a multiparameter fitting procedure constrained by passive responses of the original complex model and the somatodendritic profile following current injection at the soma.

2.3.3 *Signal Conduction in Active Cables*

Many neurons possess active dendrites and are capable of generating *dendritic spikes* which propagate toward the soma (*forward propagating dendritic spikes*) or further out into the dendritic structure (*backward propagating dendritic spikes*). As the dynamics of spikes is determined by the kinetics of ionic conductances embedded into the membrane, active signal propagation follows different rules. However, in contrast to signal conduction in passive cables, the propagation of signals in active cable structures defies a rigorous mathematical description as, in general, the resulting membrane equations are no longer analytically treatable. Computational models and numerical simulations remain, so far, the only way to assess signal propagation along active axonal or dendritic structures. A detailed introduction and excellent review of concepts can be found in Jack et al. (1975).

For that reason, we will restrict here to the estimation of the propagation speed for the simplest case. Generalizing the cable equation (2.23), for the case of active membranes one obtains the *active cable equation*

$$\frac{a}{2R_i}\frac{\partial^2 V(x,t)}{\partial x^2} = C_\mathrm{m}\frac{\partial V(x,t)}{\partial t} + I_L + I_\mathrm{ion}(x,t)\,, \tag{2.34}$$

where I_L and $I_\mathrm{ion}(x,t)$ denote the leak current and the current due to ionic conductances. Considering a dendritic cable with constant diameter, as well as location-independent ionic currents, (2.34) yields an estimate for the constant propagation speed of an AP:

$$\theta = \sqrt{\frac{Ka}{2R_i C_\mathrm{m}}}\,. \tag{2.35}$$

Here, K is a constant which was experimentally estimated to be $K = 10.47\ \mathrm{m\,s^{-1}}$. Experimentally, the conduction velocity was measured to be about $\theta = 21.2\ \mathrm{m\,s^{-1}}$ for the squid axon. The corresponding theoretical value (2.35) is $\theta = 18.8\ \mathrm{m\,s^{-1}}$.

2.4 Summary

In this chapter, some of the most important concepts and notions used in theoretical and computational neuroscience were briefly introduced. These include models of excitable membranes to describe spike generation (Sect. 2.1), models of synaptic interactions (Sect. 2.2) and models of neuronal cables, in particular signal conduction in passive cables (Sect. 2.3).

The Hodgkin–Huxley model (Sect. 2.1.3) remains to this date the most prominent biophysical model of ionic currents describing the generation of APs in neuronal membranes. Various mathematically more rigorous and general approaches, such as Markov (kinetic) models (Sect. 2.1.4), do exist which overcome the limitations of the former and provide, despite being analytically more challenging, a more accurate description of experimental observations. Such kinetic models are very general, and can also be applied to the description of synaptic interactions, as presented here for excitatory (glutamate AMPA, Sect. 2.2.1 and NMDA, Sect. 2.2.2) and inhibitory ($GABA_A$ and $GABA_B$, Sects. 2.2.3 and 2.2.4, respectively) receptors.

Finally, the electrical equivalent circuit (Sect. 2.1.2) was shown to provide a powerful model for describing neuronal dynamics in small membrane patches, in particular when used in computational studies of extended dendritic structures. This model provides the basis of the cable formalism of neuronal membranes (Sect. 2.3), which can be utilized to study signal conduction in elaborate neuronal structures, such as dendritic trees.

The concepts and notions which were outlined in this Chapter will be used throughout this book.

Chapter 3
Synaptic Noise

This chapter will review the highly irregular and seemingly noisy neuronal activity during different brain states, such as wake and sleep states, as well as different types of anesthesia. During these states, intracellular recordings in cortical neurons in vivo show a very intense and noisy synaptic activity, also called *synaptic noise*. The properties of synaptic noise are reviewed, in particular the quantitative measurements of its large conductance (*high-conductance states*).

3.1 Noisy Aspects of Extracellular Activity In Vivo

Figure 3.1 shows extracellular recordings of unit activity across different stages of the motor and premotor cortex of awake macaque monkey. This recording reveals that the activity of cerebral cortex is not silent, but instead is extremely active. It also shows that even outside of the movement, there is intense spontaneous activity, while the movement itself is associated with subtle modifications of this activity. A similar picture holds as well for the sensory cortex. Thus, the cerebral cortex is not silent with neurons only firing in relation to movements or sensory inputs, but is characterized by large amounts of spontaneous activity, which is only slightly modified in relation to sensory inputs or motor behavior. This spontaneous activity is also very irregular, as can be seen in Fig. 3.1. In this section, we review the evidence from extracellular data that the cellular and network activity in cerebral cortex is very irregular, weakly correlated and in many ways statistically similar to "noise."

3.1.1 Decay of Correlations

An example of multisite LFPs recordings in different brain states is shown in Fig. 3.2, which was obtained using a set of eight equidistant bipolar electrodes

A. Destexhe and M. Rudolph-Lilith, *Neuronal Noise*, Springer Series in Computational Neuroscience 8, DOI 10.1007/978-0-387-79020-6_3,

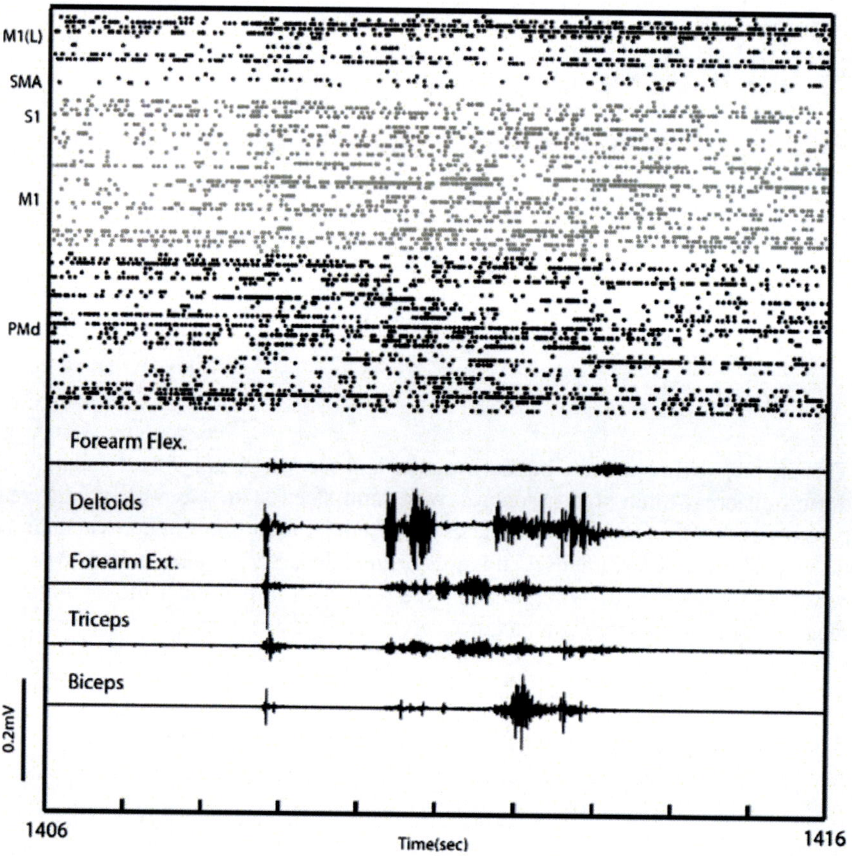

Fig. 3.1 Extracellular activity in Rhesus macaque monkey cortex during movements. The figure shows raster of unit activity recorded using 96 chronically implanted microwires, in five different cortical areas including premotor and motor cortex, as well as the simultaneously recorded muscle (EMG) activity of the animal's forelimb. The muscular command is associated with small modulations of activity, while the "spontaneous" activity seems sustained and very irregular. Courtesy of Miguel Nicolelis, Duke University

(interelectrode distance of 1 mm; see Fig. 3.2, *scheme*). From such recordings, wake and sleep states can be identified using the following criteria: *Wake*: low-amplitude fast activity in LFPs, high electrooculogram (EOG) and high electromyogram (EMG) activity; (SWS): LFPs dominated by high-amplitude slow waves, low EOG activity and EMG activity present; *Rapid Eye Movement sleep*: low-amplitude fast LFP activity, high EOG activity and abolition of EMG activity. During waking and attentive behavior, LFPs are characterized by low-amplitude fast (15–75 Hz) activity (Fig. 3.2, *Awake*), whereas during SWS, LFPs are dominated by high-amplitude slow-wave complexes occurring at a frequency of <1 Hz (Fig. 3.2, *SWS*).

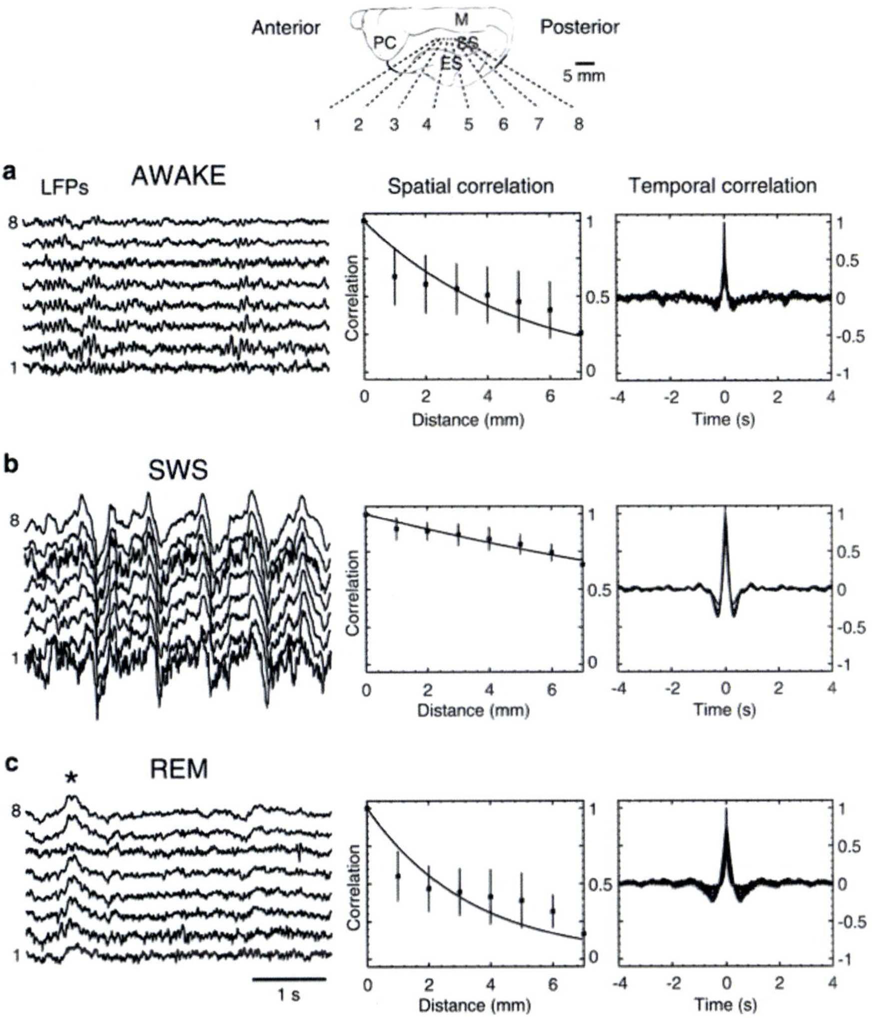

Fig. 3.2 Extracellular activity during wake and sleep states in cat parietal cortex. *Top*: scheme of cat cortex showing the implantation of eight bipolar electrodes, inserted in gray matter at different sites separated by 1 mm. *Bottom*: LFP activity during wakefulness (**a**), slow-wave sleep (**b**) and REM sleep (**c**). In each case, the signals (*left*), the spatial correlations (*middle*) and temporal correlations (*right*) are represented. During wakefulness, the spatial and temporal coherence are minimal. Modified from Destexhe et al. (1999)

Slow-wave complexes of higher frequency (1–4 Hz; * in Fig. 3.2) and spindle waves (7–14 Hz) are also present in SWS states. During periods of REM sleep, in contrast, the activity is similar to waking periods (Fig. 3.2, *REM*).

By evaluating autocorrelations over long periods (e.g., 20 s, as in Fig. 3.2) for each state of the animal, a relative steep decay toward zero is observed

(Fig. 3.2, right), indicating that the LFP activity is very irregular despite the dominant frequencies characteristic to each state. This is contrary to the pronounced rhythmicity that can appear in autocorrelations calculated for small time windows. For long periods of time, the autocorrelations show similar behavior for SWS, REM, and waking states, and therefore, cannot be used to distinguish these states.

In contrast, correlations represented as a function of distance usually display marked differences between awake/REM and SWS (Fig. 3.2, middle). SWS displays slow-wave complexes of a remarkable spatiotemporal coherence, as indicated by the high values of spatial correlations for large distances, in contrast with the steeper decline of spatial correlations with distance during waking and REM sleep. Similar spatial correlations were observed in different animals and during different wake/sleep episodes (Destexhe et al. 1999).

3.1.2 $1/f$ Frequency Scaling of Power Spectra

The irregular aspect of neuronal activity during wake and sleep states can further be analyzed by computing the PSD of the LFP in these states. The corresponding PSDs typically display a broadband structure: during wakefulness, the PSD shows two different scaling regions, according to the frequency band. For low frequencies (between 1 Hz and 20 Hz), the PSD scales approximately as $1/f$, whereas for higher frequencies (between 20 Hz and 65 Hz), the scaling is approximately of $1/f^3$ (Fig. 3.3, *Awake*). During SWS, the additional power at slow frequencies masks the $1/f$ scaling, but the same $1/f^3$ scaling is present in the high-frequency band (Fig. 3.3, *SWS*). In Bédard et al. (2006b), this behavior was observed for other electrodes in the same experiment, and in three different animals. Thus, these results confirm that the $1/f$ frequency scaling reported in the EEG (Pritchard 1992) is also present in LFPs from cat association cortex, but only during waking and for specific frequency bands. They also show that global variables, such as the LFP, display a broadband spectral structure similar to stochastic systems. It must be noted that this $1/f$ scaling of LFPs at low frequencies can be explained by the "filtering" of the electric signals through extracellular space, as shown by Bédard et al. (2006b, 2010) and Bédard and Destexhe (2009).

3.1.3 Irregular Neuronal Discharges

Single-unit activity recorded and separated using standard procedures (Destexhe et al. 1999) show that during the waking state, units tend to discharge tonically (Fig. 3.4a), similar to early observations by Hubel (1959), Evarts (1964) and Steriade (1974, 1978). However, in such recordings, the relation between units and LFP is not evident at first sight, although there is a tendency to discharge during LFP negativity (see below). During SWS, the pattern of discharge is typically more

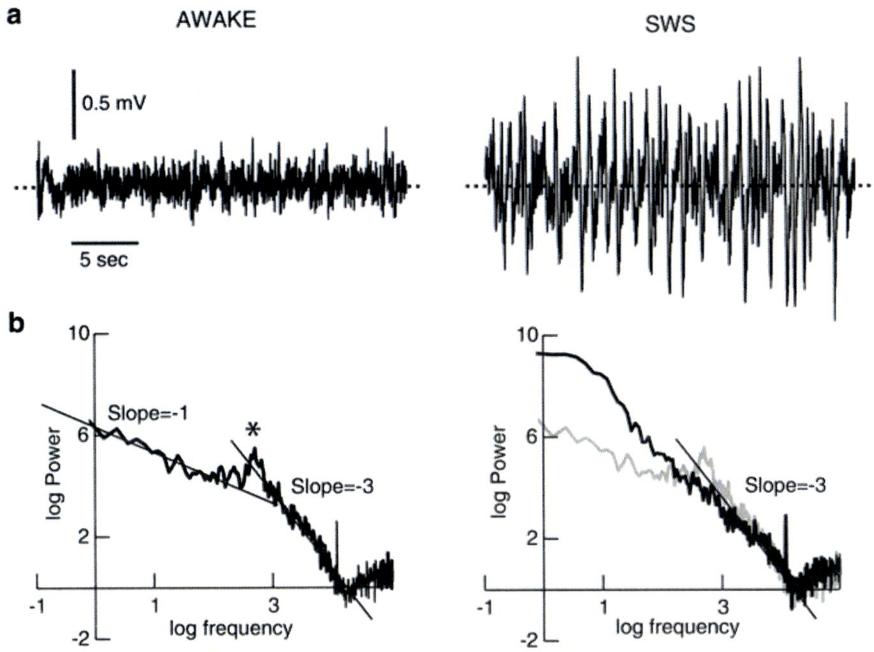

Fig. 3.3 Frequency scaling of local field potentials from cat parietal cortex. (**a**) LFPs recorded in cat parietal cortex during wake and (SWS) states. (**b**) Power spectral density of LFPs, calculated from 55 s of recordings sampled at 300 Hz, and represented in log–log scale (*thin black lines* represent $1/f^{\alpha}$ scaling). During waking (*left*), the frequency band below 20 Hz scales approximately as $1/f$ (*: peak at 20 Hz beta frequency), whereas the frequency band between 20 Hz and 65 Hz scales approximately as $1/f^3$. During slow-wave sleep (*right*), the power in the slow frequency band is increased, and the $1/f$ scaling is no longer visible, but the $1/f^3$ scaling at high frequencies remains unaffected. PSDs were calculated over successive epochs of 32 s, which were averaged over a total period of 200 s for Wake and 500 s for SWS. Modified from Bédard et al. (2006b)

phasic and characterized by periods of silence and of increased firing (Fig. 3.4b), as reported previously (Evarts 1964; Steriade et al. 1974). Positive deflections of slow-wave complexes are almost always associated with a "concerted neuronal silence" in all units, while negative deflections tend to be correlated with a brief increase of firing. REM sleep displays similar activity patterns as in awake animals (Fig. 3.4c).

Such patterns of firing are also illustrated in Fig. 3.5, where the LFP is compared to the distributed multiunit activity at 8 electrodes. During wakefulness, the activity is irregular and tonic, with no obvious sign of synchrony (Fig. 3.5, *Awake*). During SWS, however, it is apparent that a generalized pause of firing is associated with slow-wave complexes (Fig. 3.5, *SWS*). Moreover, here the activity alternates between periods where neurons fire tonically, and pauses of firing. These two periods are commonly referred to as *Up-states* and *Down-states*.

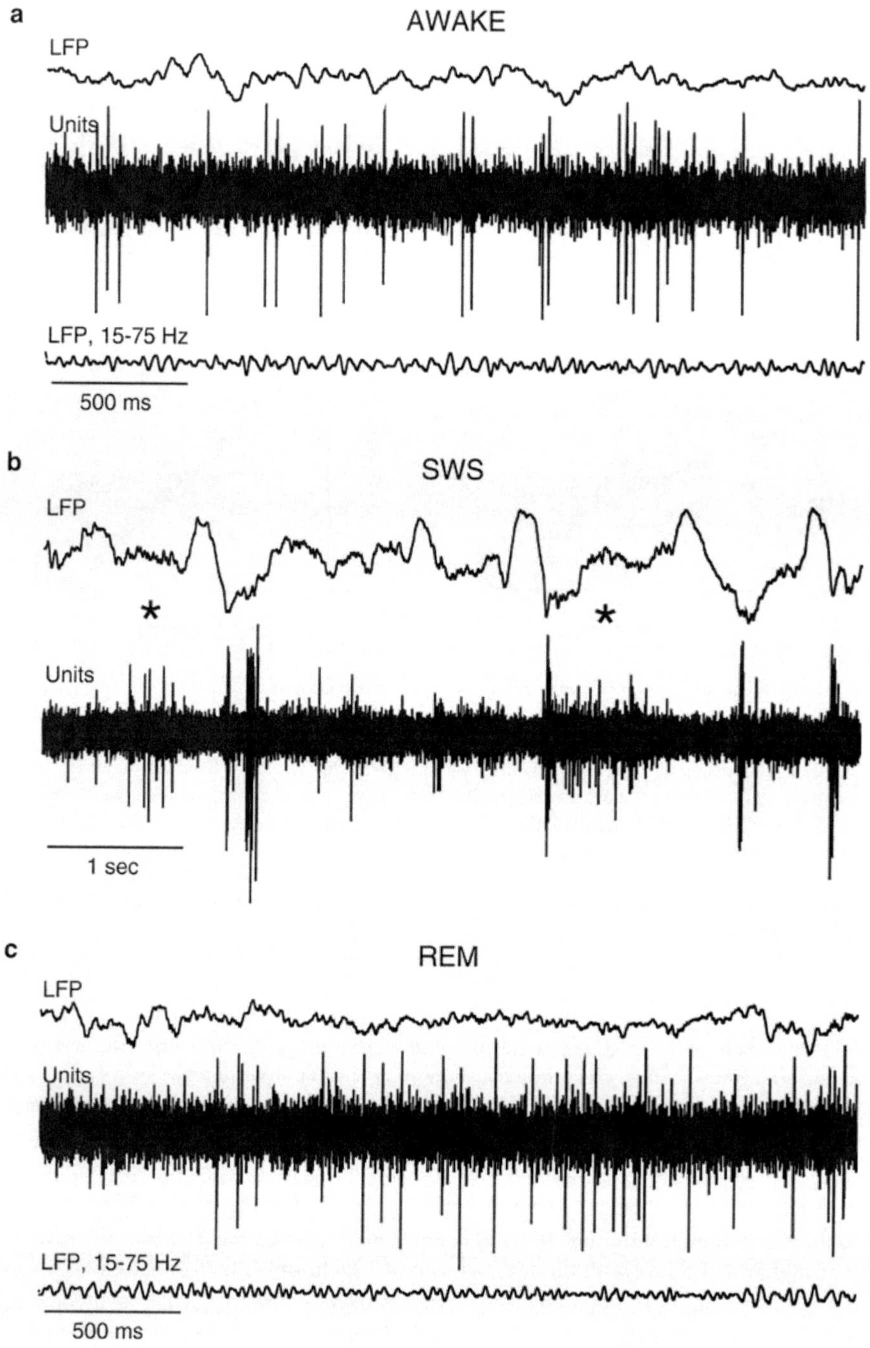

Fig. 3.4 Relation between LFP and multiunit activity during wake and sleep states. All LFP and unit activity are from the same pair of electrodes, across different states: wake (**a**), slow-wave sleep (**b**) and REM sleep (**c**). The LFP filtered between 15 Hz and 75 Hz is also shown for wake and REM sleep states. Modified from Destexhe et al. (1999)

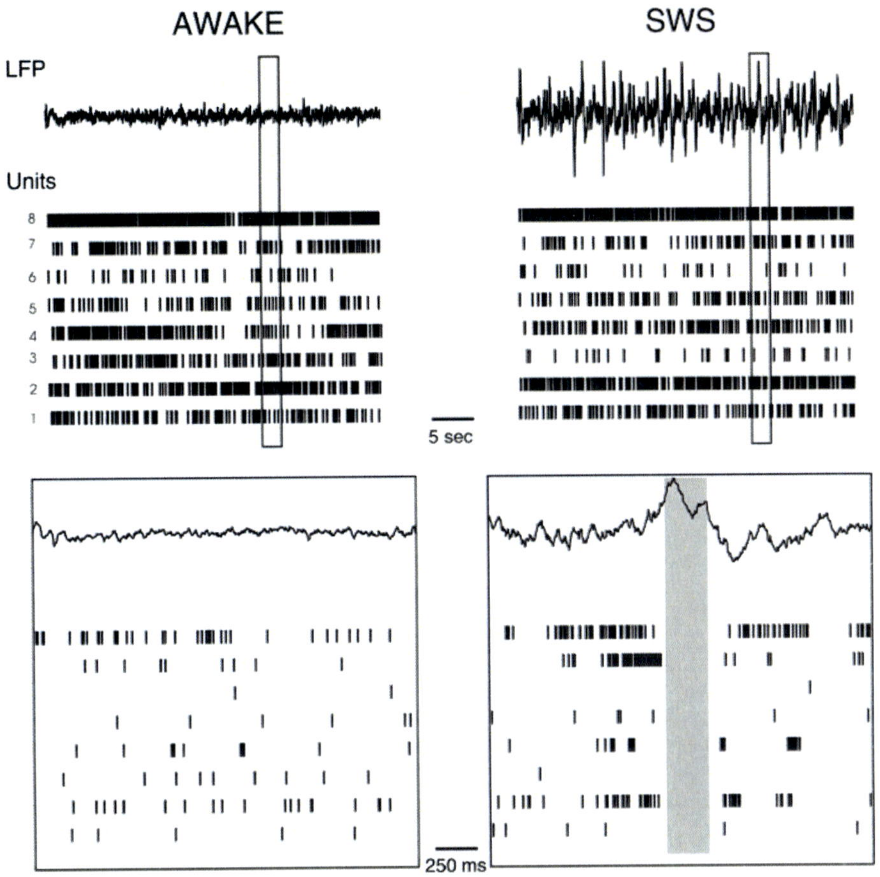

Fig. 3.5 Distributed firing activity in relation to LFPs during wake and sleep states. Irregular firing activity of eight multiunits shown at the same time as the LFP recorded in electrode 1 (same setting as in Fig. 3.2). During wakefulness, the activity is sustained and irregular (see magnification below). During slow-wave sleep (*SWS*), the activity is similar to wakefulness, except that "pauses" of firing occur in all cells, and in relation to the slow waves (one example is shown in the bottom graph, *gray bar*). The boxes in the top graphs are shown in bottom graphs at 20 times higher temporal resolution

3.1.4 Firing of Cortical Neurons is Similar to Poisson Processes

The analogy with stochastic systems can further be pursued by analyzing unit discharges. For that, the ISI distributions can be calculated from neurons recorded in wake and sleep states, and represented in log-linear scale for individual neurons (Fig. 3.6; log–log scale in insets). Such an analysis revealed that during waking, the ISI distributions were close to exponentially distributed ISIs, as generated by Poisson stochastic process with same statistics characteristics as the neurons

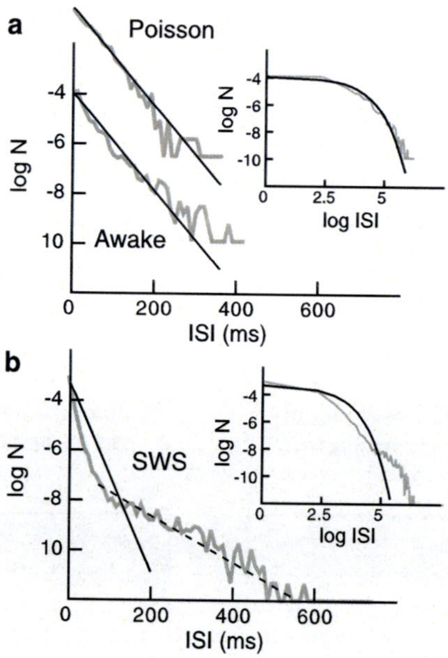

Fig. 3.6 Spike trains in awake animals are analogous to Poisson processes. The logarithm of the distribution of interspike intervals (ISI) during waking (*Awake*, **a**; 1,951 spikes) and slow-wave sleep (*SWS*, **b**; 15,997 spikes) is plotted as a function of ISI length, or log ISI length (insets). A Poisson process of the same rate and statistics is displayed in (**a**) (*Poisson*; curve displaced upwards for clarity). The exponential ISI distribution predicted by Poisson processes of equivalent rates is shown as *straight lines* (*black solid* in inset). The *dotted line* in (**b**) indicates a Poisson process with lower rate which fits the tail of the ISI distribution in SWS. Modified from Bédard et al. (2006b)

analyzed (Fig. 3.6, *Poisson*). For all neurons studied, no evidence for a power-law behavior was found in both wakefulness and SWS (Bédard et al. 2006b). For instance, in 22 neurons recorded during the wake state and analyzed in a study by Bédard and colleagues, the Pearson coefficient was of 0.91 ± 0.13 for exponential distribution fits, and of 0.86 ± 0.16 for power-law distribution fits (Bédard et al. 2006b). When taking only the subset of 7 neurons with more than 2,000 spikes, this study showed that the fit was nearly perfect for an exponential distribution (Pearson coefficient of 0.999 ± 0.001). However, during SWS, there was a marked difference between the experimental ISI and the corresponding Poisson process (Fig. 3.6b). As seen above, in this state, neurons tend to produce long periods of silences, which is visible as a prominent tail of the distribution for large ISIs. This tail can be well fit by a Poisson process of low rate as shown in Fig. 3.6b (*dashed line*).

Similar results were also obtained in other studies, such as Softky and Koch (1993). As illustrated in Fig. 3.7, the irregularity of the discharge of awake macaque monkey was quantified here by the coefficient of variation (ratio of the variance and the mean) of the discharge rate, and units tended to behave similar to Poisson processes.

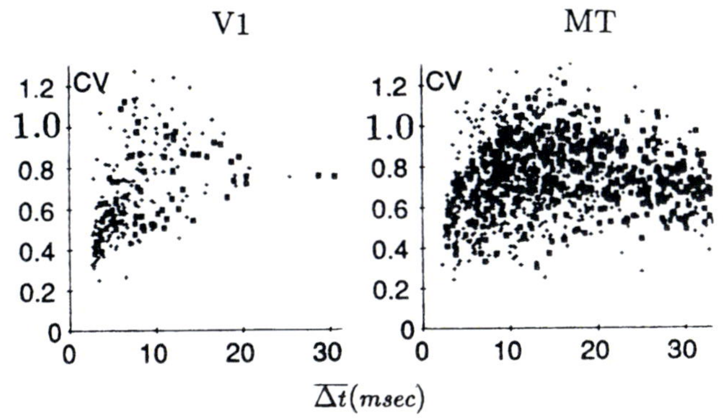

Fig. 3.7 Discharge variability of neurons in area V1 and MT of awake macaque monkey cortex. The coefficient of variation (C_V) is plotted for a large number of neurons recorded in this area (*squares* represent the most significant estimates with largest number of spikes). The C_V is represented against the mean firing rate of the neuron (represented in terms of the mean interspike interval $\bar{\Delta t}$). Modified from Softky and Koch (1993)

3.1.5 The Collective Dynamics in Spike Trains of Awake Animals

To check for the presence of collective dynamics, such as self-organized critical states, one can perform an avalanche analysis by taking into account the collective information from multisite recordings. Using the method introduced by Beggs and Plenz (2003), which amounts to detecting clusters of contiguous events separated by silences, Bédard et al. (2006b) binned their recordings in time windows of 1–16 ms (Fig. 3.8, *scheme*). In this study, there was no evidence for any recognizable event in LFPs which could be taken as avalanche (see Fig. 3.4), so the spike times among the ensemble of simultaneously recorded neurons were used. Here, the distribution of avalanche size does not follow power-law scaling (Fig. 3.8, bottom), but was found closer to an exponential distribution as predicted by Poisson processes (Fig. 3.8, *gray curve*). This analysis revealed that there is no sign of self-organized criticality or avalanche dynamics in this system, but that the cellular dynamics is rather analogous to a Poisson stochastic process (see details in Bédard et al. 2006b).

3.2 Noisy Aspects of Intracellular Activity In Vivo and In Vitro

In this section, we overview the intracellular activity of cortical neurons in different brain states, such as natural sleep and wakefulness, active states under anesthesia and activated states in vitro, and assess the different levels of *synaptic noise* in these states as found in experimental studies.

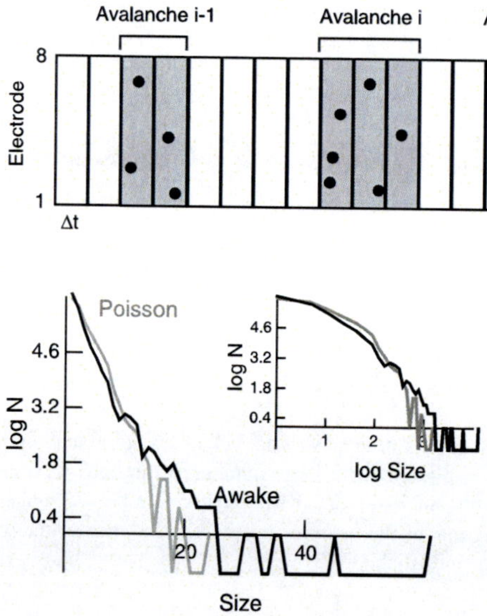

Fig. 3.8 Absence of self-organized critical states in spike trains from the parietal cortex of awake cats. *Top*: scheme of the procedure used to calculate neuronal avalanches by binning the time axis in bins of size Δt. *Bottom*: distribution of avalanche size as a function of the bin size. The distribution scales exponentially (*black solid*), similar to the same analysis performed on a Poisson process with the same statistics (*gray solid*). Inset: same curves shown in log–log scale. Modified from Bédard et al. (2006b)

3.2.1 Intracellular Activity During Wakefulness and Slow-wave Sleep

In a recent study, intracellular recordings of cortical neurons were performed in parietal cortex of awake and naturally sleeping cats (Steriade et al. 2001) and complemented with simultaneous LFP, electromyogram (EMG) and electrooculogram (EOG) recordings to identify the behavioral states of the animals. With pipettes filled with K^+-Acetate (KAc), electrophysiologically identified activities of 96 presumed excitatory neurons were recorded during the waking state. Of them, 47 neurons revealed a regular-spiking firing pattern, with significant spike-frequency adaptation in response to depolarizing current pulses, and spike width of 0.69 ± 0.20 ms (range 0.4–1.5 ms). In this study, the V_m of *regular spiking neurons* was found to vary in a range between -56 mV and -76 mV (mean -64.0 ± 5.9 mV). Among the recorded regular spiking cells, 26 were identified as *wake-active* cells, in which the firing was sustained all through the wake state, as described in other studies (Matsumura et al. 1988; Baranyi et al. 1993; Steriade et al. 2001; Timofeev et al. 2001).

In regular spiking neurons (Fig. 3.9, *Awake*), the V_m is typically depolarized (around -65 mV) and shows high-amplitude fluctuations and sustained irregular

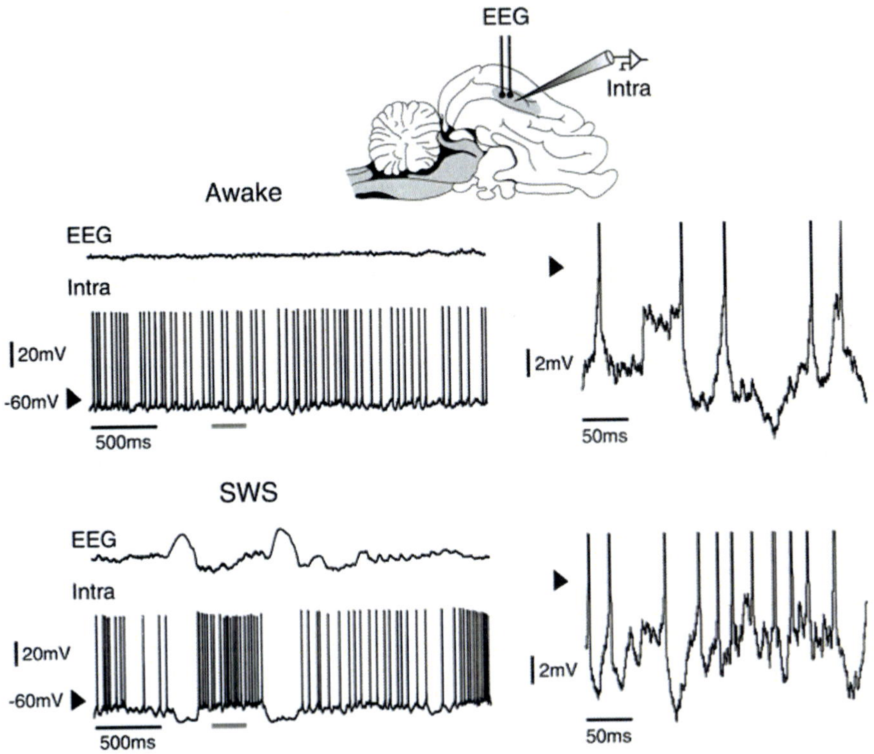

Fig. 3.9 Intracellular activity in cat parietal cortex during wakefulness and slow-wave sleep. LFP (called here EEG) and intracellular recording were obtained in area 5–7 of cat cortex (*scheme*). When the animal was awake (*top traces*), the EEG was desynchronized and the intracellular activity was sustained and irregular. During slow-wave sleep (*SWS*), the EEG displayed slow waves, which were correlated with "Down-states": brief hyperpolarizations with interruption of firing. In between slow waves, the EEG was closer to desynchronized and the activity displayed "Up-states" with sustained and irregular firing similar to wakefulness. The *right panels* show a magnification of the V_m activity in each case. Modified from Steriade et al. (2001)

firing (3.1 Hz on average; range 1–11 Hz) during wakefulness. During SWS, neurons always display Up- and Down-states in the V_m activity in phase with the slow waves (Fig. 3.9, *SWS*), a pattern corroborating previous observations that show that slow waves coincide with "concerted neuronal silences" (Fig. 3.5; Destexhe et al. 1999). Such recordings in awake animals will be shown and analyzed in more detail in Chap. 9, Sect. 9.3.

3.2.2 Similarity Between Up-states and Activated states

An intriguing feature of Up-states is also visible in Fig. 3.9: Up-states share similar features with the activated state of wakefulness. In both cases, the membrane potential is similarly depolarized, there is a comparable level of V_m fluctuations and the firing is sustained and irregular. The LFP is also locally *desynchronized* and of low amplitude during Up-states, while the Down-states are associated with high-amplitude slow-wave complexes (Fig. 3.9). These features suggest that the activity during wakefulness resembles a prolonged Up-state.

The resemblance between activated states and the SWS Up-states in cortex is not only apparent at the level of single cells, but can also be found in the EEG and LFPs. First, as described above, the typical desynchronized EEG pattern of arousal is evident locally in the EEG during Up-states. A second, and stronger, indication of the similarities comes from studies using multiple bipolar electrodes in awake and naturally sleeping cats (Destexhe et al. 1999), where it has been shown that the desynchronized EEG patterns that are present during wakefulness are not only irregular temporally, but more generally, are characterized by low and fluctuating spatiotemporal coherence in LFPs (Fig. 3.10a, b, left). The LFP signals at neighboring electrodes (1 mm distance in the aforementioned study) alternate between periods of high and low coherence, as quantified by the correlation excursions (Fig. 3.10b, left). However, these fluctuations in coherence are local, since the excursions of correlations are greatly diminished between more distant pairs of electrodes. Very similar results are obtained for the Up-states of SWS (Fig. 3.10a, b, right), indicating that the spatiotemporal dynamics of these Up-states, as viewed through the EEG and LFPs, are essentially indistinguishable from those of wakefulness.

Extracellularly recorded cortical neurons also display similar dynamics during activated states and SWS Up-states. It has been known for some time that the mean firing rate of cortical neurons during wakefulness and SWS is in the same range (Hobson and McCarley 1971), as clearly visible from Fig. 3.5. However, the dynamic similarity not only concerns the mean rate, but also the temporal patterns of discharge and the respective timing between different cells. Although such entities are more difficult to characterize, a convenient, albeit subjective, way to appreciate such information is by transforming the spike patterns into audio signals by assigning a specific note to each neuron. Allowing for the presence of "concerted neuronal silences" during SWS due to the slow oscillation Down-states, the melodies produced by wakefulness and SWS are remarkably similar (such audio material can be obtained from `http://www.archive.org/details/NeuronalTones`).

Importantly, the similarity between cortical Up-states and activated states is not only evident in EEG/LFP dynamics and single-cell firing, but also extends to the relationship between them. Performing wave-triggered averages of spiking activity, by averaging the periods of neuronal firing around the negative peaks of the LFP, reveals that the depth-negative EEG component is correlated with an increased probability of unit firing, both in the awake state and during the slow oscillation Up-states (Fig. 3.10c).

From the recent intracellular investigation of SWS (Steriade et al. 2001), the membrane potential dynamics of cortical neurons were decomposed into excitatory and inhibitory conductance components (Rudolph et al. 2007, see also Chap. 8). Such an analysis demonstrates that in the majority of cortical cells, inhibition

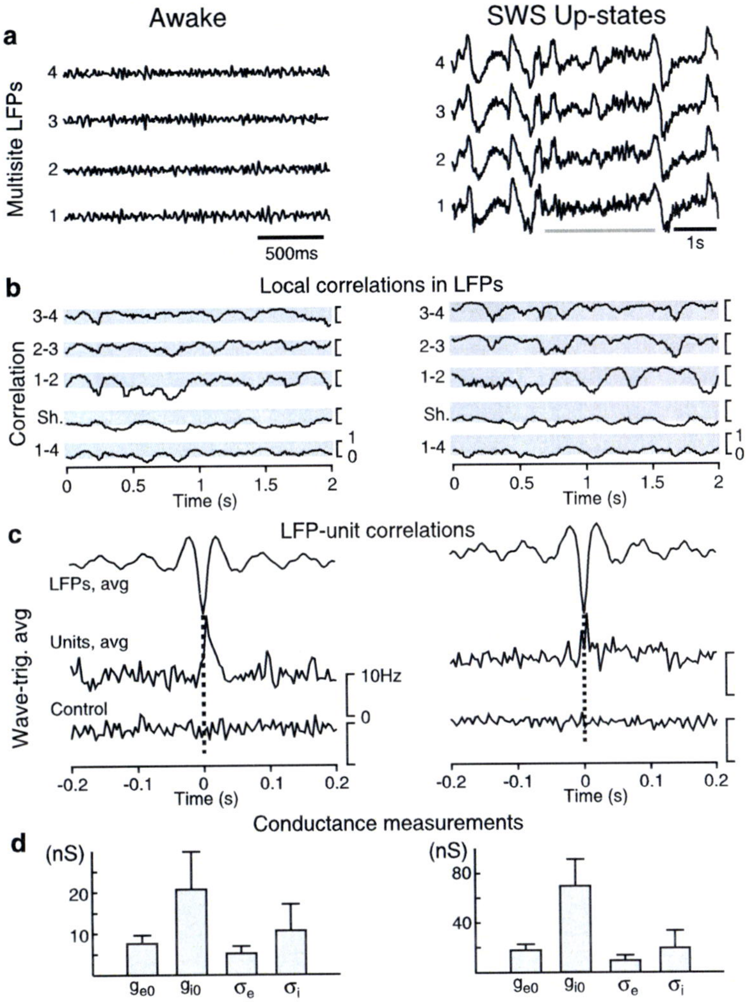

Fig. 3.10 Sleep Up-states and wakefulness are similar network states. (**a**) LFP activity in the cat parietal cortex during wakefulness (*left*) and SWS (*right*). Four electrodes (traces 1–4) were placed at cortical depth along a line in the anterior–posterior axis of the suprasylvian gyrus at an interelectrode distance of 1 mm. During SWS, "Up-states" were identified by locally desynchronized EEG activity (indicated by the *gray bar*). (**b**) Dynamics of local correlations in LFP activity. Correlations between adjacent electrodes calculated in small time windows (100 ms) show fluctuations between high and low values. Distant electrodes (1 and 4) did not show any significant correlation. (**c**) Correlation between unit firing and LFP activity. The wave-triggered average shows that the negative LFP deflections are correlated to increased firing in both wakefulness and Up-states of SWS. The control indicates the absence of relation after randomizing spike times. (**d**) Conductance measurements show similar excitatory and inhibitory components in wakefulness and Up-states of SWS. (**a–c**) modified from Destexhe et al. (1999); (**d**) modified from Rudolph et al. (2007)

is stronger than excitation, both at the level of mean conductances, as well as at the level of conductance variations (as quantified by the standard deviation σ of the conductance; see Fig. 3.10d). This pattern is seen for both wakefulness and the Up-states of SWS. However, there is a significant difference between the absolute values measured during the two states, with SWS generally showing higher conductances (Fig. 3.10d). Nevertheless, both states have a qualitatively similar ratio of excitatory-to-inhibitory conductance, in which both the mean inhibitory conductance and its associated fluctuations are larger on average compared to excitatory contributions (Fig. 3.10d). This resemblance also extends to the spike-triggered average conductance patterns in the two states (Rudolph et al. 2007).

3.2.3 Intracellular Activity During Anesthesia

The vast majority of electrophysiological studies in vivo are performed on anesthetized animals. Thus, it is important to analyze cortical dynamics during anesthesia in its relation with the activity of unanesthetized animals. Figure 3.11 shows typical examples of intracellular and EEG recordings during different states of activity, including awake animals (Fig. 3.11a), ketamine–xylazine (KX; Fig. 3.11b) and barbiturate (Fig. 3.11c) anesthesia. In all these cases, the recorded activity contrasts with the relative quiescence usually seen in intracellularly recorded neurons in cortical slices kept in vitro (Fig. 3.11d). The activity during KX anesthesia displays Up- and Down-state dynamics, very similar to that observed during SWS (Fig. 3.11b), while the activity of barbiturate anesthesia is mostly hyperpolarized (Fig. 3.11c), with neurons firing at low rates and EEG slow waves different from SWS.

The level of synaptic "noise" can be quantified by the standard deviation (SD) of membrane potential, σ_V. A comparison between different states reveals that σ_V is generally highest during barbiturate anesthesia, although the V_m distribution is not symmetric in this case (see below). Membrane potential fluctuations are around 4 mV in the Up-states, and around 2 mV in awake animals. In these cases, the V_m distributions is essentially symmetric.

Various intracellular studies of waking animals reported that cortical neurons have a low input resistance (5–40 MΩ) and a depolarized membrane potential (about -60 mV) which fluctuates markedly ($\sigma_V = 2$–6 mV), causing irregular and tonic discharges in the 5–40 Hz frequency range (Fig. 3.11a; Steriade et al. 2001). Differences in input resistance are also observed depending on the behavioral state (e.g. wakefulness, slow-wave sleep or paradoxical sleep; see Steriade et al. 2001), but in most cases, input resistance values are relatively low compared to those reported in vitro (Bindman et al. 1988, Paré et al. 1998b). These findings are the same, irrespective of the cortical area (Woody and Gruen 1978, Berthier and Woody 1988, Matsumura et al. 1988, Baranyi et al. 1993, Steriade et al. 2001). However, input resistance measurements cannot be compared from one study to the next for a number of reasons. First, they depend on the electrode shape and impedance which might significantly vary between laboratories. Second, in the case

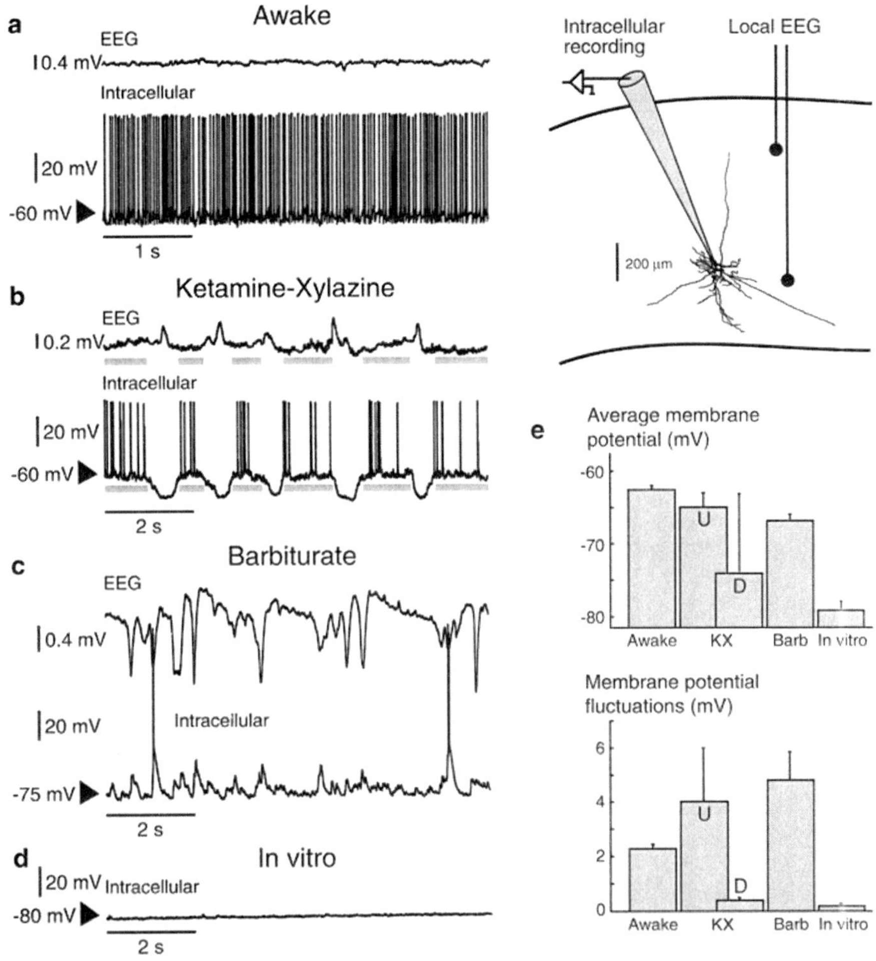

Fig. 3.11 Intracellular and electroencephalogram (EEG) activity during different brain states. Parallel intracortical EEG and intracellular recordings are compared across different cortical states. (**a**) Awake animals: the EEG is desynchronized and the intracellular recording is characterized by a depolarized and highly fluctuating membrane potential associated with irregular firing. (**b**) Under ketamine–xylazine anesthesia, the EEG oscillates between two phases: desynchronized periods (Up-states, U; bars) with fast irregular EEG oscillations; and slow waves, during which fast EEG activities are absent or strongly reduced (Down-states, D). During the desynchronized periods (*bars*), the membrane potential is depolarized and highly fluctuating, whereas it is hyperpolarized during slow waves. (**c**) Barbiturate anesthesia: the EEG displays slow waves, whereas the intracellular signal consists of depolarized bursts riding on a hyperpolarized level. (**d**) In vitro recordings obtained in cortical slices. In this case, the network activity was reduced, as shown by the quiescent intracellular signal. (**e**) Comparison of the average value and standard deviation of the membrane potential across different states. (**a**) modified from Steriade et al. (2001); (**b**) modified from Destexhe and Paré (1999); (**c, d**) modified from Paré et al. (1998b); (**e, f**) modified from Destexhe et al. (2003a)

of data from unanesthetized animals, they may depend on, or be influenced by, their behavioral state.

Findings similar to the those mentioned in the last paragraph were reported using anesthetics such as KX or urethane. In low doses, these anesthetics produce alternating periods of Up- and Down-states (Fig. 3.11b). During the Up-state (Fig. 3.11b, *bars*), which is associated with locally desynchronized fast EEG activity (as described for SWS above), cortical neurons are depolarized, fire spontaneously, and have a low input resistance (Contreras et al. 1996, Paré et al. 1998b, Destexhe and Paré 1999), similar to the observations in awake animals. During the Down state, which is accompanied by a reduction in fast EEG activity, cortical neurons have a more stable and hyperpolarized membrane potential (around -90 mV) and a higher input resistance ($39 \pm 9\text{M}\Omega$, compared with $9.3 \pm 4.3\text{M}\Omega$ for Up-states). Furthermore, there is a remarkable correspondence between the LFP (local EEG) and intracellular activity during KX anesthesia (Fig. 3.12).

3.2.4 *Activated States During Anesthesia*

A further evidence for the resemblance between Up-states and activated states comes from artificial EEG activation. Stimulation of the brainstem ascending systems responsible for the maintenance of the waking state, such as the pedonculo-pontine tegmentum (PPT), elicits a prolonged Up-state with a desynchronized EEG in animals anesthetized with KX (Fig. 3.13), as shown earlier (Steriade et al. 1993a, Steriade et al. 1993b, Rudolph et al. 2007) for KX or with urethane (Kasanetz et al. 2002). This pattern is similar to that seen in unanesthetized animals during the transition from SWS to wakefulness (see Fig. 3.9). These observations further support the notion that such Up-states represent network states similar to wakefulness.

Analysis of PPT-induced activated states indicated similar electrophysiological characteristics as seen in awake animals (Rudolph et al. 2005). The average membrane potential is slightly more hyperpolarized compared to the Up-state (average membrane potential about -75 mV $\pm$ 10 mV; Fig. 3.14a, top) and the V_m fluctuations are slightly reduced ($\sigma_V = 2.28 \pm 1.36$ mV; Fig. 3.14a, middle). Input resistance (R_{in}) measurements showed further that the R_{in} is always significantly lower during the Up-states ($R_{\text{in}} = 10.08 \pm 3.87$ MΩ; Fig. 3.14a, bottom) compared to the Down-state ($R_{\text{in}} = 14.91 \pm 2.28$ MΩ; $p < 0.006$). Following PPT stimulation, the R_{in} increases markedly by 44% $\pm$16% ($R_{\text{in}} = 18.03 \pm 9.44$ MΩ; Fig. 3.14a, bottom and b), yielding a R_{in} ratio of 2.09 ± 1.23 between Up-states and post-PPT states. In the aforementioned study, all R_{in} estimates were obtained by linear fits of the I–V-curves in the corresponding states, and gave values in good agreement with estimates obtained by injection of short current pulses (Rudolph et al. 2005).

After PPT stimulation, the SD of the membrane potential was found to be stable for up to 20 s (Fig. 3.14c, top), before increasing due to the alternating pattern of Up-states and Down-states (slow oscillations). In contrast, the R_{in} shows higher values only for shorter periods (around 7 s) before returning back to the values typical of Up- and Down-states (Fig. 3.14c, bottom).

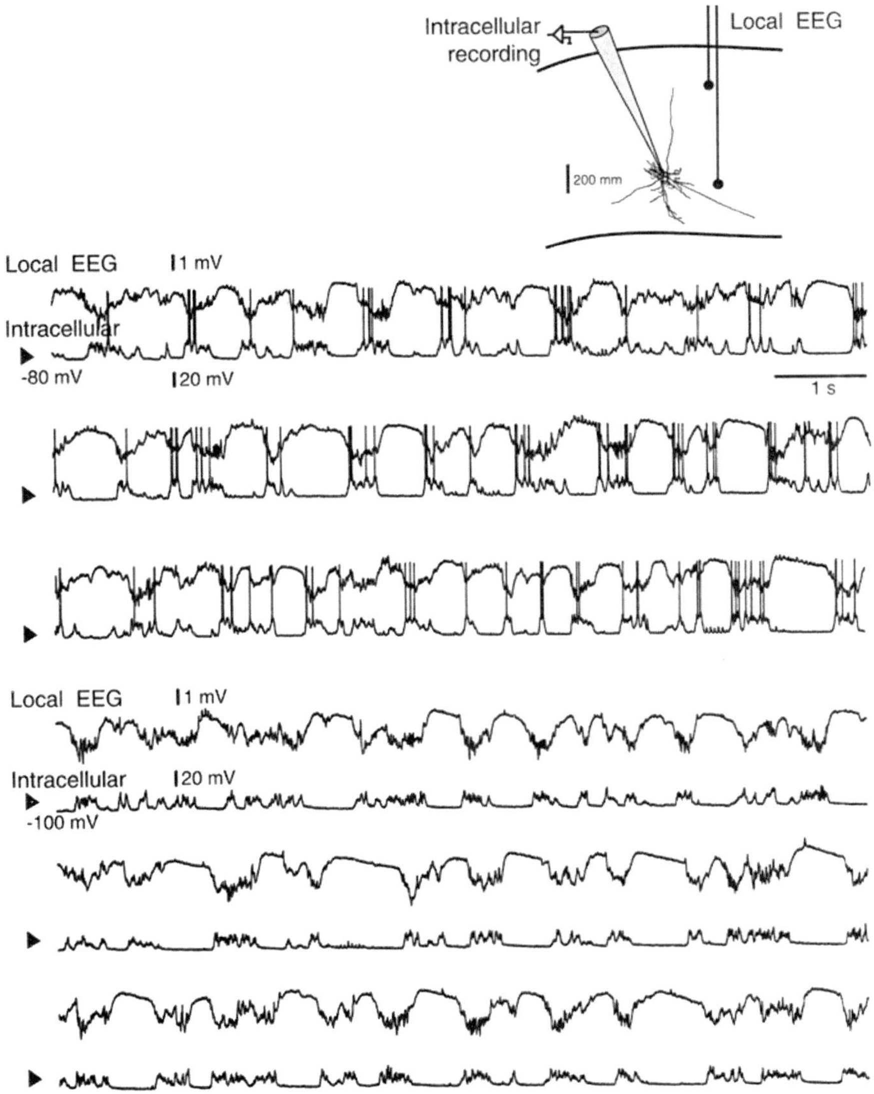

Fig. 3.12 Relation between intracellular and local electroencephalogram (EEG) activity during Up- and Down-states. Parallel intracortical EEG ("local EEG") and intracellular recordings are compared during ketamine–xylazine anesthesia. The *top* recordings are with no current injected, while the *bottom* graph shows recordings with −1 nA steady injected current, which limited action potential firing. These results reveal a remarkable coincidence between subthreshold activity and local EEG. Courtesy of Joe-Guillaume Pelletier and Denis Paré, Rutgers University; Modified from Rudolph et al. (2005)

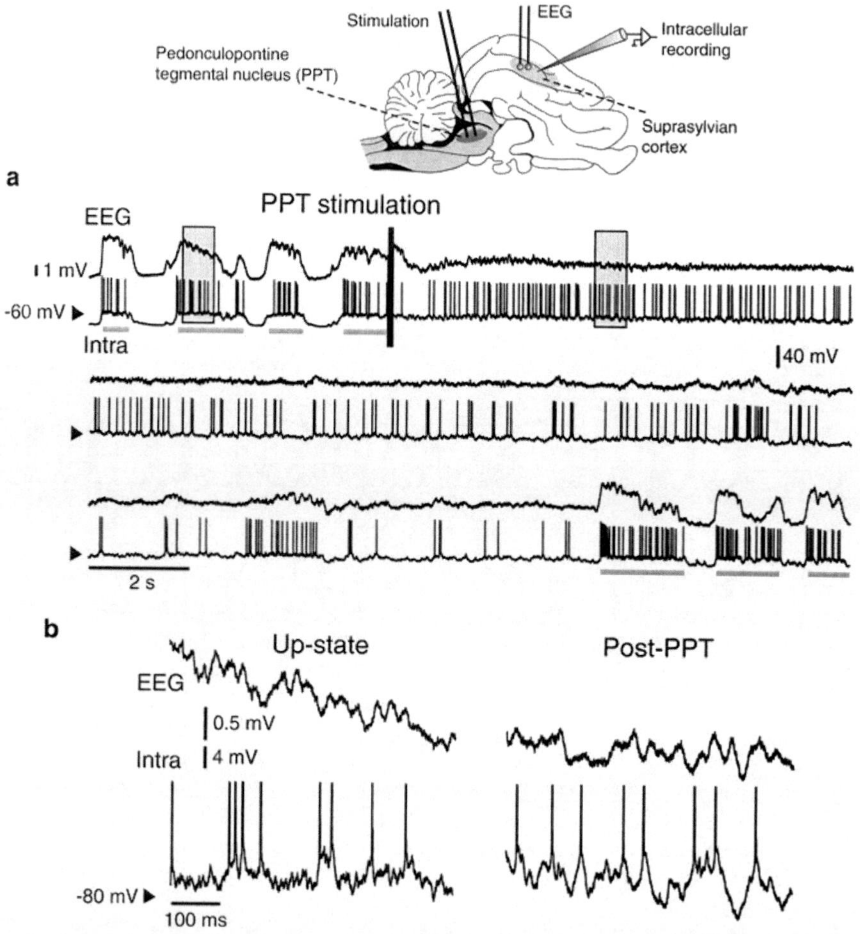

Fig. 3.13 Spontaneous and pedonculopontine tegmental (PPT)-induced electroencephalogram (EEG)-activated states under ketamine–xylazine anesthesia. (**a**) cortical neuron recorded in cat parietal cortex (areas 5–7) displays Up-states (*gray bars*) and Down-states of activity that were paralleled by slow waves in the EEG. Electrical stimulation of the PPT (100 Hz for 0.1 s; see *scheme*) produced long periods of desynchronized EEG activity. Intracellularly, PPT stimulation induced periods characterized by a depolarized membrane potential as well as membrane potential fluctuations and discharge activity similar to that seen in the Up-states. After 20–30 s, the slow waves progressively reappeared in the EEG, paralleled by the return to Up/Down-state dynamics. (**b**) Expanded views of segments indicated by shaded boxes of (**a**). Modified from Rudolph et al. (2005)

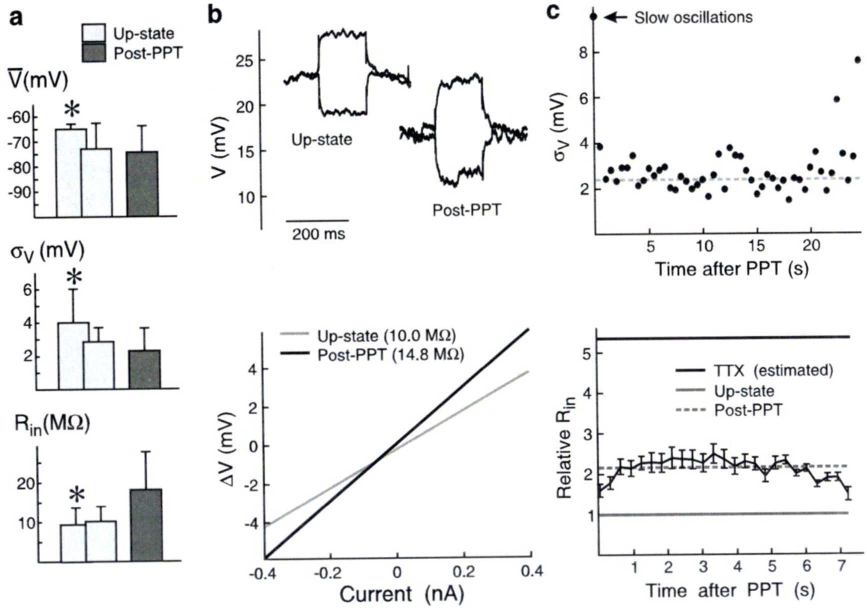

Fig. 3.14 Analysis of PPT-induced activated states during anesthesia. (**a**) Up-states and post-PPT states are characterized by a marked depolarization (*top*: average membrane potential $\bar{V}$) and large membrane potential fluctuation amplitudes (*middle*: V_m standard deviation σ_V). The input resistance R_{in} was smaller in Up-states (*bottom*) compared to post-PPT states. The *stars mark* corresponding values obtained in other measurements in the same preparation (Destexhe and Paré 1999; Paré et al. 1998b). (**b**) The input resistance was estimated by injecting brief current pulses (0.4 nA in the example shown in *top panel*) in the linear portion of current–voltage relations. The R_{in} was always smaller during the Up-states compared to the state induced by PPT stimulation, as indicated by the steeper slope in the I–V plot (*bottom*). (**c**) Standard deviation of membrane potential σ_V (*top*) and normalized R_{in} (*bottom*) as a function of time after PPT stimulation (average of seven cells; consecutive windows of 500 ms and 300 ms, respectively). The first point in the σ_V graph indicates the value calculated over a long period of slow oscillations). The reference R_{in} (*bottom*) was the average input resistance during Up-states (*gray*). Modified from Rudolph et al. (2005)

3.2.5 *Miniature Synaptic Activity In Vivo*

Miniature synaptic activity was first discovered in the neuromuscular junction (Fatt and Katz 1952) and later identified in central neurons (Martin 1977; Redman 1990). Miniature events correspond to the spontaneous, AP independent release of synaptic terminals. Although such spontaneous release is a rare event at the level of a single synaptic terminal, this activity is detectable in central neurons because of the very large number of synapses. Miniature synaptic events were also found in neocortical neurons in vivo (Paré et al. 1997). To reveal miniature synaptic activity, AP dependent release must be blocked, which can be done experimentally using microdialysis

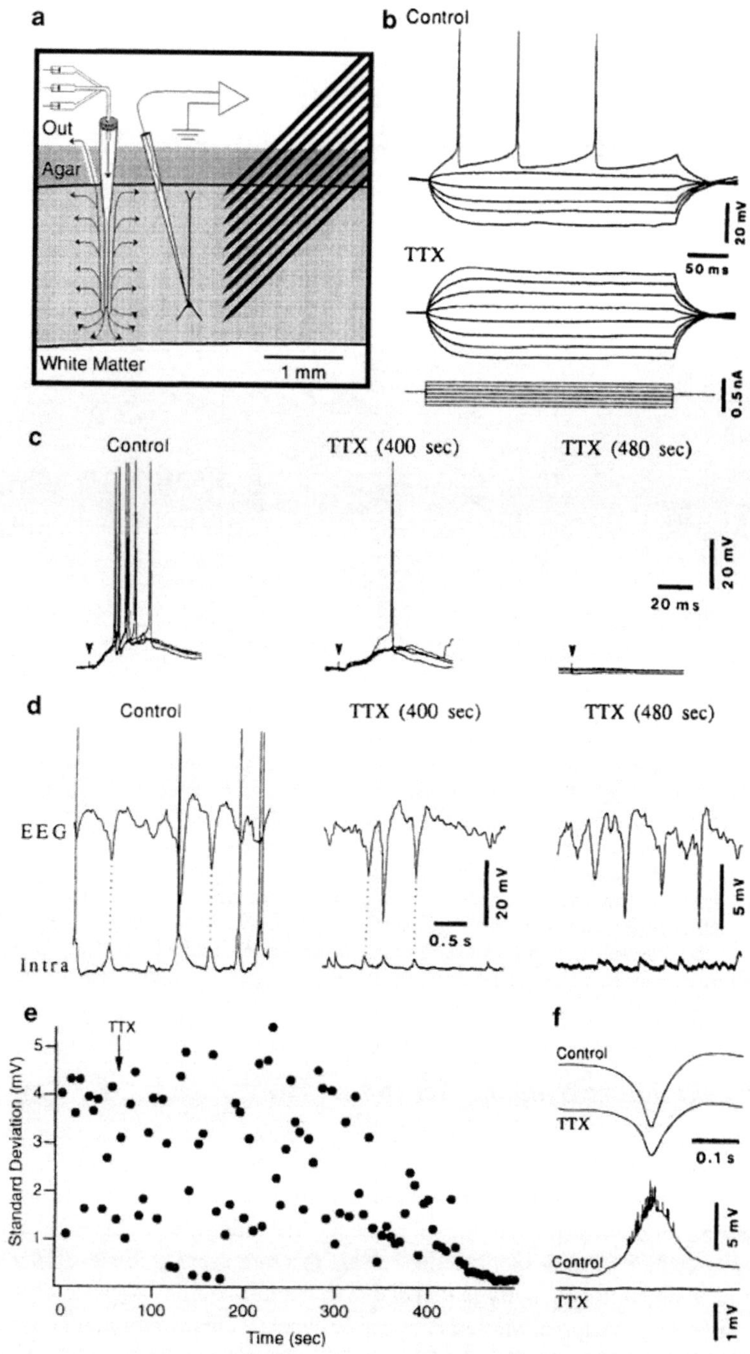
a
Out
Agar
White Matter
1 mm
b Control
20 mV
50 ms
TTX
0.5nA
c
Control
TTX (400 sec)
TTX (480 sec)
20 mV
20 ms
d
Control
TTX (400 sec)
TTX (480 sec)
EEG
Intra
20 mV
0.5 s
5 mV
e
TTX
Standard Deviation (mV)
Time (sec)
f
Control
TTX
0.1 s
5 mV
Control
TTX
1 mV

of tetrodotoxin (TTX), a sodium channel blocker (see Fig. 3.15). TTX microdialysis abolishes intracellularly evoked spikes (Fig. 3.15b), as well as synaptic potentials evoked by extracellular electrical stimulation (Fig. 3.15c). It also abolishes network activity in relation to the contralateral EEG (Fig. 3.15d). The SD of the intracellular V_m activity was observed to drop spectacularly following TTX application (Fig. 3.15e).

Although TTX microdialysis suppresses all synaptic events related to network activity, it does not abolish all synaptic activity, revealing the presence of miniature events (*minis*). Figure 3.16a illustrates the frequency distribution of minis in a sample of neurons. The mini frequencies appear to cluster around two values (9 Hz and 21 Hz; Fig. 3.16a). In a study by Paré et al. (1997), in order to analyze the cellular correlates of this distribution, neurons were divided into two groups using a frequency of 12.5 Hz as a dividing point (group 1 $<$ 12.5 Hz $<$ group 2). In agreement with previous findings (Salin and Prince 1996), variations in mini frequency were not found to be related to differences in the laminar position of the cells, because the proportion of supra- and infragranular neurons was similar in both groups. Moreover, differences in R_{in} were not significant, so that neurons with similar R_{in} could display minis at low or high frequencies. Examples of group 1 and 2 from the aforementioned recordings are shown in Fig. 3.16b, c, respectively. As exemplified in Fig. 3.16, minis are not only more frequent in group 2 recordings, but are also larger in amplitude (1.56 ± 0.88 mV, compared with 0.77 ± 0.35 mV). Further, group 2 neurons show a significantly more depolarized V_m (-61.7 ± 5.02 mV) than group 1 cells (-69.9 ± 6.03 mV).

In Paré et al. (1997), to test the pharmacological sensitivity of minis in group 1 (Fig. 3.16d) and group 2 recordings (Fig. 3.16e), a current pulse of constant amplitude was applied every 3–6 s while the cells were manually clamped at a relatively constant V_m (-70 mV). The variance of the intracellular signal was measured between the current pulses (Fig. 3.16d1, e1) and the R_{in} was estimated from the amplitude of the voltage responses to the current pulses (Fig. 3.16d2, e2). After obtaining a baseline period, the TTX solution was switched to one containing TTX plus bicuculline (200 μM) or the non-NMDA glutamatergic antagonist NBQX (200 μM).

Minis observed in group 1 recordings resisted NBQX, but were abolished by bicuculline (Fig. 3.16d). Also shown in Fig. 3.16d, the disappearance of these

Fig. 3.15 Microdialysis of tetrodotoxin (TTX) uncovers miniature synaptic potentials (minis) in neocortical neurons in vivo. (**a**) scheme of the experimental setup. Dialysis of TTX abolishes current-evoked spikes (**b**) as well as synaptic events evoked by intracortical stimuli (**c**) or occurring in relation to large-amplitude electroencephalographic (EEG) potentials ((**d**); contralateral EEG shown). (**e**) SD of the intracellular signal as a function of time. The SD was measured every 5 s, from stimulation-free epochs of 4 s. *Downward arrow*: onset of TTX dialysis. (**f**) peri-event averages ($n = 30$) of intracellular events using the negative peak of large EEG potentials as a temporal reference before (Control) and 10 min after the beginning of TTX application. *Top and bottom* 2 traces: average EEG and intracellular signal, respectively. Modified from Paré et al. (1997)

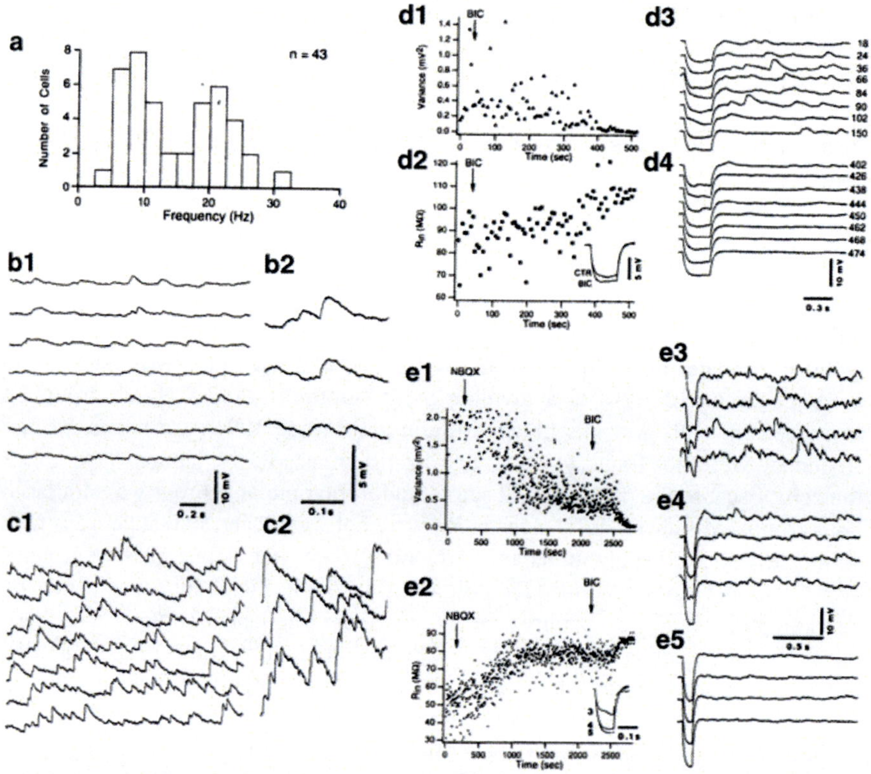

Fig. 3.16 Analysis of miniature synaptic events in cortical pyramidal neurons in vivo. (**a**) Distribution of mini frequencies in a sample of 43 neurons, 26 of which were morphologically identified as pyramidal cells. Representative examples of group 1 (**b**) and group 2 recordings (**c**) at low (1) and high gain (2). (**d, e**). contrasting pharmacological sensitivity of minis in group 1 and 2 recordings. In group 1 recordings (**d**), bicuculline (BIC) dialysis abolishes the minis ($n = 6$), thus reducing the signal variance (**d1**) and increasing the input resistance (R_{in}, **d2**). Sample traces in **d3** and **d4** illustrate the effects of bicuculline on the minis in the same cell. Numbers on the right: s, in reference to the X-axis in **d1** and **d2**. Current pulse amplitude: 0.1 nA. In group 2 recordings (**e**), most minis were abolished by NBQX dialysis, which resulted in a reduction of the signal variance (**e1**) and an augmentation of the R_{in} (**e2**). Some minis persisted after NBQX application. These events were abolished by dialysis of bicuculline. Sample traces on right: minis in control condition (**e3**), 25 min after the onset of NBQX dialysis (**e4**), and 10 min after the onset of bicuculline (**e5**). Membrane potential (V_m) changed from −55 mV in control conditions to −66 mV after NBQX. Current pulse amplitude: 0.08 nA (**d**), 0.2 nA (**e**). Modified from Paré et al. (1997)

GABA_A minis produced a 11% increase in R_{in} (Fig. 3.16d2), that coincided with a decrease of the variance of the intracellular signal (Fig. 3.16d1).

In contrast, as illustrated in Fig. 3.16, group 2 recordings were characterized by a much higher signal variance (Fig. 3.16e1) than recordings from group 1 cells (Fig. 3.16d1; 2.01 ± 0.76 mV compared with 0.16 ± 0.03 mV). Moreover, in contrast with group 1 recordings, all minis from group 2 were abolished by dialysis of 1,2,3,4-tetrahydro-6-nitro-2,3dioxo-benzo[f]quinoxaline-7-sulfonamide disodium (NBQX), a blocker of glutamate AMPA receptors. As shown in Fig. 3.16e, NBQX produced a decrease in signal variance (Fig. 3.16e1) and mini frequency (Fig. 3.16, e3 and e4) that coincided with a marked increase in R_{in} (49 ± 8 %). Further application of bicuculline abolished the remaining minis (Fig. 3.16e5), further reducing the signal variance (Fig. 3.16e1), and increasing the R_{in} by an additional 10% (Fig. 3.16e2; Paré et al. 1997).

Finally, there is substantial evidence that these group 1 and 2 recordings represent somatic and dendritic recordings, respectively. In pyramidal neurons, the soma and initial axon segment exclusively form symmetrical synaptic contacts presumed to be GABAergic, while dendrites have a higher synaptic density, with the majority of synapses (70–95%) being asymmetric and, presumably, glutamatergic (White 1989 DeFelipe and Fariñas 1992). Many group 1 and 2 recordings were obtained from morphologically identified pyramidal neurons, and the dendritic impalement was confirmed morphologically in some cases of group 2 recordings (Paré et al. 1997).

3.2.6 Activated States In Vitro

Some in vitro preparations can show activity states very similar to some anesthetized states in vivo, as illustrated in Fig. 3.17. Intracellular recordings from primary visual cortical neurons in halothane-anesthetized cats revealed (Sanchez-Vives and McCormick 2000) the rhythmic occurrence of Up- and Down-state dynamics, in which the Up-state appeared as a barrage of PSPs that could reach AP threshold (Fig. 3.17a, b). Intracellular and extracellular recordings from ferret visual and prefrontal cortical slices maintained in vitro in "traditional" slice bathing medium (2 mM Ca^{2+}, 2 mM Mg^{2+} and 2.5 mM K^{+}) did not reveal spontaneous rhythmic activities. However, changing the bath solution to more closely mimic the ionic composition of brain interstitial fluid in situ (1.0 or 1.2 mM Ca^{2+}, 1 mM Mg^{2+} and 3.5 mM K^{+}) caused the appearance of spontaneous activity with Up- and Down-state dynamics (Fig. 3.17d, e).

These dynamical characteristics were found to be very similar to those occurring in vivo (Fig. 3.17d–f). Once the slow oscillation developed, it is stable for the life of the slice (up to 12 h). Moreover, it appears as a depolarized state associated with AP activity at 2–10 Hz (in regular spiking neurons), followed by a hyperpolarized state, recurring with a periodicity of once every 3.44 ± 1.76 s (Fig. 3.17d–f). The same Up- and Down-state dynamics also appeared in fast-spiking local interneurons in phase with extracellularly recorded multiunit activity (see details in Sanchez-Vives and McCormick 2000).

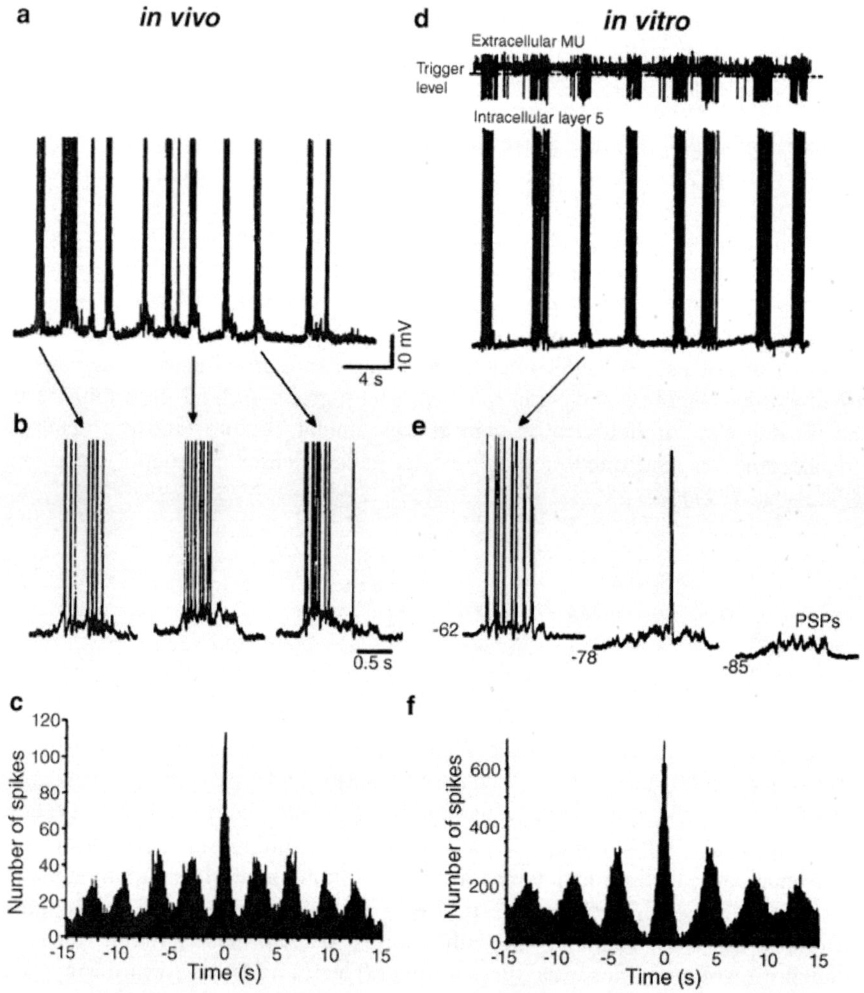

Fig. 3.17 Up/Down-states in vivo and in vitro. (**a**) Intracellular recordings in the primary visual cortex of a halothane-anesthetized cat reveal a rhythmic sequence of depolarized and hyperpolarized membrane potentials (Up- and Down-states). (**b**) Expansion of three of the Up-states for clarity. (**c**) Autocorrelogram of the intracellular recording reveals a marked periodicity of about one cycle per 3 s. (**d**) Simultaneous intracellular and extracellular recordings of a slow oscillation in ferret visual cortical slices maintained in vitro. Note the marked synchrony between the two recordings. The trigger level for the window discriminator of the extracellular multi-unit recording is indicated. (**e**) The Up-state at three different membrane potentials. (**f**) Autocorrelogram of the intracellular recording in (**d**) shows a marked periodicity of approximately once per 4 s. Modified from Sanchez-Vives and McCormick (2000)

3.3 Quantitative Characterization of Synaptic Noise

In previous sections, we have reviewed several characteristics of "synaptic noise," as seen extracellularly and intracellularly in awake and anesthetized states, as well as in vitro. In the present section, we review more detailed characterizations, such as V_m distributions, conductance measurements and power spectral analysis.

3.3.1 Quantifying Membrane Potential Distributions

The V_m distribution calculated from different states are generally symmetric, as shown in Fig. 3.18a for V_m distributions calculated from intracellular recordings in awake cats. Distributions calculated from the Up-states of SWS are also symmetric (Fig. 3.18b). The same also applies to PPT-activated states (Fig. 3.18c) and the Up-states of Ketamine–xylazine anesthesia (Fig. 3.18d). Finally, approximately symmetric V_m distributions are also observed during the Up-states in vitro (Fig. 3.18e). Thus, in all these different preparations, the V_m distribution is symmetric, unless Up- and Down-states are mixed (see Fig. 3.19a), or in barbiturate anesthesia, where the V_m distribution was found to be clearly asymmetric (Fig. 3.19b).

The quantification of these V_m distributions can be done by evaluating the SD of membrane potential, σ_V. Because V_m distributions are essentially symmetric and close to Gaussian, the average value $\bar{V}$ and SDs are sufficient. In various studies, these values were evaluated for different anesthetized states as well as during wakefulness (see Sect. 3.2.3). The level of fluctuations, as quantified by σ_V, was found to be around 2 mV in awake animals (Rudolph et al. 2007), but values of 2–6 mV were observed in anesthetized states (Paré et al. 1998b; Destexhe and Paré 1999). These values were represented graphically for different cells (see Fig. 3.22 below). In another study (Rudolph et al. 2005), values of $\sigma_V = 2.28 \pm 1.36$ mV were reported in artificially induced activated states (Fig. 3.14a). TTX microdialysis (Paré et al. 1998a) reduced the σ_V down to about 0.4 mV (see Fig. 3.21c).

3.3.2 Conductance Measurements In Vivo

As mentioned in Sect. 3.2, differences in input resistance are observed depending on the behavioral state, such as wakefulness, slow-wave sleep or paradoxical sleep (Steriade et al. 2001). However, in most of these cases, input resistance values are relatively low compared to those reported in vitro (Bindman et al. 1988, Paré et al. 1998b). These results are the same irrespective of the cortical area (Woody and Gruen 1978, Berthier and Woody 1988, Matsumura et al. 1988, Baranyi et al. 1993, Steriade et al. 2001). However, such input resistance measurements cannot be easily compared across different studies as they depend not only on the electrode

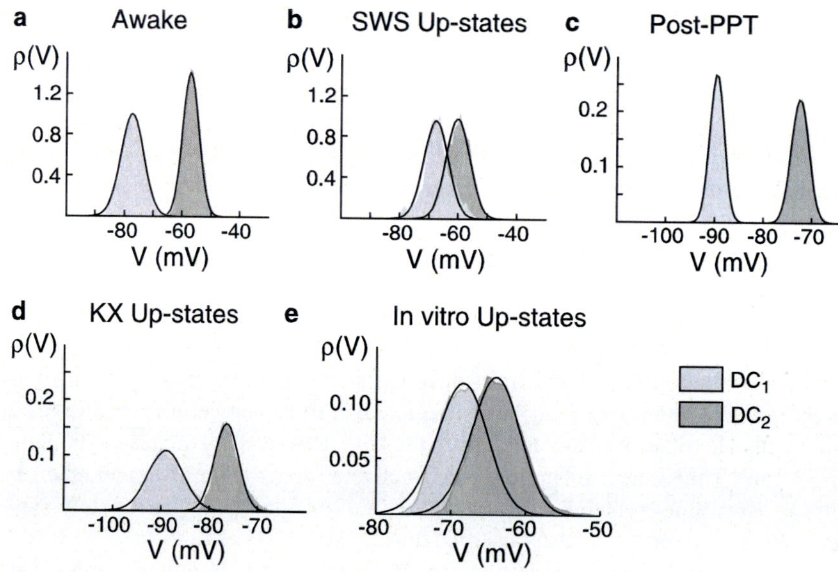

Fig. 3.18 Membrane potential distributions during activated states in vivo and in vitro. (**a**) V_m distribution calculated from intracellular recordings in awake cats. (**b**)V_m distribution during the Up-states of slow-wave sleep in cats (different cell compared to (**a**)). (**c**) V_m distribution during post-PPT activated states in cats under KX anesthesia. (**d**) V_m distribution during the Up-states of KX anesthesia (same cell as in (**a**)). (**e**) V_m distribution during Up-states in ferret visual cortex slices. In all cases, two levels of current injection (DC_1 and DC_2) are indicated. (**a, b**) modified from Rudolph et al. (2007); (**c, d**) modified from Rudolph et al. (2005); (**e**) modified from Piwkowska et al. (2008)

shape and impedance, but also, in particular in the case of data from unanesthetized animals, on the animal's behavioral state. For instance, the relatively high input resistance reported in Steriade et al. (2001) is based on data obtained during quiet wakefulness, which is presumably associated with lower levels of synaptic bombardment when compared to active waking. However, values obtained in the same laboratory (Fig. 3.11b–d; *KX*, *Barb*, and in vitro in Fig. 3.19d) are comparable if they were obtained in the same cell type (for instance, the values before and after TTX in Fig. 3.19c and e were from the same cells).

Furthermore, comparing different states of anesthesia using current injection shows that barbiturates induce a state of lower global conductance compared to KX. For instance, Fig. 3.19a, b compare the effect of constant current injection in barbiturate and KX anesthesia. In Fig. 3.19a, the voltage distribution shows two peaks, typical of the Up- and Down-states under KX. Current injection has a much larger effect on the Down-state than the Up-state, betraying a reduced conductance in the Down-state. Interestingly, the distributions obtained under barbiturate anesthesia (Fig. 3.19b) are similar to the Down-states of KX.

To more firmly investigate the contribution of synaptic activity to high-conductance states, previous studies compared the same intracellularly recorded

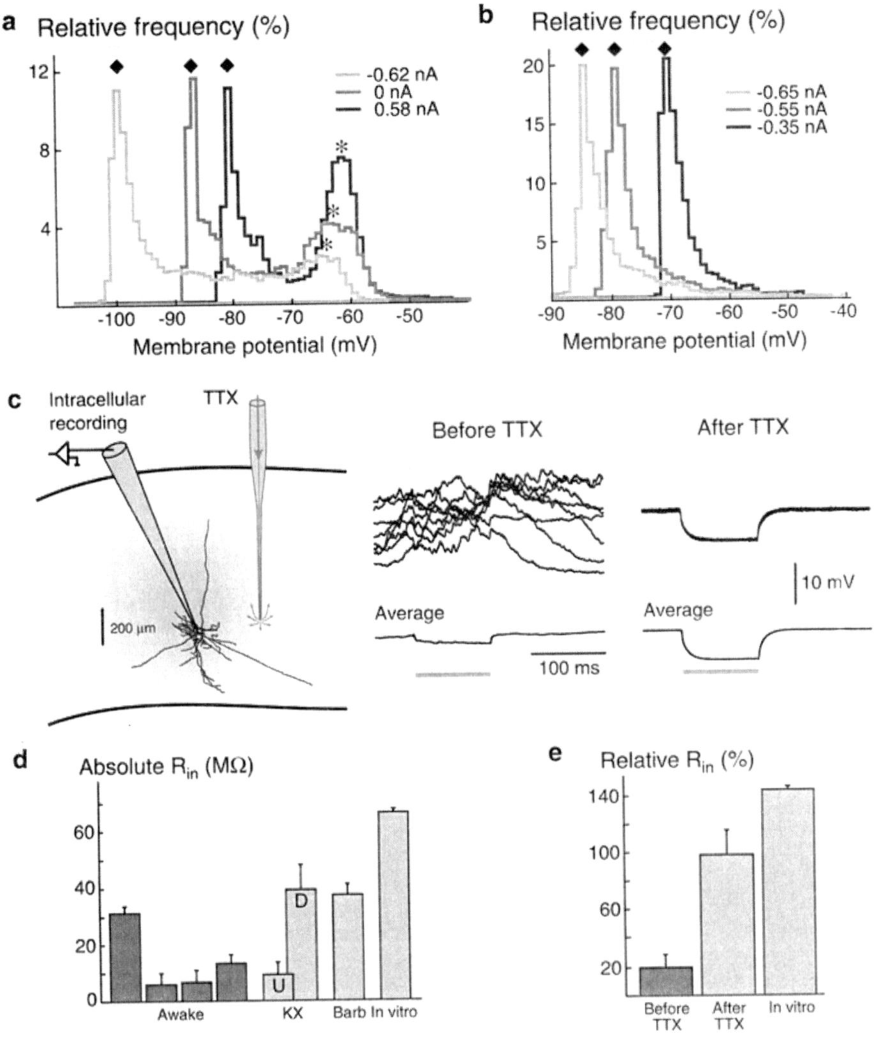

Fig. 3.19 Conductance measurements during different brain states. (**a**) Ketamine–xylazine (KX) anesthesia: membrane potential distributions at different direct current levels. The hyperpolarized (*diamonds*) and depolarized (*asterisks*) peaks correspond to the two states of the membrane (Down- and Up-states, respectively). Current injection has less effect on the peak of the Up-state than the Down-state, indicating a lower input resistance in the Up-state. (**b**) Barbiturate anesthesia: same procedure as in (**a**). In this case, the distribution of membrane potential is closer to the Down-state of KX anesthesia shown in (**a**). (**c**) Suppression of network activity using microperfusion of TTX. The scheme (*left*) illustrates the experimental setup; a microperfusion pipette was used to infuse TTX into the cortex in vivo. *Middle panel*: individual (*top*) and averaged (*bottom*) responses to injection of hyperpolarizing current pulses during the Up-state of KX. *Right panel*: responses to the same current pulse obtained in the same neuron after suppression of network activity by TTX. In this case, the input resistance and membrane time constant were about fivefold larger than in the Up-state. The post-TTX input resistance was similar to in vitro measurements using similar recording electrodes. (**d**) Absolute value of input resistance (R_{in}) measurements in different studies. In awake animals, from left to right, data from Steriade et al. (2001); Matsumura et al. (1988); Woody and Gruen (1978); Baranyi et al. (1993); (**a–c**) modified from Paré et al. (1998b)

neurons before and after suppression of network activity by microperfusion of the Na^+ channel blocker TTX in vivo (Paré et al. 1998b). TTX microperfusion produces a membrane hyperpolarization, an increased input resistance, and a remarkable stabilization of the membrane potential (Fig. 3.19c–e). After TTX application in vivo, the membrane potential and input resistance of cortical cells are similar to those seen in vitro using the same type of electrodes (Paré et al. 1998b). These findings indicate that increased cell damage by intracellular electrodes in vivo does not account for the differences between in vivo and in vitro results. Rather, these experiments suggest that the depolarized level and the low input resistance of cortical neurons in vivo are essentially due to spontaneous synaptic activity. This conclusion was strengthened using computational models, which indicate that, indeed, less than 10% of the input resistance is due to activation of voltage-dependent channels (Destexhe and Paré 1999).

An example of TTX dialysis experiment, carried out in a ketamine–xylazine anesthetized cat, is shown in Fig. 3.20 (Paré et al. 1998b). In this experiment, initially, a continuous flow of Ringer solution (1.0 μl/min) was applied through the ejection pipette. Figure 3.20b illustrates samples of evoked (Fig. 3.20b1) and spontaneous (Fig. 3.20b2) intracellular events at regular intervals (1–1.5 min) during the course of the experiment. After obtaining a baseline period, the Ringer solution was switched to one containing TTX (50 μM). Prior to the diffusion of the TTX, the injected pulse amplitude was highly variable as spontaneous synaptic events sometimes obliterated the response to the current pulse. TTX produced a gradual decline in the SD of the V_m (Fig. 3.20a1), accompanied by a reduction in the amplitude of orthodromic responses (Fig. 3.20a2) and a 54% increase in the amplitude of the voltage response to the current pulses (Fig. 3.20a3). In this comparison, the input resistance after TTX dialysis was compared to the "average" input resistance before TTX, which was mixing pulses in the Up- and Down-states. This explains why the difference is only of about twice, while it was about four times when Up-states were compared to post-TTX quiescent states (Destexhe and Paré 1999).

Figure 3.21 shows an example of the comparison between Up-states and post-TTX states. During the Up-states of KX anesthesia, neocortical neurons are characterized by highly fluctuating activity and spontaneous firing in the 5–20 Hz range (Fig. 3.21a1). After microperfusion of TTX, this activity is abolished (Fig. 3.21a2), as mentioned above, and the responses to cortical stimulation are totally suppressed (Paré et al. 1998a). In a subsequent study by Destexhe and Paré (1999), the input resistance R_{in} was estimated before and after TTX application by injection of hyperpolarizing current pulses in the linear portion of current–voltage relations. It was found that during active states, neocortical neurons had a low R_{in} (9.2 ± 4.3 MΩ; mean $\pm$ SE; $n = 26$), as evidenced by the relatively small voltage responses to intracellular current injection (Fig. 3.21b1). After TTX, the same cells showed a much larger R_{in} (46 ± 8 MΩ; $n = 9$), calculated at the same V_m (Fig. 3.21b2). Although in this state the absolute R_{in} values varied from cell to cell, nine cells recorded in this study before and after TTX showed similar *relative*

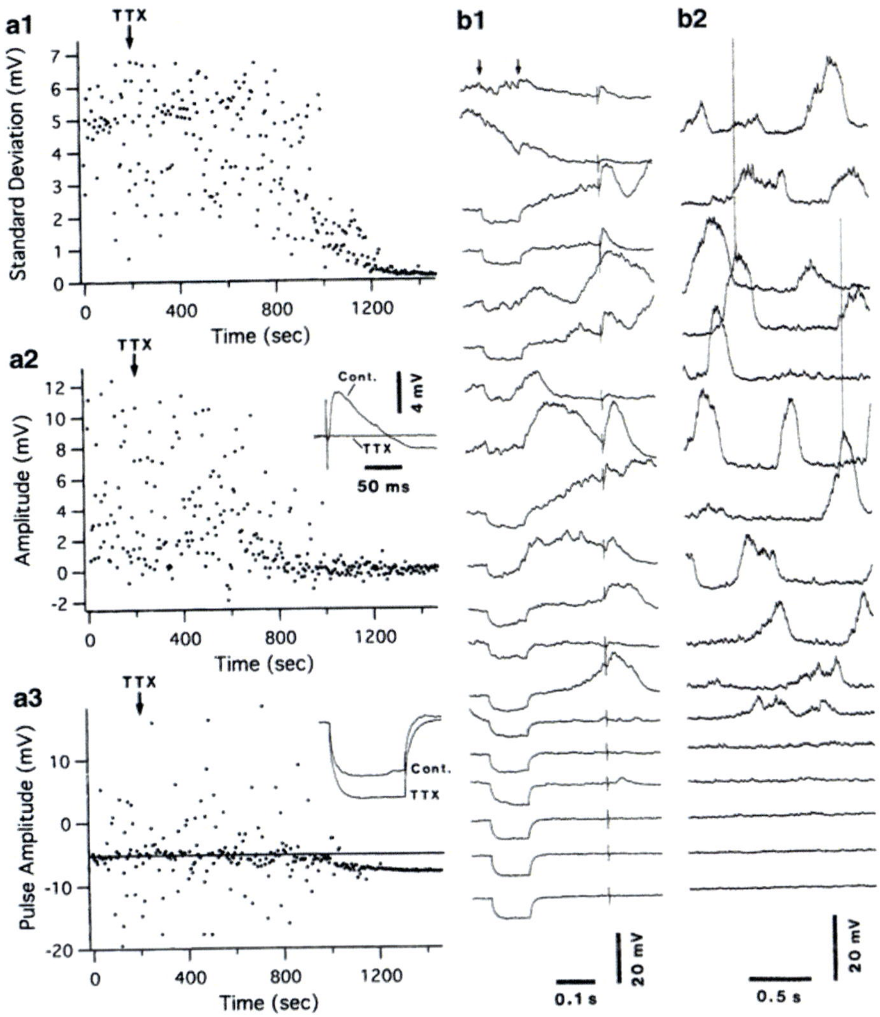

Fig. 3.20 Effect of TTX dialysis on the input resistance. (**a**) Time evolution of the SD of the intracellular signal σ_V (**a1**), of the amplitude of the response to a cortical shock (**a2**), and the voltage response to a current pulse of constant amplitude (0.2 nA; **a3**). *Arrow*: onset of TTX dialysis. The σ_V was measured every 5 s, from stimulation-free epochs of 2 s. A, 2 and 3, insets: comparison of cortically evoked response and voltage response to current pulses before and 20 min after onset of TTX dialysis. Averages of 20 sweeps, same scaling. (**b**) samples of evoked (**b1**) and spontaneous (**b2**) intracellular events at regular intervals (1–1.5 min) during the course of this experiment. Modified from Paré et al. (1998b)

changes, amassing to a fivefold (81.4 ± 3.6%) R_{in} decrease during active periods compared to quiescent conditions.

Another aspect of background activity is that cortical neurons display high-amplitude V_m fluctuations (Fig. 3.21a1). Measuring the amplitude of fluctua-

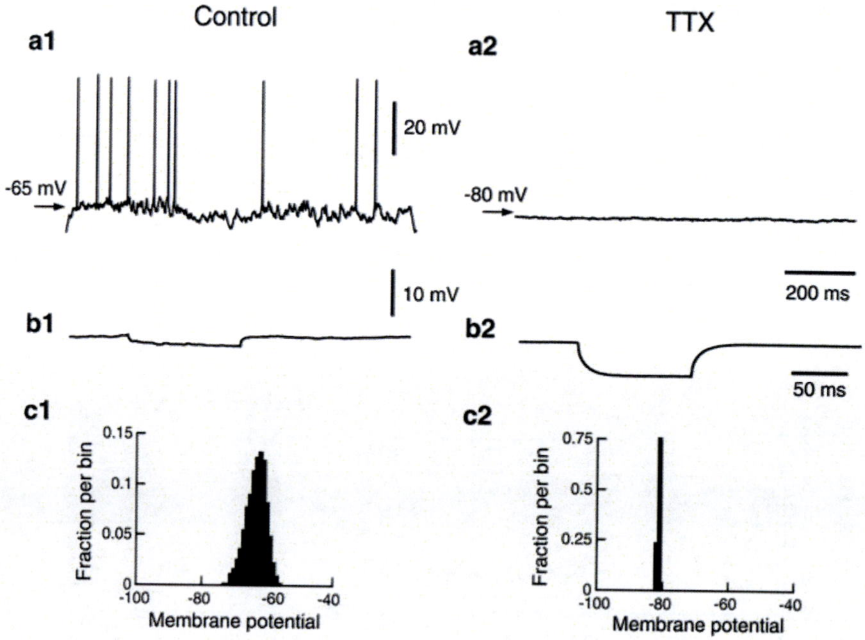

Fig. 3.21 Comparison of Up-states and quiescent states in Ketamine–xylazine anesthesia. (**a**) Intracellular recording of a neocortical neuron in cat parietal cortex (area 5–7) during a phase of desynchronized EEG activity (Up-state, Ketamine–Xylazine anesthesia). The same neuron is shown before (**a1**) and after (**a2**) microperfusion of TTX, abolishing all spontaneous activity. (**b**) Average of 50 hyperpolarizing pulses (−0.1 nA) to estimate the input resistance, during active periods (**b1**) and after TTX (**b2**). (**c**) Distribution of membrane potential before (**c1**) and after TTX (**c2**). TTX abolished most membrane potential (V_m) fluctuations, increased the input resistance R_{in} by about fivefold, and hyperpolarized the cell by about 15 mV. Modified from Destexhe and Paré (1999)

tions by calculating the SD of the V_m (i.e., σ_V), Destexhe and Paré showed that, in different cells recorded successively during Up-states and after TTX application, σ_V was reduced from 4.0 ± 2.0 mV to 0.4 ± 0.1 mV, respectively (Destexhe and Paré 1999), as illustrated by histograms of V_m distribution (Fig. 3.21c). These histograms also show that, after TTX, the V_m drops significantly to -80 ± 2 mV, as reported previously (Paré et al. 1998b). During Up-states, the average V_m was -65 ± 2 mV in control conditions (K-Acetate-filled pipettes) and -51 ± 2 mV with chloride-filled recording pipettes. These conditions correspond to chloride reversal potentials of -73.8 ± 1.6 mV and -52.0 ± 2.9 mV, respectively (Paré et al. 1998a).

In their seminal study, Paré and colleagues recorded successfully a total of nine different cells during Up-states and after TTX application (Paré et al. 1998b). Figure 3.22 represents the relative R_{in} increase obtained after TTX, which was quantified by

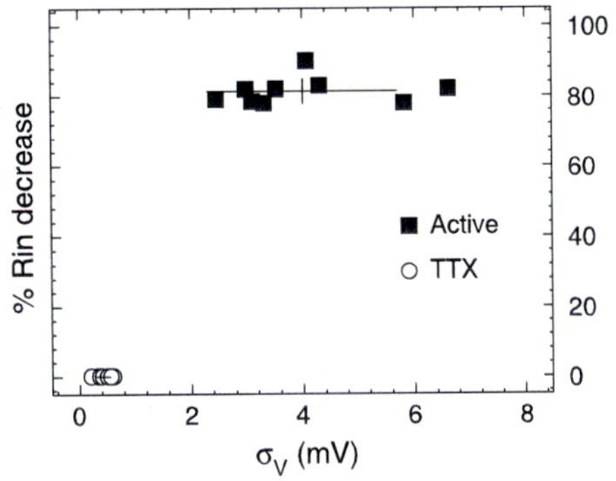

Fig. 3.22 Change of input resistance and membrane potential fluctuations between Up-states and quiescent states. Nine cells recorded before and after TTX are represented. The graph plots the percentage decrease in R_{in} (normalized to the R_{in} under TTX) as a function of σ_V during Up-states and after TTX. Modified from Destexhe and Paré (1999)

$$\frac{R_{\text{in}}(\text{TTX}) - R_{\text{in}}(\text{control})}{R_{\text{in}}(\text{TTX})}. \tag{3.1}$$

All cells clustered around the value of 80% (81.4 ± 3.6%), which represents about 5-times magnitude change in absolute R_{in}. The values of σ_V were more variable from cell to cell, ranging from about 2 mV to 6 mV (4.0 ± 2.0 mV; Fig. 3.22). These values represent important constraints for the construction of models and will be considered further in Chap. 4.

Finally, an important type of measurement is to estimate the relative contribution of excitatory and inhibitory conductances of synaptic activity. A first estimate is to consider the average of the passive membrane equation at steady state:

$$< V_m >= \frac{g_{\text{L}} E_L + < g_e > E_e + < g_i > E_i}{g_{\text{L}} + < g_e > + < g_i >}, \tag{3.2}$$

where $<>$ denotes the time average, g_{L} is the leak conductance, E_L is the leak reversal, and g_e and g_i (and their respective reversal potentials E_e and E_i) are the time-dependent global excitatory and inhibitory conductances, respectively. Including in this equation results from in vivo measurements obtained under KX in the Up-states and after TTX (Destexhe and Paré 1999; Paré et al. 1998b), namely $< V_m >= -65 \pm 2$ mV, $E_L = -80 \pm 2$ mV, $E_e = 0$ mV, $E_i = -73.8 \pm 1.6$ mV, and $R_{\text{in}}(\text{TTX})/R_{\text{in}}(\text{active}) = 5.4 \pm 1.3$, one obtains the following ratios:

$$\frac{< g_e >}{g_{\text{L}}} = 0.73 \quad \text{and} \quad \frac{< g_i >}{g_{\text{L}}} = 3.67. \tag{3.3}$$

According to these measurements, the ratio of the contributions of the average excitatory and inhibitory conductances, $< g_i > / < g_e >$, is about 5. Ratios between 4 and 5 were also obtained in cells with reversed inhibition (e.g., recorded with chloride-filled electrodes; see analysis in Destexhe and Paré 1999), or after brainstem stimulation (Rudolph et al. 2005). Other in vivo experimental studies concluded as well that inhibitory conductances are twofold to sixfold larger than excitatory conductances during sensory responses

(Hirsch et al. 1998; Borg-Graham et al. 1998, Anderson et al. 2000) or following thalamic stimulation (Contreras et al. 1997; but see Haider et al. 2006). This issue will be discussed further in Chap. 8.

Measurements have also been obtained during active states in vivo in other preparations, usually by comparing up- and down-states under various anesthetics such as ketamine–xylazine or urethane. Such estimates are very variable, ranging from up to several-fold smaller R_{in} in up-states (Contreras et al. 1996; Paré et al. 1998b, Petersen et al. 2003, Leger et al. 2005), to nearly identical R_{in} between up- and down-states or even larger R_{in} in up-states (Metherate and Ashe 1993; Zou et al. 2005, Waters and Helmchen 2006). The latter paradoxical result may be explained by voltage-dependent rectification (Waters and Helmchen 2006), or the presence of potassium currents in down-states (Zou et al. 2005). Consistent with the latter, cesium-filled electrodes have negligible effects on the up-state, but greatly abolish the hyperpolarization during the down-states (Timofeev et al. 2001). Moreover, the R_{in} of the down-state differs from that of the resting state (after TTX) by about twofold (Paré et al. 1998b). It is, thus, clear that the down-state is very different from the true resting state of the neuron. Finally, conductance measurements in awake and naturally sleeping animals have revealed a wide diversity between cell to cell in cat cortex (Rudolph et al. 2007), ranging from large (much larger than the leak conductance) to mild (smaller or equal to the leak) synaptic conductances. On average, the synaptic conductance was estimated as about three times the resting conductance, with breaks into about one third excitatory and two third inhibitory conductance (Rudolph et al. 2007). Strong inhibitory conductances were also found in artificially evoked active states using PPT stimulation (Rudolph et al. 2005). These aspects will be considered in more detail in Chap. 9.

3.3.3 Conductance Measurements In Vitro

The technique of voltage clamp can be applied to directly measure the conductances due to synaptic activity. In the experiments by Hasenstaub et al. (2005), cells were recorded with Na^+ and K^+ channel blockers (cesium and QX-314, respectively) to avoid the contamination with spikes. By clamping the cell at the apparent reversal potential of inhibition (−75 mV), a current which is exclusively due to excitatory synapses can be observed. Conversely, clamping at the reversal of excitation (0 mV) will reveal inhibitory currents. These paradigms are illustrated in Fig. 3.23. Because the membrane potential is clamped, the driving force is constant and, therefore, the conductance distributions can be directly obtained from the current distributions measured experimentally.

Such voltage-clamp experiments revealed several features. First, the conductance distributions are symmetric, both for excitation and inhibition (Fig. 3.23a, b, *histograms*). Second, they are well fit by Gaussians (Fig. 3.23a, b, *continuous lines*). As we will see in Chap. 4, this Gaussian nature of the current is important

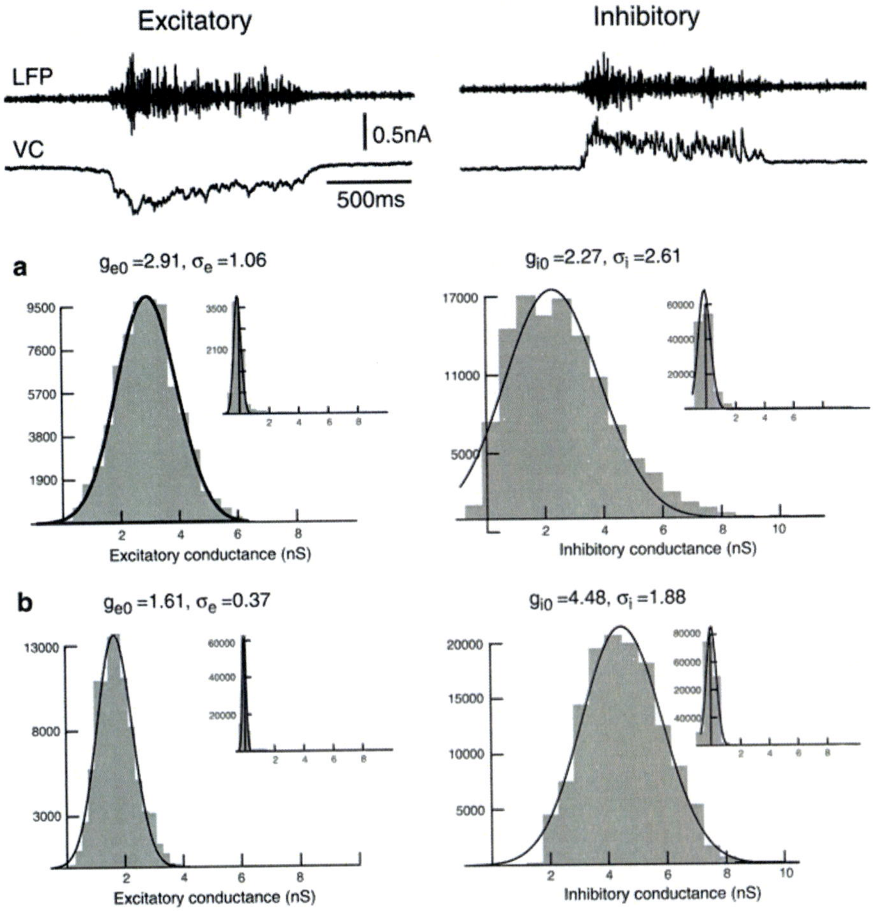

Fig. 3.23 Voltage-clamp recordings during Up-states in vitro. *Left*: Excitatory currents revealed by clamping the voltage at −75 mV. *Right*: Inhibitory currents obtained by clamping the V_m at 0 mV (experiments with Cesium and QX314 in the pipette). (**a**) and (**b**) show the distributions obtained in different cells when a large number of Up-states were pooled together. The insets show the distribution obtained in the Down-states. In all cases, the distributions were symmetric (*continuous lines* are Gaussian fits). Modified from Destexhe et al. (2003b); data from Hasenstaub et al. (2005)

for deriving simplified models. Third, the mean synaptic conductances can take different values according to different cells. This aspect is examined in more detail in Fig. 3.24a, where the analysis of various cells using this method is shown. The mean inhibitory conductances seemed to be slightly larger than excitatory conductances (Fig. 3.24b), while the variances of inhibition were always larger compared to excitatory variances (Fig. 3.24c).

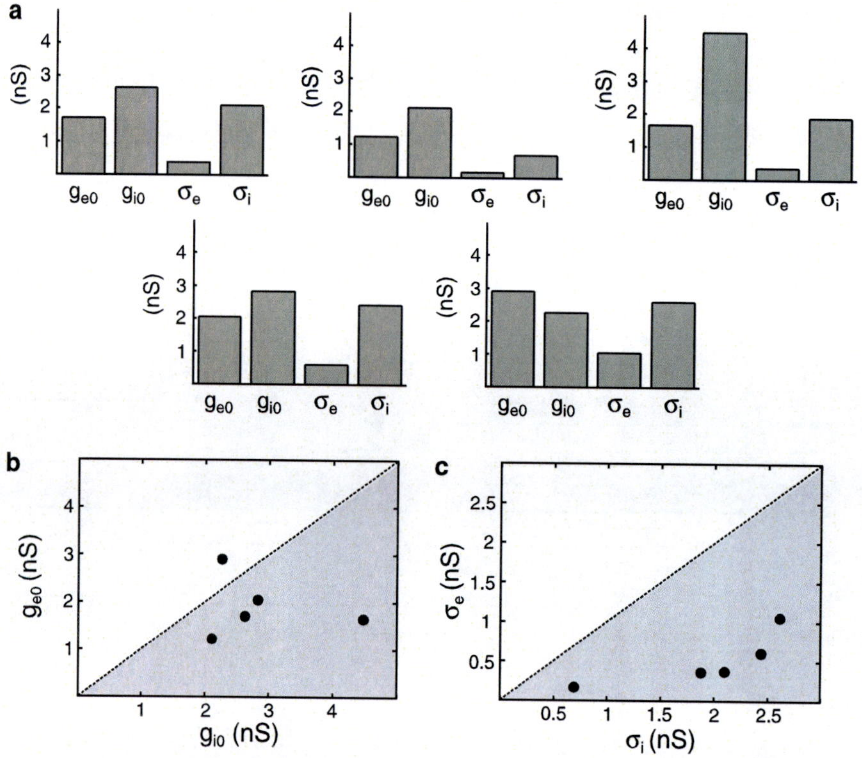

Fig. 3.24 Voltage-clamp measurements of excitatory and inhibitory conductances and their variances during Up-states in vitro. (**a**) Magnitude of the excitatory and inhibitory conductances (g_{e0}, g_{i0}), and their respective variance (σ_e, σ_i), in five cells. (**b**) Representation of excitation against inhibition showing that inhibitory conductances were generally larger. (**c**) Same representation between σ_e and σ_i Modified from Destexhe et al. (2003b); data from Hasenstaub et al. (2005)

3.3.4 Power Spectral Analysis of Synaptic Noise

Another important characteristics of synaptic noise is the PSD of the membrane potential. Similar to the PSD of LFP activity shown above (Fig. 3.3), the PSD of the V_m displays a broadband structure, both in vivo (Fig. 3.25) and in vitro (Fig. 3.26). However, these PSDs have a different structure. The PSD of V_m activity $S(\nu)$ follows a frequency-scaling behavior described by a Lorentzian

$$S(\nu) = \frac{D}{1 + (2\pi\nu\,\tau)^m}, \tag{3.4}$$

where ν is the frequency, τ is an effective time constant, D is the total spectral power at zero frequency, and m is the exponent of the frequency scaling at high frequencies. These parameters depend on various factors, such as the kinetics of synaptic currents

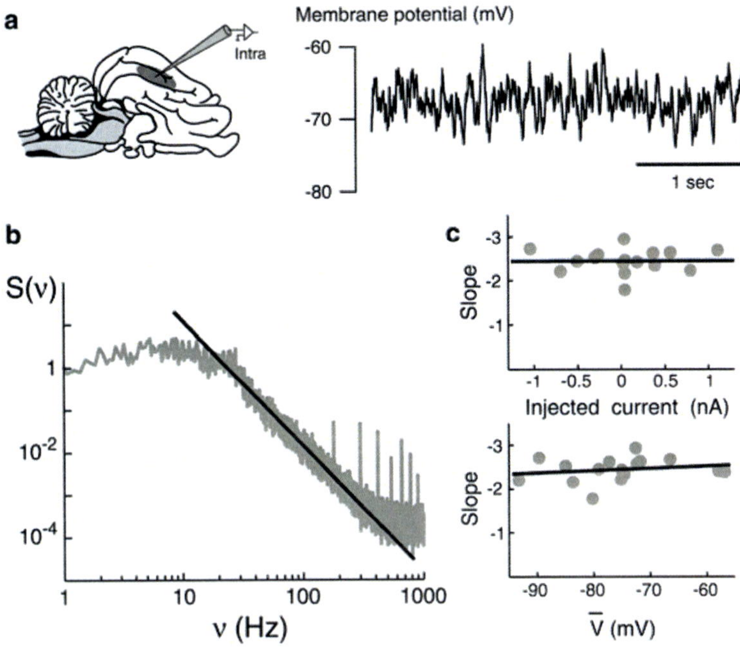

Fig. 3.25 Power spectrum of the membrane potential during post-PPT activated states in vivo. (**a**) Subthreshold intracellular activity of a cat cortical neuron recorded after stimulation of the PPT. (**b**) Example of the power spectral density (PSD) in the post-PPT state. The *black line* indicates the slope ($m = -2.76$) obtained by fitting the V_m PSD to a Lorentzian function $1/[1 + (2\pi f)^m]$. (**c**) Slope m obtained for all investigated cells as a function of the injected current (*top*) and resulting average membrane potential $\bar{V}$ (*bottom*). Modified from Rudolph et al. (2005)

(Destexhe and Rudolph 2004 see also Chap. 8) and the contribution of active membrane conductances (Manwani and Koch 1999a; Manwani and Koch 1999b). Consistent with this, the slope shows little variations as a function of the injected current (Fig. 3.25b, top), and of the membrane potential (Fig. 3.25b, bottom). It was found to be nearly identical for Up-states (slope $m = -2.44 \pm 0.31$ Hz^{-1}) and post-PPT states (slope $m = -2.44 \pm 0.27$ Hz^{-1}; see Fig. 3.25c). These results are consistent with the fact that the subthreshold membrane dynamics are mainly determined by synaptic activity, less so by active membrane conductances.

Piwkowska and colleagues also calculated the PSD of the V_m activity from Up-states recorded in neurons of the primary visual cortex of the guinea pig in vitro (Piwkowska et al. 2008), and a very similar scaling was observed (the exponent shown in Fig. 3.26 is of $m = 2.24$). Similar exponents m around 2.5 were also observed for channel noise in neurons (Diba et al. 2004; Jacobson et al. 2005), in agreement with the exponents estimated from synaptic noise in vivo (Destexhe et al. 2003a; Rudolph et al. 2005) and in vitro (Piwkowska et al. 2008). It is important to note that the traditional cable model of the neuron fails to predict these values,

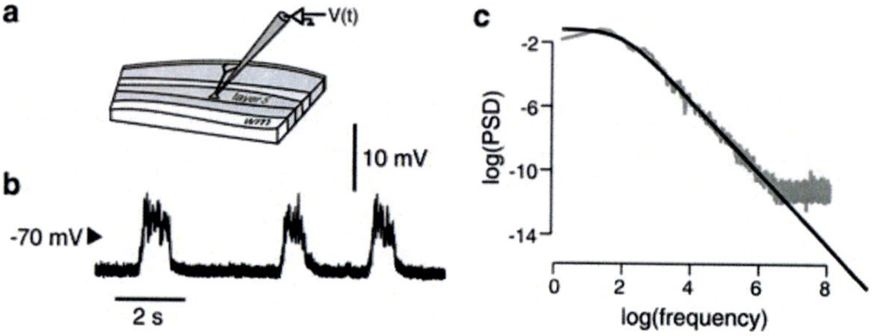

Fig. 3.26 Power spectrum of the membrane potential during Up-states in vitro. (**a**) Scheme of the recording in slices of guinea-pig primary visual cortex. (**b**) Subthreshold intracellular activity during three successive Up-states in the slice. (**c**) Power spectral density (PSD) calculated within Up-states (16 Up-states concatenated). The *black line* indicates the slope ($m = -2.24$) obtained by fitting the V_m PSD to a Lorentzian function $1/[1+(2\pi f)^m]$. Modified from Piwkowska et al. (2008)

and other mechanisms seems necessary to explain these observations, such as the nonideal character of the membrane capacitance proposed by Bédard and Destexhe (Bédard and Destexhe 2008; 2009; Bédard et al. 2010 see also Chap. 4).

3.4 Summary

In this chapter, we have reviewed the measurement of background activity in neurons. One of the aspects that tremendously progressed in the last years is that this measurement was made quantitative in the intact brain, as detailed here.

In the first part of this chapter (Sects. 3.1 and 3.2), we have shown that the activity of cerebral cortex is highly "noisy" in aroused brain states, as well as during sleep or anesthetized states. In particular, the activity during SWS consists of Up- and Down-states, with the Up-states displaying many characteristics in common with the aroused brain, at both the population level, the EEG which is desynchronized, and the fine structure of correlations in LFPs. The similarities extent to the intracellular level, where the membrane potential activity is very similar.

Section 3.3 reviewed the quantitative measurement of synaptic noise, which was first achieved for different network states under anesthesia in vivo in a seminal study by Paré et al. (1998b). In this work, the impact of background activity could be measured, for the first time, by comparing the same cortical neurons recorded before and after total suppression of network activity. This was done globally for two types of anesthesia, barbiturates and ketamine–xylazine. In a subsequent study (Destexhe and Paré 1999), this analysis was refined by focusing specifically on the "Up-states" of ketamine–xylazine anesthesia, with locally desynchronized EEG. These analyses evidenced a very strong impact of synaptic background activity on increasing the membrane conductance of the cell into "high-conductance states" and provided measurements of this conductance.

The contribution of excitatory and inhibitory synaptic conductances were later measured in awake and naturally sleeping animals (Rudolph et al. 2007). The availability of such measurements, as will be detailed later in Chap. 9, can be considered as an important cornerstone, because they allow the construction of precise models and dynamic-clamp experiments to evaluate their consequences on the integrative properties of cortical neurons. This theme will be followed in the next chapters.

Chapter 4
Models of Synaptic Noise

In this chapter, we build models of "synaptic noise" in cortical neurons based on the experimental characterization reviewed in Chap. 3. We first consider detailed models, which incorporate a precise morphological representation of the cortical neuron and its synapses. Next, we review simplified models of synaptic "noise." Both type of models will be used in the next chapters to investigate the integrative properties of neurons in the presence of synaptic noise.

4.1 Introduction

How neurons integrate synaptic inputs under the noisy conditions described in the last chapter, is a problem which was identified in early work on motoneurons (Barrett and Crill 1974; Barrett 1975), followed by studies in Aplysia (Bryant and Segundo 1976) and cerebral cortex (Holmes and Woody 1989). This early work motivated further studies using compartmental models in cortex (Bernander et al. 1991) and cerebellum (Rapp et al. 1992; De Schutter and Bower 1994), which pointed out that the integrative properties of neurons can be drastically different in such noisy states. However, at that time, no precise experimental measurements were available to characterize and quantify the noise sources in neurons.

Around the same time, researchers also began to use detailed biophysical models to investigate the consequences of synaptic background activity in cortical neurons, starting with the first investigation of this kind by Bernander et al. (1991). These studies revealed that the presence of background activity is able to change several features of the integrative properties of the cell, such as coincidence detection or its responsiveness to synaptic inputs. In this chapter, we will present a number of models at various levels of computational complexity that were directly constrained from the quantitative measurements of synaptic noise detailed in Chap. 3.

A. Destexhe and M. Rudolph-Lilith, *Neuronal Noise*, Springer Series in Computational Neuroscience 8, DOI 10.1007/978-0-387-79020-6_4,

4.2 Detailed Compartmental Models of Synaptic Noise

In this first section, we will describe a particular class of neuronal models, namely detailed compartmental models with 3-dimensional dendritic arborization based on morphological studies, as well as biophysically plausible ion channel and synaptic dynamics. Although computationally very demanding, this class of models is interesting and enjoys much attention, as it allows to investigate in great detail the behavior and response of neurons under a variety of conditions. Here, we focus on models of cortical neurons, whose biophysical parameters, such as ion channel kinetics and density, synaptic kinetics and distribution as well as synaptic activity, are constrained by experimental measurements and observations. In Chap. 5, these models will be used to characterize the integrative properties of cortical neurons in conditions of intense synaptic activity resembling those found in awake and naturally sleeping animals.

4.2.1 Detailed Compartmental Models of Cortical Pyramidal Cells

One of the greatest challenges in constructing detailed biophysical neuronal models is the confinement of their vast parameter space. Typically, depending on the level of detail of the morphological reconstruction, such models are described by huge systems of coupled differential equations, each describing the biophysical dynamics in a small equipotential membrane patch (see Sect. 2.3). This description, in turn, is accompanied by a number of passive parameters, parameters which characterize active properties and parameters which describe the dynamics at synaptic terminals. In the following pages, we will provide a coarse overview over the resulting immense parameter space, and how these parameters should and can be constrained.

Morphology

A great number of cellular morphologies for use in computational studies exists, obtained from 3-dimensional reconstructions of stained neurons. In Chap. 5, we will make particular use of cat layer II–III, layer V and layer VI neocortical pyramidal cells, obtained from two previous studies (Contreras et al. 1997; Douglas et al. 1991). The cellular geometries can be incorporated into software tools for neuronal simulations, such as the NEURON simulation environment (Hines and Carnevale 1997; Carnevale and Hines 2006).

Due to the limited spatial resolution of the reconstruction technique, the dendritic surface must typically be corrected for spines, assuming that spines represent about 45% of the dendritic membrane area (DeFelipe and Fariñas 1992). This surface correction is achieved by rescaling the membrane capacitance and conductances by 1.45 as described previously (Bush and Sejnowski 1993; Paré et al. 1998a).

Also, commonly, an axon has to be added to the reconstructed morphology. Here, one can restrict to the most simple axon morphology, consisting of an initial segment of 20 μm length and 1 μm diameter, followed by ten segments of 100 μm length and 0.5 μm diameter each.

Passive Properties

The passive properties of the model should be adjusted to experimental recordings in the absence of synaptic activity. In order to block synaptic events mediated by glutamate α-amino-3-hydroxy-5-methyl-4-isoxazolepropionic acid (AMPA) and γ-aminobutyric acid type-A ($GABA_A$) receptors, one can use a microperfusion solution containing a mixture of ringer + TTX (50 μM) + 1,2,3,4-tetrahydro-6-nitro2,3-dioxo-benzo[f]quinoxaline-7-sulfonamide disodium (NBQX, 200 μM) + bicuculline (200 μM). This procedure suppresses all miniature synaptic events, as demonstrated previously (Paré et al. 1997, see also Chap. 3).

Fitting of the model to passive responses obtained in the condition of absence of synaptic activity can be performed using a simplex algorithm (Press et al. 1993). Such fits provide values for the leak conductance and reversal potential, while other passive parameters are fixed (membrane capacitance of 1 μF/cm^2 and axial resistivity of 250 Ωcm). Other combinations of passive parameters can also be considered, including a supplementary leak in the soma (10 ns) due to electrode impalement, combined with a larger membrane resistance of 0.015 ms cm^{-2} (Pongracz et al. 1991; Spruston and Johnston 1992) and/or a lower axial resistivity of 100 Ωcm.

Also a nonuniform distribution of leak parameters can be considered, based, for instance, on estimations in layer V neocortical pyramidal cells (Stuart and Spruston 1998). As estimated by these authors, the axial resistance is low (80 Ωcm), the leak conductance is low ($g_L = 0.019$ ms cm^{-2}) in the soma but high ($g_L = 0.125$ ms cm^{-2}) in distal dendrites. Moreover, in this study, g_L is given by a sigmoid distribution

$$\frac{1}{g_L} = 8 + \frac{44}{1 + \exp\left[\frac{1}{50}(x - 406)\right]},$$

where x is the distance to soma. The exact form of this distribution, depending on the cellular morphology used, can be obtained by fitting the model to passive responses as described above.

Synaptic Inputs

Synaptic inputs can be simulated by using densities of synapses in different regions of the cell as estimated from morphological studies of neocortical pyramidal cells (Mungai 1967; White 1989; Fariñas and DeFelipe 1991a,b; Larkman 1991; DeFelipe and Fariñas 1992). These densities (per 100 μm^2 of membrane) are: 10–20

GABAergic synapses in soma, 40–80 GABAergic synapses in axon initial segment, 8–12 GABAergic synapses and 55–65 glutamatergic (AMPA) synapses in dendrites.

The kinetics of AMPA and $GABA_A$ receptor types are commonly simulated using two-state kinetic models (Destexhe et al. 1994a,b):

$$I_{syn} = \bar{g}_{syn} m (V - E_{syn})$$

$$\frac{dm}{dt} = \alpha[T](1-m) - \beta m , \tag{4.1}$$

where I_{syn} is the PSC, $\bar{g}_{syn}$ is the maximal conductance, m is the fraction of open receptors, E_{syn} is the reversal potential, $[T]$ is the transmitter concentration in the cleft, α and β are forward and backward binding rate constants of T to open the receptors. Here, typically, $E_{syn} = 0$ mV, $\alpha = 1.1 \times 10^6\ M^{-1}\ s^{-1}$, $\beta = 670\ s^{-1}$ for AMPA receptors, $E_{syn} = -80$ mV, $\alpha = 5 \times 10^6\ M^{-1}\ s^{-1}$, $\beta = 180\ s^{-1}$ for $GABA_A$ receptors. When a spike occurs in the presynaptic compartment, a pulse of transmitter is triggered such that $[T] = 1$ mM during 1 ms. These kinetic parameters were obtained by fitting the model to PSCs recorded experimentally (see Destexhe et al. 1998b). N-methyl-D-aspartate (NMDA) receptors are blocked by ketamine and are usually not included in models of cortical neurons.

Correlation of Release Events

In some simulations used in later sections of this book, N Poisson-distributed random presynaptic trains of APs are generated according to a correlation coefficient c. This correlation is applied to any pair of the presynaptic trains, irrespective of the proximity of synapses on the dendritic tree, and correlations are treated independently for excitatory and inhibitory synapses for simplicity. To generate correlated presynaptic trains, a set of N_0 independent Poisson-distributed random variables is generated and distributed randomly among the N presynaptic trains. This procedure is repeated at every integration step such that the N_0 random variables are constantly redistributed among the N presynaptic trains. Correlations arise from the fact that $N_0 \leq N$ and the ensuing redundancy within the N presynaptic trains. There is a complex and highly nonlinear relation between N_0, N as well as c and the more commonly used Pearson correlation coefficient. Typically, for $N = 16{,}563$ and $N_0 = 400$, a c value of 0.7 yields a Pearson correlation of 0.0005.

For more details about the generation of correlated Poisson-distributed presynaptic activity, we refer to Appendix B.

Voltage-Dependent Currents

Active currents are usually inserted into the soma, dendrites and axon with different densities in accordance with available experimental evidence in neocortical and hippocampal pyramidal neurons (Stuart and Sakmann 1994; Magee and Johnston 1995a,b; Hoffman et al. 1997; Magee et al. 1998). Active currents are usually expressed in the generic form

$$I_i = \bar{g}_i m^M h^N (V - E_i)\,,$$

where $\bar{g}_i$ is the maximal conductance of current I_i and E_i its reversal potential. The current activates according to M activation gates, represented by the gating variable m. It inactivates with N inactivation gates represented by the gating variable h. Here, m and h obey to first-order kinetic equations (see Sect. 2.1.3).

Voltage-dependent Na^+ current can be described by Traub and Miles (1991):

$$\begin{aligned}
I_{\mathrm{Na}} &= \bar{g}_{\mathrm{Na}} m^3 h (V - E_{\mathrm{Na}}) \\
\frac{\mathrm{d}m}{\mathrm{d}t} &= \alpha_m(V)(1-m) - \beta_m(V)m \\
\frac{\mathrm{d}h}{\mathrm{d}t} &= \alpha_h(V)(1-h) - \beta_h(V)h \\
\alpha_m &= \frac{-0.32\,(V - V_T - 13)}{\exp\left[-\frac{1}{4}(V - V_T - 13)\right] - 1} \\
\beta_m &= \frac{0.28\,(V - V_T - 40)}{\exp\left[\frac{1}{5}(V - V_T - 40)\right] - 1} \\
\alpha_h &= 0.128 \exp\left[-\frac{1}{18}(V - V_T - V_S - 17)\right] \\
\beta_h &= \frac{4}{1 + \exp\left[-\frac{1}{5}(V - V_T - V_S - 40)\right]}\,,
\end{aligned}$$

where $V_T = -58$ mV is adjusted to obtain a spiking threshold of around -55 mV, and the inactivation is shifted by 10 mV toward hyperpolarized values ($V_S = -10$ mV) to match the voltage dependence of Na^+ currents in neocortical pyramidal cells (Huguenard et al. 1988). The density of Na^+ channels in cortical neurons is similar to that suggested in a previous study in hippocampal pyramidal cells (Hoffman et al. 1997). Specifically, the channel density is comparable low in the soma and dendrites (120 ps/μm^2), but about 10 times higher in the axon.

The "delayed-rectifier" K^+ current can be described by Traub and Miles (1991):

$$\begin{aligned}
I_{\mathrm{Kd}} &= \bar{g}_{\mathrm{Kd}} n^4 (V - E_{\mathrm{K}}) \\
\frac{\mathrm{d}n}{\mathrm{d}t} &= \alpha_n(V)(1-n) - \beta_n(V)n \\
\alpha_n &= \frac{-0.032\,(V - V_T - 15)}{\exp\left[-\frac{1}{5}(V - V_T - 15)\right] - 1} \\
\beta_n &= 0.5 \exp\left[-\frac{1}{40}(V - V_T - 10)\right].
\end{aligned}$$

K^+ channel densities are 100 ps/μm^2 in soma and dendrites, and 1,000 ps/μm^2 in the axon.

A noninactivating K^+ current is described by Mainen et al. (1995):

$$I_M = \bar{g}_M n(V - E_K)$$

$$\frac{dn}{dt} = \alpha_n(V)(1-n) - \beta_n(V)n$$

$$\alpha_n = \frac{0.0001\,(V+30)}{1-\exp\left[-\frac{1}{9}(V+30)\right]}$$

$$\beta_n = \frac{-0.0001\,(V+30)}{1-\exp\left[\frac{1}{9}(V+30)\right]}.$$

This current is typically present in soma and dendrites (density of 2–5 ps/μm^2) and is responsible for spike frequency adaptation, as detailed previously (Paré et al. 1998a).

It was reported that some pyramidal cell have a hyperpolarization-activated current termed I_h (Spain et al. 1987; Stuart and Spruston 1998). However, most cortical cells recorded in studies such as Destexhe and Paré (1999) had no apparent I_h (see passive responses in Fig. 4.1), therefore this current is usually omitted.

It is important to note that the models considered here do not include the genesis of broad dendritic calcium spikes, which were shown to be present in thick-tufted Layer 5 pyramidal cells (Amitai et al. 1993). Such calcium spikes are generated in the distal apical dendrite (Amitai et al. 1993), and they are believed to strongly influence dendritic integration in these cells (Larkum et al. 1999, 2009) (for a recent modeling study, see Hay et al. 2001). In the present book, the integrative properties are considered in cortical neurons with the "classic" dendritic excitability, mediated by Na^+ and K^+ voltage-dependent currents (Stuart et al. 1997a,b). The inclusion of dendritic calcium spikes, and how they interact with synaptic background activity, should be done in future work.

4.2.2 *Calibration of the Model to Passive Responses*

The first step in adjusting the parameter space of a constructed model is to calibrate its passive properties in accordance with experiments. As both somatic and dendritic recordings are critical to constrain the simulations of synaptic activity (see below), passive responses from both types of recordings (Paré et al. 1997) need to be used to set the passive parameters of a model. Typically, a fitting procedure is performed, such that the same model reproduces both somatic and dendritic recordings obtained, for example, in deep pyramidal cells in the absence of synaptic activity (TTX and synaptic blockers; see Sect. 4.2.1). In that case, the same model can fit both traces (Fig. 4.1a) when the following optimal passive parameters are

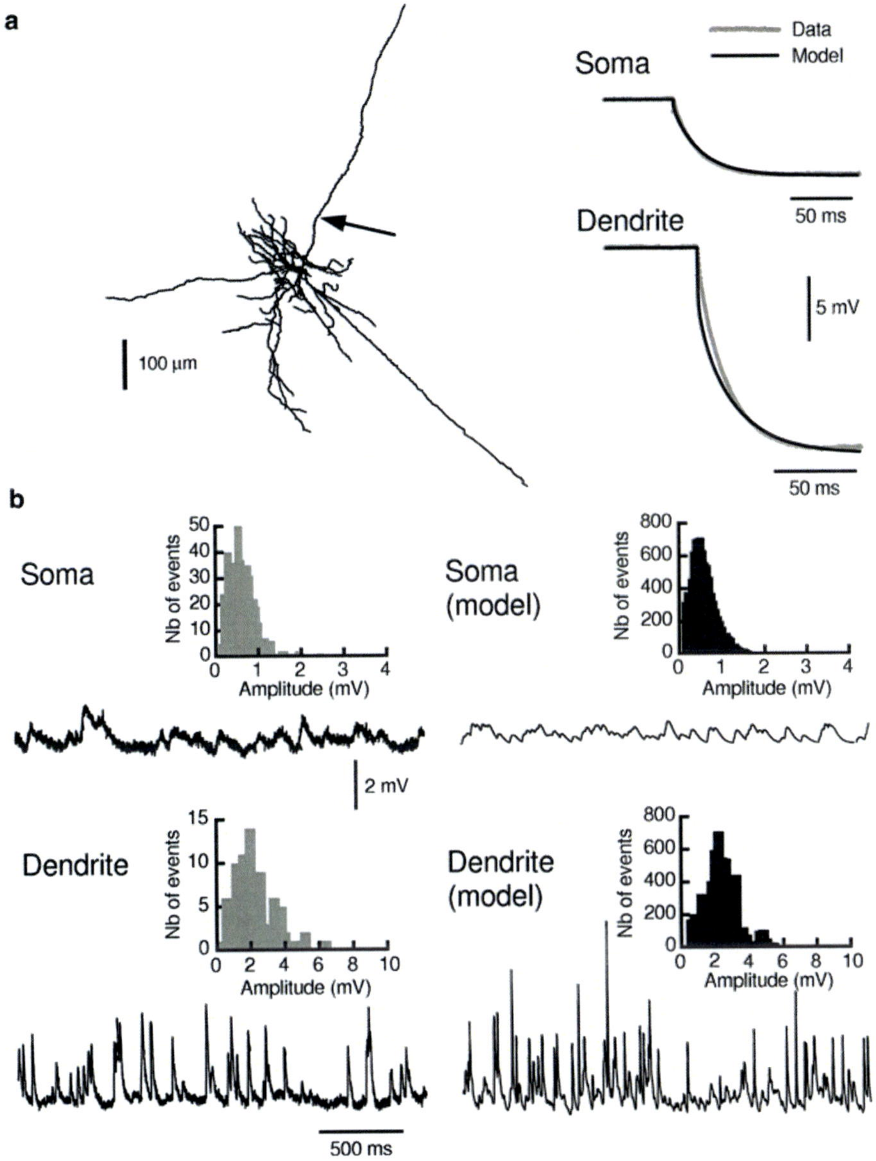

Fig. 4.1 Calibration of the model to passive responses and miniature synaptic events recorded intracellularly in vivo. (**a**) Morphology of a layer VI neocortical pyramidal cell from cat cerebral cortex, which was reconstructed and incorporated into computational models. Passive responses of the model were adjusted to somatic (*Soma*; −0.1 nA current pulse) and dendritic recordings (*Dendrite*; −0.2 nA current pulse) obtained in vivo in the presence of TTX and synaptic blockers (see Sect. 4.2.1). (**b**) Miniature synaptic potentials in neocortical pyramidal neurons. *Left*: TTX-resistant miniature events in somatic (*Soma*) and dendritic (*Dendrite*) recordings. Histograms of mini amplitudes are shown in the insets. *Right*: simulated miniature events. About 16,563 glutamatergic and 3,376 GABAergic synapses were simulated with Poisson-distributed spontaneous release. Quantal conductances and release frequency were estimated by matching simulations to experimental data. Best fits were obtained with an average release frequency of 0.01 Hz and conductances of 1,200 and 600 ps at glutamatergic and GABAergic synapses, respectively. Modified from Destexhe and Paré (1999)

chosen: $g_L = 0.045$ ms cm^{-2}, $C_m = 1\ \mu$F/cm^2 and an axial resistance $R_i = 250\ \Omega$cm (see Sect. 4.2.1).

Another fit can be performed by forcing R_i to 100 Ωcm ($C_m = 1\ \mu$F/cm^2 and $g_L = 0.039$ ms cm^{-2}). Although the latter set of values is not optimal, they can be used to check for the dependence of the results on axial resistance. A passive fit can also be performed with high membrane resistance, based on whole-cell recordings (Pongracz et al. 1991; Spruston and Johnston 1992), and a somatic shunt due to electrode impalement. In this case, the parameters are: 10 ns somatic shunt, $g_L = 0.0155$ ms cm^{-2}, $C_m = 1\ \mu$F/cm^2 and R_i is of either 250 Ωcm or 100 Ωcm.

A nonuniform leak conductance, low in the soma and a high leak in distal dendrites according to Stuart and Spruston (1998), is also often used in detailed models (see Sect. 4.2.1). Although such fitting procedures usually depend on the particularities of the cellular morphologies utilized, they typically ensure that the constructed models have an input resistance and time constant consistent with both somatic and dendritic recordings free of synaptic activity.

4.2.3 Calibration to Miniature Synaptic Activity

A second type of calibration consists of simulating TTX-resistant miniature synaptic potentials occurring in the same neurons. These miniature events can be characterized in somatic and dendritic intracellular recordings after microperfusion of TTX in vivo (Paré et al. 1997) (Fig. 4.1b, left). To simulate such minis, a plausible range of parameters must be determined, based on in vivo experimental constraints. Then, a search within this parameter range is performed to find an optimal set which is consistent with all of the constraints. These constraints are: (a) the densities of synapses in different regions of the cell, as derived from morphological studies of neocortical pyramidal cells (Mungai 1967; Fariñas and DeFelipe 1991a,b; Larkman 1991; DeFelipe and Fariñas 1992; White 1989; see Sect. 4.2.1); (b) the quantal conductance at AMPA and GABA$_A$ synapses, as determined by whole-cell recordings of neocortical neurons (Stern et al. 1992; Salin and Prince 1996; Markram et al. 1997); (c) the value of membrane potential fluctuations during miniature events following TTX application in vivo (about 0.4 mV for somatic recordings and 0.6–1.6 mV for dendritic recordings; see Paré et al. 1997); (d) the change in R_{in} due to miniature events, as determined in vivo (about 8–12% in soma and 30–50% in dendrites Paré et al. 1997); (e) the distribution of mini amplitudes and frequency, as obtained from in vivo somatic and dendritic recordings (Fig. 4.1b, *insets*).

With these constraints, an extensive search in the model's parameter space can be performed and, usually, a narrow region is found. Such an optimal set of values is: (a) a density of 20 GABAergic synapses per 100 μm^2 in the soma, 60 GABAergic synapses per 100 μm^2 in the initial segment, 10 GABAergic synapses and 60 glutamatergic (AMPA) synapses per 100 μm^2 in the dendrites; (b) a rate of spontaneous release (assumed uniform for all synapses) of 0.009–0.012 Hz; (c) quantal conductances of 1,000–1,500 ps for glutamatergic and 400–800 ps for GABAergic synapses. In these conditions, simulated miniature events are consistent with experiments (Fig. 4.1b, right), with σ_V of 0.3–0.4 mV in soma and 0.7–1.4 mV in dendrites, and R_{in} changes of 8–11% in soma and 25–37% in dendrites.

4.2.4 *Model of Background Activity Consistent with In Vivo Measurements*

The next step in the construction of detailed biophysical computational models is to simulate the intense synaptic activity consistent with in vivo recordings. Here, one can hypothesize that miniature events and active periods are generated by the same population of synapses, with different conditions of release for GABAergic and glutamatergic synapses. Thus, the model of miniature events detailed in the previous section can be utilized in order to simulate active periods by simply increasing the release frequency at all synaptic terminals.

A commonly used synaptic activity pattern is a Poisson-distributed random release, simulated with identical release frequency for all excitatory synapses (ν_e) as well as for inhibitory synapses (ν_i). The release frequencies ν_e and ν_i do affect the R_{in} and average V_m (Fig. 4.2a). The release frequencies can, thus, be constrained by comparing the models's R_{in} and average V_m under synaptic bombardment with experimental measurements (see preceding section): (a) the R_{in} change produced by TTX should be about 80%; (b) the V_m should be around -80 mV without synaptic activity; (c) the V_m should be about -65 mV during active periods ($E_{\text{Cl}} = -75$ mV); (d) the V_m should be around -51 mV during active periods recorded with chloride-filled electrodes ($E_{\text{Cl}} = -55$ mV). Here, again, an extensive search in this parameter space can be performed and several combinations of excitatory and inhibitory release frequencies will reproduce correct values for the R_{in} decreases and V_m differences between active periods and after TTX (Fig. 4.2a). Typical values of release frequencies are $\nu_e = 1$ Hz (range 0.5–3 Hz) for excitatory synapses and $\nu_i = 5.5$ Hz (range 4–8 Hz) for inhibitory synapses.

An additional constraint is the large V_m fluctuations experimentally observed during active periods, as quantified by σ_V (see Chap. 3). As shown in Fig. 4.2b, increasing the release frequency of excitatory or inhibitory synapses produces the correct R_{in} change but always yields too small values for σ_V. High release frequencies lead to membrane fluctuations of small amplitude, due to the large number of summating random events (Fig. 4.2b4). Variations within 50–200% of the optimal value of different parameters, such as synapse densities, synaptic conductances, frequency of release, leak conductance, and axial resistance, do yield approximately correct R_{in} changes and correct V_m, but fail to account for values of σ_V observed during active periods (Fig. 4.2c, *crosses*). This failure suggests that an additional fundamental property is missing in the model to account for the amplitude of V_m fluctuations observed experimentally.

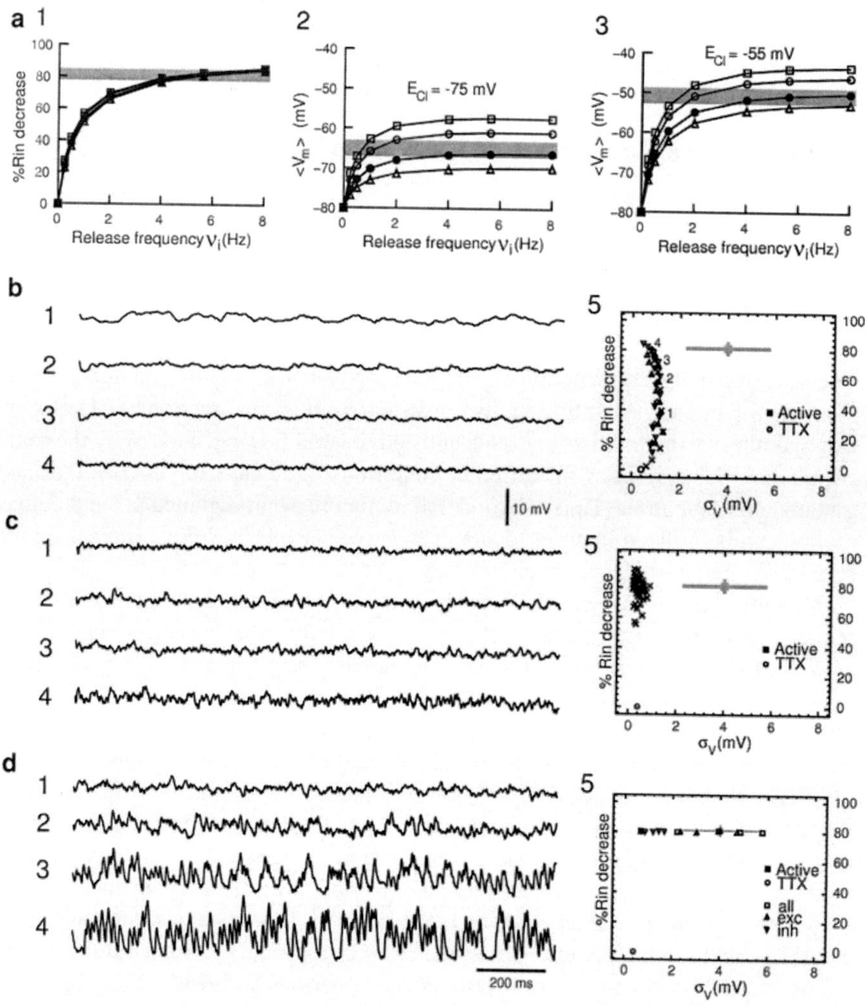

Fig. 4.2 Constraining the release parameters of the model to simulate periods of intense synaptic activity. (**a**) Effect of release frequencies on R_{in} (**a1**) and average V_m ($< V_m >$) for two values of chloride reversal potential E_{Cl} (**a2–a3**). Both excitatory (ν_e) and inhibitory (ν_i) release frequencies were varied; each curve represents different ratios between them: $\nu_e = 0.4\ \nu_i$ (*squares*), $\nu_e = 0.3\ \nu_i$ (*open circles*), $\nu_e = 0.18\ \nu_i$ (*filled circles*), $\nu_e = 0.1\ \nu_i$ (*triangles*). *Shaded areas* indicate the range of values observed during in vivo experiments using either KAc- or KCl-filled pipettes. The optimal value was $\nu_e = 1$ Hz and $\nu_i = 5.5$ Hz. (**b**) Increasing the release frequency can account for the experimentally observed R_{in} decrease but not for the standard deviation of V_m (σ_V). **b1–b4** show the effect of increasing the release frequency up to $\nu_e = 1$ Hz, $\nu_i = 5.5$ Hz (**b4**). Different symbols in the graph (**b5**) indicate different combinations of release frequencies, synaptic conductances and densities. (**c**) Several combinations of conductance and release frequencies could yield correct R_{in} decrease but failed to reproduce σ_V. **c1–c4** show different parameter combinations giving the highest σ_V. All parameters were varied within 50–200% of their value in **b4** and are shown by crosses in **c5**. (**d**) Introducing a correlation between release events led to correct R_{in} and σ_V values. **d1–d4** correspond to $\nu_e = 1$ Hz and $\nu_i = 5.5$ Hz, as in **b4**, with increasing values of correlation (0.025, 0.05, 0.075, 0.1 from **d1** to **d4**). *Open symbols* in graph (**d5**) indicate R_{in} and σ_V obtained with different values of correlation (between 0 and 0.2) when all inputs (*squares*), only excitatory inputs (*triangles*) or only inhibitory inputs (*reversed triangles*) were correlated. Modified from Destexhe and Paré (1999)

Table 4.1 Membrane parameters of neocortical neurons during intense synaptic activity and after TTX

Parameter measured	Experiments	Model (passive)	Model (I_{Na}, I_{Kd}, I_M)
σ_V	4.0 ± 2.0 mV	3.6–4.0 mV	3.8–4.2 mV
$< V_m >$ (KAc)	-65 ± 2 mV	-63.0 to -66.1 mV	-64.3 to-66.4 mV
$< V_m >$ (KCl)	-51 ± 2 mV	-50.1 to -52.2 mV	-50.7 to -53.1 mV
σ_V (TTX)	0.4 ± 0.1 mV	0.3–0.4 mV	–
$< V_m >$ (TTX)	-80 ± 2 mV	-80 mV	–
R_{in} change	$81.4 \pm 3.6\%$	79–81%	80–84%

Experimental values were measured in intracellularly recorded pyramidal neurons in vivo (*Experiments*). The average value ($< V_m >$) and SD (σ_V) of the V_m are indicated, as well as the R_{in} change, during active periods and after TTX application. The values labeled "TTX" correspond to somatic recordings of miniature synaptic potentials. Experimental values are compared to the layer VI pyramidal cell model where active periods were simulated by correlated high-frequency release on glutamatergic and GABAergic receptors. The model is shown without voltage-dependent currents (*Model (passive)*; same model as in Fig. 4.2d4) and with voltage-dependent currents distributed in soma and dendrites (Na^+ and K^+ currents; same model as in Fig. 4.2a). The range of values indicate different combinations of release frequency (from 75% to 150% of optimal values). Miniature synaptic events were simulated by uncorrelated release events at low frequency (same model as in Fig. 4.2b). See text for more details

One additional assumption has to be made in order to reproduce experimental measurements of V_m fluctuations in vivo. In the cortex, action potential-dependent release is clearly not independent at different synapses, as single axons usually establish several contacts in pyramidal cells (Markram et al. 1997; Thomson and Deuchars 1997). More importantly, the presence of high amplitude fluctuations in the EEG implies correlated activity in the network. It is this correlation which, therefore, needs to be included in the release of different synapses (see Sect. 4.2.1). For the sake of simplicity, one can assume that the correlation is irrespective of the proximity of synapses on the dendritic tree and correlations can be treated independently for excitatory and inhibitory synapses. Figure 4.2d (*open symbols*) shows simulations of random synaptic bombardment similar to Fig. 4.2b4 but using different correlation coefficients. The horizontal alignment of open symbols in Fig. 4.2d suggests that the degree of correlation has a negligible effect on the R_{in}, because the same amount of inputs occurred on average. However, the degree of correlation affects significantly the SD of the signal. Here, several combinations of excitatory and inhibitory correlations, within the range of 0.05–0.1 (see Appendix B), give rise to V_m fluctuations with comparable σ_V as that observed experimentally during active periods (Fig. 4.2d4; see also Table 4.1). Introducing correlations among excitatory or inhibitory inputs alone shows that excitatory correlations are most effective in reproducing the V_m fluctuations (Fig. 4.2d, right). Similar results are also obtained with oscillatory correlations (Destexhe and Paré 2000).

Finally, it is important to assess to which extent the above parameter fits are dependent on the specific morphology of the studied cell. In Fig. 4.3a, four different cellular geometries are compared, ranging from small layer II–III cells to large layer

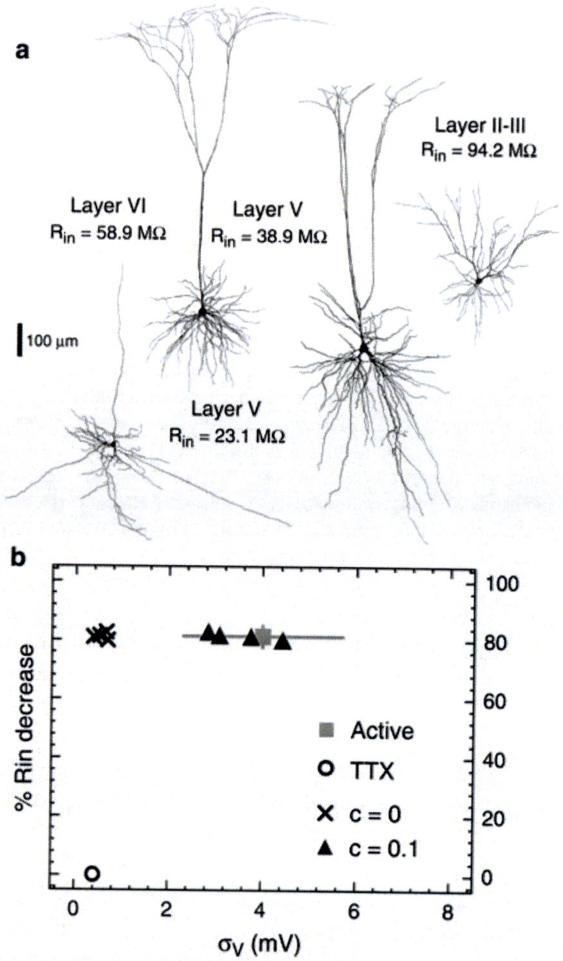

Fig. 4.3 Effect of dendritic morphology on membrane parameters during intense synaptic activity. (**a**) Four cellular reconstructions from cat neocortex used in simulations. All cells are shown at the same scale and their R_{in} was measured in the absence of synaptic activity (identical passive parameters for all cells). (**b**) Graph plotting the R_{in} decrease as a function of the standard deviation of the signal for these four cells. For all cells, the release frequency was the same ($\nu_e = 1$ Hz, $\nu_i =$ 5.5 Hz). Results obtained without correlation (*crosses*) and with a correlation of $c =$ 0.1 (*triangles*) are indicated. Modified from Destexhe and Paré (1999)

V pyramidal cells. In experiments, the absolute R_{in} values will vary from cell to cell. However, the relative R_{in} change produced by TTX was found to be similar irrespective of the cell recorded (Paré et al. 1998b). Similarly, in the constructed models, the absolute R_{in} values will depend on the cellular geometry: using identical passive parameters, the R_{in} values of the four neurons shown, for example, in Fig. 4.3a ranged from 23 MΩ to 94 MΩ. However, high-frequency release conditions will have a similar impact on their membrane properties. Using identical synaptic densities, synaptic conductances, and release conditions as detailed above leads to a decrease in R_{in} of around 80% for all cells (Fig. 4.3b).

Similarly, V_m fluctuations will also depended critically on the degree of correlation between the release of different synapses. However, uncorrelated events will in all cases produce too small σ_V (Fig. 4.3b, *crosses*), whereas a correlation of 0.1

is able to reproduce both the R_{in} change and σ_V (Fig. 4.3b, *triangles*) observed experimentally. In contrast, the value of σ_V displays a strong correlation with cell size, and the variability of σ_V values are relatively high compared to that of R_{in} decreases.

4.2.5 Model of Background Activity Including Voltage-Dependent Properties

In order to assess how the model of background activity is affected by the presence of voltage-dependent currents, one first has to estimated the voltage-dependent currents present in cortical cells from their current–voltage (I–V) relationship. The I–V curve of a representative neocortical cell after TTX microperfusion is shown in Fig. 4.4a. In this example, the I–V curve is approximately linear at a membrane potential more hyperpolarized than -60 mV, but displays an important outward rectification at more depolarized potentials, similar to in vitro observations (Stafstrom et al. 1982). The R_{in} is of about 57.3 MΩ at values around rest (~ -75 mV) and 30.3 MΩ at more depolarized V_m (> -60 mV), which represents a relative R_{in} change of 47%.

This type of I–V relation can be simulated by including two voltage-dependent K^+ currents, I_{Kd} and I_M (see Sect. 4.2.1). In the presence of these two currents, a constructed model will display a rectification comparable to that observed in cells showing the strongest rectification in experiments under TTX (Fig. 4.4b; the straight lines indicate the same linear fits as in Fig. 4.4a for comparison).

After such a confinement, the model can then be used to estimate the release conditions in the presence of voltage-dependent currents. First, the model including I_{Kd} and I_M must again be fit to passive traces obtained in the absence of synaptic activity to estimate the leak conductance and leak reversal (similar to Fig. 4.1a). Second, the release rate required to account for the σ_V and R_{in} change produced by miniature events must be estimated as in Fig. 4.1b. Third, the release rates that could best reproduce the R_{in}, average membrane potential and σ_V (see Fig. 4.1) must be estimated.

In the presence of voltage-dependent currents, the fitting of the model usually yields small, but detectable, changes in the optimal release conditions. In the example above, the fit yield release rates of $\nu_e = 0.92$ Hz and $\nu_i = 5.0$ Hz in the presence of I_{Kd} and I_M, which is about 8–9% lower compared to a solely passive model ($\nu_e = 1$ Hz and $\nu_i = 5.5$ Hz for the same R_{in} change, $< V_m >$ and σ_V). Both models give, however, nearly identical σ_V values for the same value of correlation. A similar constraining procedure can also be performed using a nonuniform leak distribution with high leak conductances in distal dendrites (see Sect. 4.2.1), in addition to I_{Kd} and I_M. Here, for the example described above, nearly identical results are obtained. This suggests that leak and voltage-dependent K^+ currents have a small contribution to the R_{in} and σ_V of active periods, which are mostly determined by synaptic activity.

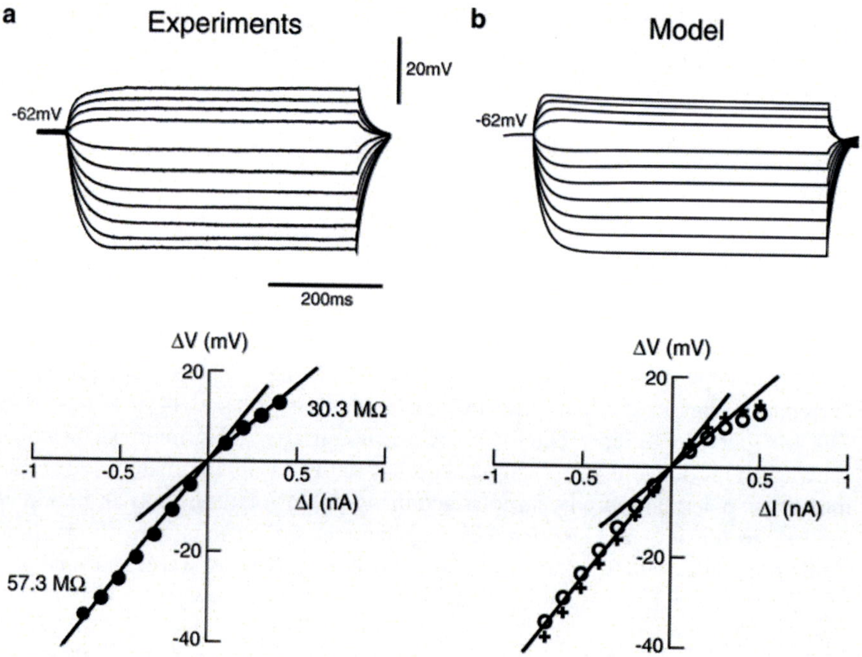

Fig. 4.4 Outward rectification of neocortical pyramidal neurons. (**a**) Current–voltage relation of a deep pyramidal neuron after microperfusion of TTX. The cell had a resting V_m of -75 mV after TTX and was maintained at -62 mV by DC current injection. The current–voltage (I–V) relation was obtained by additional injection of current pulses of different amplitudes. The I–V relation revealed a significant reduction of R_{in} at depolarized levels (*straight lines* indicate the best linear fits). (**b**) Simulation of the same protocol in the model pyramidal neuron. The model had two voltage-dependent K^+ currents, I_{Kd} (100 ps/μm^2) and I_M (2 ps/μm^2). The I–V relation obtained in the presence of both currents (*circles*) is compared with the same model with I_M removed (+). The model displayed a comparable rectification as experiments, although more pronounced (*straight lines* indicate the same linear fits as in (**a**) for comparison). Modified from Destexhe and Paré (1999)

In the presence of voltage-dependent Na^+ currents, in addition to I_{Kd} and I_M, models usually display pronounced spike frequency adaptation due to I_M, similar to regular spiking pyramidal cells in vitro (Connors et al. 1982). However, spike frequency adaptation is much less apparent in the presence of correlated synaptic activity (Fig. 4.5b), probably due to the very small conductance of I_M compared to synaptic conductances. Nevertheless, the presence of I_M affects the firing behavior of the cell, as suppressing this current enhances the excitability of the cell (Fig. 4.5b, *No* I_M). This is consistent with the increase of excitability demonstrated in neocortical slices (McCormick and Prince 1986) following application of acetylcholine, a blocker of I_M.

The *spontaneous* firing is also greatly affected by Na^+ current densities, which impact on the firing threshold. For instance, setting the threshold at about -55 mV in the soma leads to a sustained firing rate of around 10 Hz (Fig. 4.6a), with all

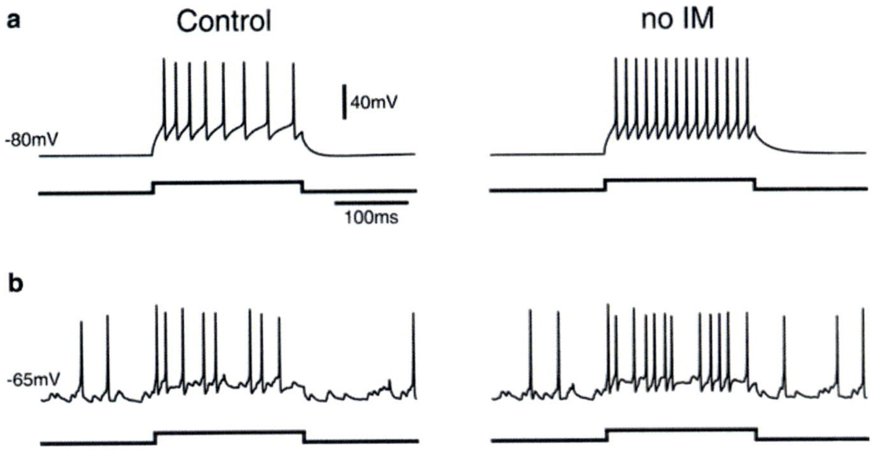

Fig. 4.5 Model of spike frequency adaptation with and without synaptic activity. (**a**) In the absence of synaptic activity, the slow voltage-dependent K^+ current I_M caused spike frequency adaptation in response to depolarizing current pulses (*Control*; 1 nA injected from resting V_m of -80 mV). The same stimulus did not elicit spike frequency adaptation in the absence of I_M (*No* I_M). (**b**) In the presence of correlated synaptic activity, spike frequency adaptation was not apparent although the same I_M conductance was used (*Control*; depolarizing pulse of 1 nA from resting V_m of ≈ -65 mV). Without I_M, the excitability of the cortical pyramidal cell was enhanced (*No* I_M). Modified from Destexhe and Paré (1999)

other features consistent with the model described above. In particular, the R_{in} reductions and values of σ_V produced by synaptic activity are minimally affected by the presence of voltage-dependent currents (Table 4.1). However, this observation is only valid for R_{in} calculated in the linear region of the I–V relation (< -60 mV; see Fig. 4.4). Thus, at membrane potentials more negative than -60 mV, the membrane parameters are essentially determined by background synaptic currents, with a minimal contribution from intrinsic voltage-dependent currents.

The sensitivity of the firing rate of the model cell to the release frequency is shown in Fig. 4.6b, c. A threefold increase in release frequencies leads to a proportional increase in firing rate (from ~10 Hz to ~30 Hz; Fig. 4.6b, gray solid). Indeed, if all release frequencies are increased by a given factor, the firing rate increases by about the same factor (Fig. 4.6c, *squares*). This shows that, within this range of release frequencies, the average firing rate of the cell reflects the average firing rate of its afferents. However, this relationship is broken if the release frequency is changed only at excitatory synapses: doubling the excitatory release frequency with no change in inhibition triples in this specific model the firing rate (Fig. 4.6c, *circles*).

Finally, an interesting observation is that sharp events of lower amplitude than APs are also visible in Fig. 4.6a, b. These events are likely to be dendritic spikes that do not reach AP threshold in the soma/axon region, similar to the fast prepotentials described by Spencer and Kandel (1961). Similar events were reported in intracellular recordings of neocortical pyramidal cells in vivo (Deschênes 1981).

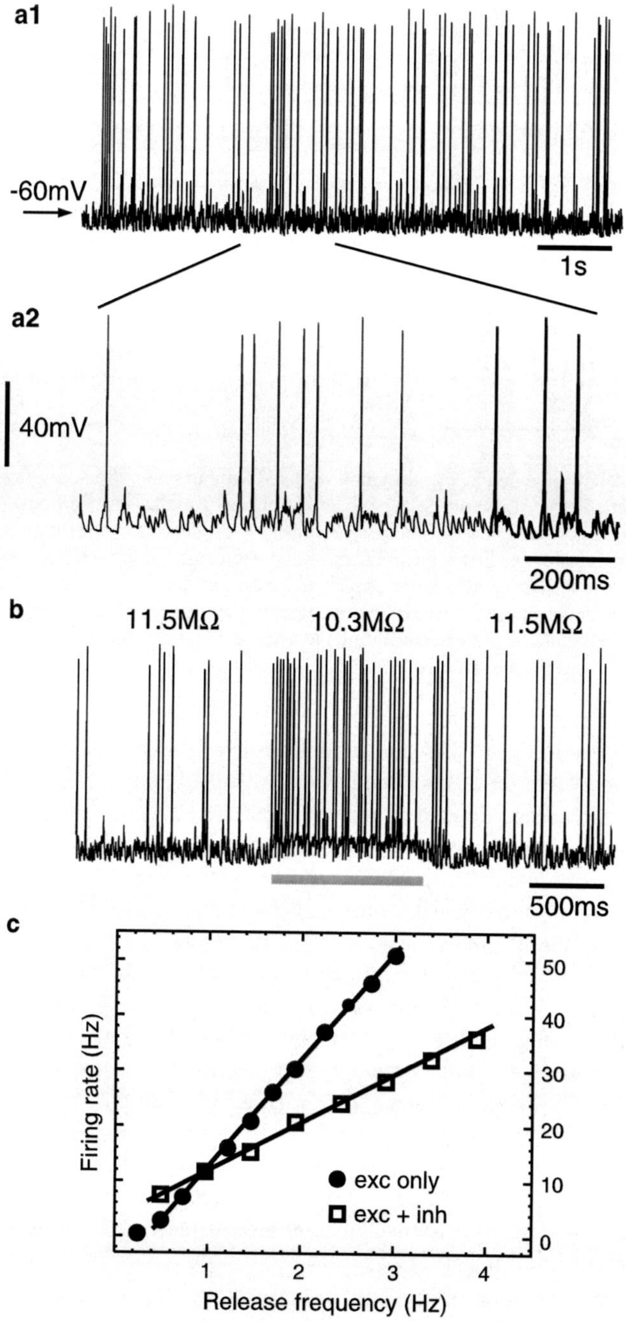
a1
-60mV
1s
a2
40mV
200ms
b
11.5MΩ
10.3MΩ
11.5MΩ
500ms
c
Firing rate (Hz)
50
40
30
20
10
0
exc only
exc + inh
1
2
3
4
Release frequency (Hz)

4.3 Simplified Compartmental Models

Simplified models are necessary to perform large-scale network simulations, because of the considerable computational requirement and complexity of morphologically realistic models. Systematic approaches for designing simplified models with electrical behavior equivalent to more detailed representations started with the concept of the *equivalent cylinder* introduced by Rall (1959, 1995). The equivalent cylinder representation of a neuron is, however, only possible under specific morphological constraints: all dendrites must end at the same electrotonic distance to soma and branch points must follow the 2/3 power rule (see Sect. 2.3.2), which makes it applicable only to a small subset of cellular morphologies.

Another approach was proposed later by Stratford et al. (1989) and consists of drawing a "cartoon model" in which the pyramidal cell morphology is reduced to 24 compartments. A related approach was also proposed to design simplified models based on the conservation of axial resistance (Bush and Sejnowski 1993). The latter type of reduced model had the same total axial resistance as the original model, but the membrane area was not conserved. A rescaling factor had to be applied to capacitance and conductances, to compensate for the different membrane area and yield correct input resistance.

Finally, a method to obtain reduced models that preserve membrane area and voltage attenuation with as few as three compartments was introduced by Destexhe (2001). Complex morphologies are collapsed into equivalent compartments to yield the same total membrane area. The axial resistances of the simplified model are adjusted by fitting, such that passive responses and voltage attenuation are identical between simplified and detailed models. The reduced model, thus, has the same membrane area, input resistance, time constant, and voltage attenuation as the detailed model.

In what follows, we illustrate the latter approach, and how well it accounts for situations such as the attenuation of synaptic potentials and the electrotonic structure in the presence of synaptic background activity.

Fig. 4.6 Tonic firing behavior in simulated active periods. (**a1**) In the presence of voltage-dependent Na^+ and K^+ conductances distributed in axon, soma, and dendrites, the simulated neuron produced tonic firing at a rate of ∼10 Hz (action potential threshold of −55 mV, $< V_m >= -65$ mV, $\sigma_V = 4.1$ mV). (**a2**) Same simulation as (**a1**) shown with a faster time base. (**b**) Effect of a threefold increase in release frequency at excitatory and inhibitory synapses. The firing rate of the simulated cell also increased threefold (*gray bar*). The $R_{\rm in}$ is indicated before, during and after the increase of release frequency. (**c**) Relation between firing rate and release frequency when the frequency of release was increased at both excitatory and inhibitory synapses (*squares*) or solely at excitatory synapses (*circles*). Modified from Destexhe and Paré (1999)

4.3.1 *Reduced 3-Compartment Model of Cortical Pyramidal Neuron*

To obtain a reduced model, the first step is to identify different functional regions in the dendritic morphology. A possibility is to chose these regions based on morphology and/or the distribution of synapses in the cell. For example, in neocortical pyramidal cells, these regions can be: (a) the soma and the first 40 μm of dendrites, which are devoid of spines (Jones and Powell 1969; Peters and Kaiserman-Abramof 1970) and mostly form inhibitory synapses (Jones and Powell 1970); (b) all dendrites laying between 40 μm and about 240 μm from the soma, which includes the vast majority of basal dendrites and the proximal trunk of apical dendrite; (c) all dendrites laying farther from 240 μm from the soma, which contains the major part of the apical dendritic tree. The latter two regions (b–c) contain nearly all excitatory synapses (DeFelipe and Fariñas 1992; White 1989). They also contain inhibitory synapses, although at lower density than in the soma. Thus, in the case of pyramidal cells, these considerations lead to three different functional regions that can be used to build a reduced model.

The second step is to identify each functional region with a single compartment in the reduced model, and to calculate its length and diameter. To allow comparison between the models, the length of the equivalent compartment is chosen as the typical physical length of its associated functional region, and the diameter of the equivalent compartment is chosen such that the total membrane area is the same as the ensemble of dendritic segments it represents. For example, in the layer VI pyramidal cell considered earlier (Fig. 4.7a), one can identify three compartments: the soma and first 40 μm of dendrites ("soma/proximal" compartment), the dendritic region between 40 μm and 240 μm ("middle" compartment), containing the majority of basal dendritic branches, and the dendritic segments from 240 μm to about 800 μm ("distal" compartment), which includes the apical dendrites and the distal ends of the longest basal branches. The lengths and diameters in this specific example are: $L = d = 34.8$ μm (*Soma/proximal*), $L = 200$ μm and $d = 28.8$ μm (*Middle*), $L = 515$ μm and $d = 6.20$ μm (*Distal*), yielding a reduced model with the same total membrane area as the detailed model (Fig. 4.7c).

Finally, the third step is to adjust the reduced model such that it produces correct passive responses and voltage attenuation. To this end, the same passive parameters as in the detailed model are used, except for the axial resistivities, which must be adjusted by a multiple fitting procedure (Press et al. 1993) constrained by passive responses (Fig. 4.7d) and somatodendritic profile following current injection (Fig. 4.7e). The optimal values in the given example are 4.721 kΩcm (*Soma/proximal*), 3.560 kΩcm (*Middle*) and 0.896 kΩcm (*Distal*). These values are high compared to recent estimates (Stuart and Spruston 1998) because each equivalent compartment represents here a large number of dendritic branches that are collapsed together. Thus, in these conditions, the reduced model has a total membrane area, input resistance, time constant, and voltage attenuation consistent with the detailed model.

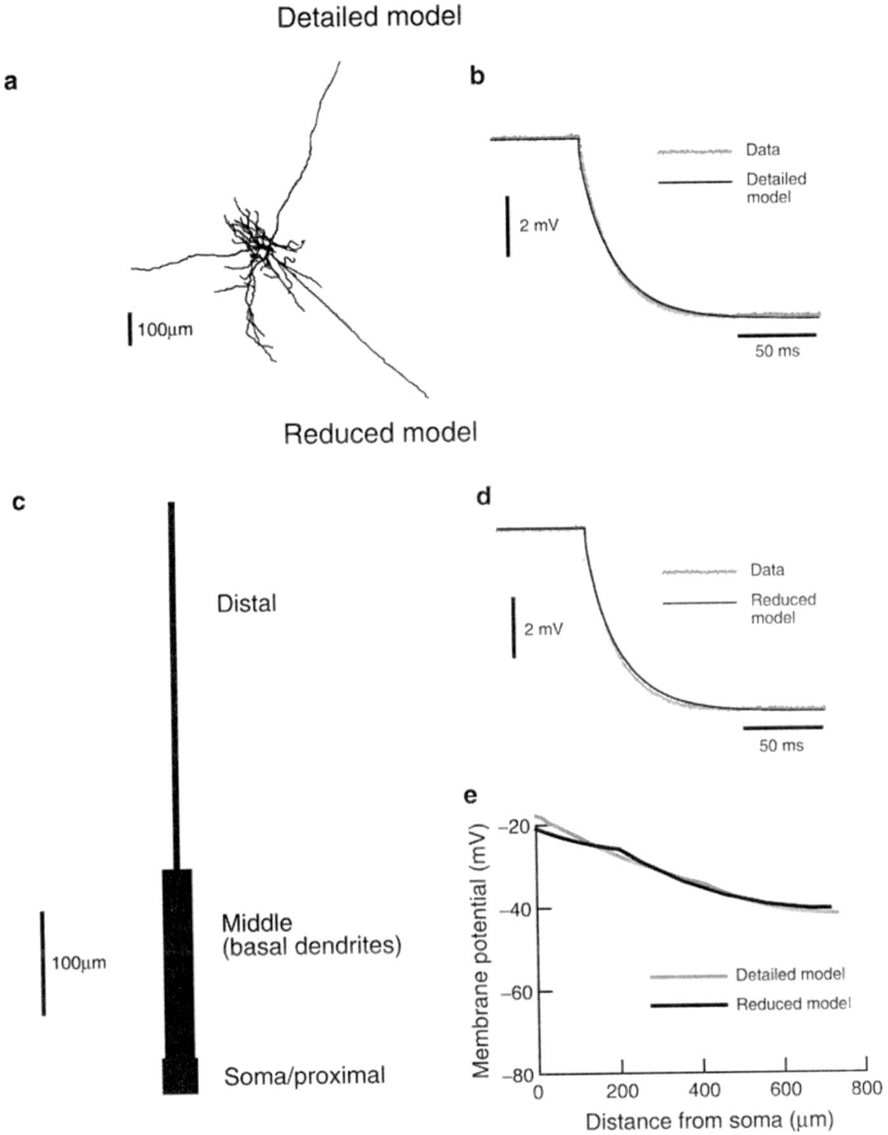

Fig. 4.7 Simplified 3-compartment model of layer VI neocortical pyramidal cell. (**a**) Dendritic morphology of a layer VI pyramidal cell from cat parietal cortex (from Contreras et al. 1997). (**b**) Best fit of the model to passive responses obtained experimentally ($g_L = 0.045$ ms cm^{-2}, $C_m = 1\ \mu$F/cm^2 and $R_i = 250\ \Omega$cm; see Destexhe and Paré 1999). (**c**) Simplified 3-compartment model obtained from the layer VI morphology in (**a**). The length and area of each compartment were calculated based on length and total area of the parent dendritic segments. (**d**) Adjustment of the 3-compartment model to passive responses (same experimental data as in (**b**)). (**e**) Adjustment of the 3-compartment model (*continuous line*) to the profile of voltage attenuation at steady state in the original layer VI model (*gray line*; each curve is an average over 10 s of activity with 0.8 nA injection in soma). Axial resistivities were adjusted by the fitting procedure, which was constrained by (**d**) and (**e**) simultaneously. Modified from Destexhe (2001)

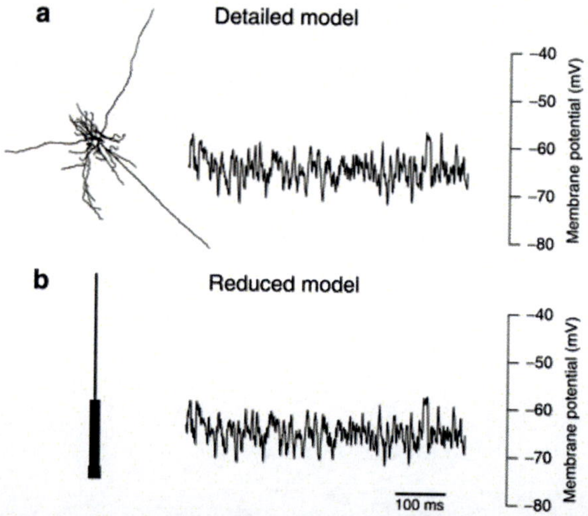

Fig. 4.8 Simulation of synaptic background activity using the 3-compartment model. (**a**) Synaptic background activity in the Layer VI pyramidal neuron described in Fig. 4.7a. A large number of randomly occurring synaptic inputs (16,563 glutamatergic and 3,376 GABAergic synapses) were simulated in all compartments of the model. (**b**) Same simulation using the same number of synaptic inputs in the reduced model. Both models gave membrane potential fluctuations of comparable fine structure. Modified from Destexhe (2001)

4.3.2 Test of the Reduced Model

To assess the performance of the 3-compartment model, one first simulates synaptic background activity, with a large number of randomly occurring synaptic inputs distributed in all compartments of the model (Fig. 4.8). Synaptic currents are modeled by kinetic models of glutamatergic and GABAergic receptors which distribution was based on morphological studies in pyramidal neurons (DeFelipe and Fariñas 1992; White 1989, see details in Destexhe and Paré 1999). Simulating background activity with a total of 16,563 glutamatergic and 3,376 GABAergic synapses distributed in dendrites (with same number of synapses in corresponding locations in the reduced model) and releasing randomly according to Poisson processes (Fig. 4.8, *Detailed model*), leads to membrane potential fluctuations of similar fine structure at the soma (Fig. 4.8, *Reduced model*).

The reduced model can further be tested by considering the variations of input resistance ($R_{\rm in}$) and average somatic membrane potential ($< V_m >$) due to the presence of synaptic background activity. In the layer VI pyramidal cell model example, the $R_{\rm in}$ and $< V_m >$ vary as a function of the release frequency at synaptic terminals (Fig. 4.9a, *symbols*; see details of this model in Destexhe and Paré 1999, see also Sect. 4.2). The reduced model behaves remarkably similarly to the layer VI model when the same densities of synaptic conductances are used (Fig. 4.9a, *black solid*), a result which is not achievable with a single-compartment model.

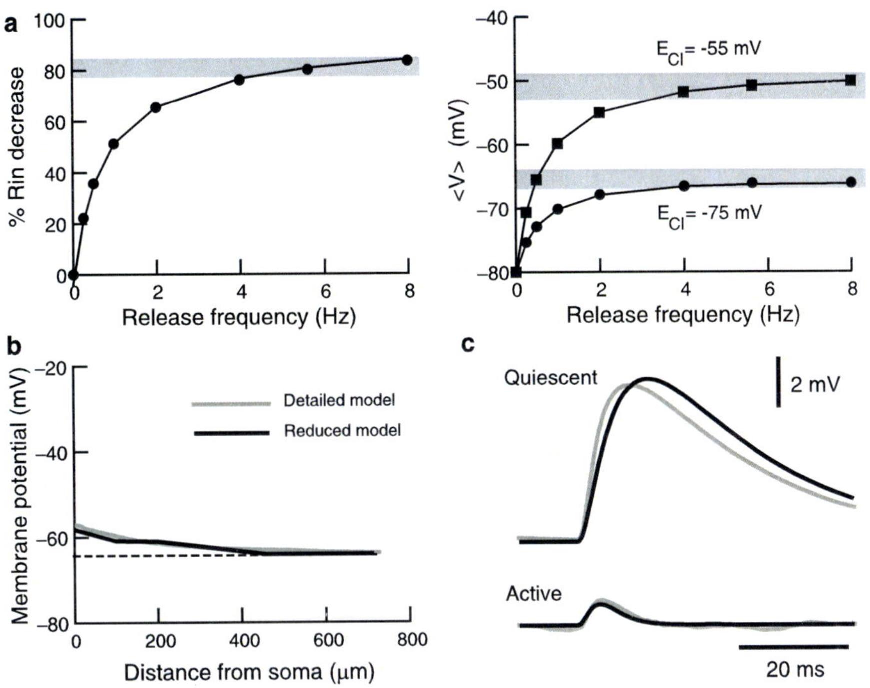

Fig. 4.9 Behavior of the 3-compartment model in the presence of synaptic background activity. (**a**) Decrease in input resistance (R_{in}) and average membrane potential ($< V_m >$) in the presence of synaptic background activity (E_{Cl} is the chloride reversal potential). The behavior of the 3-compartment model (*continuous line*) is remarkably similar to that of the layer VI pyramidal cell model (*symbols*; data from Destexhe and Paré 1999). The values measured intracellularly in cat parietal cortex in vivo (Paré et al. 1998b) are shown in gray for comparison. (**b**) Voltage attenuation in the presence of synaptic background activity. The 3-compartment model (*black solid*) is compared to the layer VI pyramidal cell model (*gray solid*; 0.8 nA injection in soma) in the presence of background activity (same scale as in Fig. 4.7e). (**c**) Attenuation of distal EPSPs. A 100 ns AMPA-mediated EPSP was simulated in the distal dendrite of the simplified model (*black solid*) and is compared to synchronized stimulation of the same synaptic conductance distributed in distal dendrites of the layer VI pyramidal cell model (*gray solid*). The EPSP was of 5.6 mV amplitude without background activity (*Quiescent*) and dropped to 0.68 mV in the presence of background activity (*Active*). Both models generated EPSPs of similar amplitude. Modified from Destexhe (2001)

Two additional properties are correctly captured by the reduced model compared to the detailed layer VI model. First, the profile of voltage attenuation in the presence of background activity is similar in both models (Fig. 4.9b). The simplified model, thus, reproduces the observation that background activity is responsible for an about fivefold increase in voltage attenuation (Destexhe and Paré 1999, compare Figs. 4.9b with 4.7e). Second, stimulation of synaptic inputs distributed in distal dendrites yield similar somatic EPSPs amplitudes in both models, in the presence of background activity (Fig. 4.9c, *Active*) and in quiescent conditions (Fig. 4.9c, *Quiescent*). These properties will be investigated in more detail in Chap. 5.

In conclusion, the reduced model introduced here can be obtained by fitting the axial resistances constrained by passive responses and voltage attenuation. The model obtained preserves the membrane area, input resistance, time constant, and voltage attenuation of more detailed morphological representations. It also captures the electrotonic properties of the neurons in the presence of synaptic background activity. However, this model is one to two orders of magnitude faster to simulate when compared to the corresponding detailed model.

4.4 The Point-Conductance Model of Synaptic Noise

The fluctuating activity present in neocortical neurons in vivo and its high conductance can also be modeled using single-compartment models. We review here one of the most simplified versions, the "point-conductance" model, which can be used to represent the conductances generated by thousands of stochastically releasing synapses. In this model, synaptic activity is represented by two independent fast glutamatergic and GABAergic conductances described by stochastic random-walk processes. As we will show in this section, an advantage of this approach is that all the model parameters can be determined from voltage-clamp experiments. Finally, the point-conductance model also captures the amplitude and spectral characteristics of the synaptic conductances during background activity. The point-conductance model has many advantages, which allow its efficient application in dynamic-clamp experiments (Chap. 6), the possibility of a mathematical treatment (Chap. 7), as well as its use as an analysis technique to extract conductances and other properties from experimental recordings (Chap. 8).

4.4.1 The Point-Conductance Model

In the *point-conductance model*, synaptic background activity (Destexhe et al. 2001) is given by the following membrane equation:

$$C\frac{\mathrm{d}V}{\mathrm{d}t} = -g_{\mathrm{L}}(V - E_{\mathrm{L}}) - I_{\mathrm{syn}} + I_{\mathrm{ext}}\,, \tag{4.2}$$

where C is the membrane capacitance, I_{ext} a stimulation current, g_{L} the leak conductance and E_{L} the leak reversal potential. I_{syn} is the total synaptic current, which is decomposed into the sum of two independent terms:

$$I_{\mathrm{syn}} = g_e(t)(V - E_e) + g_i(t)(V - E_i), \tag{4.3}$$

where $g_e(t)$ and $g_i(t)$ represent the time-dependent excitatory and inhibitory conductances , respectively. Their respective reversal potentials, $E_e = 0$ mV and $E_i = -75$ mV, are identical to those used in the detailed biophysical model.

The conductances $g_e(t)$ and $g_i(t)$ are described by a one-variable stochastic process similar to the Ornstein–Uhlenbeck (OU) process (Uhlenbeck and Ornstein 1930):

$$\begin{aligned}\frac{\mathrm{d}g_e(t)}{\mathrm{d}t} &= -\frac{1}{\tau_e}[g_e(t) - g_{e0}] + \sqrt{D_e}\chi_1(t) \\ \frac{\mathrm{d}g_i(t)}{\mathrm{d}t} &= -\frac{1}{\tau_i}[g_i(t) - g_{i0}] + \sqrt{D_i}\chi_2(t)\,,\end{aligned} \tag{4.4}$$

where g_{e0} and g_{i0} are average conductances, τ_e and τ_i are time constants, D_e and D_i are noise "diffusion" coefficients, $\chi_1(t)$ and $\chi_2(t)$ are Gaussian white noise processes of unit SD (see Sect. 4.4.4 for a formal derivation of this model).

The numerical scheme for integration of these stochastic differential equations (SDEs) takes advantage of the fact that these stochastic processes are Gaussian, which leads to an exact update rule (Gillespie 1996):

$$\begin{aligned}g_e(t+h) &= g_{e0} + (g_e(t) - g_{e0})\mathrm{e}^{-\frac{h}{\tau_e}} + A_e N_1(0,1) \\ g_i(t+h) &= g_{i0} + (g_i(t) - g_{i0})\mathrm{e}^{-\frac{h}{\tau_i}} + A_i N_2(0,1)\,,\end{aligned} \tag{4.5}$$

where $N_1(0,1)$ and $N_2(0,1)$ are normal distributed random numbers with zero average and unit SD, and A_e, A_i are amplitude coefficients given by:

$$\begin{aligned}A_e &= \sqrt{\frac{D_e\tau_e}{2}\left[1 - \mathrm{e}^{-\frac{2h}{\tau_e}}\right]} \\ A_i &= \sqrt{\frac{D_i\tau_i}{2}\left[1 - \mathrm{e}^{-\frac{2h}{\tau_i}}\right]}\,.\end{aligned}$$

This update rule provides a stable integration procedure for Gaussian stochastic models, which guarantees that the statistical properties of the variables $g_e(t)$ and $g_i(t)$ are not dependent on the integration step h. The point-conductance model is then inserted into a single-compartment that includes voltage-dependent conductances described by Hodgkin and Huxley (1952d) type models:

$$\begin{aligned}C_\mathrm{m}\frac{\mathrm{d}V}{\mathrm{d}t} &= -g_\mathrm{L}(V - E_\mathrm{L}) - I_\mathrm{Na} - I_\mathrm{Kd} - I_M - \frac{1}{a}I_\mathrm{syn} \\ I_\mathrm{Na} &- \bar{g}_\mathrm{Na}m^3h(V - E_\mathrm{Na}) \\ I_\mathrm{Kd} &= \bar{g}_\mathrm{Kd}n^4(V - E_\mathrm{K}) \\ I_M &= \bar{g}_M p(V - E_\mathrm{K})\,,\end{aligned} \tag{4.6}$$

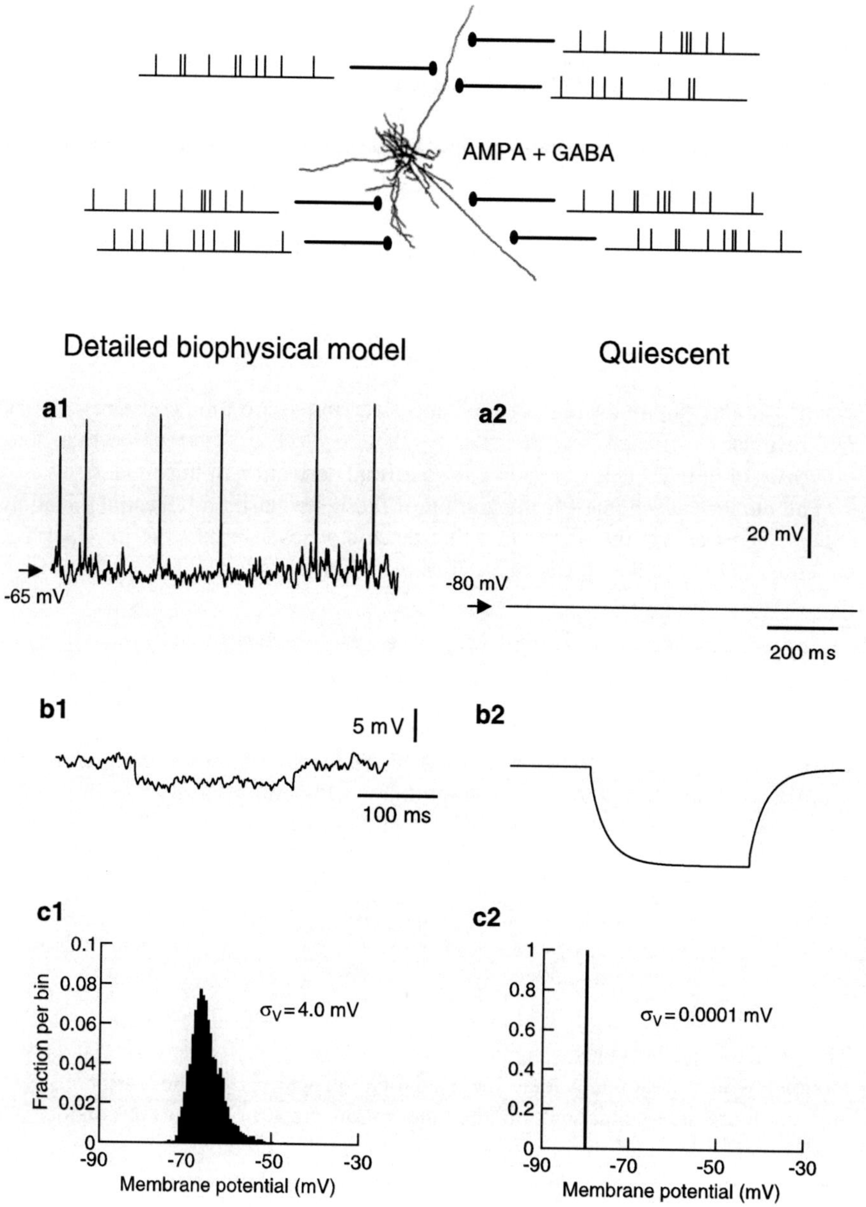
AMPA + GABA
Detailed biophysical model
Quiescent
a1
-65 mV
a2
20 mV
-80 mV
200 ms
b1
5 mV
100 ms
b2
c1
Fraction per bin
0.1
0.08
0.06
0.04
0.02
0
-90
-70
-50
-30
Membrane potential (mV)
σV=4.0 mV
c2
1
0.8
0.6
0.4
0.2
0
-90
-70
-50
-30
Membrane potential (mV)
σV=0.0001 mV

where $C_m = 1\ \mu F/cm^2$ denotes the specific membrane capacitance, $g_L = 0.045\ ms/cm^2$ is the leak conductance density, and $E_L = -80$ mV is the leak reversal potential. I_{Na} is the voltage-dependent Na^+ current and I_{Kd} is the "delayed-rectifier" K^+ current responsible for action potentials. I_M is a noninactivating K^+ current responsible for spike frequency adaptation. In order to allow comparison with the biophysical model described earlier (see also Destexhe and Paré 1999), the currents and their parameters are the same in both models. Furthermore, in 4.6, a denotes the total membrane area, which is, for instance, 34,636 μm^2 for the layer VI cell described in Fig. 4.10.

4.4.2 Derivation of the Point-Conductance Model from Biophysically Detailed Models

Figure 4.10 shows an example of in vivo-like synaptic activity simulated in a detailed layer VI compartmental model (see Sect. 4.2). The membrane potential fluctuates around −65 mV and spontaneously generates APs at an average firing rate of about 10 Hz (Fig. 4.10a1), which is within the range of experimental measurements in cortical neurons of awake animals (Hubel 1959; Evarts 1964; Steriade 1978; Matsumura et al. 1988; Steriade et al. 2001). Without background activity, the constructed cell is resting at −80 mV (Fig. 4.10a2), similar to experimental measurements after microperfusion of TTX (Paré et al. 1998b). The model further reproduces the ∼80% decrease of input resistance (Fig. 4.10b), as well as the distribution of membrane potential (Fig. 4.10c) typical of in vivo activity (see Paré et al. 1998b; Destexhe and Paré 1999).

A model of synaptic background activity must include the large conductance due to synaptic activity and also its large V_m fluctuations, as both aspects are important and determine cellular responsiveness (Hô and Destexhe 2000, see also Chap. 5). To build such a model, one first needs to characterize these fluctuations as outlined in Chap. 3. Briefly, one records the total synaptic current resulting from synaptic background activity using an "ideal" voltage clamp (e.g., with a series resistance of 0.001 MΩ) in the soma. This current (I_{syn} in (4.3)) is then decomposed into a sum

Fig. 4.10 Properties of neocortical neurons in the presence of background activity simulated using a detailed biophysical model. *Top*: Layer VI pyramidal neuron reconstructed and incorporated in simulations. (**a**) Membrane potential in the presence of background activity (**a1**) and at rest (**a2**). Background activity was simulated by random release events described by weakly correlated Poisson processes of average releasing frequency of 1 Hz and 5 Hz for excitatory and inhibitory synapses, respectively (Destexhe and Paré 1999). (**b**) Effect on input resistance. A hyperpolarizing pulse of −0.1 nA was injected at −65 mV in both cases (average of 100 pulses in **a1**). The presence of background activity (**b1**) was responsible for a ∼ fivefold decrease in input resistance compared to rest (**b2**). (**c**) Membrane potential distribution in the presence (**c1**) and in the absence (**c2**) of background activity. Modified from Destexhe et al. (2001)

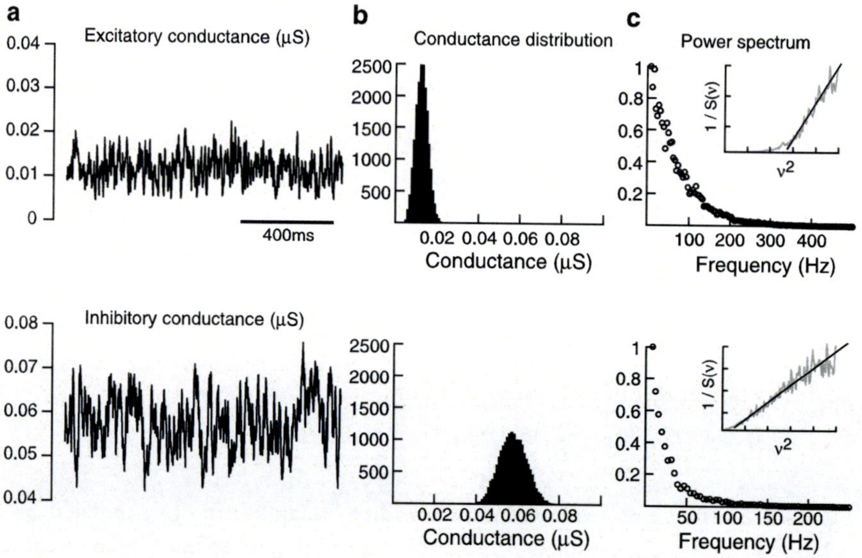

Fig. 4.11 Statistical properties of the conductances underlying background activity in the detailed biophysical model. (**a**) Time course of the total excitatory (*top*) and inhibitory (*bottom*) conductances during synaptic background activity. (**b**) Distribution of values for each conductance, calculated from (**a**). (**c**) Power spectral density of each conductance. The insets show the inverse of the power spectral density ($1/S(\nu)$) represented against squared frequency (ν^2; same scale used). Modified from Destexhe et al. (2001)

of two currents, associated respectively with an excitatory (g_e) and an inhibitory (g_i) conductance. The latter are calculated by running the model twice at two different clamped voltages (-65 and -55 mV), which leads to two equations similar to (4.3). The time course of the conductances are then obtained by solving these equations for g_e and g_i, at each time step.

The time course of g_e and g_i during synaptic background activity is illustrated in Fig. 4.11a. As expected from the stochastic nature of the release processes, these conductances are highly fluctuating and have a broad, approximately symmetric distribution (Fig. 4.11b). It is also apparent that the inhibitory conductance g_i accounts for most of the conductance seen in the soma, similar to voltage-clamp measurements in cat visual cortex in vivo (Borg-Graham et al. 1998). This is not surprising here, because the release frequency of GABAergic synapses is five times larger than that of glutamatergic synapses, their decay time constant is slower, and the perisomatic region contains exclusively GABAergic synapses (DeFelipe and Fariñas 1992). The average values and SDs are, respectively, 0.012 μs and 0.0030 μs for g_e, and of 0.057 and 0.0066 μs for g_i.

The PSD of g_e and g_i (Fig. 4.11c) shows a broad spectral structure, as expected from their apparent stochastic behavior, but also there is a clear decay at larger frequencies, suggesting that these processes are analogous to *colored noise*. Interestingly, the power spectrum clearly seems to decay in $1/\nu^2$ for inhibitory synapses, as shown by representing the inverse of the power spectrum as a function of ν^2 (Fig. 4.11c, *insets*). However, for excitatory synapses, the ν^2 decay is only true for high frequencies.

In order to confine the model's parameter space, one next searches for stochastic representations that will capture the amplitude of the conductances, their SD and their spectral structure. Such a representation is provided by the OU process (Uhlenbeck and Ornstein 1930), described by the differential equation

$$\frac{\mathrm{d}x}{\mathrm{d}t} = -\frac{x}{\tau} + \sqrt{D}\chi(t), \tag{4.7}$$

where x is the random variable, D is the amplitude of the stochastic component, $\chi(t)$ is a normally distributed (zero-mean) noise source, and τ is the time constant. Here, $\tau = 0$ gives white noise, whereas $\tau > 0$ gives "colored" noise.

Indeed, the integration of the activity at thousands of individual synaptic terminals results in global excitatory and inhibitory conductances which fluctuate around a mean value (Fig. 4.12a, left). These global conductances are characterized by nearly Gaussian amplitude distributions $\rho(g)$ (Fig. 4.12b, *gray*), and their power spectral density $S(\nu)$ follows a Lorentzian behavior (Destexhe et al. 2001) (Fig. 4.12c, *gray*). The OU process is known as one of the simplest stochastic processes which captures these statistical properties. Introduced at the beginning of the last century to describe Brownian motion (Uhlenbeck and Ornstein 1930; Wang and Uhlenbeck 1945; Hänggi and Jung 1994), this model has long been used as a effective stochastic model of synaptic noise (Ricciardi and Sacerdote 1979; Lánský and Rospars 1995).

Indeed, it can be shown that incorporating synaptic conductances described by OU processes into single-compartment models provides a compact stochastic representation that captures in a broad parameter regime their temporal dynamics (Fig. 4.12a, right), amplitude distribution (Fig. 4.12b, *black*) and spectral structure (Fig. 4.12c, *black*; Destexhe et al. 2001), as well as subthreshold and response properties of cortical neurons characteristic for in vivo conditions (Fellous et al. 2003; Prescott and De Koninck 2003).

The advantage of using this type of stochastic model is that the distribution of the variable x (see (4.7)) and its spectral characteristics are known analytically (see details in Gillespie 1996). The stochastic process x is Gaussian, its variance is given by

$$\sigma^2 = D\tau/2, \tag{4.8}$$

and its PSD is

$$S(\nu) = \frac{2D\tau^2}{1 + (2\pi\nu\tau)^2}, \tag{4.9}$$

where ν denotes the frequency.

The Gaussian nature of the OU process, and its spectrum in $1/\nu^2$, qualitatively match the behavior of the conductances underlying background activity in the

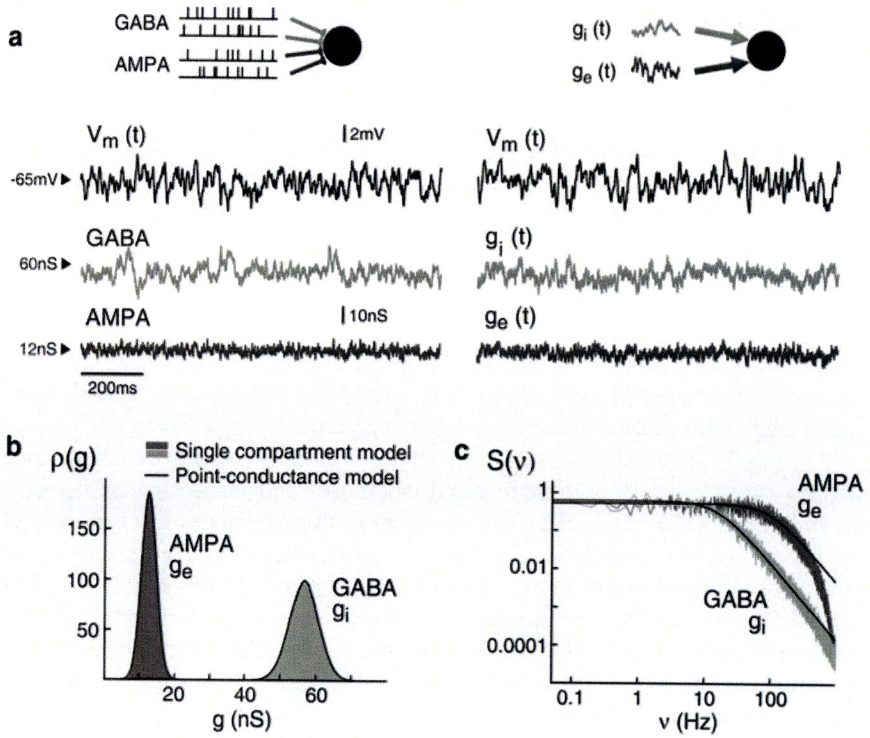

Fig. 4.12 Simplified models of synaptic background activity. (**a**) Comparison between detailed and simplified models of synaptic background activity. *Left*: biophysical model of a cortical neuron with realistic synaptic inputs. Synaptic activity was simulated by the random release of a large number of excitatory and inhibitory synapses (4,472 and 3,801 synapses, respectively; see scheme on top) in a single-compartment neuron. Individual synaptic currents were described by 2-state kinetic models of glutamate (AMPA) and GABAergic ($GABA_A$) receptors. The traces show, respectively, the membrane potential (V_m) as well as the total excitatory (*AMPA*) and inhibitory (*GABA*) conductances. *Right*: simplified "point-conductance" model of synaptic background activity produced by two global excitatory and inhibitory conductances ($g_e(t)$ and $g_i(t)$ in scheme on top), which were simulated by stochastic processes (Destexhe et al. 2001). The traces show the V_m as well as the global excitatory ($g_e(t)$) and inhibitory ($g_i(t)$) conductances. (**b**) Distribution of synaptic conductances $\rho(g)$ (*gray*: biophysical model; *black*: point-conductance model). (**c**) Comparison of the power spectral densities $S(\nu)$ of global synaptic conductances obtained in these two models (*gray*: biophysical model; *black*: point-conductance model). The two models share similar statistical and spectral properties, but the point-conductance model is more than two orders of magnitude faster to simulate. Modified from Rudolph and Destexhe (2004)

detailed biophysical model. Moreover, the variance and the spectral structure of this model can be manipulated by only two variables (D and τ), which is very convenient for fitting this model, for instance, to experimental data (using (4.8) and (4.9)).

The above outlined procedure can be applied to obtain a simple point-conductance representation of synaptic background activity. Figure 4.13 shows the result of fitting the model described by (4.4) to the detailed biophysical simulations shown in Fig. 4.11. Here, (4.9) provides excellent fits to the PSD (Fig. 4.13a), both

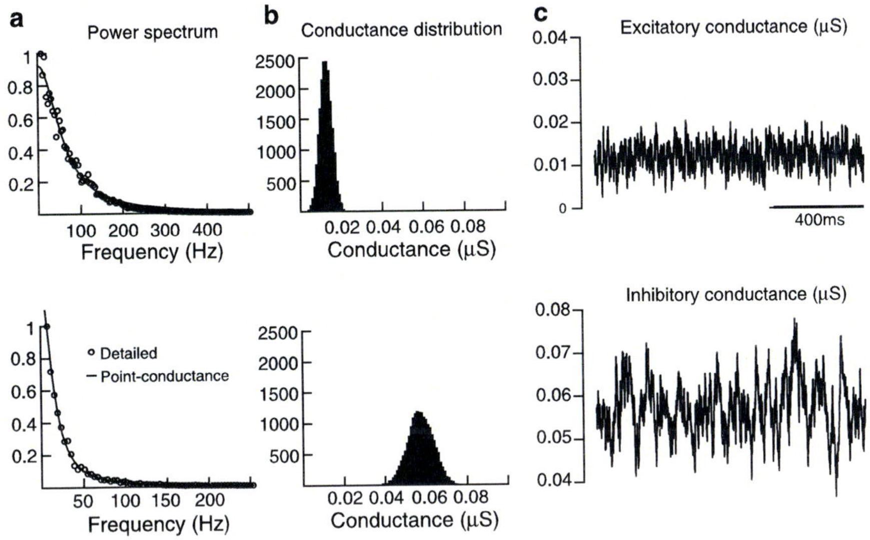

Fig. 4.13 Fit of a point-conductance model of background synaptic activity. (**a**) Power spectral density of the conductances from the biophysical model (*top*: excitatory, *bottom*: inhibitory). The *continuous lines* show the best fits obtained with the stochastic point-conductance model. (**b**) Distribution of conductance values for the point-conductance model. (**c**) Time course of the excitatory and inhibitory conductances of the best stochastic model. The same data length as in Fig. 4.11 were used for all analyses. Modified from Destexhe et al. (2001)

for excitatory and inhibitory conductances, and by using the average values and standard deviations estimated from the distribution of conductances in Fig. 4.11b, the other parameters of the point-conductance model can be calculated. A summary of the so-estimated parameters is given in Table 4.2. The resulting stochastic model has a distribution (Fig. 4.13b) and temporal behavior (Fig. 4.13c) consistent with the conductances estimated from the corresponding detailed model (compare with Fig. 4.11). Most importantly, the so-constructed point-conductance model is more than two orders of magnitude faster to simulate compared to the corresponding detailed biophysical model.

In order to use the point-conductance approach to build a single-compartment model of synaptic background activity in neocortical neurons, one investigates the basic membrane properties of the model, including the effect of background activity on the average V_m and input resistance, as well as the statistical properties of subthreshold voltage fluctuations (Fig. 4.14). In the absence of background activity, the cell is resting at a V_m of around -80 mV (Fig. 4.14a2), similar to measurements in vivo in the presence of TTX (Paré et al. 1998b). In the presence of synaptic background activity, intracellular recordings indicate that the V_m of cortical neurons fluctuates around -65 mV, a feature that is reproduced by the point-conductance model (Fig. 4.14a1; compare with the biophysical model in Fig. 4.10a1). The point-conductance model also captures the effect of background activity on input

Table 4.2 Effect of cellular morphology

Cell	Layer VI	Layer III	Layer Va	Layer Vb
Area (μm^2)	34,636	20,321	55,017	93,265
R_{in} (MΩ)	58.9	94.2	38.9	23.1
g_{e0} (μs)	0.012	0.006	0.018	0.029
σ_e (μs)	0.0030	0.0019	0.0035	0.0042
τ_e (ms)	2.7	7.8	2.6	2.8
g_{i0} (μs)	0.057	0.044	0.098	0.16
σ_i (μs)	0.0066	0.0069	0.0092	0.01
τ_i (ms)	10.5	8.8	8.0	8.5

Synaptic background activity was simulated in four reconstructed neurons from cat cerebral cortex, using several thousand glutamatergic and GABAergic inputs distributed in the soma and dendrites (Destexhe and Paré 1999). The same respective densities of synapses were used in all four cells. The table shows the parameters of the best fit of the point-conductance model using the same procedure. These parameters generally depended on the morphology, but their ratios (g_{e0}/g_{i0} and σ_e/σ_i) were approximately constant

resistance (Fig. 4.14b; compare with Fig. 4.10b), as well as on the amplitude of voltage fluctuations (Fig. 4.14c; compare with Fig. 4.10c). The SD calculated from the V_m distributions (see Fig. 4.14b) is about $\sigma_V = 4$ mV for both models.

4.4.3 *Significance of the Parameters*

Figure 4.15 illustrates changes of the parameters of background activity in the detailed model, and how these changes affect the mean and variance of the conductances. As shown, during a sudden change in the correlation (Fig. 4.15, *gray box*), the variance of excitatory and inhibitory conductances do significantly change, while their average value remains nearly unaffected.

Input signals consisting in the simultaneous firing of a population of cells occur in vivo on a background of random synaptic noise. In order to assess how correlated synaptic events are reflected at the soma, Destexhe and Paré (1999) used a reconstructed multicompartmental cell (Fig. 4.16a) from the rat prefrontal cortex that received 16,563 AMPA synapses and 3,376 GABA synapses discharging in a Poisson manner at 1 Hz and 5.5 Hz, respectively. At the soma, such an intense synaptic bombardment yields voltage fluctuations that depended on the amount of correlations introduced among the synaptic inputs. Figure 4.16a displays sample traces in cases of low (0.1) and high (0.9) correlations in the excitatory synaptic inputs, and the relationship between the standard deviation of the membrane potential measured at the soma and the synaptic correlation (right panel). Figure 4.16b also shows that for the point-conductance model (one compartment) it is possible to find a unique value of the SD σ_e of the excitatory stochastic conductance g_e that results in a simulated somatic synaptic current yielding membrane voltage fluctuations

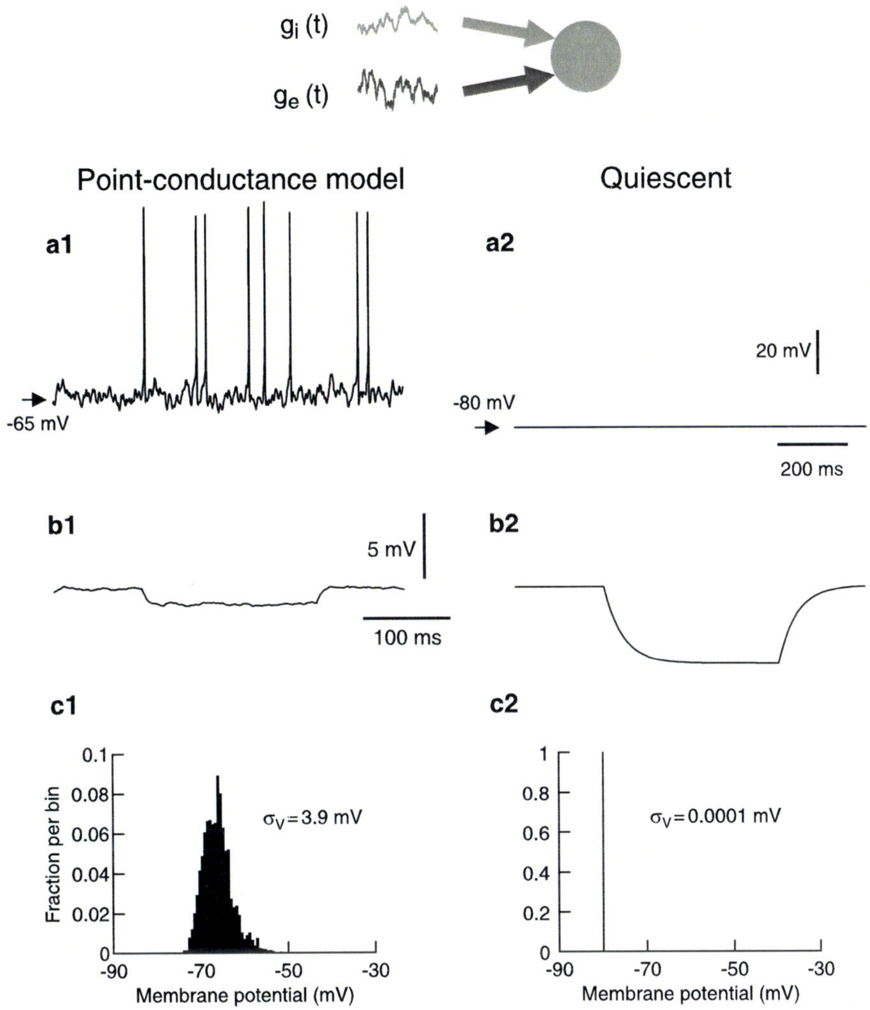

Fig. 4.14 Membrane potential and input resistance of the point-conductance model. *Top*: scheme of the point conductance model, where two stochastically varying conductances determine the V_m fluctuations through their (multiplicative) interaction. (**a**) Membrane potential in the presence (**a1**) and in the absence (**a2**) of synaptic background activity represented by two fluctuating conductances. The point-conductance model was inserted in a single-compartment with only a leak current (same conductance density as in the detailed model). (**b**) Effect on input resistance (same description and hyperpolarizing pulse as in Fig. 4.10b). (**c**) V_m distribution in the presence (**c1**) and in the absence (**c2**) of the fluctuating conductances. Modified from Destexhe et al. (2001)

equivalent to the ones observed in the detailed model. Moreover, Fig. 4.16 gives the corresponding curves obtained with the reconstructed model of a cat pyramidal cell which was extensively used in other studies, and for which parameters have been

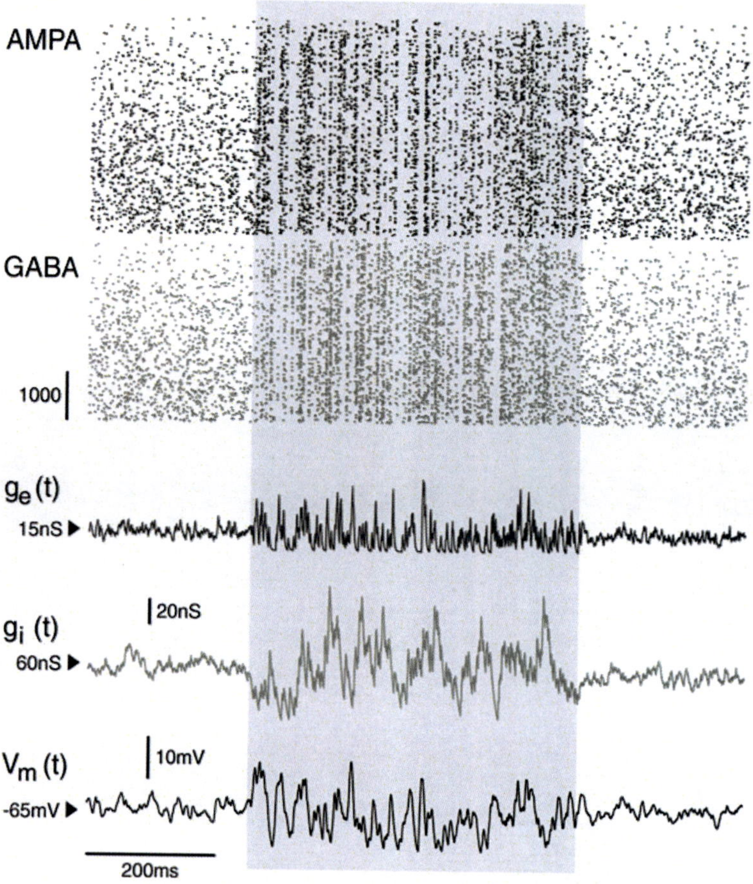

Fig. 4.15 Correspondence between network activity and global synaptic conductances. A biophysical model of cortical neuron in high-conductance state was simulated (same model as in Fig. 4.12, with 4,472 AMPA and 3,801 GABAergic synapses releasing randomly). The graph shows, from top to bottom: raster plots of the release times at excitatory (AMPA) and inhibitory (GABA) synapses, the global excitatory ($g_e(t)$) and inhibitory ($g_i(t)$) synaptic conductances, and the membrane potential ($V_m(t)$). The synapses were initially uncorrelated for the first 250 ms, resulting in rasters with no particular structure (*top*), and global average conductances which were about 15 ns for excitation and 60 ns for inhibition (*bottom*). At $t = 250$ ms, a correlation was introduced between the release of individual synapses for a period of 500 ms (*gray box*). The correlation increases the probability of different synapses to co-release, which appears as vertical bands in the raster. Because the release rate was not affected by correlation, the average global conductance was unchanged. However, the amount of fluctuations (i.e., the variance) of the global conductances increased as a result of this correlation change, and led to a change in the fluctuation amplitude of the membrane potential. Such changes in the V_m can affect spiking activity, thus allowing cortical neurons to detect changes in the release statistics of their synaptic inputs. Modified from Rudolph and Destexhe (2004)

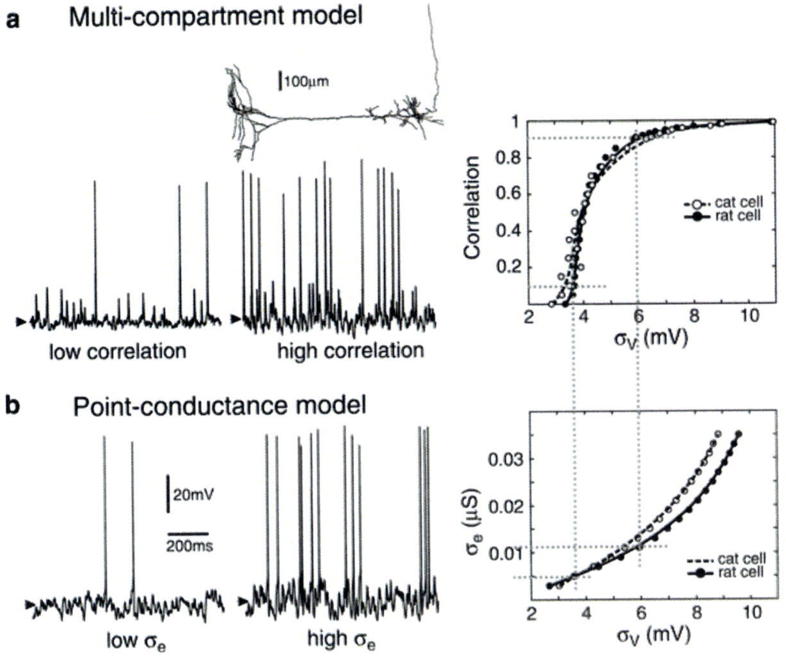

Fig. 4.16 Relationship between the variance and correlation of synaptic inputs. (**a**) Detailed model: The *left panels* show sample voltage traces for low ($c = 0.1$; average membrane potential of −65.5 mV; *arrow*) and high ($c = 0.9$; average membrane potential of −65.2 mV; *arrow*) AMPA synaptic correlations. The *right panel* shows the relationship between the amount of synaptic correlations and the resulting standard deviation (SD) of the membrane voltage. *Horizontal dashed lines* correspond to the sample traces shown on the left. The correlation among inhibitory synapses was fixed ($c = 0$). The inset shows the detailed morphology of the rat cell used in this study. (**b**) Point conductance model: The *left panels* show sample voltage traces for low ($\sigma_e = 5$ ns; average membrane potential of −64.8 mV; *arrow*) and high ($\sigma_e = 11$ ns; average membrane potential of −64.9 mV; *arrow*) SD of the stochastic variable (σ_e) representing excitatory inputs to the one compartment model. The *right panel* shows the relationship between σ_e and the resulting SD of the membrane voltage of the point-conductance model. The *dashed lines* show that there is a one-to-one correspondence between the value of correlation and σ_e. The SD of the stochastic variable representing inhibitory inputs was fixed ($\sigma_i = 15$ ns). The *dashed lines* correspond to the sample traces shown on the left. Low (5 ns) and high (11 ns) σ_e yield membrane potential fluctuations and firing rate equivalent to correlations those obtained in the detailed model for synaptic correlations $c = 0.1$ and $c = 0.9$, respectively. Modified from Fellous et al. (2003)

directly constrained by in vivo recordings (Destexhe and Paré 1999). In both cases, the correlation of synaptic inputs translates directly into the variance of synaptic conductances.

Finally, comparing the firing behavior of the two types of model in the presence of background activity, one observes irregular firing in both cases (see Figs. 4.10a1 and 4.14a1). Moreover, the variation of the average firing rate as a function of the conductance parameters is remarkably similar (Fig. 4.17). Although the absolute

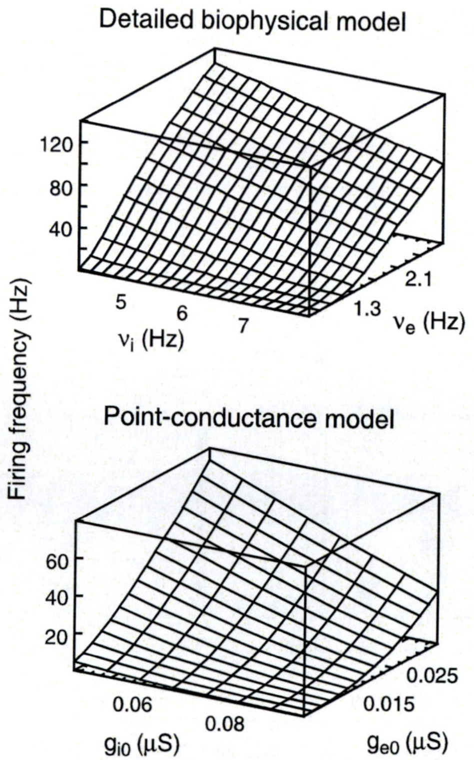

Fig. 4.17 Comparison of the mean firing rate point-conductance and detailed biophysical models. The firing rate is shown as a function of the strength of excitation and inhibition. The biophysical model was identical to that used in Fig. 4.10, while the point-conductance model was inserted in a single-compartment containing the same voltage-dependent currents as the biophysical model. The strength of excitation/inhibition was changed in the detailed model by using different release frequencies for glutamatergic (ν_e) and GABAergic (ν_i) synapses, while it was changed in the point-conductance model by varying the average excitatory (g_{e0}) and inhibitory (g_{i0}) conductances. g_{e0} and g_{i0} were varied within the same range of values (relative to a reference value), compared to ν_e and ν_i in the detailed model. Modified from Destexhe et al. (2001)

firing rates are different, the shape of 3-dimensional plot is correctly captured by the point-conductance model, and a rescaling of the absolute value can be achieved by changing the excitability of the cell by manipulating Na^+/K^+ channel densities.

4.4.4 Formal Derivation of the Point-Conductance Model

The derivation of the point-conductance model in the previous sections was done numerically, but more formal derivations are possible, as we will show in this section (for details, see Destexhe and Rudolph 2004).

We start from the simplest model of postsynaptic receptors, which consists of a two-state scheme of the binding of transmitter (T) to the closed form of the channel (C), leading to its open form (O):

$$C+T \underset{\beta}{\overset{\alpha}{\rightleftharpoons}} O\,. \tag{4.10}$$

Here, α and β are the forward and backward rate constants, respectively. These rate constants are assumed to be independent of voltage. Assuming further that the form C is in excess, and that the transmitter occurs as very brief pulses (Dirac δ-functions), one obtains for the scheme, (4.10), the kinetic equation

$$\frac{\mathrm{d}c}{\mathrm{d}t} = \alpha \sum_j \delta(t-t_j) - \beta c\,, \tag{4.11}$$

where c is the fraction of channels in the open state, and the sum runs over all presynaptic spikes j, with Dirac delta functions $\delta(t-t_j)$ occurring at the times of each presynaptic spike (t_j). The synaptic conductance of the synapse is given by $g_{\mathrm{syn}} = g_{\mathrm{max}} c(t)$, where g_{max} is the maximal conductance.

This equation is solvable by Fourier or Laplace transform. The Fourier transform of (4.11) reads:

$$C(\omega) = \frac{\alpha \sum_j \mathrm{e}^{-i\omega t_j}}{\beta + i\omega}\,, \tag{4.12}$$

where $\omega = 2\pi f$ is the angular frequency and f is the frequency. Assuming uncorrelated inputs, this expression leads to the following PSD for the variable c:

$$P_c(\omega) = \frac{\alpha^2 \lambda}{\beta^2 + \omega^2}\,. \tag{4.13}$$

Here, λ denotes the average rate of presynaptic spikes. The explicit solution of (4.11) is then given by

$$c(t) = \alpha \sum_j \mathrm{e}^{-\beta(t-t_j)}\,. \tag{4.14}$$

As seen from the latter expression, this model consists of a linear summation of identical exponential waveforms with an instantaneous rise of α at each time t_j, followed by an exponential decay with time constant $1/\beta$. This *exponential synaptic current* model is widely used to represent synaptic interactions (Dayan and Abbott 2001; Gerstner and Kistler 2002).

We now examine the case of exponential synapses submitted to a high-frequency train of random presynaptic inputs. If the synaptic conductance time course is the same for each event (e.g., nonsaturating exponential synapse, (4.11)), and if successive events are triggered according to a Poisson-process, then the conductance of the synapse is equivalent to a *shot-noise process*. In this case, for high enough presynaptic average frequency, it can be demonstrated that the synaptic conductance

g will approach a Gaussian distribution (Papoulis 1991), which is characterized by its mean g_0 and standard deviation σ_g. In general, these values can be calculated using Campbell's theorem (Campbell 1909a; Papoulis 1991): for a shot-noise process of rate λ and described by

$$s(t) = \sum_j p(t - t_j), \tag{4.15}$$

where $p(t)$ is the time course of each event occurring at time t_j, the average value s_0 and variance σ_s are given by

$$\begin{aligned} s_0 &= \lambda \int_{-\infty}^{\infty} p(t)\mathrm{d}t \\ \sigma_s^2 &= \lambda \int_{-\infty}^{\infty} p^2(t)\mathrm{d}t, \end{aligned} \tag{4.16}$$

respectively. Applying this theorem to the conductance of exponential synapses, one obtains

$$\int_{-\infty}^{\infty} g(t)\mathrm{d}t = \int_{-\infty}^{\infty} g_{\max}\alpha \mathrm{e}^{-\beta t}\mathrm{d}t = \frac{g_{\max}\alpha}{\beta}, \tag{4.17}$$

which combined with (4.16) gives the following expressions for the mean and variance of the synaptic conductance

$$\begin{aligned} g_0 &= \frac{\lambda g_{\max}\alpha}{\beta} \\ \sigma_g^2 &= \frac{\lambda g_{\max}^2 \alpha^2}{2\beta}, \end{aligned} \tag{4.18}$$

respectively.

From the last equations we see that, for a large number of synaptic inputs triggered by a Poisson process, the mean and variance of the conductance distribution depend linearly on the rate λ of the process and increase linear or with the square of the maximum quantal conductance $g_{\max}$, respectively. Figure 4.18 illustrates this behavior by simulating numerically an exponential synapse triggered by a high-frequency Poisson process (Fig. 4.18a). The distribution of the conductance is in general asymmetric (Fig. 4.18b, *histogram*), although it approaches a Gaussian (Fig. 4.18b, *continuous curve*; the exact solution is given in (4.30)). This Gaussian curve was obtained using the mean and the variance estimated from (4.18), which, therefore, provides a good approximation of the conductance distribution.

One can now compare this shot-noise process to the Gaussian stochastic process

$$\frac{\mathrm{d}g(t)}{\mathrm{d}t} = -\frac{g(t) - g_0}{\tau} + \sqrt{D}\chi(t), \tag{4.19}$$

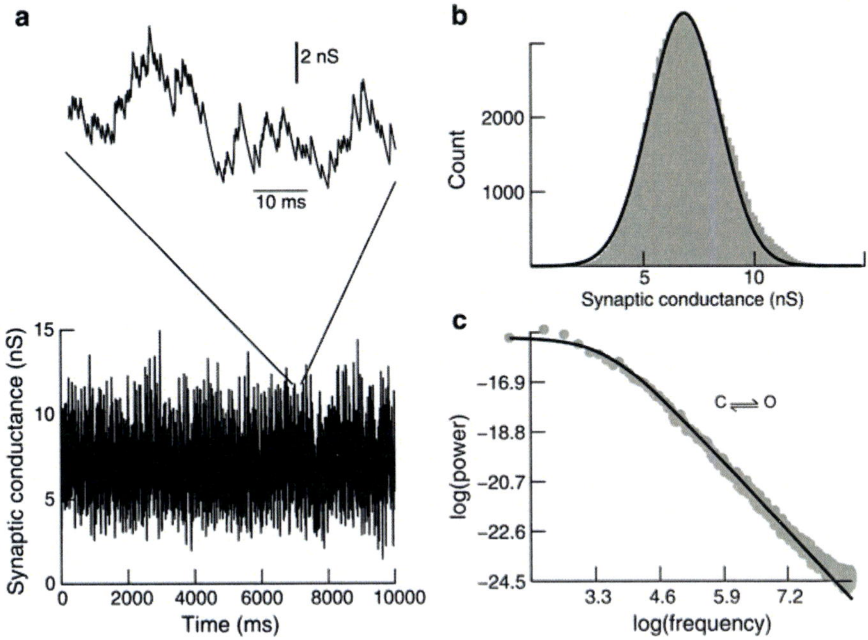

Fig. 4.18 Random synaptic inputs using two-state kinetic models. (**a**) Synaptic conductance resulting from high-frequency random stimulation of the nonsaturating exponential synapse model (4.11; $\alpha = 0.72\ \mathrm{ms}^{-1}$, $\beta = 0.21\ \mathrm{ms}^{-1}$, $g_{\max} = 1$ ns). Presynaptic inputs were Poisson distributed (average rate of $\lambda = 2{,}000\ \mathrm{s}^{-1}$). The conductance time course (bottom; inset at higher magnification on top) shows the irregular behavior resulting from these random inputs. (**b**) Distribution of conductance. The histogram (*gray*) was computed from the model shown in (**a**) (bin size of 0.15 ns). The conductance distribution is asymmetric, although well approximated by a Gaussian (*black*), with mean and variance calculated from Campbell's theorem (see (4.16) and (4.18)). (**c**) Power spectrum of the conductance. The power spectral density (PSD, represented here in log–log scale) was calculated from the numerical simulations of the model in (**a**) (*circles*) and was compared to the theoretical value of the PSD of an Ornstein–Uhlenbeck stochastic process (*black*). Modified from Destexhe and Rudolph (2004)

where $g(t) = g_{\max}c(t)$, D is the amplitude of the stochastic component, $\chi(t)$ is a normally distributed (zero-mean) noise source, and τ is the time constant. This process is Gaussian-distributed, with a mean value of g_0 and a variance given by $\sigma_g^2 = D\tau/2$. To relate both models, one observes the expression of the PSD for the OU process

$$P_g(\omega) = \frac{2D\tau^2}{1+\omega^2\tau^2}\,. \tag{4.20}$$

This Lorentzian form is equivalent to (4.13), from which one can deduce that $\tau = 1/\beta$. The relation $\sigma_g^2 = D\tau/2$, combined with (4.18), gives for the amplitude of the noise

$$D = \lambda g_{\max}^2 \alpha^2\,. \tag{4.21}$$

Using this expressions, the predicted PSD was found to match very well with numerical simulations of Poisson-distributed synaptic currents described by 2-state kinetic models, as illustrated in Fig. 4.18c.

These results suggest that the model of nonsaturating exponential synaptic currents, (4.11), is very well described by an OU process in the case of a large number of random inputs. The decay time of the synaptic current gives the *correlation time* τ of the noise, whereas the other parameters of the OU process can be deduced directly from the parameters of the kinetic model of the synaptic conductance. The point-conductance model considered in this section is an excellent approximation for a shot-noise process with exponential synapses. However, this model assumes that the presynaptic activity is uncorrelated, and analogous to a Poisson process. In the next section, we will present a generalization of this approach to the case of correlated presynaptic activity.

4.4.5 A Model of Shot Noise for Correlated Inputs

In the previous section, Campbell's theorem (Campbell 1909a,b; Rice 1944, 1945, see (4.16)) was used to deduce explicit expressions for the first and second cumulant (i.e., the mean and variance, respectively) of single-channel, or multiple uncorrelated, Poisson-driven shot-noise processes with exponential quantal response function, which we will denote in the following by $h(t)$. However, as was shown earlier in Sect. 4.2, temporal correlations among the synaptic input channels, which result in changes of the statistical properties of the synaptic noise impinging on the cellular membrane, do have a significant impact on the neuronal response (see Fig. 4.15). Such changes in the effective stochastic noise process due to multiple correlated synaptic inputs can be mathematically described by a generalization of Campbell's theorem (Rudolph and Destexhe 2006a), as we will outline below.

Using the model of correlated activity outlined in Appendix B, it can be shown that correlations among multiple synaptic input channels are covered by the definition of a new shot-noise process. The latter can be constructed with two assumptions. First, the time course of k co-releasing identical quantal events $h_k(t)$ equals the sum of k quantal time courses: $h_k(t) = kh_1(t)$, $h_1(t) \equiv h(t)$. Second, the output $s(t)$ due to N correlated Poisson processes equals the sum over the time course of k ($k = 0, \ldots, N$) co-releasing identical quantal events stemming from N_0 independent Poisson trains. For each of the N_0 independent Poisson trains, this sum is weighted according to a binomial distribution

$$\rho_k(N, N_0) = \binom{N}{k} \left(\frac{1}{N_0}\right)^k \left(1 - \frac{1}{N_0}\right)^{N-k}, \tag{4.22}$$

where $k = 0, \ldots, N$. Mathematically, this process is equivalent to a shot-noise process $s(t) = \sum_j A_j h(t - t_j)$ with amplitude A_j given by an independent random variable with a distribution given in (4.22), and t_j arising from a Poisson process with total rate $N_0 \lambda$.

With this, the cumulants C_n, $n \geq 1$, of a multichannel shot-noise process take the form

$$C_n := \lambda N_0 \sum_{k=0}^{N} \rho_k(N,N_0) \int_{-\infty}^{\infty} h_k^n(t)\mathrm{d}t\,, \tag{4.23}$$

with C_1 and C_2 defining the mean s_0 and variance σ_s^2:

$$s_0 = \lambda N_0 \sum_{k=0}^{N} \rho_k(N,N_0) \int_{-\infty}^{\infty} \mathrm{d}t h_k(t)$$

$$\sigma_s^2 = \lambda N_0 \sum_{k=0}^{N} \rho_k(N,N_0) \int_{-\infty}^{\infty} \mathrm{d}t h_k(t)^2\,. \tag{4.24}$$

The last two equations generalize (4.16) of the previous section. For exponential quantal response function $h(t) = h_0 \mathrm{e}^{-t/\tau}$, $t \geq 0$ ($h(t) \equiv 0$, $t < 0$), where τ denotes the time constant and h_0 the maximal response for each channel, (4.24) yields

$$\begin{aligned} s_0 &= \lambda N h_0 \tau\,, \\ \sigma_s^2 &= \frac{1}{2} \lambda N h_0^2 \tau \left(1 + \frac{N-1}{N+\sqrt{c}(1-N)}\right) \end{aligned} \tag{4.25}$$

and, generally, for the cumulants

$$C_n = \frac{\lambda \tau h_0^n}{n} \sum_{k=0}^{N} \binom{N}{k} k^n \frac{\left((N-1)(1-\sqrt{c})\right)^{N-k}}{\left(N+\sqrt{c}(1-N)\right)^{N-1}}\,. \tag{4.26}$$

It is interesting to note that due to the model of correlation in the multichannel input pattern, the total release rate λN is preserved. This directly translates into the independence of the mean on the correlation measure c, whereas s_0 is linearly dependent on λ and N (Fig. 4.19b). The variance σ_s^2 shows a monotonic but nonlinear dependence on c and N, being proportional to

$$\lambda N \left(1 + \frac{N-1}{N+\sqrt{c}(1-N)}\right).$$

For vanishing correlation ($c = 0$), σ_s^2 approaches a value proportional to $2\lambda N$ for large N. Moreover, for zero correlation the system is still equivalent to a shot-noise process of rate $\lambda N = \lambda N_0$, as can be inferred from the mean s_0, but a factor $2\lambda N$, resulting from the used shuffling algorithm (see Appendix B), enters now the variance σ_s^2. On the other hand, for maximal correlation ($c = 1$), there is only one independent input channel, in which case $\sigma_s^2 \sim \lambda N^2$ (Fig. 4.19b).

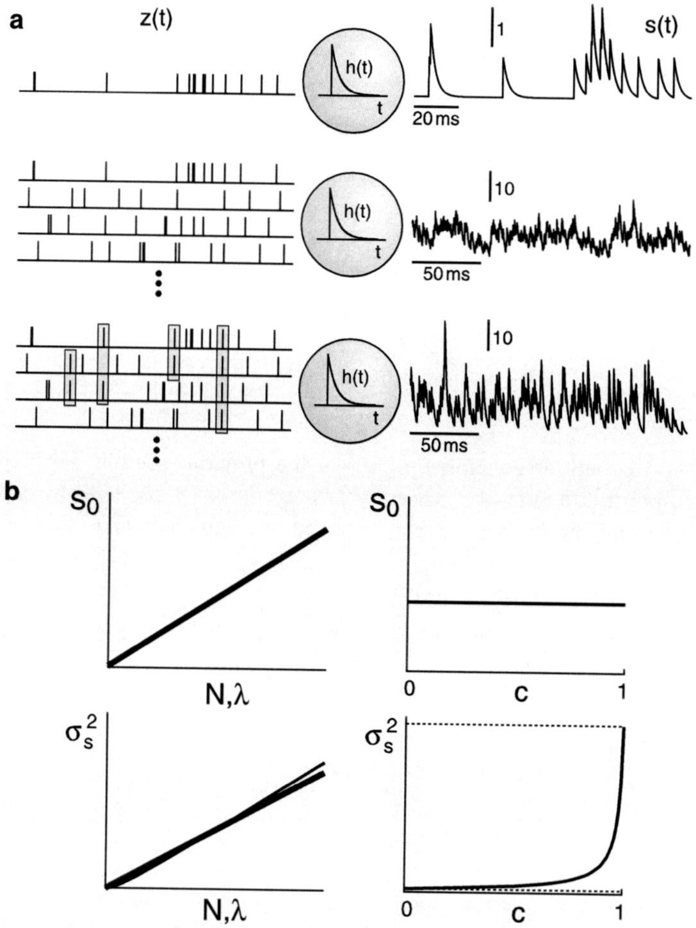

Fig. 4.19 (**a**) Shot-noise process for single (*top*), multiple uncorrelated (*middle*) and multiple correlated (*bottom*) input channels. The output $s(t)$ of the dynamical system equals the sum of the quantal responses $h(t)$ triggered by the arrival of a sequence of impulses $z(t)$ occurring at random times according to a Poisson distribution. Whereas $s(t)$ triggered by multiple but uncorrelated channels (*middle*) can be described by a single channel with a rate equivalent to the product of the number of channels and the individual channel rate, the output differs when temporal correlations (*bottom, left, gray bars*) are introduced. Parameters: $N = 1$, $\lambda = 100$ Hz (A); $N = 100$, $\lambda = 50$ Hz, $c = 0$ (B); $c = 0.7$ (C); $h_0 = 1$, $\tau = 2$ ms in all cases. (**b**) The mean s_0 shows a linear dependence on N and λ, and is independent on c (*top*). The variance σ_s^2 is linear in λ (*bottom left, gray*) but depends nonlinearly on the number of input channels N (*bottom left, black*) and correlation c (*bottom right*) of the multichannel inputs. *Dashed lines* show the asymptotic values of σ_s^2 for $N \to \infty$ in the case of $c = 0$ and $c = 1$. Modified from Rudolph and Destexhe (2006a)

The approach outlined above allows to obtain analytic expressions for other statistical parameters as well. The correlation function $C(t_1,t_2)$ and autocorrelation function $C(T) := C(T,0)$ of a multichannel shot-noise process $s(t)$ take the form

$$\begin{aligned} C(t_1,t_2) &:= \lambda N_0 \sum_{k=0}^{N} p_k(N,N_0) \int_{-\infty}^{\infty} h_k(t-t_1)h_k(t-t_2)\mathrm{d}t \\ &= \frac{1}{2}\lambda \tau h_0^2 N \left(1+\frac{N-1}{N+\sqrt{c}(1-N)}\right) \mathrm{e}^{\frac{|t_2-t_1|}{\tau}} . \end{aligned} \tag{4.27}$$

In a similar fashion, the moments M_n, $n \geq 1$, of the shot-noise process $s(t)$, defined by $M_n = \int_{-\infty}^{\infty} s(t)^n \mathrm{d}t$, can be deduced, yielding the finite sum

$$M_n = \sum_{k=1}^{n} \sum_{(n_1,\dots,n_k)} \frac{n!}{k!n_1!\cdots n_k!} C_{n_1}\cdots C_{n_k} , \tag{4.28}$$

where the second sum denotes the partition of n into a sum over k integers $n_i \geq 1$ for $1 \leq i \leq k$.

From the real part of the Fourier transform of the moment generating function

$$Q_s(u) := \exp\left[-\lambda N_0 \sum_{k=0}^{N} p_k(N,N_0) \int_{-\infty}^{\infty} \{1-\mathrm{e}^{-uh_k(t)}\}\mathrm{d}t \right] , \tag{4.29}$$

the amplitude probability distribution $\rho_s(h)$ can be calculated. A lengthy but straightforward calculation yields

$$\begin{aligned} \rho_s(h) = & \frac{1}{\sqrt{2\pi C_2}} \mathrm{e}^{-\frac{(C_1-h)^2}{2C_2}} + \sum_{m=1}^{\infty}\sum_{k=0}^{\infty} \frac{(-2)^{m-k+1}}{\sqrt{2\pi}k!} \frac{\Gamma[3/2+m+k]}{\Gamma[1/2+k]} \\ & \times \frac{(C_1-h)^{2k}}{C_2^{3/2+m+k}} \left(\frac{C_1-h}{1+2k} d_{2m+1} + d_{2m+2} \right) . \end{aligned} \tag{4.30}$$

Here,

$$d_n = \sum_{k=1}^{[\frac{n}{3}]} \sum_{(n_1,\dots,n_k)} \frac{C_{n_1}\cdots C_{n_k}}{k!n_1!\cdots n_k!} ,$$

where the second sum runs over all partitions of n into a sum of k terms $n_i \geq 3$, $1 \leq i \leq [\frac{n}{3}]$, $n \geq 3$. Finally, the PSD of a multichannel shot-noise process with exponential quantal response function

$$S_s(\nu) = \lambda N_0 \sum_{k=0}^{N} p_k(N,N_0)|H_k(\nu)|^2 , \tag{4.31}$$

where $H_k(\nu) = \int_{-\infty}^{\infty} h_k(t) e^{-2\pi i \nu t} dt$, is given by

$$S_s(\nu) = \lambda N \left(1 + \frac{N-1}{N+\sqrt{c}(1-N)}\right) \frac{h_0^2 \tau^2}{1+(2\pi\nu\tau)^2}. \tag{4.32}$$

Both the equation for the amplitude distribution, (4.30), and PSD, (4.31), can be used to justify from a more general setting the use of the OU stochastic process as an effective model for synaptic noise. Indeed, in lowest orders, (4.30) yields

$$\rho_s(h) = \frac{1}{\sqrt{2\pi C_2}} e^{-\frac{(C_1-h)^2}{2C_2}} \left(1 - \frac{C_3(C_1-h)(C_1^2-3C_2-2C_1h+h^2)}{6C_2^3}\right), \tag{4.33}$$

with the second-order correction, and all higher order terms, vanishing with powers of $C_1/\sqrt{C_2} \equiv s_0/\sigma_s$. Thus, even in the presence of correlations, the Gaussian amplitude distribution characterizing the OU stochastic process is an excellent approximation as long as $s_0 > \sigma_s$, a condition which is easily observed for neuronal noise in the cortex. Similarly, the power spectral density shows a Lorentzian behavior

$$S(\nu) = \frac{2D\tau^2}{1+(2\pi\tau\nu)^2}, \tag{4.34}$$

(Fig. 4.20b) with the frequency dependence unaltered by the correlation c, whereas the maximal power D is a nonlinear monotonic function of c and N:

$$D = \frac{1}{2}\lambda N h_0^2 \left(1 + \frac{N-1}{N+\sqrt{c}(1-N)}\right). \tag{4.35}$$

The relations given above allow to characterize multichannel shot-noise processes from experimental recordings. As shown, the mean $s_0 \equiv C_1$ and variance $\sigma_s^2 \equiv C_2$ of the amplitude distribution (4.25) are monotonic functions of c and λ (see Fig. 4.20b), thus allowing to estimate those parameters from experimentally obtained distributions for which s_0 and σ_s^2 are known. Here, in order to obtain faithful estimates, the total number of input channels N as well as the quantal decay time constant and amplitude, τ and h_0, respectively, need to be known. Average values for N can be obtained from detailed morphological studies. Estimates for the quantal conductance h_0 and synaptic time constant τ can be obtained using whole cell recordings of miniature synaptic events. Average values for the synaptic time constants are also accessible through fits of the PSD obtained from current-clamp recordings. Finally, the explicit expression for $S_s(\nu)$ (4.32), can be used to fit experimental power spectral densities obtained from voltage-clamp recordings, yielding values for the time constant τ and the power coefficient D.

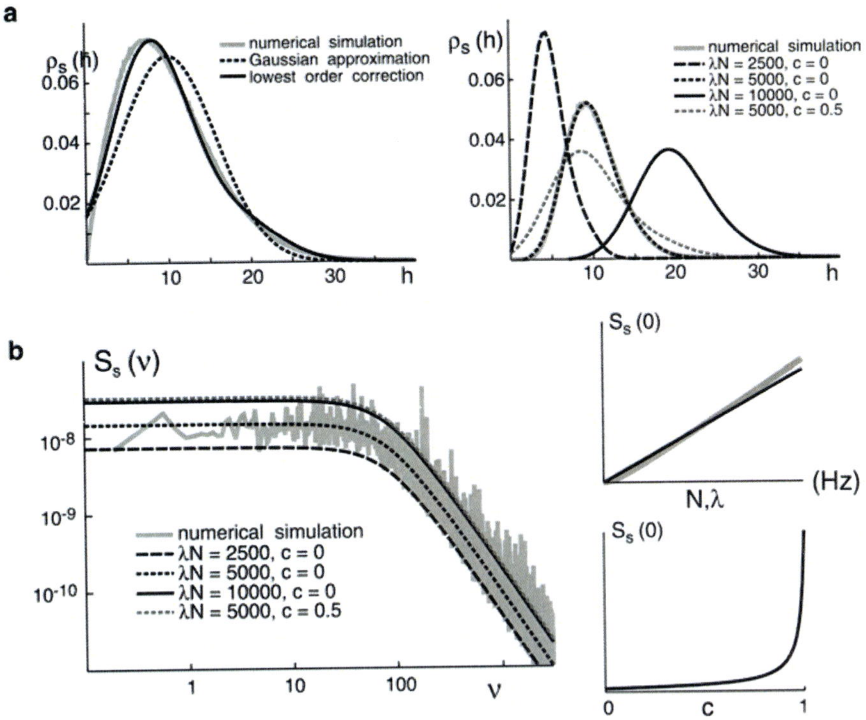

Fig. 4.20 (**a**) Amplitude probability distribution for multiple correlated input channels. $\rho_s(h)$, (4.30), shows a generally asymmetric behavior (*left, gray*), which can be approximated by the lowest order correction (*left, black solid*) to the corresponding Gaussian distribution (*left, black dashed*). The amplitude distribution depends on c and the total rate, and approaches a Gaussian for high total input rates or small correlations (*right*). Parameters: $N = 100$, $\lambda = 50$ Hz, $c = 0.7$ (A); $h_0 = 1$, $\tau = 2$ ms in all cases. (**b**) The power spectral density of resulting from shot-noise process with multiple correlated input channels and exponential quantal response function shows a Lorentzian behavior (*left*). The total power depends on both, the number of input channels N (*right, black*), rate λ (*right, gray*) and the level of correlation c of the multichannel inputs. Parameters: $h_0 = 1$, $\tau = 2$ ms. Modified from Rudolph and Destexhe (2006a)

4.5 Summary

In this chapter, we have covered different modeling approaches to directly incorporate the quantitative measurements of synaptic noise described in Chap. 3.

In the first part of the chapter (Sects. 4.2 and 4.3), we have reviewed the building and testing of detailed biophysical models of synaptic noise, using compartmental models with dendrites (Destexhe and Paré 1999). In Sect. 4.3, these models were simplified to a few compartments to keep the essence of the soma–dendrite interactions, but still use realistic parameter values for the synaptic conductances.

One of the main conclusions of these models is that, in order to match the experimental measurements quantitatively, it is necessary to include a correlation between the stochastic release of the different synapses in soma and dendrites. Only using correlated stochastic release at synaptic terminals, it is possible to match all of the intracellular and conductance measurements made in vivo (see details in Sect. 4.2 and Destexhe and Paré 1999).

We also reviewed another important modeling aspect, which is to obtain simplified representations that capture the main properties of the synaptic "noise." The approach presented in Sect. 4.4 (see Sect. 4.4.2 for a formal derivation) leads to a simplified stochastic model of synaptic noise, called the "point-conductance model" (Destexhe et al. 2001). This advance is important, because simple models have enabled real-time applications such as the dynamic-clamp injection of synaptic noise in cortical neurons (see Chap. 6). The main interest of the simplified model is that several fundamental properties of synaptic noise, such as the mean and variance of conductances, can be changed independently, which is difficult to achieve with complex or even shot-noise type models. We will see in the next chapters that this feature is essential to understand many effects of synaptic noise on neurons.

Finally, simple models also have enabled a number of mathematical applications, some of which resulted in methods to analyze experiments, as we will outline in Chaps. 7 and 8. Before that, in the next chapter, we will demonstrated that the point-conductance stochastic model of synaptic noise provides a very useful tool in the investigation of the effect of synaptic noise on neurons.

Chapter 5
Integrative Properties in the Presence of Noise

In the previous chapter, we have reviewed models of synaptic noise which were directly based on in vivo measurements. In the present chapter, we use these models to investigate the type of consequences that this noisy synaptic activity can have on the fundamental behavior of neurons, such as their responsiveness and integrative properties. We start by reviewing early results of the effect of noise on neurons, before detailing, using computational models, a number of beneficial effects of the presence of intense synaptic background activity present in neocortical pyramidal neurons in vivo.

5.1 Introduction

As reviewed in the preceding chapter, the impact of synaptic noise on the integrative properties is a classic theme of modeling studies. It was noted already in early models that the effectiveness of synaptic inputs is highly dependent on the membrane properties, and, thus, is also dependent on other synaptic inputs received by the dendrites, as observed first in motoneurons (Barrett and Crill 1974; Barrett 1975), Aplysia (Bryant and Segundo 1976) and cerebral cortex (Holmes and Woody 1989). Figure 5.1 shows an example of a model simulation of the effect of synaptic background activity on the effectiveness of synaptic inputs (Barrett 1975). This study evidenced that the synaptic efficacy is greatly dependent on the level of synaptic background activity.

This theme was later investigated by models using detailed dendritic morphologies, in cerebral cortex (Bernander et al. 1991) and cerebellum (Rapp et al. 1992; De Schutter and Bower 1994). These studies revealed as well that the effectiveness of synaptic inputs is highly dependent on the level of background activity. Moreover, the presence synaptic background activity was found to also alter the dynamics of synaptic integration in neurons, rendering pyramidal neurons to behave as better coincidence detectors (Bernander et al. 1991).

A. Destexhe and M. Rudolph-Lilith, *Neuronal Noise*, Springer Series in Computational Neuroscience 8, DOI 10.1007/978-0-387-79020-6_5,

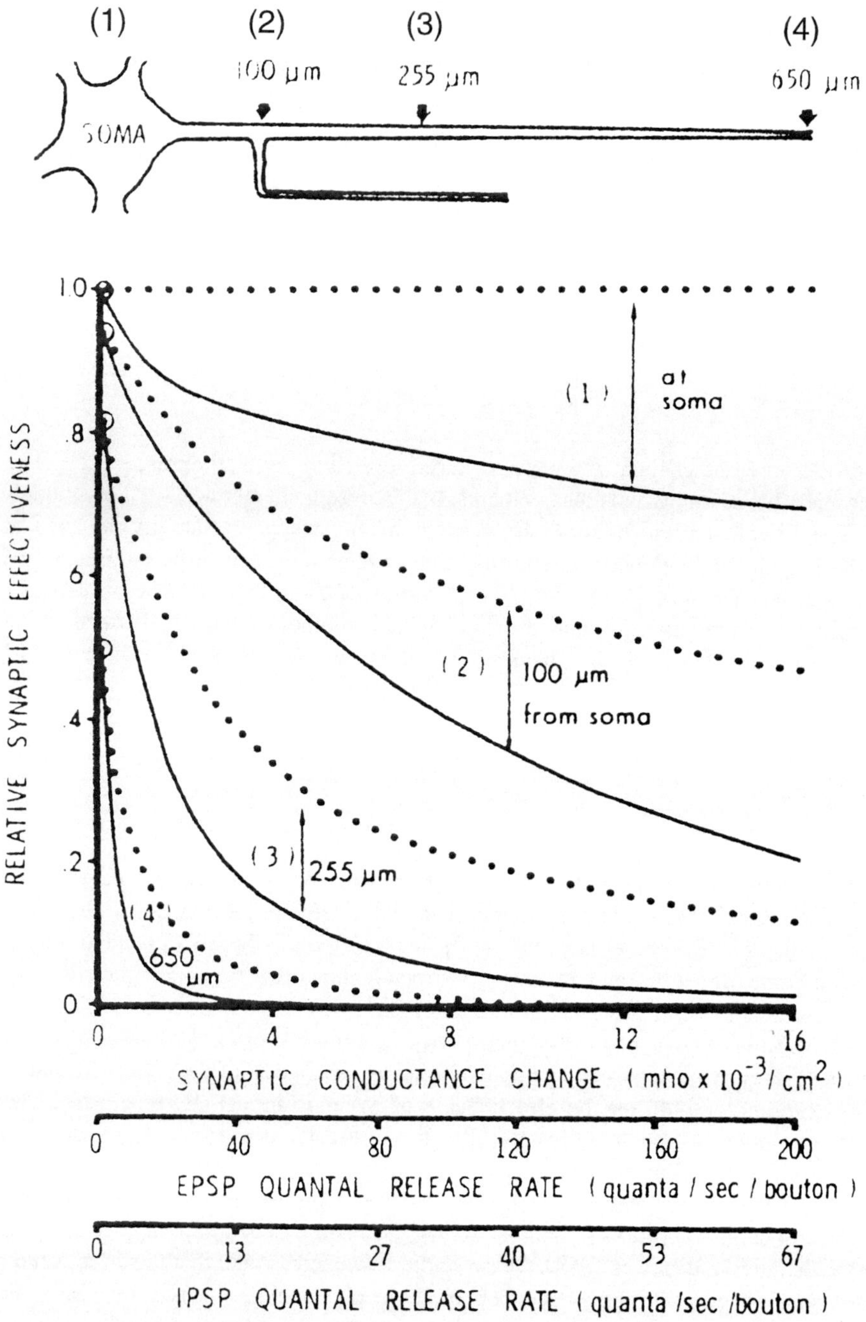
(1)
(2)
(3)
(4)
100 μm
255 μm
650 μm
SOMA
1.0
.8
.6
.4
.2
0
RELATIVE SYNAPTIC EFFECTIVENESS
(1)
at soma
(2)
100 μm from soma
(3)
255 μm
(4)
650 μm
0
4
8
12
16
SYNAPTIC CONDUCTANCE CHANGE (mho x 10⁻³/cm²)
0
40
80
120
160
200
EPSP QUANTAL RELEASE RATE (quanta / sec / bouton)
0
13
27
40
53
67
IPSP QUANTAL RELEASE RATE (quanta /sec /bouton)

Following these pioneering studies, and incorporating the first experimental measurements of synaptic background activity (as reviewed in Chap. 3), a third generation of models appeared (Destexhe and Paré 1999), in which model parameters were directly constrained by experiments (Chap. 4). With this approach, one is, finally, in a position to make quantitative and biophysically relevant predictions, which are directly verifiable experimentally (for example using dynamic-clamp experiments; see Chap. 6). The use of such models to investigate the integrative properties of neurons is the subject of the present chapter.

5.2 Consequences on Passive Properties

A first and most obvious consequence of the presence of background activity is that it will necessarily have a major impact on various passive properties, such as the input resistance (as shown experimentally), but also on properties critical for integration, such as the attenuation of the signal with distance along dendrites. These properties were recently investigated using detailed biophysical models (Sect. 4.2), and will be reviewed in this section.

The experimental evidence for a $\sim$80% decrease in R_{in} due to synaptic bombardment betrays a massive opening of ion channels. In the compartmental model introduced in Chap. 4, the total conductance due to synaptic activity is 7 to 10 times larger than the leak conductance. The impact of this massive increase in conductance on dendritic attenuation can be investigated by comparing the effect of current injection in active periods and synaptic quiescence (Fig. 5.2). In the absence of synaptic activity (Fig. 5.2b, *smooth traces*), somatic current injection (Fig. 5.2b, left) elicits large voltage responses in dendrites, and reciprocally (Fig. 5.2b, right), shows a moderate electrotonic attenuation. By contrast, during simulated active periods (Fig. 5.2b, *noisy traces*), voltage responses to identical current injections are markedly reduced, betraying a greatly enhanced electrotonic attenuation. In these conditions, the relative amplitude of the deflection induced by the same amount of current with and without synaptic activity, as well as the difference in time constant, are in agreement with experimental observations (compare Fig. 5.2b, *Soma*, with Fig. 3.19c,d). Similarly, in these models, the effect of synaptic bombardment on the time constant is also in agreement with previous models (Holmes and Woody 1989; Bernander et al. 1991; Koch et al. 2002).

Fig. 5.1 Dependence of the effectiveness of synaptic inputs on the level of synaptic background activity. The relative synaptic efficacy of a test excitatory quantal conductance change is shown at four different synaptic locations in dendrites, as a function of the conductance change due to synaptic background activity. *Dotted lines* are with a background activity's reversal potential of −70 mV (producing no net change of V_{m} in the cell), while for *solid curves* the reversal was 0 mV. The lower abscissas show the excitatory and inhibitory conductance release rates required to produce the conductance change showed in the main abscissa. Modified from Barrett (1975)

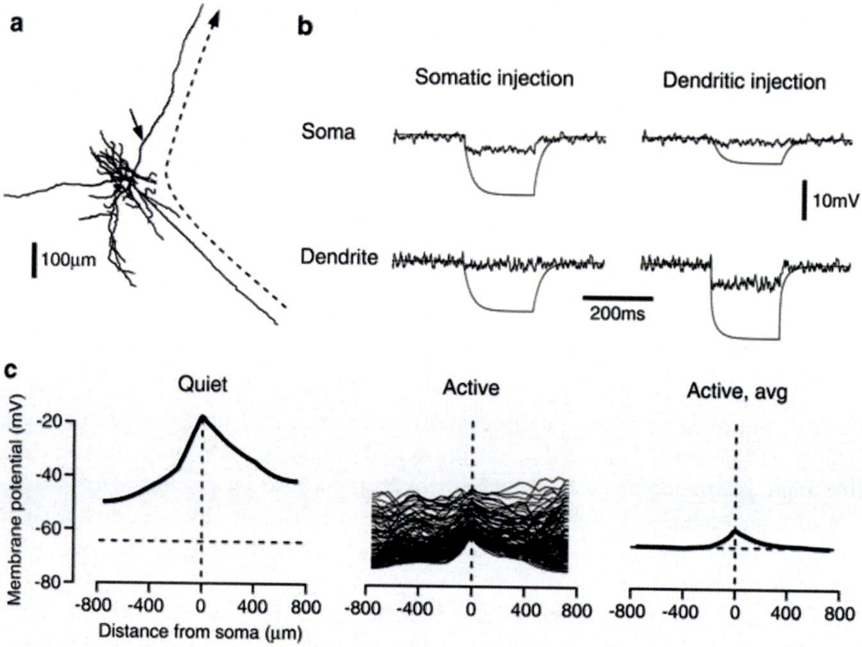

Fig. 5.2 Passive dendritic attenuation during simulated active periods. (**a**) Layer VI pyramidal cell used for simulations. (**b**) Injection of hyperpolarizing current pulses in the soma (*left*, −0.1 nA) and a dendritic branch (*right*; −0.25 nA). The dendritic voltage is shown at the same site as the current injection (indicated by a *small arrow* in (**a**)). The activity during simulated active periods (*noisy traces*; average of 50 pulses) is compared to the same simulation in the absence of synaptic activity (*smooth lines*). (**c**) Somatodendritic V_m profile along the path indicated by a *dashed line* in (**a**). The V_m profile is shown at steady-state following injection of current in the soma (+0.8 nA). In the absence of synaptic activity (*Quiet*) there was moderate attenuation. During simulated active periods (*Active*) the profile was fluctuating (100 traces shown) but the average of 1,000 sweeps (*Active, avg*) revealed a marked attenuation of the steady-state voltage. Modified from Destexhe and Paré (1999)

Dendritic attenuation can be further characterized by computing somatodendritic profiles of V_m with steady current injection in the soma: in the absence of synaptic activity (Fig. 5.2c, *Quiet*), the decay of V_m following somatic current injection is typically characterized by large space constants (e.g., 515 to 930 μm in the layer VI pyramidal cell depicted in Fig. 5.2, depending on the dendritic branch considered), whereas the space constant is reduced by about fivefold (105–181 μm) during simulated active periods (Fig. 5.2c, *Active*).

This effect on voltage attenuation is also present when considering different parameters for passive properties. Using a model adjusted to whole-cell recordings in vitro (Stuart and Spruston 1998), a relatively moderate passive voltage attenuation can be observed (Fig. 5.3, *Quiescent*; 25–45% attenuation for distal events). Taking into account the high conductance and more depolarized conditions of in vivo-like activity shows a marked increase in voltage attenuation (Fig. 5.3, *In vivo-like*;

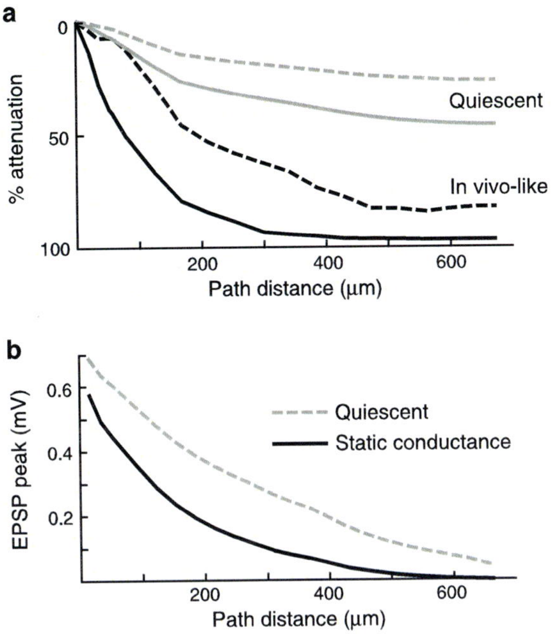

Fig. 5.3 Impact of background activity on passive voltage attenuation for different passive parameters. (**a**) somatodendritic membrane potential profile at steady state after current injection at the soma (+0.4 nA; layer VI cell). Two sets of passive properties were used, *solid*: from Destexhe and Paré (1999), *dashed*: from Stuart and Spruston (1998). (**b**) peak EPSP at the soma as a function of path distance for AMPA-mediated 1.2 nS stimuli at different dendritic sites (dendritic branch shown in Fig. 5.2. Peak EPSPs in quiescent conditions are compared with EPSPs obtained with a high static conductance. Modified from Rudolph and Destexhe (2003b)

80–90% attenuation). Furthermore, computing the EPSP peak amplitude in these conditions reveals an attenuation with distance (Fig. 5.3, lower panel), which is more pronounced if background activity is represented by an equivalent static (leak) conductance. Thus, the high-conductance component of background activity enhances the location-dependent impact of EPSPs, and leads to a stronger individualization of the different dendritic branches (London and Segev 2001; Rhodes and Llinás 2001).

In order to estimate the convergence of synaptic inputs necessary to evoke a significant somatic depolarization during active periods, a constant density of excitatory synapses can be stimulated synchronously in "proximal" and "distal" regions of dendrites (as indicated in Fig. 5.4a). In the absence of synaptic activity, simulated EPSPs have large amplitudes (12.6 mV for proximal and 6.0 mV for distal; Fig. 5.4b, *Quiet*). By contrast, during simulated active periods, the same stimuli give rise to EPSPs that are barely distinguishable from spontaneous V_m fluctuations (Fig. 5.4b, *Active*). In the model shown in Fig. 5.4, the average EPSP amplitude is 5.4 mV for proximal and 1.16 mV for distal stimuli (Fig. 5.4b, *Active, avg*), suggesting that EPSPs are attenuated by a factor of 2.3 to 5.2 in this case,

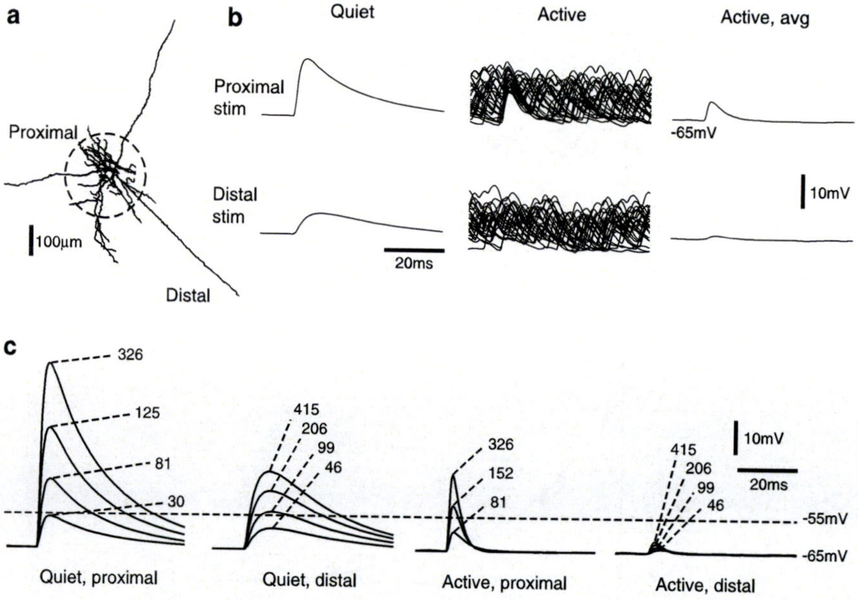

Fig. 5.4 Somatic depolarization requires the convergence of a large number of excitatory inputs during simulated active periods. (**a**) The layer VI pyramidal cell was divided into two dendritic regions: *Proximal* included all dendritic branches laying within 200 μm from the soma (*circle*) and *Distal* referred to dendritic segments outside this region. (**b**) Attenuation following synchronized synaptic stimulation. A density of 1 excitatory synapse per 200 μm^2 was stimulated in proximal (81 synapses) and distal regions (46 synapses). Responses obtained in the absence of synaptic activity (*Quiet*) are compared to those observed during simulated active periods (*Active*; 25 traces shown). In the presence of synaptic activity (*Active*), the evoked EPSP was visible only when proximal synapses were stimulated. Average EPSPs (*Active, avg*; n = 1,000) showed a marked attenuation compared to the quiescent case. (**c**) Averaged EPSPs obtained with increasing numbers of synchronously-activated synapses. Protocols similar to (**b**) were followed for different numbers of synchronously-activated synapses (indicated for each trace). The *horizontal dashed line* indicates a typical value of action potential threshold. Modified from Destexhe and Paré (1999)

with the maximal attenuation occurring for distal EPSPs. Figure 5.4c also shows the effect of increasing the number of synchronously activated synapses. In quiescent conditions, less than 50 synapses on basal dendrites are sufficient to evoke a 10 mV depolarization at the soma (*Quiet, proximal*) and the activation of about 100 distal synapses is needed to achieve a similar depolarization (*Quiet, distal*). In contrast, during simulated active periods, over 100 basal dendritic synapses are necessary to reliably evoke a 10 mV somatic depolarization (*Active, proximal*) while the synchronous excitation of up to 415 distal synapses does evoke only depolarization of a few millivolts (*Active, distal*).

This effect of synaptic activity on dendritic attenuation is also mostly independent of the cell geometry: in models of active states with various cellular morphologies it was found that the space constant is reduced by about fivefold.

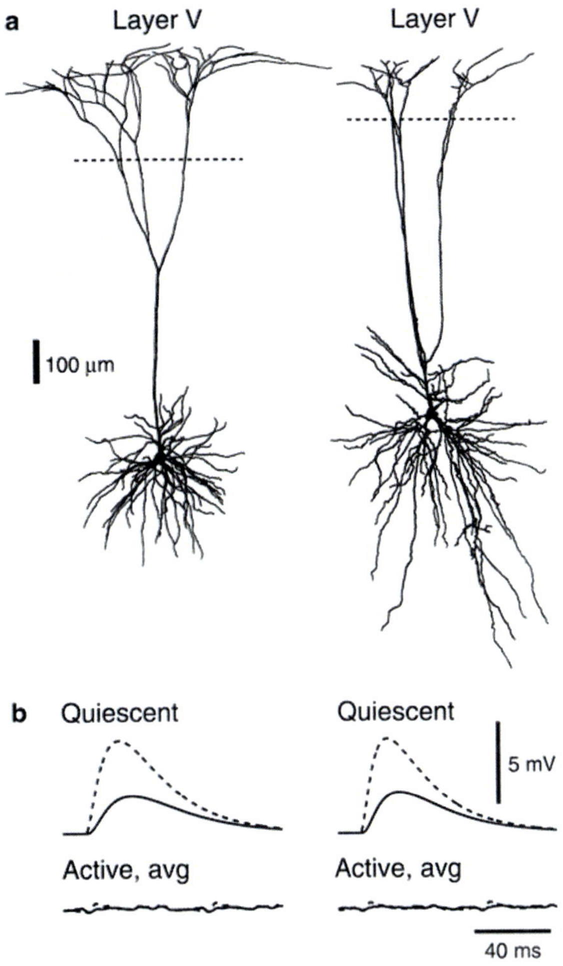

Fig. 5.5 Attenuation of distal EPSPs in two layer V cells. (**a**) Morphologies of the two pyramidal cells, where excitatory synapses were stimulated synchronously in the distal apical dendrite (>800 μm from soma; indicated by *dashed lines* in each morphology). (**b**) The EPSP resulting from the stimulation of 857 (*left*) and 647 AMPA synapses (*right*, 23.1 MΩ cell) are shown for quiescent (*Quiescent*) and active conditions (*Active*). These EPSPs were about 2-3 mV in amplitude without synaptic activity but were undetectable during active periods. The same simulation, performed with low axial resistance (100 Ωcm; *dashed lines*), gave qualitatively identical results. Modified from (Destexhe and Paré 1999)

Moreover, stimulating layer V neurons with several hundreds of synapses at a distance of over 800 μm from the soma shows undetectable effects during active periods (Fig. 5.5). These results can also be reproduced using low axial resistivities (Fig. 5.5c, *dashed lines*).

The above findings show that intense synaptic activity does have a drastic effect on the attenuation of distal synaptic inputs. However, it must also be noted that voltage-dependent currents in dendrites may amplify EPSPs (Cook and Johnston 1997) or trigger dendritic spikes that propagate towards the soma (Stuart et al. 1997a). Therefore, the attenuation of EPSPs must be re-examined in models that include active dendritic currents. This will be done in Sect. 5.7.

5.3 Enhanced Responsiveness

Another major consequence of the presence of background activity is the modulation of the responsiveness of the neuron. As mentioned previously, the synaptic background activity can be decomposed into two components: a tonically active conductance and voltage fluctuations. Modeling studies have mostly focused on the conductance effect, revealing that background activity is responsible for a decrease in responsiveness, which, in turn, imposes severe conditions of coincidence of inputs necessary to discharge the cell (see, e.g., Bernander et al. 1991; see also Sect. 5.1). Here, it will be shown that, in contrast, responsiveness is enhanced if voltage fluctuations are taken into account. In this case, the model can produce responses to inputs that would normally be subthreshold. This effect is called *enhanced responsiveness*.

To investigate this effect, we start by illustrating the procedure to calculate the response of model pyramidal neurons in the presence of background activity. We then investigate the properties of responsiveness and which parameters are critical to explain them. Finally, we illustrate a possible consequence of these properties at the network level (see details in Hô and Destexhe 2000).

5.3.1 Measuring Responsiveness in Neocortical Pyramidal Neurons

In the layer VI pyramidal cell shown in Fig. 5.6a, which will serve as an example here, synaptic background activity is simulated by Poisson-distributed random release events at glutamatergic and GABAergic synapses (see Sect. 4.2). As before, model is constrained by intracellular measurements of the V_{m} and input resistance before and after application of TTX (Destexhe and Paré 1999; Paré et al. 1998b). A random release rate of about 1 Hz for excitatory synapses and 5.5 Hz for inhibitory synapses is necessary to reproduce the correct V_{m} and input resistance. In addition, it is necessary to include a correlation between release events to reproduce the amplitude of V_{m} fluctuations observed experimentally (Fig. 5.6b, *Correlated*). This model, thus, reproduces the electrophysiological parameters measured intracellularly in vivo: a depolarized V_{m}, a reduced input resistance and high-amplitude V_{m} fluctuations.

To investigate the response of the model cell in these conditions, a set of excitatory synapses is activated in dendrites, in addition to the synapses involved in generating background activity (see details in Hô and Destexhe 2000). Simultaneous activation of these additional synapses, in the presence of background activity, evokes APs with considerable variability in successive trials (Fig. 5.6c), as expected from the random nature of the background activity. A similar high variability of synaptic responses is typically observed in vivo (Arieli et al. 1996; Contreras et al. 1996; Nowak et al. 1997; Paré et al. 1998b; Azouz and Gray 1999; Lampl et al. 1999). The evoked response, expressed as a probability of evoking a spike in

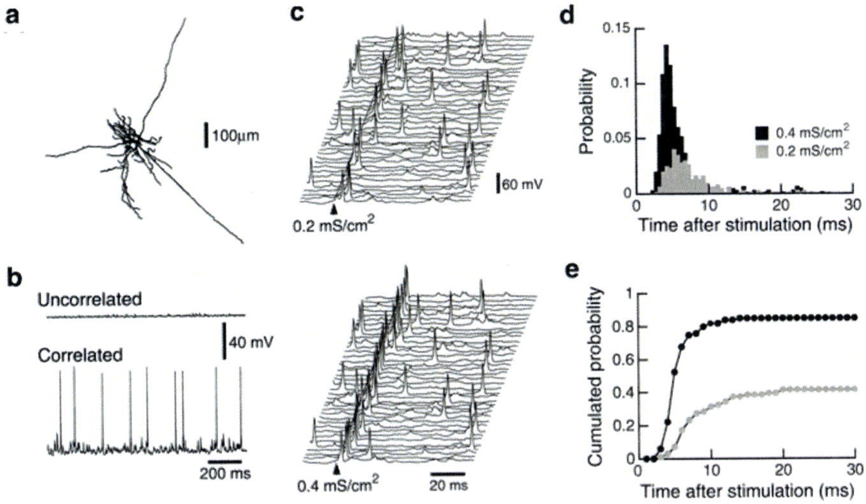

Fig. 5.6 Method to calculate the response to synaptic stimulation in neocortical pyramidal neurons in the presence of synaptic background activity. (**a**) Layer VI pyramidal neuron reconstructed and incorporated in simulations. (**b**) Voltage fluctuations due to synaptic background activity. Random inputs without correlations (*Uncorrelated*) led to small-amplitude V_m fluctuations. Introducing correlations between release events (*Correlated*) led to large-amplitude V_m fluctuations and spontaneous firing in the 5–20 Hz range, consistent with experimental measurements. (**b**) Evoked responses in the presence of synaptic background activity. The response to a uniform AMPA-mediated synaptic stimulation is shown for two values of maximum conductance density (0.2 and 0.4 mS/cm^2). The *arrow* indicates the onset of the stimulus and each graph shows 40 successive trials in the presence of correlated background activity. (**d**) Probability of evoking a spike. The spikes specifically evoked by the stimulation were detected and the corresponding probability of evoking a spike in successive 0.5 ms bins was calculated over 600 trials. (**e**) Cumulative probability obtained from (**d**). Modified from Hô and Destexhe (2000)

successive 0.5 ms intervals, is shown in Fig. 5.6d (cumulative probability shown in Fig. 5.6e). The variability of responses depends on the strength of synaptic stimuli, with stronger stimuli leading to narrower probabilities of evoking a spike (Fig. 5.6d).

Thus, the most appropriate measure of synaptic response in the presence of highly fluctuating background activity is to compute probabilities of evoking a spike. In the following, we use this measure to characterize the responsiveness of pyramidal neurons in different conditions.

5.3.2 *Enhanced Responsiveness in the Presence of Background Activity*

In order to characterize the responsiveness, one defines the *input–output response function* of the neuron. This relation (also called *transfer function*) is defined as

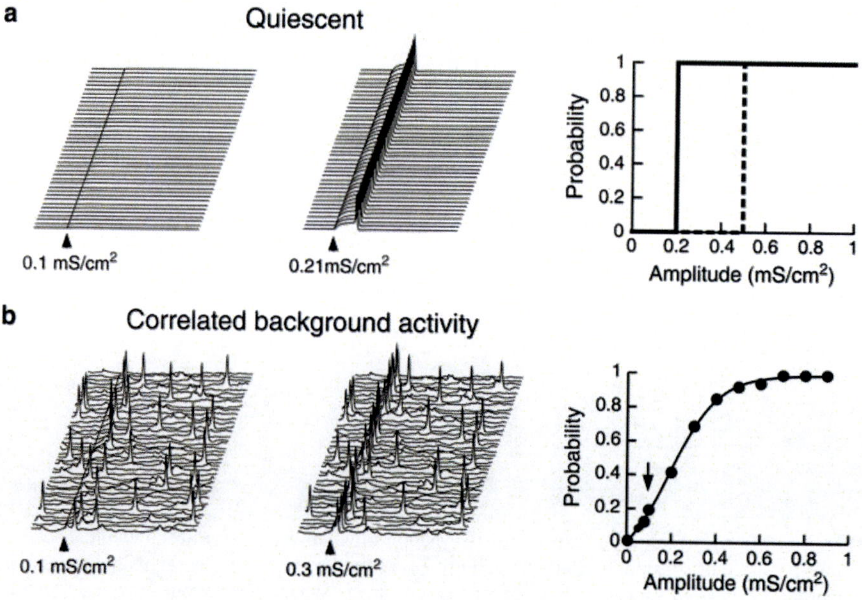

Fig. 5.7 Synaptic background activity enhances the responsiveness to synaptic inputs. *Left*: successive trials of synaptic stimulation for two different stimulus amplitudes (curves arranged similarly to Fig. 5.6c). *Right*: input–output response function expressed as the cumulative probability of evoking a spike (calculated over 100 trials) as a function of stimulation amplitude (in mS/cm^2; same procedures as in Fig. 5.6). (**a**) Response to synaptic stimulation in the absence of background activity (*Quiescent*). The neuron had a relatively high input resistance (R_{in} = 46.5 MΩ) and produced an all-or-none response. The response is compared to the same model in the presence of shunt conductances equivalent to background activity (*dashed line*; R_{in} = 11.1 MΩ). (**b**) Response in the presence of correlated synaptic background activity. In this case, the neuron had a relatively low input resistance (R_{in} = 11.2 MΩ) but produced a different response over a wide range of input strengths. In particular, the probability of evoked spikes was significant for inputs that were subthreshold in the quiescent model (*arrow*). All simulations were done at the same average resting V_m of −65 mV. Modified from Hô and Destexhe (2000)

the cumulated probability of firing a spike, computed for increasing input strength. This input–output response function is similar to the *frequency-current* (F-I) response often studied in neurons. It is also equivalent to the *transfer function* of the neuron, which expresses the firing rate as a function of input strength (the latter function can be normalized to yield the probability of spiking per stimulus). The slope of the input–output response function is commonly called the *gain* of the neuron.

In quiescent conditions, the cell typically responds in an all-or-none manner. In this case, the response function is a simple step function (Fig. 5.7a), reflecting the presence of a sharp threshold for APs. The response function can also be calculated by adding a constant conductance, equivalent to the total conductance due to synaptic background activity (usually, this equivalent conductance should be calculated for each compartment of the neuron). In the presence of this additional dendritic shunt, the response function is shifted to higher input strength

(Fig. 5.7a, *dashed line*). Thus, consistent with the overall decrease of responsiveness evidenced in previous studies (Barrett 1975; Holmes and Woody 1989; Bernander et al. 1991), the conductance of background activity imposes strict conditions of convergence to discharge the cell.

However, in the presence of correlated background activity, the response is qualitatively different (Fig. 5.7b). Here, the cell is more responsive, because small excitatory inputs that were subthreshold in quiescent conditions (e.g., 0.1 mS/cm^2 in Fig. 5.7a,b) can generate APs in the presence of background activity. More importantly, the model cell produces a different response to a wide range of input strength, thus generating a different response to inputs that were previously indistinguishable in the absence of background activity.

These simulations suggest that the presence of background activity at a level similar to in vivo measurements (Destexhe and Paré 1999; Paré et al. 1998b) is responsible for a significant effect on the responsiveness of pyramidal neurons. The specific role of the different components of background activity is investigated next.

5.3.3 *Enhanced Responsiveness is Caused by Voltage Fluctuations*

To investigate the role of voltage fluctuations, one first compares two models with background activity of equivalent conductance but different V_m fluctuations. By using uncorrelated and correlated background activities (Fig. 5.6b), the neuron receives the same amount of random inputs, but individual inputs are combined differently, resulting in equivalent average conductance but different amplitudes of V_m fluctuations (see Appendix B). With uncorrelated background activity, small inputs become subthreshold again (e.g., 0.1 mS/cm^2 in Fig. 5.8a). The response function is typically steeper (Fig. 5.8a, right) and closer to that observed in the case of equivalent leak conductance (compare with Fig. 5.7a, *dashed line*). Thus, comparing correlated and uncorrelated activity, it appears that the presence of high-amplitude V_m fluctuations significantly affects cellular responsiveness. The persistence of small-amplitude V_m fluctuations in the uncorrelated case is presumably responsible for the sigmoidal shape in Fig. 5.8a.

To dissociate the effect of V_m fluctuations from that of shunting conductance, background activity is now replaced by injection of noisy current waveforms at all somatic and dendritic compartments. To that end, the total net currents due to background activity is recorded at each compartment and injected in the same locations in a model without background activity. This "replay" procedure leads to V_m fluctuations similar to those produced by synaptic background activity (Fig. 5.8b, left), but without the important tonically activated synaptic conductance, thus allowing to dissociate these two factors. With noisy current injection, the input resistance is comparable to that of quiescent conditions (e.g., $R_{in} = 45.5\,\mathrm{M}\Omega$ vs. $46.5\,\mathrm{M}\Omega$), but the cell is more responsive, with subthreshold inputs in quiescent conditions evoking a significant response in the presence of V_m fluctuations (e.g., 0.05 mS/cm^2 in Fig. 5.8b).

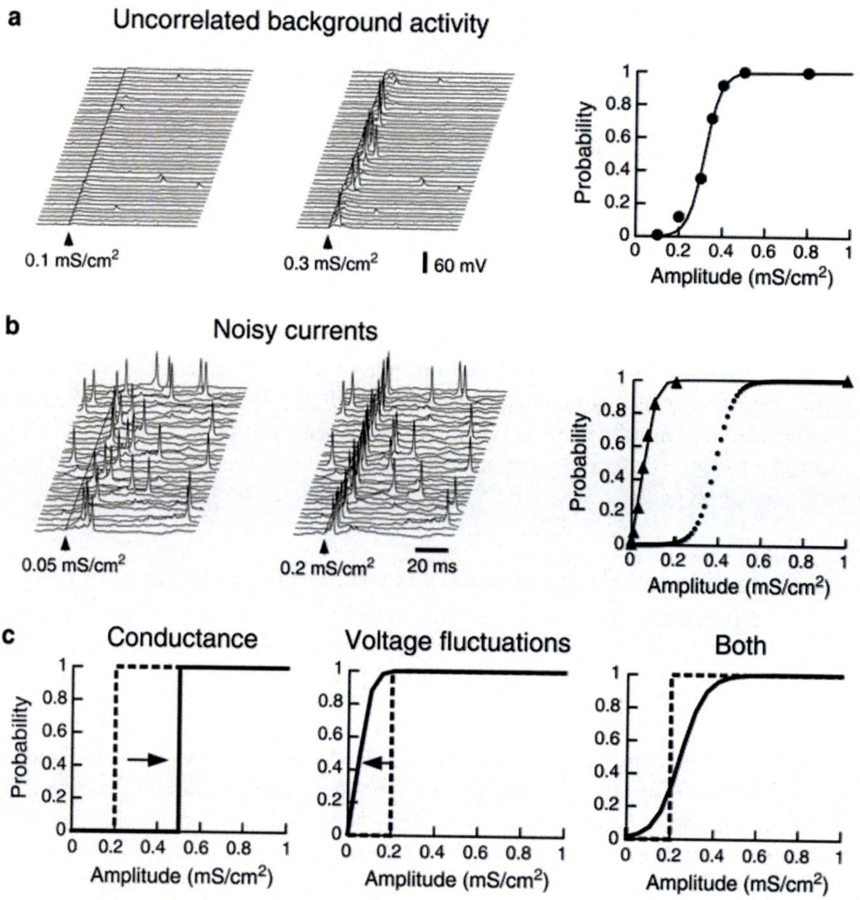

Fig. 5.8 Enhanced responsiveness is due to voltage fluctuations. (**a**) Synaptic responses in the presence of uncorrelated background activity. This simulation was the same as in Fig. 5.7b, but without correlations in background activity, resulting in similar input resistance ($R_{in} = 10.0\,\mathrm{M}\Omega$) but smaller amplitude V_m fluctuations (see Fig. 5.6b). The response function was steeper. (**b**) Simulation in the presence of fluctuations only. Noisy current waveforms were injected in soma and dendrites, leading to similar V_m fluctuations as in Fig. 5.7b, but with a high input resistance ($R_{in} = 45.5\,\mathrm{M}\Omega$). The response function (*continuous line*) showed enhanced responsiveness. Fluctuating leak conductances without V_m fluctuations did not display a significant enhancement in responsiveness (*dotted curve*; $R_{in} = 11\,\mathrm{M}\Omega$). Panels in (**a**, **b**) were arranged similarly as in Fig. 5.7. (**c**) Reconstruction of the response function. *Conductance*: effect of adding a leak conductance equivalent to synaptic bombardment (*continuous line*; $R_{in} = 11.1\,\mathrm{M}\Omega$), compared to a quiescent model (*dashed line*; $R_{in} = 46.5\,\mathrm{M}\Omega$). *Voltage fluctuations*: effect of adding noisy current waveforms (*continuous line*; $R_{in} = 45.5\,\mathrm{M}\Omega$), compared to the quiescent model (*dashed line*; $R_{in} = 46.5\,\mathrm{M}\Omega$). *Both*: combination of noisy current waveforms and the equivalent shunt (*continuous line*; $R_{in} = 11.1\,\mathrm{M}\Omega$), compared to the quiescent model (*dashed line*). The response function was qualitatively similar to that in the presence of correlated background activity (Fig. 5.7b). All simulations correspond to the same average resting V_m of $-65\,\mathrm{mV}$. Modified from Hô and Destexhe (2000)

The case of a fluctuating conductance without V_m fluctuations can also be tested. For this, the total conductance is recorded in each compartment during correlated background activity, and later assigned to the leak conductance in each compartment of a model without background activity. This procedure leads to a relatively steep response function (Fig. 5.8b, *dotted curve*). Although these conductance fluctuations slightly enhances responsiveness, the observed effect is small compared to that of V_m fluctuations.

To assess the importance of these different factors, their impact is compared in Fig. 5.8c. The effect of conductance is to decrease responsiveness, as shown by the shift of the response function toward larger input strength (Fig. 5.8c, *Conductance*). The effect of voltage fluctuations is to increase responsiveness by shifting the response function to the opposite direction (Fig. 5.8c, *Voltage fluctuations*). Combining these two factors leads to a response function (Fig. 5.8c, *Both*) which is qualitatively similar to the correlated background activity (compare with Fig. 5.7b, right).

One can, therefore, conclude that the behavior of the neocortical cell in the presence of correlated background activity can be understood qualitatively by a combination of two opposite influences: a tonically active conductance, which decreases responsiveness by *shifting* the response function to higher thresholds, and voltage fluctuations, which increase responsiveness by modifying the *shape* of the response function (Hô and Destexhe 2000). The effect of these parameters will be further investigated in the following sections, as well as verified in dynamic-clamp experiments (see Chap. 6).

5.3.4 Robustness of Enhanced Responsiveness

The robustness of this finding is examined by performing variations in the configuration of the model. Simulations using four different reconstructed pyramidal cells from cat neocortex, including a layer II-III cell and two layer V cells, show a similar enhancement in responsiveness for all cases (Fig. 5.9; Hô and Destexhe 2000). For each cell, correlated background activity (*continuous curves*) was compared to the equivalent shunt conductance (*dashed curves*), showing that the presence of background activity significantly enhances responsiveness, an observation which is remarkably robust against changes in cellular morphology.

To test the influence of dendritic excitability, the densities of Na^+ and K^+ channels are varied in dendrites, soma, and axon (Hô and Destexhe 2000). It was found that rescaling these densities by the same factor results in a different global excitability of the cell, and gives rise to a shift in the response functions, as expected (Fig. 5.10a). However, in all cases, comparing correlated background activity (Fig. 5.10, *continuous curves*) to models with equivalent shunt (*dashed curves*), an enhancement in responsiveness irrespective of the exact position of the response functions can be observed.

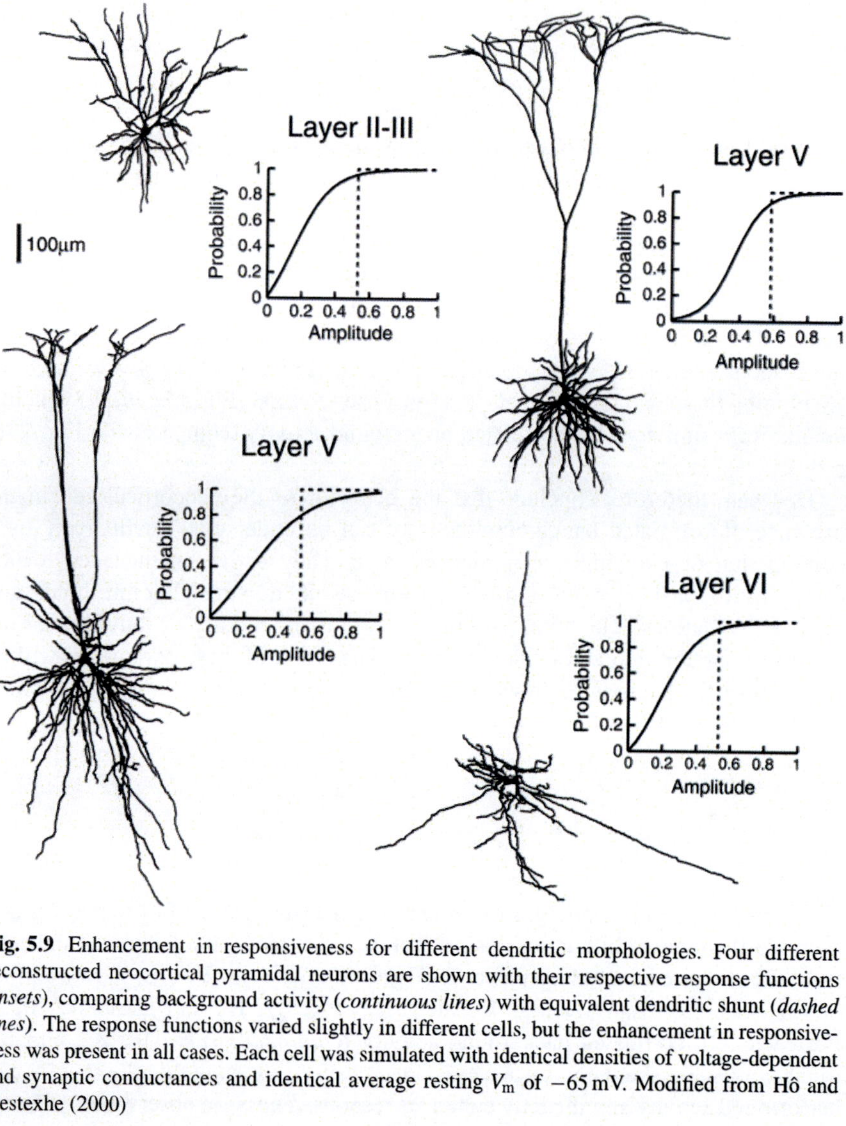

Fig. 5.9 Enhancement in responsiveness for different dendritic morphologies. Four different reconstructed neocortical pyramidal neurons are shown with their respective response functions (*insets*), comparing background activity (*continuous lines*) with equivalent dendritic shunt (*dashed lines*). The response functions varied slightly in different cells, but the enhancement in responsiveness was present in all cases. Each cell was simulated with identical densities of voltage-dependent and synaptic conductances and identical average resting V_m of -65 mV. Modified from Hô and Destexhe (2000)

The same phenomenon is present for variations of other parameters, such as the distribution of leak currents (Fig. 5.10b), different axial resistivities (Fig. 5.10b), different sets and densities of voltage-dependent currents (Fig. 5.10c), different combinations of synaptic receptors and different release frequencies (Fig. 5.10d). In all these cases, variations of parameters have an expected effect of shifting the response function, but the presence of background activity always leads to a significant enhancement in responsiveness similar to Fig. 5.10a.

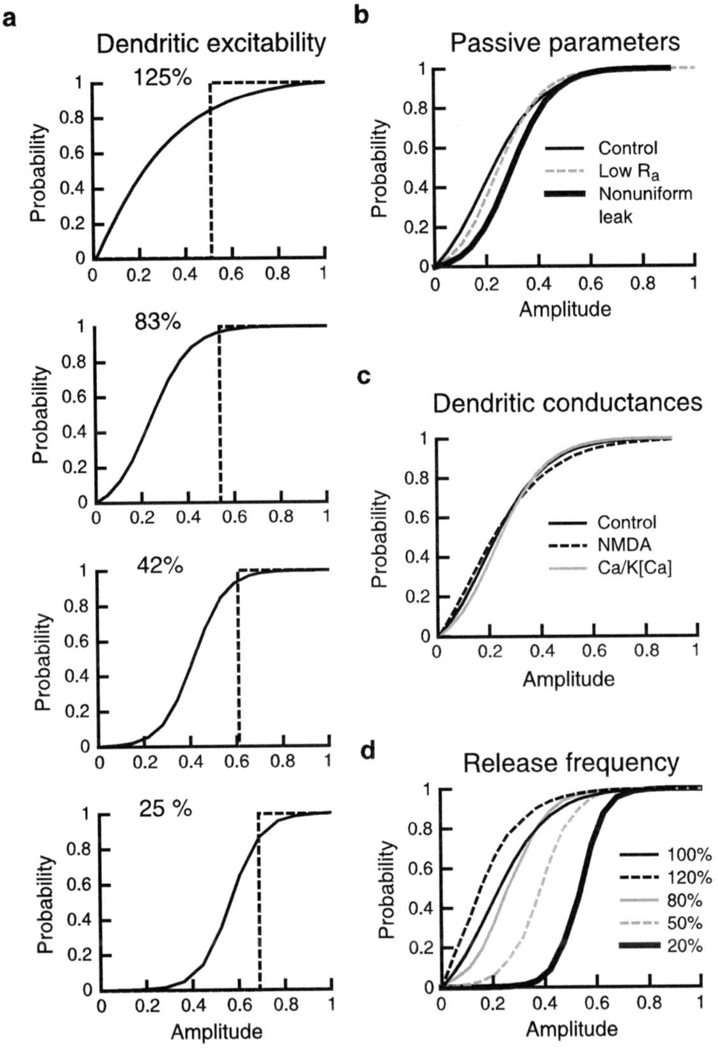

Fig. 5.10 Enhancement in responsiveness for different distributions of conductances. (**a**) Modulating dendritic excitability by using different densities of Na^+/K^+ conductances shifts the response function, but the enhancement in responsiveness was present in all cases (the relative Na^+/K^+ conductance densities are indicated with respect to control values). (**b**) Effect of different variations of the passive parameters Stuart and Spruston (low axial resistivity of $R_a = 80\ \Omega$cm; nonuniform leak conductances, from 1998). (**c**) Addition of NMDA receptors or dendritic Ca^{2+} currents and Ca^{2+}-dependent K^+ currents. (**d**) Modulation of the intensity of background activity (values indicate the excitatory release frequency relative to control). Similar to (**a**), the parameters in (**b–d**) affected the position of the response function, but the enhancement in responsiveness was present in all cases. All simulations were obtained with the layer VI cell and correspond to the same average resting V_m of −65 mV. Modified from Hô and Destexhe (2000)

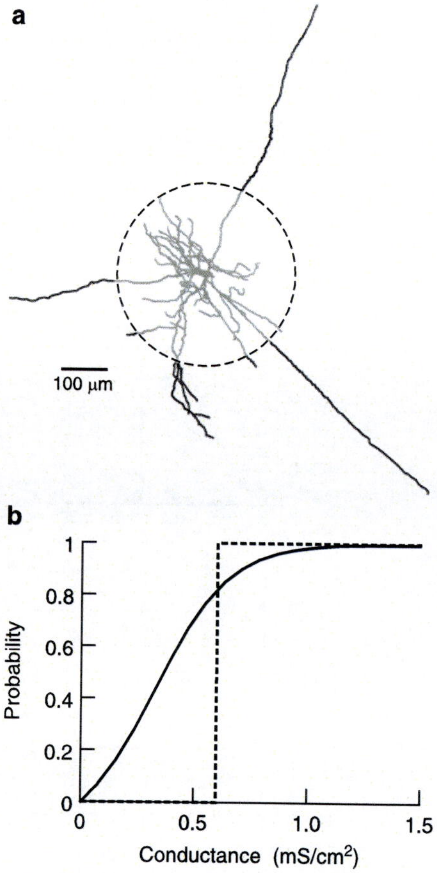

Fig. 5.11 Enhancement in responsiveness for inputs localized in distal dendrites. (**a**) Subdivision of the distal dendrites in the layer VI cell. The distal dendrites (*black*) were defined as the ensemble of dendritic segments laying outside 200 μm from the soma (*dashed circle*). The unstimulated dendrites are shown in *light gray*. (**b**) Probability of evoking a spike as a function of the strength of synaptic stimulation in distal dendrites. The response obtained in the presence of correlated background activity (*continuous line*) is compared to that of a model including the equivalent dendritic shunt (*dashed line*). Background activity enhanced the responsiveness in a way similar to uniform stimulation. Modified from Hô and Destexhe (2000)

It is important to test whether the enhancement in responsiveness is sensitive to the proximity of the excitatory inputs to the somatic region. To this end, Hô and Destexhe (2000) investigated the synchronized stimulation of increasing densities of AMPA-mediated synapses located exclusively in the distal region of dendrites (>200 μm from soma; see Fig. 5.11a). The response function, following stimulation of distally located AMPA-mediated inputs, was computed similarly as for uniform stimulation. Here again, the presence of background activity leads to an enhancement of the responsiveness of the cell (Fig. 5.11b), showing that the mechanisms described above also apply to distally located inputs as well.

5.3.5 *Optimal Conditions for Enhanced Responsiveness*

To evaluate the range of voltage fluctuations at which responsiveness is optimally enhanced, the response probability has to be computed for subthreshold inputs

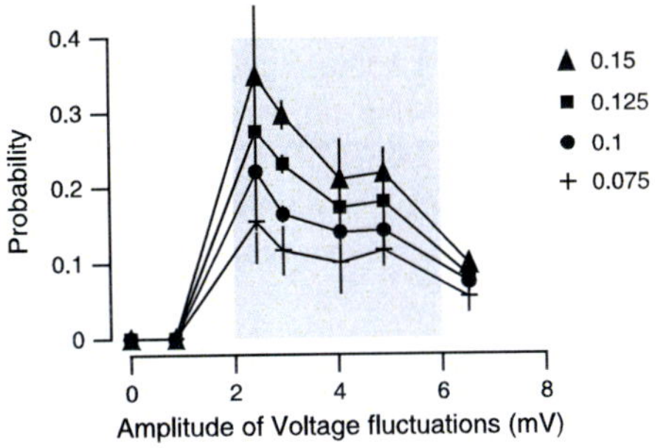

Fig. 5.12 Enhancement in responsiveness occurs for levels of background activity similar to in vivo measurements. The probability of evoking a spike was computed for subthreshold inputs in the presence of different background activities of equivalent conductance but different amplitudes of voltage fluctuations, indicated by their standard deviation of the V_m, σ_V. These different conditions were obtained by varying the value of the correlation. The different symbols indicate different subthreshold input amplitudes ($+ = 0.075\,\mathrm{mS/cm^2}$, *circles* $= 0.1\,\mathrm{mS/cm^2}$, *squares* $= 0.125\,\mathrm{mS/cm^2}$, *triangles* $= 0.15\,\mathrm{mS/cm^2}$; *vertical bars* = standard error). In all cases, the enhanced responsiveness occurred at the same range of voltage fluctuations, which also corresponded to the range measured intracellularly in vivo (*gray area*; $\sigma_V = 4.0 \pm 2.0\,\mathrm{mV}$; from Destexhe and Paré 1999). All simulations correspond to the same average V_m of $-65\,\mathrm{mV}$. Modified from Hô and Destexhe (2000)

at different conditions of V_m fluctuations. One of these conditions is obtained by varying the value of the correlation, leading to background activities of identical mean conductance and average V_m, but different amplitudes of V_m fluctuations (see Destexhe and Paré 1999). The probability of spikes specifically evoked by subthreshold stimuli can, thus, be represented as a function of the amplitude of V_m fluctuations in Fig. 5.12 (*symbols*).

Figure 5.12 shows that there are no spikes evoked without background activity, or with background activity with fluctuations which are too small in amplitude. However, for V_m fluctuations larger than about $\sigma_V = 2\,\mathrm{mV}$, the response probability displays a steep increase and stays above zero for background activities with larger fluctuation amplitudes. The responsiveness is, therefore, enhanced for a range of V_m fluctuations of σ_V between 2 and 6 mV or more. Interestingly, this optimal range approximately matches the level of V_m fluctuations measured intracellularly in cat parietal cortex in vivo ($\sigma_V = 4.0 \pm 2.0\,\mathrm{mV}$ in Destexhe and Paré 1999, indicated by a gray area in Fig. 5.12). This suggests that the range of amplitude of V_m fluctuations found in vivo is precisely within the "noise" level effective to enhance the responsiveness of cortical neurons, as found by models.

5.3.6 *Possible Consequences at the Network Level*

The above results show that enhanced responsiveness is present at the single-cell level, in which case the high variability of responses makes it necessary to perform averages over a large number of successive stimuli. Because the nervous system does not perform such temporal averaging, the physiological meaning of enhanced responsiveness would therefore be unclear if it relied exclusively on performing a large number of trials. However, as illustrated below, this averaging can also be performed at the population level by employing spatial averaging, leading to an instantaneous enhancement in responsiveness for single-trial stimuli.

In order to address this possibility, a simple case of a feedforward network of pyramidal neurons can be considered, and its dynamics compared with and without synaptic background activity. This simple paradigm is illustrated in Fig. 5.13a. One thousand identical presynaptic pyramidal neurons received simultaneous afferent AMPA-mediated inputs with conductance randomized from cell-to-cell. The differences in afferent input, thus, created variations in the amplitude of the excitation and on the timing of the resulting spike in the presynaptic cells. The output of this population of cells was monitored through the EPSP evoked in a common postsynaptic cell (Fig. 5.13a).

In quiescent conditions, i.e., the absence of background activity, the EPSP evoked in the postsynaptic cell was roughly all-or-none (Fig. 5.13b), reflecting the AP threshold in the presynaptic cells (similar to Fig. 5.7a). When the presynaptic cells received correlated synaptic background activity (which was different in each cell), the EPSPs were more graded (Fig. 5.13c,d), compatible with the sigmoid response function in Fig. 5.7b, left. Perhaps the most interesting property is that the smallest inputs, which were subthreshold in quiescent conditions, led to a detectable EPSP in the presence of background activity (0.1–0.15 mS/cm^2 in Fig. 5.13d). This shows that the network, indeed, transmits some information about these inputs, while the latter are effectively filtered out in quiescent conditions.

Although this paradigm is greatly simplified (identical presynaptic cells, independent background activities), it nevertheless illustrates the important possibility that the enhanced responsiveness shown in Fig. 5.7b may be used in populations of pyramidal neurons to *instantaneously* detect a single afferent stimulus. In these conditions, the network can detect a remarkably wide range of afferent input amplitudes. Similar to an effect previously reported in neural networks with additive noise (Collins et al. 1995a), networks of pyramidal cells with background activity can detect inputs that are much smaller than the AP threshold of individual cells (Hô and Destexhe 2000). This is another clear example of beneficial property conferred by the presence of noise.

Finally, the phenomenon of enhanced responsiveness is not specific to detailed biophysical models, but is also found in simplified models. Figure 5.14 shows that the point-conductance model (introduced in Sect. 4.4) displays an effect which is basically identical to that observed in a biophysically detailed model. In this case, the probability of response is computed following activation of a single AMPA-mediated synapse. Using different values of σ_e and σ_i, yielding different σ_V, gives rise to different values of responsiveness (Fig. 5.14b, symbols), with no effect on the average V_m. This effect of fluctuations is qualitatively similar to that observed in the detailed model (see details in Destexhe et al. 2001).

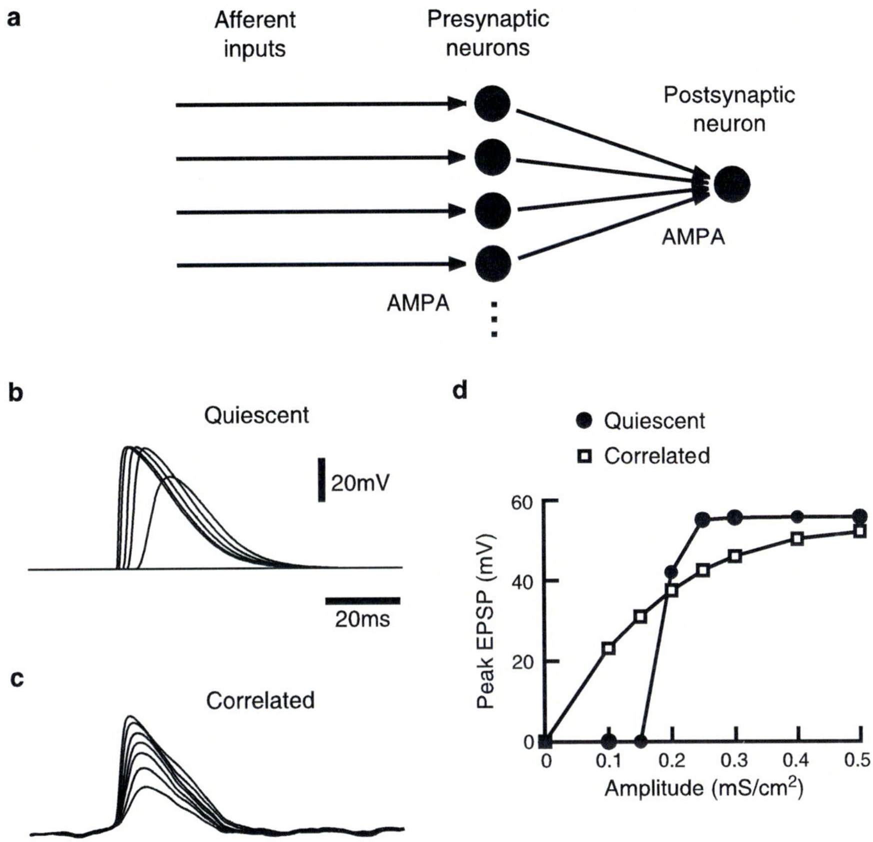

Fig. 5.13 Synaptic background activity enhances the detection of synaptic inputs at the network level. (**a**) Feedforward network consisting of 1,000 presynaptic neurons identical to Fig. 5.6a. All presynaptic neurons connected the postsynaptic cell using AMPA-mediated glutamatergic synapses. The presynaptic neurons were excited by a simultaneous AMPA-mediated afferent input with randomly distributed conductance (normal distribution with standard deviation of 0.02 mS/cm^2; other parameters as in Fig. 5.6b). (**b**) EPSPs evoked in the postsynaptic cell in quiescent conditions (average afferent conductances of 0.1, 0.15, 0.2, 0.25, 0.3, 0.4, and 0.5 mS/cm^2). The EPSP was approximately all-or-none, with smallest inputs evoking no EPSP and strongest inputs leading to EPSPs of constant amplitude. (**c**) Same simulations in the presence of correlated synaptic background activity. The same conductance densities led to detectable EPSPs of progressively larger amplitude. (**d**) Peak EPSP from (**b**) and (**c**) plotted as a function of the average afferent conductance. The response was all-or-none in control conditions (*Quiescent*) and was graded in the presence of background activity (*Correlated*), showing a better detection of afferent inputs. Modified from Hô and Destexhe (2000)

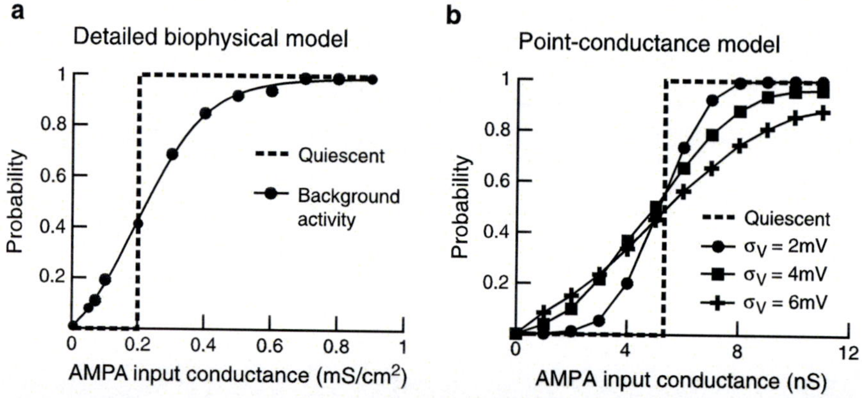

Fig. 5.14 Comparison of the responsiveness of point conductance and detailed biophysical models. (**a**) an AMPA-mediated input was simulated in the detailed model, and the cumulated probability of spikes specifically evoked by this input was computed for 1,000 trials. The curves show the probabilities obtained when this procedure was repeated for various values of AMPA conductance. (**b**) same paradigm in the point-conductance model. Four conditions are compared, with different values of standard deviation of the V_m (σ_V). In both models, there was a nonnull response for subthreshold inputs in the presence of background activity. Modified from Destexhe et al. (2001)

5.3.7 *Possible Functional Consequences*

An interesting observation is that the enhanced responsiveness is obtained for a range of V_m fluctuations comparable to that measured intracellularly during activated states in vivo (Fig. 5.12). This suggests that the level of background activity present in vivo represents conditions close to optimal for enhancing the responsiveness of pyramidal neurons. It is possible that the network maintains a level of background activity whose functional role is to keep its cellular elements in a highly responsive state.

In agreement with this view, Fig. 5.13 illustrated that, in a simple feedforward network of pyramidal neurons, the presence of background activity allows the network to instantaneously detect synaptic events that would normally be subthreshold (0.1–0.15 mS/cm^2 in Fig. 5.13d). In this case, background activity sets the population of neurons into a state of more efficient and more sensitive detection of afferent inputs, which are then transmitted to the postsynaptic cells, while the same inputs are filtered out in the absence of background activity.

These results should be considered in parallel with the observation that background activity is particularly intense in intracellularly recorded cortical neurons of awake animals (Matsumura et al. 1988; Steriade et al. 2001). In the light of this model, one can interpret the occurrence of intense background activity as a factor that facilitates information transmission. It is therefore conceivable that background activity is an active component of arousal or attentional mechanisms, as proposed theoretically (Hô and Destexhe 2000) and in dynamic-clamp experiments (Fellous et al. 2003; Shu et al. 2003b). Such a link with attentional processes is a very interesting direction which should be explored by future studies.

5.4 Discharge Variability

As already mentioned above, cortical neurons in vivo were found to show a highly irregular discharge activity, both during sensory stimuli (e.g., Dean 1981; Tolhurst et al. 1983; Softky and Koch 1993; Holt et al. 1996; Shadlen and Newsome 1998; Stevens and Zador 1998; Shinomoto et al. 1999) and during spontaneous activity (e.g., Smith and Smith 1965; Noda and Adey 1970; Burns and Webb 1976). To quantify the variability of a neuronal spike train, one commonly uses the *coefficient of variation* (C_V), defined as

$$C_V = \frac{\sigma_{ISI}}{<ISI>}, \tag{5.1}$$

where $<ISI>$ and σ_{ISI} are, respectively, the average value and the SD of ISIs. In experiments, this quantity was found to be higher than 0.5 for firing frequencies above 30 Hz in cat and macaque V1 and MT neurons (Softky and Koch 1993). A C_V of 0.8 was reported as the lower limit under in vivo conditions by investigating the responses of individual MT neurons of alert macaque monkeys driven with constant-motion stimuli (Stevens and Zador 1998).

Much theoretical work has since been devoted to find neuronal mechanisms responsible for the observed high firing irregularity. However, neither the integration of random EPSPs by a simple leaky integrate-and-fire (IAF) neuron model, nor a biophysically more realistic model of a layer V cell with passive dendrites were able to generate the high C_V as observed in vivo (Softky and Koch 1993).

To solve this apparent discrepancy, balanced or "concurrent" inhibition and excitation was proposed as a mechanism producing a discharge activity with Poisson-type variability in IAF models (Shadlen and Newsome 1994; Usher et al. 1994; Troyer and Miller 1997; Shadlen and Newsome 1998; Feng and Brown 1998, 1999), or in single compartment Hodgkin–Huxley type models (Bell et al. 1995). Later, it was demonstrated that, using a leaky integrator model with partial reset mechanism or physiological gain, Poisson-distributed discharge activity at high frequencies can also be obtained without a fine tuning of inhibitory and excitatory inputs (Troyer and Miller 1997; Christodoulou and Bugmann 2000, 2001), indicating a possible role of nonlinear spike-generating dynamics for cortical spike train statistics (Gutkin and Ermentrout 1998).

Finally, the "noisy" aspect of network dynamics was emphasized as a possible mechanism driving cortical neurons to fire irregularly (Usher et al. 1994; Hansel and Sompolinsky 1996; Lin et al. 1998; Tiesinga and José 1999). In this context, it was shown that (temporal) correlation in the inputs can produce a high C_V in the cellular response (Stevens and Zador 1998; Sakai et al. 1999; Feng and Brown 2000; Salinas and Sejnowski 2000, for a review see Salinas and Sejnowski 2001).

The consensus which emerged from these studies is that neurons operating in an excitable or noise-driven regime are capable of showing highly irregular responses. In this subthreshold regime, the membrane potential is close to spike threshold and APs are essentially triggered by fluctuations of the membrane potential. In this framework, the irregularity of the discharge and, thus, the C_V value, can be increased by either bringing the membrane closer to firing threshold (e.g., by balancing the mean of excitatory and inhibitory drive; see, e.g., Bell et al. 1995; Shadlen and Newsome 1998; Feng and Brown 1998, 1999), or by increasing the noise amplitude (e.g., by correlating noisy synaptic inputs; see, e.g., Feng and Brown 2000; Salinas and Sejnowski 2001). However, the underlying conditions for the appearance of this subthreshold regime, as well as its dependence on various electrophysiological parameters or the characteristics of the driving inputs, remain mostly unclear in such models.

5.4.1 *High Discharge Variability in Detailed Biophysical Models*

To investigate the conditions under which high discharge variability is generated, the spontaneous discharge of Hodgkin–Huxley type models of morphologically-reconstructed cortical neurons was studied by Rudolph and Destexhe (2003a). These models incorporated in vivo-like synaptic background activity, simulated by random release events at excitatory and inhibitory synapses constrained by in vivo intracellular measurements in cat parietal cortex (Paré et al. 1998b; Destexhe and Paré 1999).

It was found that neither the synaptic strength (as determined by quantal conductance and release rates; Fig. 5.15a), the balance between excitation and inhibition, nor the membrane excitability (Fig. 5.15b,c) or presence of specific ion channels (Fig. 5.16) are stand-alone factors determining a high C_V ($C_V \approx 0.8$) for physiologically relevant firing rates. Moreover, provided the neuron model is within the limits of biophysically plausible parameters regimes (e.g. ion channel kinetics and distribution, membrane excitability, morphology), no significant changes of the irregularity of spiking over that expected from a renewal process with (effective) refractory period are observed.

5.4.2 *High Discharge Variability in Simplified Models*

The results obtained with such detailed biophysical models suggest that the high-conductance state of cortical neurons is essential and, at the same time, provides natural conditions for maintaining an irregular firing activity in neurons

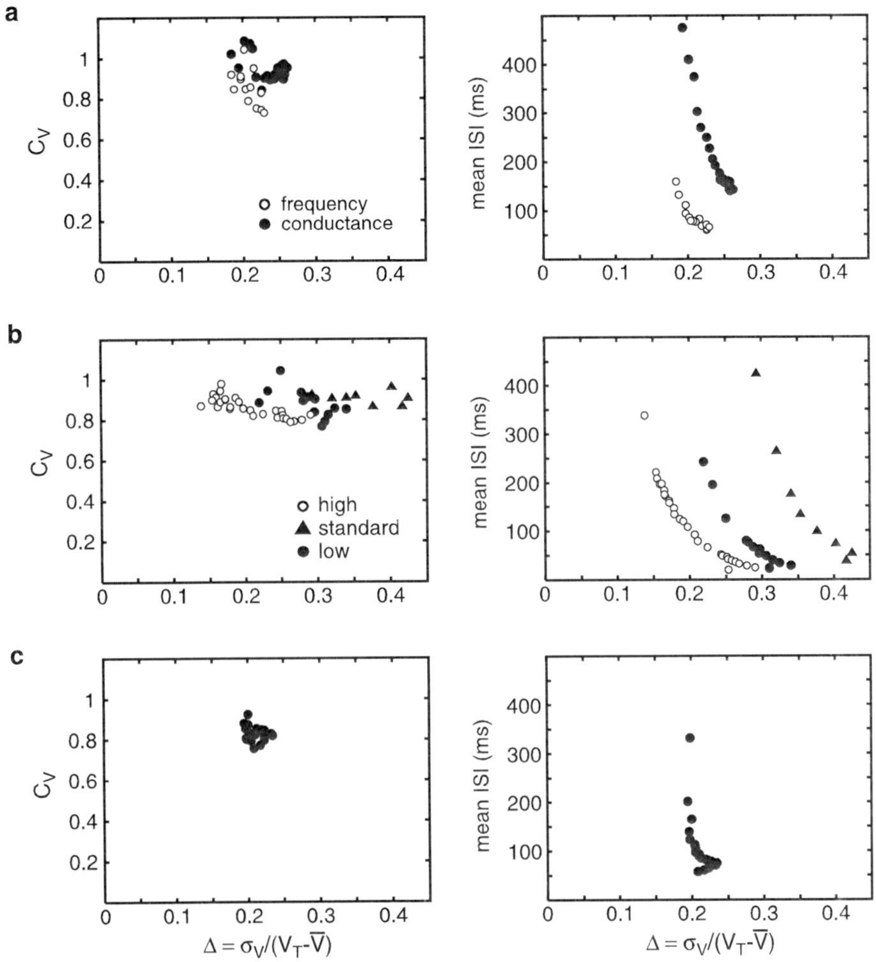

Fig. 5.15 Discharge variability C_V and mean ISI in a detailed biophysical model of cortical neurons as a function of the threshold accessibility, defined as $\Delta = \sigma_V/(V_T - \overline{V})$, where $\overline{V}$, σ_V and V_T denote the membrane potential mean, standard deviation and firing threshold, respectively. (**a**) Results for different levels of background activity, obtained by changes in the quantal synaptic conductances or the release rate synaptic terminals. (**b**) Results for the high and low excitable membranes. The firing rate was changed by altering the correlation in the synaptic background. (**c**) Results for different levels of membrane excitability in the presence of fixed correlated background (Pearson correlation coefficient ~0.1). Modified from Rudolph and Destexhe (2003a)

receiving irregular synaptic inputs. The main support for this proposition is provided by simplified models, in which it is possible to manipulate the excitatory and inhibitory conductances of background activity, and to explore their parameter space in detail.

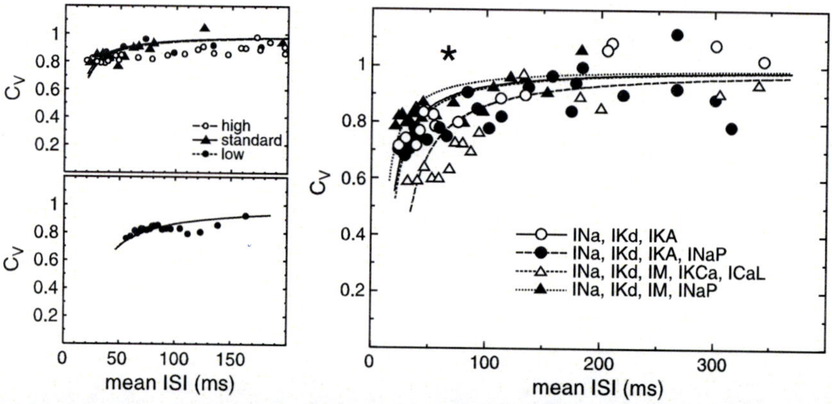

Fig. 5.16 Discharge variability for various ion channel settings. *Top, left*: Change of peak conductances of sodium, delayed-rectifier and voltage-dependent potassium channels by ±40% around experimentally observed values. *Bottom, left*: As *Top*, but in the presence of correlated background activity (Pearson correlation coefficient ~0.1). *Right*: Various ion channel settings and kinetics. *White dots*: voltage-dependent conductances including sodium I_{Na} and delayed-rectifier potassium channels I_{Kd} as well as A-type potassium channels I_{KA} according to (Migliore et al. 1999). *Black dots*: I_{Na}, I_{Kd} and I_{KA} conductances (Migliore et al. 1999) with additional persistent sodium current I_{NaP} (French et al. 1990; Huguenard and McCormick 1992; McCormick and Huguenard 1992). *White triangles*: I_{Na}, I_{Kd} and I_{M} with additional Ca^{2+}-dependent potassium current (C-current) I_{KCa} (Yamada et al. 1989) and a high-threshold Ca^{2+}-current (L-current) I_{CaL} (McCormick and Huguenard 1992). *Black triangles*: I_{Na}, I_{Kd} and I_{M} with additional persistent sodium current I_{NaP}. Modified from Rudolph and Destexhe (2003a)

By mapping the physiologically meaningful regions of parameters producing (a) highly irregular spontaneous discharges ($C_{\mathrm{V}} > 0.8$); (b) spontaneous firing rates between 5 and 20 Hz; (c) Poisson-distributed ISI intervals; (d) input resistance and voltage fluctuations consistent with in vivo estimates, it was possible to explore the impact of the high-conductance state on the generation of irregular responses (Rudolph and Destexhe 2003a). In the so constructed simplified models driven by effective stochastic excitatory and inhibitory conductances, high (>0.8) C_{V} values are observed in parameter regimes for the effective conductance ranging between 20 and 250% of the values obtained by fitting the model to experimental observations (Fig. 5.17a).

Moreover, using such simplified models also allows to compare fluctuating conductance models with fluctuating current models with comparable voltage fluctuations, but low input resistance. The latter type is often used to represent synaptic background activity in models (Bugmann et al. 1997; Sakai et al. 1999; Shinomoto et al. 1999; Svirskis and Rinzel 2000) or in experiments (Holt et al. 1996; Hunter et al. 1998; Stevens and Zador 1998). It was found that current-based models lead, in general, to a more regular firing ($C_{\mathrm{V}} \sim 0.6$ for 5–20 Hz firing frequencies) compared to conductance-based models. Intermediate models (high mean conductance with current fluctuations or mean current with conductance fluctuations) displayed the highest C_{V} when the high-conductance component was

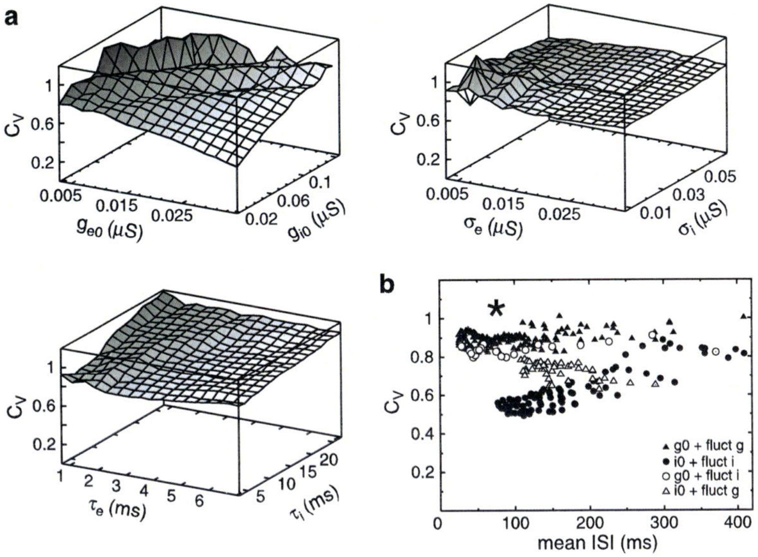

Fig. 5.17 Irregular firing activity in a single-compartment "point-conductance" model. (**a**) The C_V as a function of the mean (g_{e0} and g_{i0}, respectively), variance (σ_e and σ_i) and time constant (τ_e and τ_i) of inhibitory and excitatory effective conductances. In all cases, C_V values above 0.8 were observed. Only models with a strong dominance of excitation led to more regular firing due to a combination of high firing rates ($\gg$200 Hz) and the presence of a refractory period in the model neurons. (**b**) Evidence that high firing variability is linked to high-conductance states. In the point-conductance model (g_0 + fluct g), excitatory and inhibitory conductances were changed around a mean according to a Ornstein–Uhlenbeck process, leading to a high C_V around unity. In the fluctuating current model (i_0 + fluct i), random currents around a mean described by an Ornstein–Uhlenbeck process were injected into the cell, leading to a lower variability in the spontaneous discharge activity. In two other models, using a constant current with fluctuating conductance around zero mean (i_0 + fluct g), and a constant conductance with fluctuating current around zero mean (g_0 + fluct i), higher C_V values were obtained, showing that high-conductance states account for the high discharge variability. The *star* indicates a spontaneous firing rate of about 17 Hz. Modified from Rudolph and Destexhe (2003a)

present (Fig. 5.17b). This analysis indicates that the most robust way to obtain irregular firing consistent with in vivo estimates is to use neuron models in a high-conductance state.

5.4.3 High Discharge Variability in Other Models

The findings reported above are not in conflict with earlier results suggesting that the OU process does not reproduce cortical spiking statistics (Shinomoto et al. 1999). The latter model consists of white noise injected directly as a current to the membrane, and is, therefore, equivalent to fluctuating current models. Injection of current as colored noise (Brunel et al. 2001), however, may lead to high C_V values, although not with Poisson statistics and only for particular values of the noise time constant.

The high discharge variability reported in detailed biophysical as well as conductance-based simplified models (Rudolph and Destexhe 2003a) is, however, not undisputed. For instance, in the single-compartment Hodgkin–Huxley type model of cortical neurons investigated by Bell et al. (1995), a number of cellular and synaptic input parameters were identified which, if correctly combined, yield a balanced or "sensitive" neuronal state. Only in this state, which is characterized by a rather narrow parameter regime, the cell converts Poisson synaptic inputs into irregular output-spike trains. This fine tuning causes the cell to operate close to the threshold for firing, which, in turn, leads to an input (noise) driven cellular response. In addition, in these models, an increase in the variability for stronger inputs can be observed. In this case, there is a net decrease of the membrane time constant, which was identified as the cause of the irregularity (Bell et al. 1995). This is consistent with the necessity of a high-conductance state, but the fine tuning required in the above study contrasts with the high robustness seen in the models of Sect. 5.4.2.

A single-compartment model with Hodgkin–Huxley type Na^+ and K^+ currents of hippocampal interneurons subject to Gaussian current noise and Poisson-distributed conductance noise was investigated in Tiesinga and José (1999). For Poisson-distributed inputs, these authors report a net increase in the C_V for fixed mean ISIs for increasing noise variance, and a shift of the C_V versus mean ISI to lower mean ISIs for increasing noise average. However, noise average and variance were quantified in terms of the net synaptic current, leaving a direct link between synaptic conductance and spiking statistics open for further investigations.

The impact of correlation in the synaptic background on the neuronal response was also subject of a study by Salinas and Sejnowski (2000) using a conductance-based IAF neuron. Here, it was found that, in accordance with the above findings, the variability of the neuronal response in an intact microcircuit is mostly determined by the variability of its inputs. In addition, it was shown that using a smaller time constant (similar to that found in a high-conductance state) leads to higher C_V values, which is also in agreement with the findings described in Sect. 5.4.2. Finally, a decrease in the variability for an increase in the effective refractory period (relative to the timescale of changes in the postsynaptic conductances) caused by smaller synaptic time constants and larger maximal synaptic conductances was reported (Salinas and Sejnowski 2000). These results hold for a "balanced" model. However, a marked decrease of the C_V is obtained when the "balanced" model is replaced by an "unbalanced" one. Although the sensitivity to the balance in the synaptic inputs was not investigated in detail in the aforementioned study, the model suggests a peak in the C_V only for a narrow parameter range (see also Bell et al. 1995). The reported C_V values of 1.5 for firing rates of 75 Hz are markedly higher than that found in conductance-based simplified models studied in Rudolph and Destexhe (2003a), presumably because of the occurrence of bursts and significant deviations from the Poisson distribution, as found for high firing rates (see the decrease in irregularity for dominant excitation in Fig. 5.17a, top left).

Finally, in characterizing how fluctuations impact on the discharge variability, the representation of the C_V as a function of a measure of "threshold accessibility" Δ (see caption Fig. 5.15) reveals differences between low-conductance and high-conductance states. In low-conductance states, the C_V is found to be dependent on Δ (Rudolph and Destexhe 2003a), consistent with IAF models (see, e.g., Troyer and Miller 1997). On the other hand, for high-conductance states, the C_V was mostly independent of threshold accessibility, which is in overall agreement with the finding that the high discharge variability is highly robust with respect to the details of the model, provided it is operating in a high conductance state.

In summary, studies with biophysically realistic models, including detailed compartmental models and simplified point neurons, suggest that the genesis of highly variable discharges ($C_V > 0.8$ and Poisson distributed) by membrane potential fluctuations is highly robust only in high-conductance states. In Sect. 6.2.2, we will show results in which this relation between high-conductance states and the variability of cellular discharges was evidenced in dynamic-clamp experiments.

5.5 Stochastic Resonance

The enhanced responsiveness property investigated in detail in the preceding section is reminiscent of a more general phenomenon of enhancement of information processing capabilities that can occur in some physical systems in the presence of noise. Phenomena like the amplification of weak signals or the improvement of signal detection and reliability of information transfer in nonlinear dynamical systems in the presence of a certain nonzero noise level were intensely studied, both theoretically and experimentally. These phenomena became well-known, and are now well-established, under the term *stochastic resonance* (Wiesenfeld and Moss 1995), and have since been shown to be an inherited property of many physical, chemical, and biological systems (Gammaitoni et al. 1998). In Sect. 5.3, we have seen that the enhancement of responsiveness presents a nonmonotonic behavior as a function of noise amplitude (see Fig. 5.12). Below we will detail whether this phenomenon can be attributed to stochastic resonance.

Neurons provide particularly favorable conditions for displaying stochastic resonance due to their strongly excitable nature, nonlinear dynamics, and embedding in noisy environments. Sensory systems were among the first in which experimentalists looked for a possibly beneficial role of noise, as neurons in these systems need to detect signals superimposed on an intrinsically noisy background. First experimental evidence for the enhancement of the quality of neuronal responses in the presence of nonzero noise levels date back to the late 1960s of the last century (Rose et al. 1967; Buño et al. 1978). First direct proof for the beneficial modulation of oscillating signals by noise and, thus, the presence of stochastic resonance phenomena in biological neuronal systems was found in sensory afferents in the early 1990s. In cells of the dogfish *ampullae* of Lorenzini, Braun et al. showed that the generation of spike impulses not just depends on intrinsic subthreshold oscillations, but crucially on the superimposed noise: the frequency of the oscillations determines

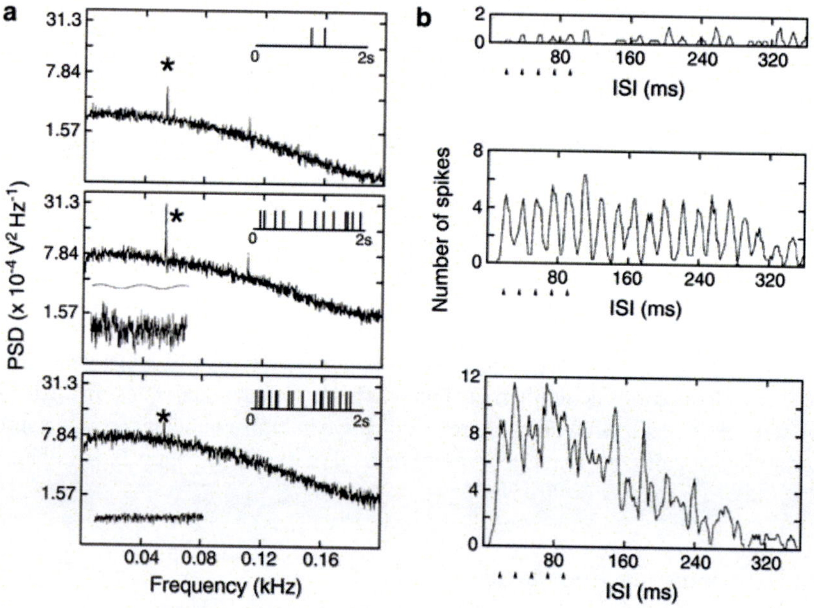

Fig. 5.18 Response of crayfish mechanoreceptors to noisy periodic inputs. (**a**) Power spectral density computed from the spiking activity, stimulated with a weak periodic signal (frequency 55.2 kHz) plus three external noise intensities (*top*: 0; *middle*: 0.14, *bottom*: 0.44 V r.m.s.). *Insets* show examples of the spike trains from which the PS were computed. The peaks at the stimulus frequency (*star*) and multiples of it are most prominent at intermediate noise levels (*middle panel*). (**b**) Interspike interval histograms (ISIHs) for the same stimulus and noise condition as in (**a**). The *arrowheads* mark the first five integer multiples of the stimulus period. For no or low noise amplitudes, the spike rate is very low and, hence, the ISIH shows only small peaks (*top*). In contrast, for large noise amplitudes, spikes occur increasingly random, leading to a masking of the periodic response and no clear peaks at the stimulus period and their integer multiples (*bottom*). The largest coherence between stimulus and response was observed at medium noise levels, with the ISIHs displaying clear peaks at the stimulus period and multiples of it, without exponential-like decay of the peak amplitudes (*middle*). Modified from Douglass et al. (1993)

the base impulse rhythm, but whether a spike is actually triggered was determined by the noise amplitude (Braun et al. 1994). Moreover, the presence of noise added a new dimension to the encoded information, as Braun et al. demonstrated that dual sensory messages can be conveyed in a single-spike train.

Among the most widely studied systems belong mechanoreceptors. Using a combination of external noise superimposed on a periodic mechanical stimulus, Douglas and collaborators demonstrated the presence of stochastic resonance in crayfish *Procambarus clarkii* near-field mechanoreceptor cells (Douglass et al. 1993). In order to quantify the response behavior, both power spectra and interspike interval histograms (ISIHs) were obtained. In the power spectrum, periodic stimuli led to narrow peaks at the fundamental stimulus frequency above a noisy floor (Fig. 5.18a). As it was shown, the amplitude of this peak depends on the noise

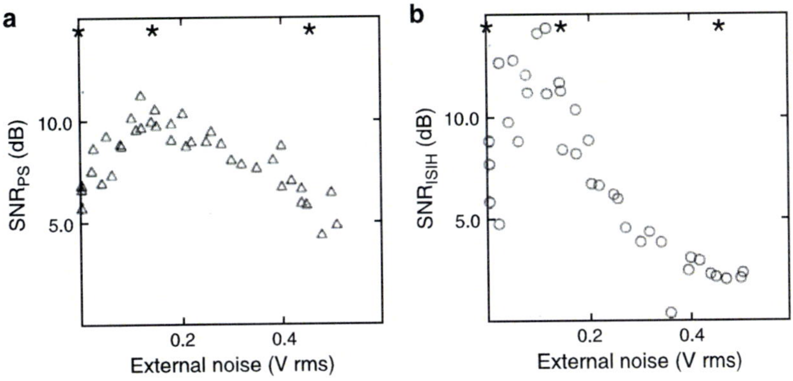

Fig. 5.19 Stochastic resonance in crayfish mechanoreceptors. (**a**) Signal-to-noise ratio (SNR) calculated from the power spectra of the spiking response according to (5.2). (**b**) SNR calculated from the ISIHs according to (5.3). *Stars* indicate the noise levels shown in Fig. 5.18. In both cases, maximum coherence between the stimulus and response is observed for intermediate noise levels. Modified from Douglass et al. (1993)

amplitude and was markedly pronounced for intermediate noise levels (Fig. 5.18a, middle) while degrading for higher noise amplitudes (Fig. 5.18a, bottom). Similarly, the ISIHs reveal peaks at the stimulus frequency and integer multiples of it, indicating a periodic response to the periodic stimulus (Fig. 5.18b). Whereas these peaks are small due to a small average rate and skipping of responses at low noise levels (Fig. 5.18b, top), the amplitude of these peaks rises markedly when the noise amplitude is increased (Fig. 5.18b, middle). For high levels of noise, the periodicity of the response is lost due to the increasing stochasticity of the response, leading to an increasing randomization of the peaks in the ISIHs (Fig. 5.18b, bottom).

To quantify the coherence of the response to periodic stimuli, the signal-to-noise ratio (SNR) is commonly considered. For the power spectrum,

$$SNR_{\mathrm{PS}} = 10 \cdot \log_{10}(S \cdot N(\nu_0)), \tag{5.2}$$

where $N(f_0)$ denotes the amplitude of the broad-band noise at the signal frequency ν_0 and S is the area under the signal peak above the noise floor in the power spectrum. Similarly, for the ISIH, a SNR can be defined by considering the integral under the peaks caused by the periodic stimulus (Fig. 5.19):

$$SNR_{\mathrm{ISIH}} = 10 \cdot \log_{10} \left(\frac{N_{\mathrm{max}}}{N_{\mathrm{min}}} \right)^2 . \tag{5.3}$$

Here, N_{max} and N_{min} denote the sum over the intervals around peaks in the ISIH and the sum over intervals around the troughs, respectively. In both cases, a clear enhancement of the coherence for intermediate noise levels can be observed (Fig. 5.18), thus demonstrating the beneficial role of noise in the investigated system.

Utilizing similar SNR coherence measures, further experimental evidence for the presence of stochastic resonance in mechanoreceptor neurons was obtained in the cercal system of the cricket *Acheta domestica* (Levin and Miller 1996) and the tibial nerve of the rat (Ivey et al. 1998). In the latter study, spiking activity after mechanical stimulation of the receptive field with sine-waves of different frequencies superimposed on varying levels of white noise was recorded. The coherence was quantified by using the correlation coefficient between the periodic input signal the nerve response. In all cases, the addition of noise improved the signal transmission. Whereas the response to the noisy stimulus tended to be rate modulated at low frequencies of the periodic stimulus component, it followed nonlinear stochastic resonance behavior at higher frequencies. Whereas most of these studies used noise external to the system superimposed on an external signal in order to study the detection of noisy sensory inputs, also system-intrinsic noise, such as the Brownian motion of hair cells in the *maculae sacculi* of leopard frogs *Rana pipiens* (Jaramillo and Wiesenfeld 1998), was shown to beneficially shape sensory responses.

However, the stochastic resonance phenomenon is not restricted to periodic input signals. Using the cross-correlation function between a stimulus and the systems response, Collins et al. demonstrated the presence of aperiodic stochastic resonance in mammalian cutaneous mechanoreceptors of rat (Collins et al. 1995b, 1996). Here, in contrast to the studies mentioned above, aperiodic stimuli in the presence of nonzero levels of noise were used. Indeed, as the input noise was increased, the stimulus response coherence, defined here as the normalized power norm (i.e., the maximum value of the normalized cross-correlation function)

$$C_1 = \frac{C_0}{\left(\overline{S^2(t)}\right)^{1/2} \left(\overline{(R(t) - \overline{R(t)})^2}\right)^{1/2}}, \tag{5.4}$$

where $C_0 = \overline{S(t)R(t)}$ denotes the power norm, $S(t)$ and $R(t)$ the aperiodic input signal and mean firing rate constructed from the output, respectively, increased to a peak and then slowly decreased (Fig. 5.20b). This was in stark contrast to the average firing rate, which increased monotonically as a function of the input noise (Fig. 5.20a) and demonstrates that the presence of noise adds an additional coding dimension to signal processing as the firing rate and the temporal characteristics of the response, quantified by its coherence to the input signal, are merely independent of each other.

Mechanoreceptors are not the only sensory afferents which exhibit the stochastic resonance phenomena. Experimental studies involving other sensory modalities, such as electrosensory afferent (Greenwood et al. 2000) or visual systems (Pei et al. 1996; Simonotto et al. 1997) complemented studies which show that noise has a beneficial impact on sensory perception. Most of these studies utilized white noise with varying amplitude superimposed on a sensory signal. However, in nature,

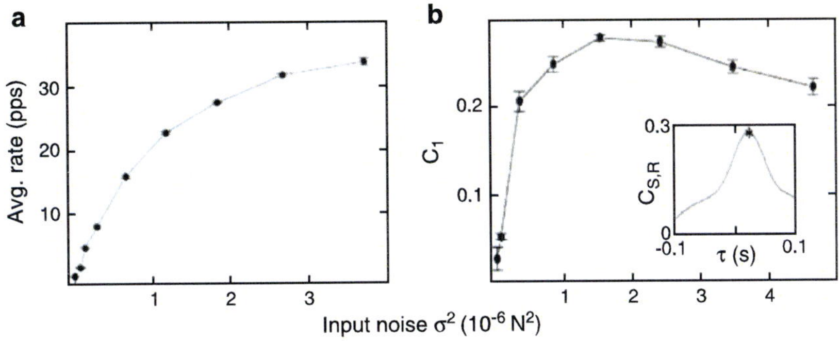

Fig. 5.20 Stochastic resonance in rat SA1 cutaneous mechanoreceptors. (**a**) Mean and SD of the average firing rate as a function of the input noise variance σ^2 for a population of three cells. The rate monotonically increases as a function of σ^2. (**b**) Normalized power norm C_1 (5.4) as a function of σ^2 for one cell. In contrast to the output rate, the power norm shows a nonmonotonic behavior, indicating an optimal coherence between input signal and output for intermediate noise levels. The *inset* shows the cross-correlation function between the aperiodic stimulus $S(t)$ and the response $R(t)$ for $\sigma^2 = 1.52 \times 10^{-6}$ N^2. Modified from Collins et al. (1996)

stochastic processes are rarely ideal white Gaussian. Using rat cutaneous afferents, Nozaki and colleagues could demonstrate the presence of the stochastic resonance phenomena also in settings with a more natural statistical signature of the stochastic component (Nozaki et al. 1999). Using colored noise with a PSD following a $1/f$ or $1/f^2$ behavior, interesting differences could be deduced: the optimal noise amplitude is lowest and the SNR highest for white noise (Fig. 5.21). However, under certain circumstances, the output SNR for $1/f$ noise can be much larger than that for white noise at the same noise intensity (Fig. 5.21), thus rendering $1/f$ noise better suited for enhancing the neuronal response to weak periodic stimuli.

As mentioned above, sensory afferents were among the first systems in which the beneficial role of noise could be demonstrated. But what about more central systems? As detailed in Chap. 3, the neuronal activity, for instance in the cortex, closely resembles a Poissonian, hence stochastic, process. Indeed, individual neurons receive a barrage of random inputs, rendering their intrinsic dynamics stochastic. Could it be that also here the stochastic resonance phenomena is manifest and allows to enhance the detection of signals embedded into this seemingly random background?

Experimentally, the activity in central systems is far more difficult to control compared to sensory afferents. One possibility to assess the role of noise here is to use psychophysical setups (Chialvo and Apkarian 1993). Simonotto and colleagues demonstrated that the stochastic resonance phenomenon indeed modifies the ability of humans to interpret a perceived stationary visual images contaminated by noise (Simonotto et al. 1997). That the influence of noise on the perception of sensory signals by humans is not restricted to one modality was demonstrated by Richardson and colleagues (Richardson et al. 1998): using electrical noise, these researches showed that the ability of an individual to detect subthreshold mechanical

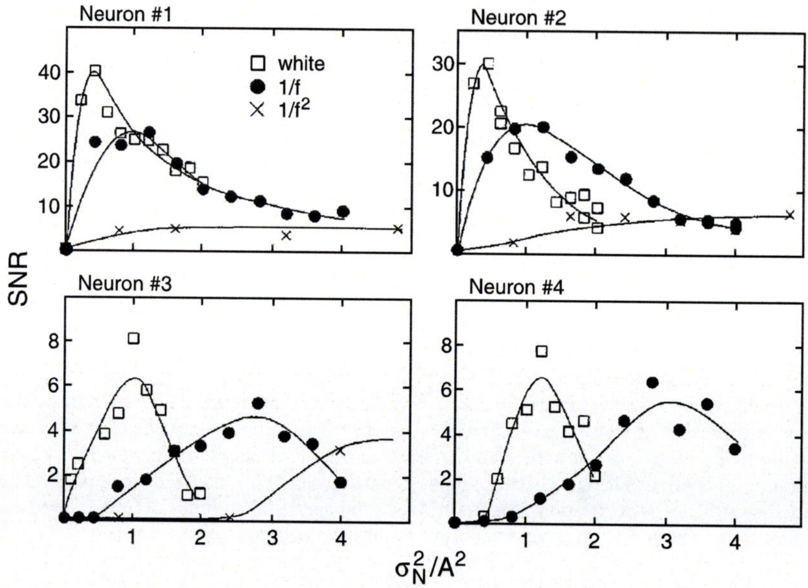

Fig. 5.21 Stochastic resonance in rat cutaneous afferents. Signal-to-noise ratio (SNR) as a function of the input noise variance (represented in units of the squared amplitude A^2 of the sinusoidal input signal) for different statistics of the input noise (*white*, $1/f$ and i/f^2) for four different neurons. *Solid lines* interpolate the experimental results. The SNR is defined as the ratio between the peak amplitude and the noise floor in the power spectrum at the frequency of the periodic input. Modified from Nozaki et al. (1999)

cutaneous stimuli was enhanced. This cross-modality stochastic resonance effect thus demonstrates that for stochastic resonance type effects in human perception to occur, the noise and stimulus need not be of the same modality.

Also direct electrophysiological studies of the stochastic resonance phenomena at the network level using electrical field potential recordings exist (e.g., Srebo and Malladi 1999). The first study demonstrating the existence of the stochastic resonance phenomena in neuronal brain networks was conducted by Gluckman and colleagues (Gluckman et al. 1996). Delivering signal and noise directly to a network of neurons in hippocampal slices from the rat temporal lobe through a time-varying electrical field, it was observed that the response of the network to weak periodic stimuli could be enhanced when the stochastic component was at an intermediate level. Finally, numerical simulations and their experimental verification showed that the stochastic resonance phenomenon is responsible for the detection of distal synaptic inputs in CA1 neurons in in vitro rat hippocampal slices (Fig. 5.22; Stacey and Durand 2000, 2001).

As experimental investigations provided overwhelming evidence for the beneficial action of noise on all levels of neuronal activity, from peripheral and sensory systems to more central systems, at the single-cell level up to networks

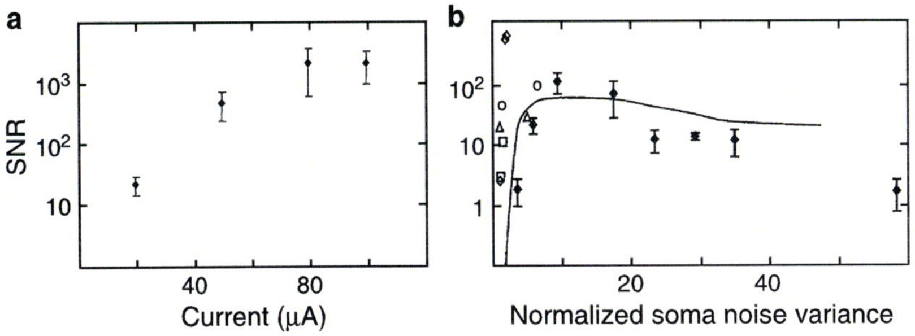

Fig. 5.22 Stochastic resonance in hippocampal CA1 neurons. (**a**) Signal-to-noise ratio (SNR) as a function of the amplitude of the input current noise. The mean and SD for a population of 13 cells are shown. The results show a clear improvement of signal detection at increasing noise levels. (**b**) SNR for four individual cells (*clear marks*) compared to simulation results (*black*). The *black line* shows the fit to the equation $SNR \propto (\varepsilon \Delta U/D)^2 e^{-(\Delta U/D)}$ which characterizes the SNR for a periodic input to a monostable system (Stocks et al. 1993; Wiesenfeld et al. 1994). In both cases, the SNR was defined as the ratio of the peak amplitude to the noise floor in the power spectra at the stimulus frequency. Modified from Stacey and Durand (2001)

and sensory perception, theoretical studies investigated the conditions on which this stochastic resonance phenomenon rests. Most prominently featured here the Fitzhugh–Nagumo (FHN) model, which serves as an idealized model for excitable systems (Fitzhugh 1955, 1961; Nagumo et al. 1962; Fitzhugh 1969). In a number of numerical studies in which the FHN model (or the Hindmarsh–Rose neuronal model as a modification of it; see Wang et al. 1998; Longtin 1997) was subjected to both periodic (Chow et al. 1998; Longtin and Chialvo 1998; Longtin 1993) and aperiodic inputs superimposed on a noisy background (Collins et al. 1995a,b), it was shown that the an optimal nonzero noise intensity maximizes the signal transmission already in low-dimensional excitable systems (Chialvo et al. 1997; Balázsi et al. 2001).

One advantage of using simplified neuronal models, such as the FHN model, driven by stochastic inputs, is that the prerequisites necessary for the stochastic resonance phenomenon to occur can be studied analytically. The stochastic FHN model is defined by a set of two coupled differential equations

$$\varepsilon \frac{V(t)}{dt} = V(t)(V(t)-a)(1-V(t)) - w(t) + A + S(t) + \xi(t),$$
$$\frac{w(t)}{dt} = V(t) - w(t) - b, \tag{5.5}$$

where $V(t)$ denotes the membrane voltage, $w(t)$ is a slow recovery variable, ε, a, and b are constant model parameters, A a constant (tonic) activation signal, $S(t)$ a time variable (periodic or aperiodic) input current, and $\xi(t)$ Gaussian white noise with zero mean and an autocorrelation of $<\xi(t)\xi(s)> = 2D\delta(t-s)$. For a certain parameter regime, (5.5) corresponds to a double-well barrier-escape problem

(Collins et al. 1995b). Defining the ensemble averages of the power norm as $< C_0 >=< \overline{S(t)R(t)} >$ and normalized power norm $< C_1 >$, (5.4), it can be shown (Collins et al. 1995b) that for the FHN model, these coherence measures take the form

$$< C_0 > \propto \frac{1}{D} \exp\left(\frac{-\sqrt{3}B^3\varepsilon}{D} \right) \overline{S^2(t)},$$

$$< C_1 > \propto \frac{< C_0 >}{N\left(\overline{S^2(t)}\right)^{1/2}}, \tag{5.6}$$

where

$$N^2 = \exp(\Theta + 2\Delta^2\overline{S^2(t)}) - \exp(\Theta + \Delta^2\overline{S^2(t)}) + \sigma(D) \tag{5.7}$$

with

$$\Theta = -2\sqrt{3}B^3\frac{\varepsilon}{D},$$

$$\Delta = 3\sqrt{3}B^2\frac{\varepsilon}{D}$$

and B denoting a constant parameter corresponding to the signal-to-threshold distance. Moreover, $\sigma(D)$ denotes the time-averaged square of the stochastic component of the response $R(t)$, which is a monotonically increasing function of the noise variance D.

This analytical description of the coherence between input signal and cellular response as a function of the input noise amplitude fits remarkably well to the corresponding numerical results (Fig. 5.23), and shows a significant increase in the normalized power norm for intermediate noise amplitudes, as found in experimental studies (Fig. 5.20b). Other studies obtained similar analytical solutions for noise-induced response enhancement, either using different neuronal models (e.g. Neiman et al. 1999) or more general input noise models (Nozaki et al. 1999). In the latter study, using the linearized FHN model

$$\varepsilon\frac{V(t)}{dt} = -\gamma V(t) - w(t) + A\sigma_{\text{in}}(2\pi\nu_0 t) + \xi(t),$$

$$\frac{w(t)}{dt} = V(t) - w(t), \tag{5.8}$$

with refractory period T_{R} and threshold potential θ driven by a periodic input of amplitude A and $1/f^{\beta}$, it was shown that the SNR obeys the analytical relation

$$SNR = \frac{2A_{\gamma}^2\theta^2 r_0^2}{h^2(\beta)\sigma_N^2(1+2T_{\text{R}}r_0)^3}. \tag{5.9}$$

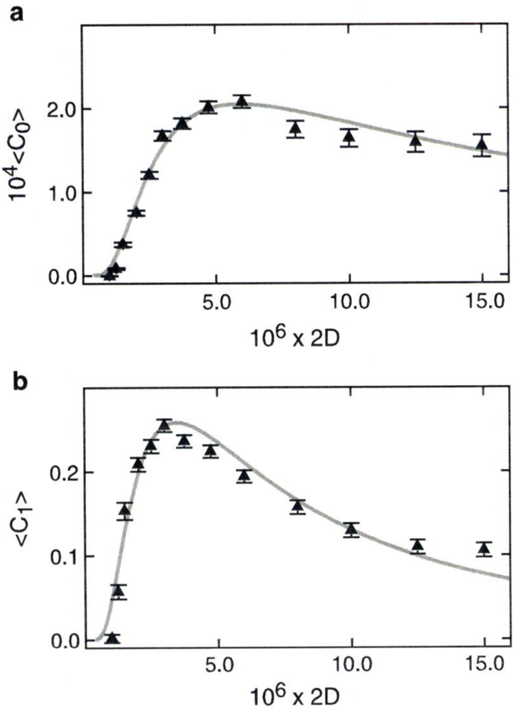

Fig. 5.23 Numerical and analytical prediction of the stochastic resonance phenomena in the FHN model driven by Gaussian white noise. (**a**) Ensemble average of the power norm (*triangles*: mean $\pm$ SD; *gray*: (5.6)) as a function of the noise intensity. (**b**) Ensemble average of the normalized power norm (*triangles*: mean $\pm$ SD; *gray*: (5.6)) as a function of the noise intensity. In both cases, the theoretical predictions fit the numerical result (obtained by averaging over 300 trials), and show qualitatively the same behavior as seen in experiments (Fig. 5.20b). Model parameters: $B = 0.07$, variance of input signal $S(t)$: 1.5×10^{-5}, in (**b**) $\sigma(D) = 1.7 \times D + 3.5 \times 10^9 D^2$. Modified from Collins et al. (1995b)

Here, $A_\gamma = A/(1+\gamma)$, σ_N is the noise variance and r_0 is the rate with which the membrane potential crosses the threshold, given by

$$r_0 = g(\beta)\exp\left[-\frac{\theta^2}{2h(\beta)\sigma_N^2}\right]. \tag{5.10}$$

In the last equation, $g(\beta)$ and $h(\beta)$ denote the lower and upper limit of the noise bandwidth. Figure 5.24 shows the analytical prediction for white, $1/f$ and $1/f^2$ noise and different noise bandwidths. These results fit qualitatively to results obtained for the SNR in experiments (Fig. 5.21).

Beyond this idealized model of neuronal excitability, stochastic resonance was also evidenced in numerical simulations using the IF neuronal model (Mar et al. 1999; Shimokawa et al. 1999; Balázsi et al. 2001) or Hodgkin–Huxley neurons (Lee et al. 1998; Lee and Kim 1999; Rudolph and Destexhe 2001a,b) driven by additive and multiplicative white or colored (Ornstein–Uhlenbeck) noise. However, due to the higher mathematical complexity of these models, at least when compared with the idealized neuronal models mentioned above, analytical descriptions of the stochastic resonance phenomenon are difficult or not at all available.

An intriguing consequence of the stochastic resonance-driven amplification of weak signals in noisy settings at the single-cell level is the amplification of distal

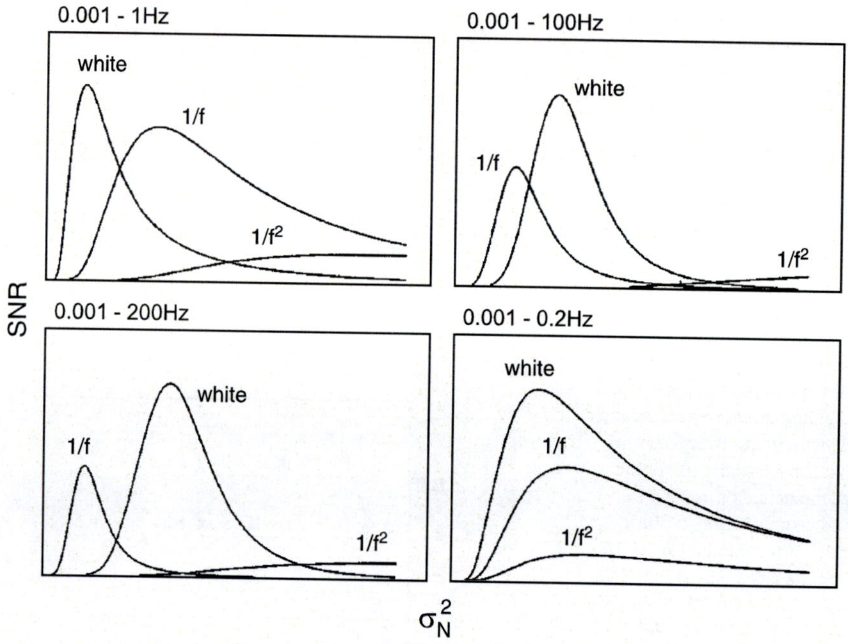

Fig. 5.24 Theoretical prediction of the SNR, (5.9) for the linearized FHN model subject to white and colored noise. The numerically calculated bandwidth is given in the upper right corners. The axes are displayed in arbitrary units. Model parameters: $\varepsilon = 0.005$, $\gamma = 0.3$, $\theta = 0.03$ and $T_R = 0.67$. Modified from Nozaki et al. (1999)

synaptic inputs. Especially in the cortex, where neurons receive an intense barrage with synaptic inputs, this could provide a fast mechanism with which individual neurons could adjust or tune in a transient manner to specific informational aspects in the received inputs, and serve as an alternative explanation besides the proposed amplification due to active channels in the dendritic tree (Stuart and Sakmann 1994; Magee and Johnston 1995b; Johnston et al. 1996; Cook and Johnston 1997; Segev and Rall 1998).

This possibility was investigated in detailed biophysical models of spatially extended Hodgkin–Huxley models of cortical neurons receiving thousands of distributed and realistically shaped synaptic inputs (Rudolph and Destexhe 2001a,b). In this study, the signal to be detected, a subthreshold periodic stimulation, was added by introducing a supplementary set of excitatory synapses uniformly distributed in the dendrites and firing with a constant period (see details in Hô and Destexhe 2000). To quantify the response to this additional stimulus, besides the SNR introduced above, a special coherence measure, COS, based on the statistical properties of spike trains was used. This measure is defined by

$$COS = \frac{N_{\mathrm{ISI}}}{N_{\mathrm{spikes}}}, \tag{5.11}$$

where N_{ISI} denotes the number of inter-spike-intervals of length equal to the stimulus period and N_{spikes} denotes the total number of spikes within a fixed time interval. In contrast to other well-known measures, such as the SNR or the synchronization index (Goldberg and Brown 1969; Pfeiffer and Kim 1975; Young and Sachs 1979; Tass et al. 1998), this coherence measure reflects in a direct way the threshold nature of the response and, thus, is well-suited to capture the response behavior of spiking systems with simple stimulation patterns.

To vary the strength of the noise in this distributed system, the release frequency of excitatory synapses generating the background activity can be changed. Such a frequency change directly impacts on the amplitude of the internal membrane voltage fluctuations (Fig. 5.25a, top). The response coherence shows a resonance peak when plotted as a function of the background strength or the resulting internal noise level (Fig. 5.25a, bottom). Quantitatively similar results are obtained in simulations in which the noise strength is altered by changing the conductance of individual synapses. Interestingly, in both cases, maximal coherence is reached for comparable amplitudes of membrane voltage fluctuations, namely $3\,\text{mV} \leq \sigma_V \leq 4\,\text{mV}$, a range which is covered by the amplitude of fluctuations measured experimentally in vivo ($\sigma_V = 4.0 \pm 2.0$ mV). Moreover, qualitatively similar results are obtained for different measures of coherence, such as the computationally more expensive SNR. These results demonstrate that for neurons with stochastically driven synaptic inputs distributed across dendritic branches, varying the noise strength is capable of inducing resonance comparable to *classical stochastic resonance* paradigms.

However, the type of model used in these studies allows to modify other statistical properties of the noisy component as well, such as the temporal correlation in the activity of the synaptic channels. Taking advantage of the distributed nature of noise sources, a redundancy and, hence, correlation in the Poisson-distributed release activity at the synaptic terminals can be introduced and quantified by a correlation parameter c (see Appendix B). As found in Rudolph and Destexhe (2001a,b), increasing the correlation leads to large-amplitude membrane voltage fluctuations as well as an increased rate of spontaneous firing (Fig. 5.25b, top). Similar to the stochastic resonance phenomena found through changes of the noise amplitude, also here the COS measure shows a clear resonance peak, this time as a function of the correlation parameter c or the resulting internal noise level σ_V (Fig. 5.25b, bottom). This suggests the existence of an optimal temporal statistics of the distributed noise sources to evoke coherent responses in the cell (see also Capurro et al. 1998; Mato 1998; Mar et al. 1999), a phenomenon which can be called *nonclassical stochastic resonance* . The optimal value of the correlation depends on the overall excitability of the cell, and is shifted to larger values for low excitability. Interestingly, also here the peak coherence is reached at $\sigma_V \sim 4\,\text{mV}$ for an excitability which compares to that found experimentally in adult hippocampal pyramidal neurons.

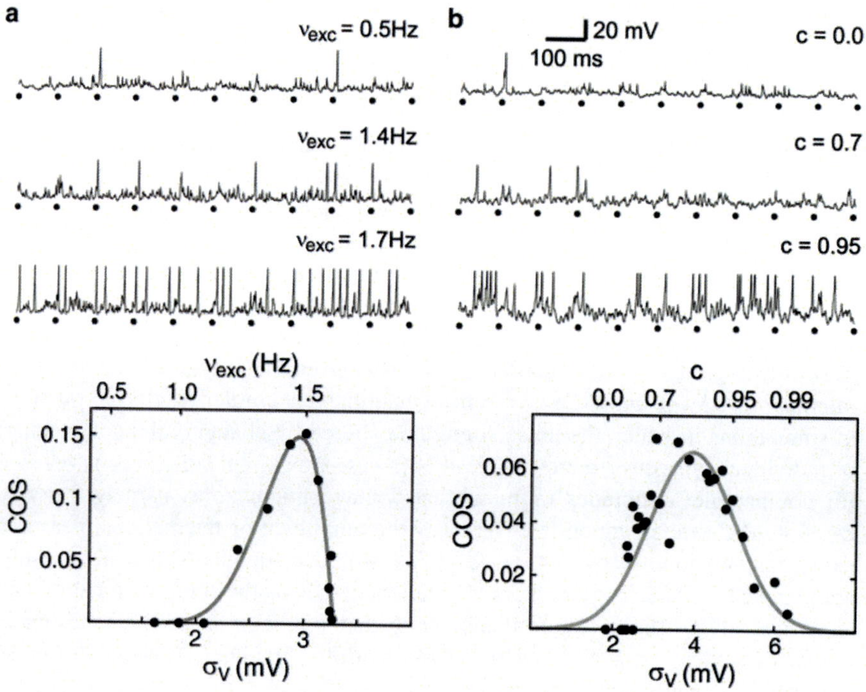

Fig. 5.25 Stochastic resonance in model pyramidal neurons. (**a**) Classical SR. The strength of the synaptic background activity ("noise") was changed by varying the average release frequency at excitatory synapses ν_{exc}, which directly impacts on the membrane voltage fluctuations (σ_V) and the firing activity of the cell (*top*, spike are truncated). Evaluating the response of the cell to a subthreshold periodic signal (*black dots*) by the coherence measure COS (5.11) reveals a resonance peak (*bottom*), showing that the detection of this signal is enhanced in a narrow range of fluctuation amplitudes ($\sigma_V \sim$ 2–3.3 mV). (**b**) Resonance behavior as a function of the correlation c between distributed random inputs. Increasing levels of correlation led to higher levels of membrane voltage fluctuation and spontaneous firing rate of the cell (*top*, spike are truncated). The response to subthreshold stimuli (*dots*) was evaluated using the coherence measure COS. A resonance peak can be observed for a range of fluctuation values of $\sigma_V \sim$ 2–6 mV (*bottom*). Optimal detection was achieved for weakly correlated distributed random inputs ($c \sim 0.7$; Pearson correlation coefficient of $\sim$0.0005). Modified from Rudolph and Destexhe (2001a)

5.6 Correlation Detection

As shown in previous sections, due to the decisive impact of neuronal noise on the cellular membrane, weak periodic signals embedded in a noisy background can be amplified and detected, hence efficiently contribute in shaping the cellular response. Interestingly, the noise levels which optimizes such responses can also be achieved by altering the temporal statistics in form of correlated releases at the synaptic terminals (nonclassical stochastic resonance). This hints at a dynamical component in the optimal noise conditions for enhanced responsiveness. In particular, it suggests the possibility of detecting temporal correlations embedded as signal into the synaptic noise background itself, without the presence of weak periodic stimuli. Mathematically, this idea is supported by the direct relation between the level of correlation and the variance of the total membrane conductance resulting from synaptic inputs, with the latter determining the fluctuations of the membrane potential and, thus, the spiking response of the cell.

Numerical simulations (Rudolph and Destexhe 2001a) showed that, indeed, both changes in the average rate of the synaptic activity as well as changes in their temporal correlations are efficient to evoke a distinguished cellular response (Fig. 5.26a,b). In the former case, the response is evoked primarily through changes in the average membrane potential resulting from a change in the ratio between inhibitory and excitatory average conductances. If such a change is excitatory in nature, hence shifts the average membrane potential closer to its threshold for spike generation, the probability for observing a response increases (Fig. 5.26a). Here, the temporal relation between the onset of the change in the synaptic activity and the evoked response is determined by the speed of membrane depolarization, hence time constant of the membrane.

In contrast, a change in the correlation of the synaptic background activity leaves the average membrane potential mostly unaffected, but instead changes its variance (Fig. 5.26b). As the latter has a determining impact on the spike-generation probability, one can expect a response of the cell to changes in the correlation which remains mostly independent of the membrane time constant. Indeed, in numerical simulations, it was shown that subjecting a detailed model of cortical neurons to fast but weak (corresponding to a Pearson correlation coefficient of not more than 0.0005) transient changes in the temporal correlation of its synaptic inputs could be detected (Rudolph and Destexhe 2001a). Not just was a clear response observable for step changes in correlation down to 2 ms, which is comparable to the timescale of single APs (Fig. 5.26c), but also occurred the response within 5 ms after the correlation onset as a result of the spatial extension of the dendritic structure.

These results demonstrate that neurons embedded in a highly active network are able to monitor very brief changes in the correlation among the distributed noise sources without major change in their average membrane potential. Moreover, these results complement experimental studies which have suggested that the presence of noise allows for an additional dimension in coding neuronal information, as the input driven responses modulated by the noise correlation become independent on the average rate, driven by the mean membrane potential.

5.7 Stochastic Integration and Location Dependence

In this section, we will investigate another important property resulting from the presence of intense background activity in neurons. As we saw in the previous sections, the presence of synaptic noise profoundly affects the efficacy of synaptic inputs. How this efficacy depends on the position of the input on the dendrite, a characteristic which we call *location dependence* of synaptic efficacy, will be described on the following pages.

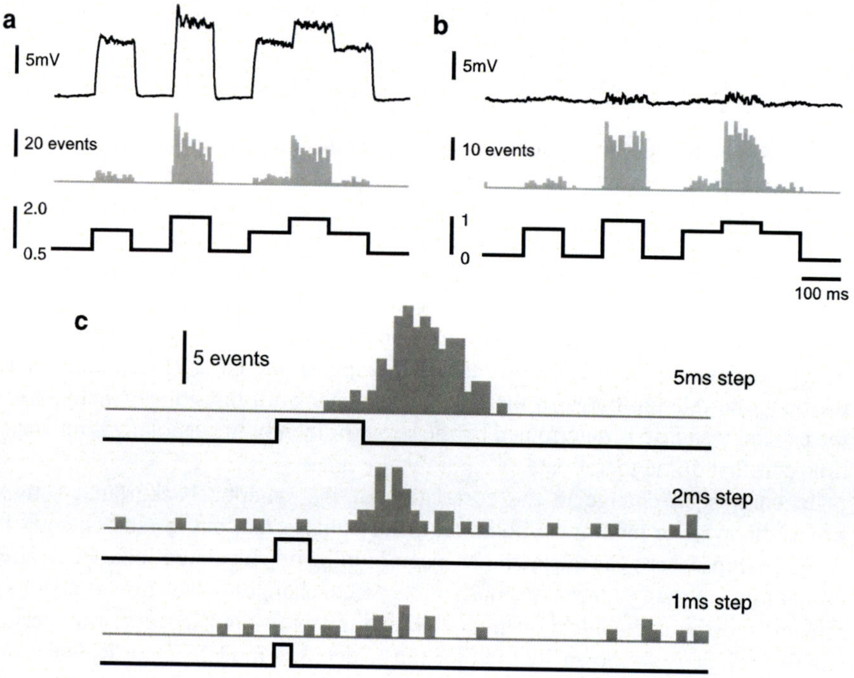

Fig. 5.26 Correlation detection in model pyramidal neurons. (**a**) Step-like changes in the amplitude of the background activity, caused by altering the excitatory frequency ν_{exc} (*bottom trace*), led to changes in firing activity (*middle trace*) but, in contrast to the correlation case, to a significant effect on the average membrane voltage (*top trace*). (**b**) Step-like changes of correlation (*bottom trace*) induced immediate changes in firing activity (*middle trace*) while the effect on average membrane potential was minimal (*top trace*). (**c**) Correlation detection can occur within timescales comparable to that of single action potentials. Brief changes in correlation (between $c = 0.0$ and $c = 0.7$) were applied by using steps of different durations. Although these steps caused negligible changes in the membrane potential, they led to a clear increase in the number of fired spikes down to steps of 2 ms duration. In all cases, the response started within 5 ms after the onset of the step. Modified from Rudolph and Destexhe (2001a)

5.7.1 *First Indication that Synaptic Noise Reduces Location Dependence*

Several computational studies examined the dendritic attenuation of EPSPs in the presence of voltage-dependent Na^+ and K^+ dendritic conductances . In a detailed biophysical model by Destexhe and Paré (1999), using a stimulation paradigm similar to that in Fig. 5.4b, proximal or distal synapses reliably evoke a cellular response in quiescent conditions (Fig. 5.27a, *Quiet*). It was observed that the stimulation of distal synapses elicits dendritic APs that propagate toward the soma, in agreement with an earlier model by the same authors (Paré et al. 1998a). However, during active periods (Fig. 5.27a, *Active*), proximal or distal stimuli do not trigger spikes reliably, although the clustering of APs near the time of the stimulation (∗)

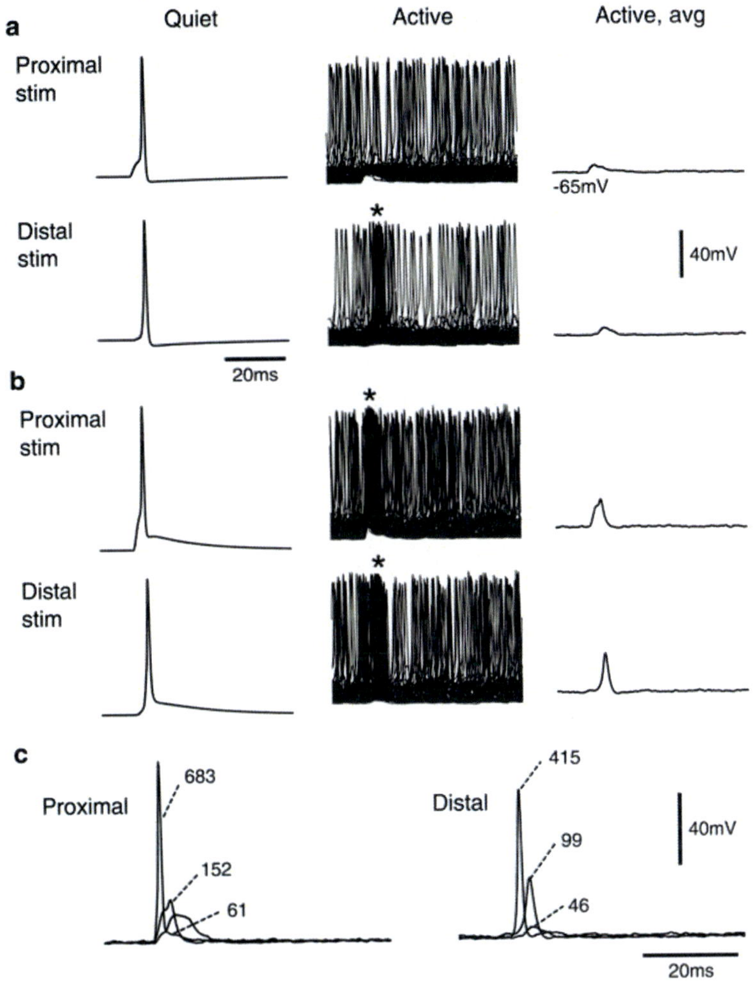

Fig. 5.27 Attenuation of EPSPs in the presence of voltage-dependent conductances. (**a**) The same stimulation paradigm as in Fig. 5.4b was performed in the presence of Na^+ and K^+ currents inserted in axon, soma, and dendrites. Excitatory synapses were synchronously activated in basal ($n = 81$) and distal dendrites ($n = 46$; $>200\,\mu m$ from soma). In the absence of spontaneous synaptic activity (*Quiet*), these stimuli reliably evoked action potentials. During simulated active periods (*Active*; 100 traces shown), the EPSP influenced action potential generation, as shown by the tendency of spikes to cluster for distal stimuli ($*$) but not for proximal stimuli. The average responses (*Active, avg*; $n = 1{,}000$) show that action potentials were not precisely timed with the EPSP. (**b**) With larger numbers of activated synapses (152 proximal, 99 distal), spike clustering was more pronounced ($*$) and the average clearly shows spike-related components. (**c**) Average response obtained with increasing numbers of synchronously activated synapses. Several hundreds of synapses were necessary to elicit spikes reliably. Modified from Destexhe and Paré (1999)

shows that EPSPs affect the firing probability. Further analysis of this behavior reveals that the evoked response, averaged from 1,000 sweeps under intense synaptic activity (Fig. 5.27a, *Active, avg*), shows similar amplitude for proximal or distal inputs. Average responses do not reveal any spiky waveform, indicating that APs are not precisely timed with the EPSP in both cases.

It is interesting to note that, in Fig. 5.27a, distal stimuli evoke AP clustering whereas proximal stimuli do not, despite the fact that a larger number of proximal synapses are activated. Distal stimuli evoke dendritic APs, some of which reach the soma and lead to the observed cluster. Increasing the number of simultaneously activated excitatory synapses enhances spike clustering for both proximal and distal stimuli (Fig. 5.27b, $*$). This observation is also evidenced by the spiky components in the average EPSP (Fig. 5.27b, *Active, avg*). Comparison of responses evoked by different numbers of activated synapses (Fig. 5.27c) shows that the convergence of several hundred excitatory synapses is necessary to evoke spikes reliably during intense synaptic activity. It is remarkable that with active dendrites, similar conditions of convergence are required for proximal or distal inputs, in sharp contrast to the case with passive dendrites, in which there is a marked difference between proximal and distal inputs (Fig. 5.27c).

The magnitude of the currents active at rest may potentially influence these results. In particular, the significant rectification present at levels more depolarized than -60 mV may affect the attenuation of depolarizing events. To investigate this aspect, Destexhe and Paré estimated the conditions of synaptic convergence using different distributions of leak conductances (Destexhe and Paré 1999): (a) although suppressing the I_{M} conductance enhanced the excitability of the cell (see above), it does not affect the convergence requirements in conditions of intense synaptic activity (compare Fig. 5.28a and Fig. 5.28b); (b) using a different set of passive parameters based on whole-cell recordings, with a low axial resistance and a high membrane resistivity (Pongracz et al. 1991; Spruston and Johnston 1992), also gives similar results (Fig. 5.28c); (c) using a nonuniform distribution of leak conductance with strong leak in distal dendrites (Stuart and Spruston 1998) also leads to similar convergence requirements (Fig. 5.28d). In addition, even the presence of a 10 nS electrode shunt in soma with a larger membrane resistivity in dendrites leads to nearly identical results. This shows that, under conditions of intense synaptic activity, synaptic currents account for most of the cell's input conductance, while intrinsic leak and voltage-dependent conductances have a comparatively small contribution. It also suggests that hundreds of synaptic inputs are required to fire the neuron reliably, and that this requirement seems independent of the location of the synaptic inputs.

Similar results are obtained when subdividing the dendritic tree into three regions, proximal, middle, and distal, as shown in Fig. 5.29. The activation of the same number of excitatory synapses distributed in either of the three regions leads to markedly different responses in a quiescent neuron, whereas they give similar responses in a simulated active state (Fig. 5.29b). The computed full response functions are also almost superimposable (Fig. 5.30), suggesting that the three dendritic regions seem equivalent with respect to their efficacy on AP generation, but that this is true only in the presence of in vivo-like synaptic noise.

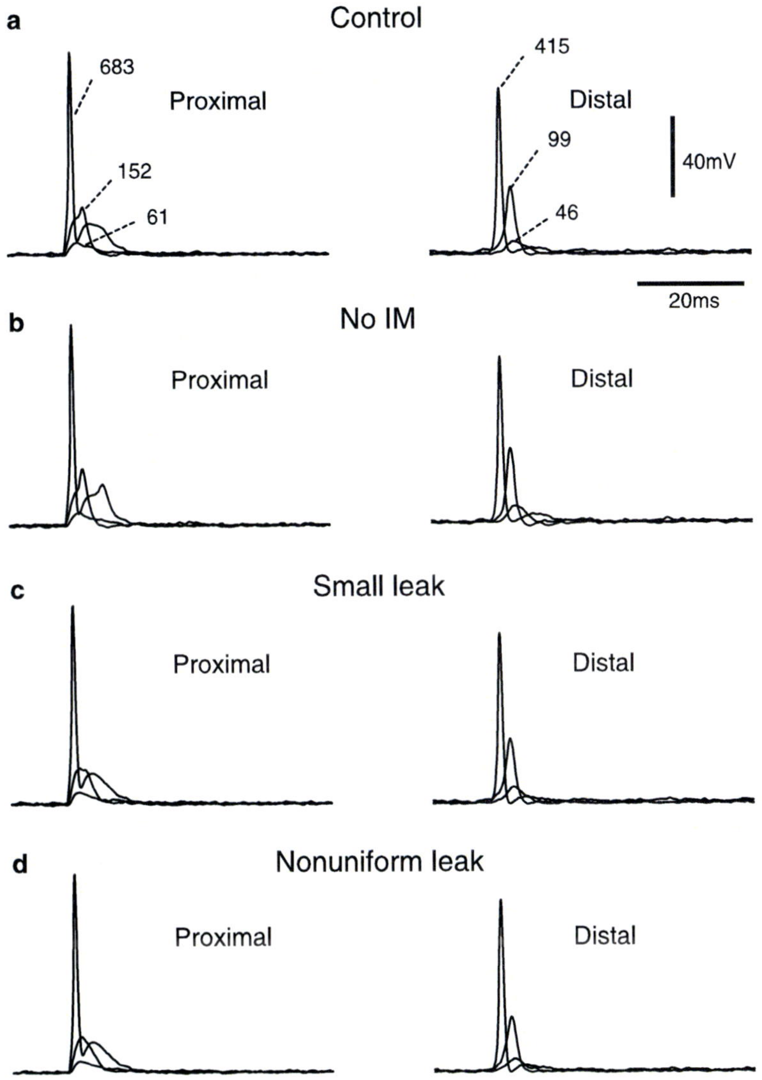

Fig. 5.28 Synaptic bombardment minimizes the variability due to input location. Average responses to synchronized synaptic stimulation are compared for proximal and distal regions of the dendritic arbor. The same stimulation paradigm as in Fig. 5.27 was repeated for different combination of resting conductances. (**a**) Control: average response obtained with increasing numbers of synchronously activated synapses (identical simulation as Fig. 5.27c). (**b**) Same simulation with I_M removed. (**c**) Same simulation as in (**a**) but with a lower axial resistance (100 Ωcm) and three times lower leak conductance ($g_L = 0.015\,\text{mS}\,\text{cm}^{-2}$). (**d**) Same simulation as in (**a**) with high leak conductance nonuniformly distributed, and low axial resistance (80 Ωcm). Modified from Destexhe and Paré (1999)

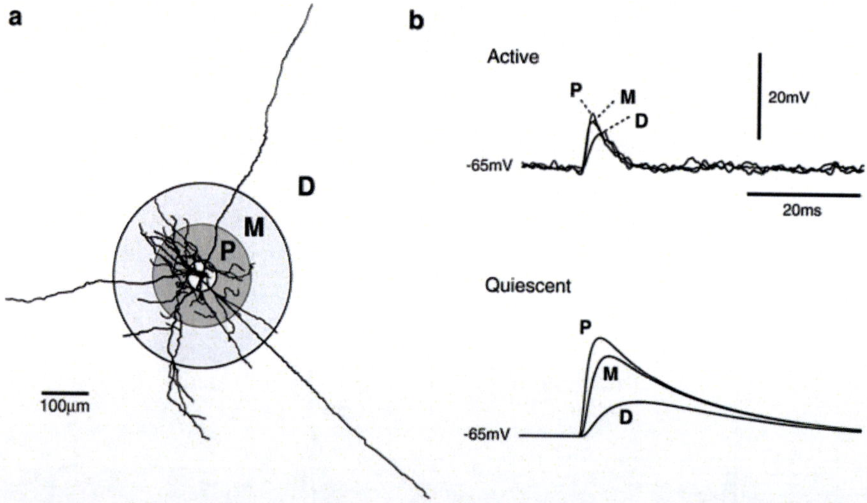

Fig. 5.29 Synaptic inputs are independent on dendritic location in the presence of synaptic background activity. (**a**) Subdivision of the layer VI cell into three regions (*P*, *proximal*: from 40 to 131 μm from soma; *M*, *middle*: from 131 to 236 μm; *D*, *distal*: >236 μm) of roughly equivalent total membrane area. 83 excitatory synapses were synchronously activated in each of these three regions. (**b**) Responses to synaptic stimulation. *Active*: average response computed over 1,000 trials in the presence of background activity. *Quiescent*: response to the same stimuli obtained in the absence of background activity at the same membrane potential (−65 mV)

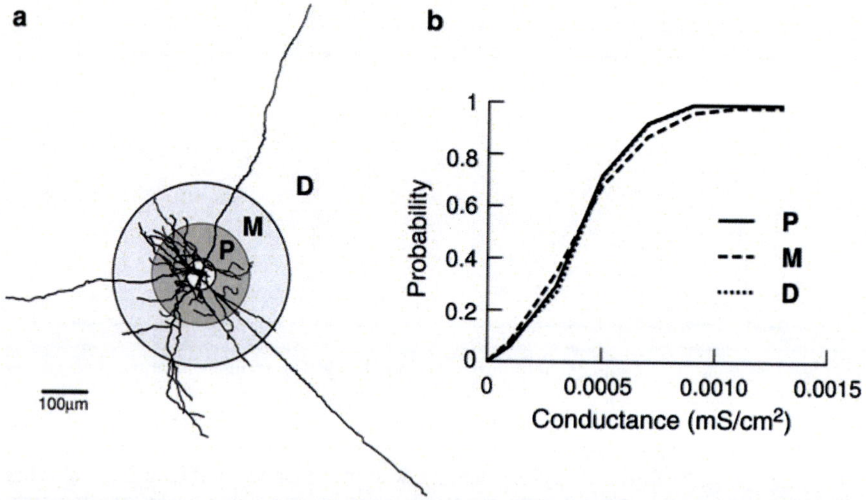

Fig. 5.30 Location independence of the whole spectrum of response probability to synaptic stimulation. The same protocol as in Fig. 5.29 was followed for stimulation, but the total (cumulated) probability of evoking a spike was computed for different input amplitudes in the three different regions in (**a**) (same description as in Fig. 5.29). Panel (**b**) shows that the response functions obtained are nearly superimposable, which demonstrates that the whole spectrum of response to synaptic stimulation does not depend on the region considered in the dendritic tree

5.7.2 *Location Dependence of Synaptic Inputs*

In this section, we explicitly investigate the effect of single, focused dendritic locations in the presence of synaptic background activity. We start by showing that background activity induces a stochastic dynamics which affects dendritic action potential initiation and propagation. We next investigate the impact of individual synapses at the soma in this stochastic state, as well as how synaptic efficacy is modulated by different factors such as morphology and the intensity of background activity itself. Finally, we present how this stochastic state affects the timing of synaptic events as a function of their position in the dendrites.

5.7.2.1 A Stochastic State with Facilitated Action Potential Initiation

Since dendrites are excitable, it is important to first determine how synaptic background activity affects the dynamics of AP initiation and propagation in dendrites. Dendritic AP propagation can be simulated in computational models of morphologically reconstructed cortical pyramidal neurons which included voltage-dependent currents in soma, dendrites, and axon (Fig. 5.31a, top). In quiescent conditions, backpropagating dendritic APs are reliable up to a few hundred microns from the soma (Fig. 5.31a, bottom, *Quiescent*), in agreement with dual soma/dendrite recordings in vitro (Stuart and Sakmann 1994; Stuart et al. 1997b). In the presence of synaptic background activity, backpropagating APs are still robust, but propagate over a more limited distance in the apical dendrite compared to quiescent states (Fig. 5.31a, bottom, In vivo-like), consistent with the limited backwards invasion of apical dendrites observed with two-photon imaging of cortical neurons in vivo (Svoboda et al. 1997).

APs can also be initiated in dendrites following simulated synaptic stimuli. In quiescent conditions, the threshold for dendritic AP initiation is usually high (Fig. 5.31b, left, *Quiescent*), and the dendritic-initiated APs propagate forward only over limited distances (100–200 μm; Fig. 5.31c, *Quiescent*), in agreement with recent observations (Stuart et al. 1997a; Golding and Spruston 1998; Vetter et al. 2001). Interestingly, background activity tends here to facilitate forward-propagating APs. Dendritic AP initiation is highly stochastic due to the presence of random fluctuations, but computing the probability of AP initiation reveals a significant effect of background activity (Fig. 5.31b, left, *In vivo-like*). The propagation of initiated APs is also stochastic, but it was found that a significant fraction (see below) of dendritic APs can propagate forward over large distances and reach the soma (Fig. 5.31c, *In vivo-like*), a situation which usually does not occur in quiescent states with low densities of Na^+ channels in dendrites.

To further explore this surprising effect of background activity on dendritic APs, one can compare different background activities with equivalent conductance but different amplitudes of voltage fluctuations. Figure 5.31b (right) shows that the probability of AP initiation, for fixed stimulation amplitude and path distance, is zero in the absence of fluctuations, but steadily rises for increasing fluctuation

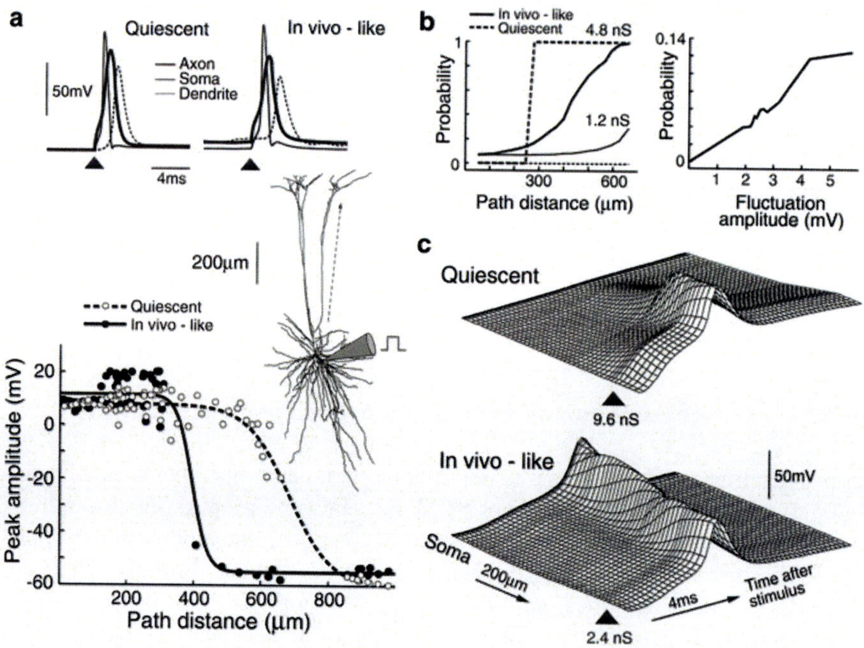

Fig. 5.31 Dendritic action potential initiation and propagation under in vivo-like activity. (**a**) Impact of background activity on action potential (AP) backpropagation in a layer V cortical pyramidal neuron. *Top*: the respective timing of APs in soma, dendrite (300 μm from soma), and axon is shown following somatic current injection (*arrow*). *Bottom*: backpropagation of the AP in the apical dendrite for quiescent (*open circles*) and in vivo-like (*filled circles*) conditions. The backwards invasion was more restricted in the latter case. (**b**) Impact of background activity on dendritic AP initiation. *Left*: probability for initiating a dendritic AP shown as a function of path distance from soma for two different amplitudes of AMPA-mediated synaptic stimuli (*thick line*: 4.8 nS; *thin line*: 1.2 nS). *Right*: probability of dendritic AP initiation (100 μm from soma) as a function of the amplitude of voltage fluctuations (1.2 nS stimulus). (**c**) Impact of background activity on dendritic AP propagation. A forward-propagating dendritic AP was evoked in a distal dendrite by an AMPA-mediated EPSP (*arrow*). *Top*: in quiescent conditions, this AP only propagated within 100–200 μm, even for high-amplitude stimuli (9.6 nS shown here). *Bottom*: under in vivo-like conditions, dendritic APs could propagate up to the soma, even for small stimulus amplitudes (2.4 nS shown here). (**b**) and (**c**) were obtained using the layer VI pyramidal cell shown in Fig. 5.32a. Modified from Rudolph and Destexhe (2003b)

amplitudes (and equivalent base-line membrane potential in the different states). This shows that subthreshold stimuli are occasionally boosted by depolarizing fluctuations. Propagating APs can also benefit from this boosting to help their propagation all the way up to the soma. In this case, the AP itself must be viewed as the stimulus which is boosted by the presence of depolarizing fluctuations. The same picture is observed for different morphologies, passive properties and for various densities and kinetics of voltage-dependent currents (see below): in vivo-like activity induced a stochastic dynamics in which backpropagating APs are minimally affected, but forward-propagating APs are facilitated. Thus, under in vivo-like conditions, subthreshold EPSPs can be occasionally boosted by depolarizing fluctuations, and have a chance to initiate a dendritic AP, which itself has a chance to propagate and reach the soma.

5.7.2.2 Location Independence of the Impact of Individual or Multiple Synapses

To evaluate quantitatively the consequences of this stochastic dynamics of dendritic AP initiation, the impact of individual EPSPs at the soma was investigated (Rudolph and Destexhe 2003b). In quiescent conditions, with a model adjusted to the passive parameters estimated from whole-cell recordings in vitro (Stuart and Spruston 1998), a relatively moderate passive voltage attenuation (25–45% attenuation for distal events) is observed (see Fig. 5.3a, *Quiescent*). Taking into account the high conductance and more depolarized conditions of in vivo-like activity shows a marked increase in voltage attenuation (80–90% attenuation; see Fig. 5.3a, *in vivo-like*). Computing the EPSP peak amplitude in these conditions further reveals an attenuation with distance (Fig. 5.3b, lower panel), which is more pronounced if background activity is represented by an equivalent static (leak) conductance. Thus, the high-conductance component of background activity enhances the location-dependent impact of EPSPs, and leads to a stronger individualization of the different dendritic branches (London and Segev 2001; Rhodes and Llinás 2001).

A radically different conclusion is reached if voltage fluctuations are taken into account. In this case, responses are highly irregular and the impact of individual synapses can be assessed by computing the poststimulus time histogram (PSTH) over long periods of time with repeated stimulation of single or groups of colocalized excitatory synapses. The PSTHs obtained for stimuli occurring at different distances from the soma (Fig. 5.32a) show that the "efficacy" of these synapses is roughly location independent, as calculated from either the peak (Fig. 5.32b) or the integral of the PSTH (Fig. 5.32c). The latter can be interpreted as the probability that a somatic spike is specifically evoked by a synaptic stimulus. Using this measure of synaptic efficacy, one can conclude that, under in vivo-like conditions, the impact of individual synapses on the soma is nearly independent on their dendritic location, despite a severe voltage attenuation.

5.7.2.3 Mechanisms Underlying Location Independence

To show that this location-independent mode depends on forward-propagating dendritic APs, one can select, for a given synaptic location, all trials which evoke a somatic spike. These trials represented a small portion of all trials. In the model investigated in Rudolph and Destexhe (2003b), this portion ranged from 0.4 to 4.5%, depending on the location and the strength of the synaptic stimuli. For these "successful" selected trials, it was found that the somatic spike is always preceded by a dendritic spike evoked locally by the stimulus. In the remaining "unsuccessful"

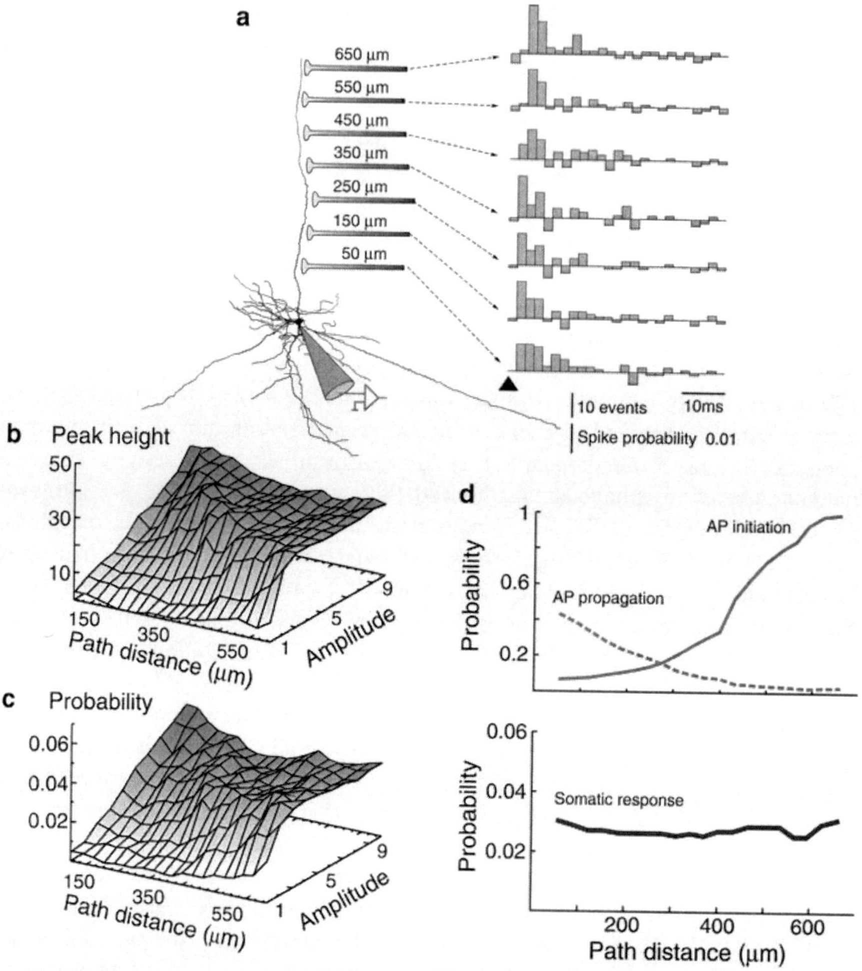

Fig. 5.32 Independence of the somatic response to the location of synaptic stimulation under in vivo-like conditions. (**a**) Poststimulus time histograms (PSTHs) of responses to identical AMPA-mediated synaptic stimuli (12 nS) at different dendritic locations (cumulated over 1,200 trials after subtraction of spikes due to background activity). (**b**) Peak of the PSTH as a function of stimulus amplitude (from 1 to 10 co-activated AMPA synapses; conductance range: 1.2 to 12 nS) and distance to soma. (**c**) Integrated PSTH (probability that a somatic spike was specifically evoked by the stimulus) as a function of stimulus amplitude and distance to soma. Both (**b**) and (**c**) show reduced location dependence. (**d**) *Top*: comparison of the probability of evoking a dendritic spike (*AP initiation*) and the probability that an evoked dendritic spike translated into a somatic/axonal spike (*AP propagation*). Both were represented as a function of the location of the stimulus (AMPA-mediated stimulus amplitudes of 4.8 nS). *Bottom*: probability of somatic spike specifically evoked by the stimulus, which was obtained by multiplying the two curves above. This probability was nearly location independent. Modified from Rudolph and Destexhe (2003b)

trials, there is a proportion of stimuli (55–97%) which evokes a dendritic spike but fails to evoke somatic spiking. This picture is the same for different stimulation sites: a fraction of stimuli evokes dendritic spikes and a small fraction of these dendritic spikes successfully evokes a spike at the soma/axon.

The latter aspect can be further analyzed by representing the probabilities of initiation and propagation along the distance axis (Fig. 5.32d). There is an asymmetry between these two measures: the chance of evoking a dendritic AP is lower for proximal stimuli and increased with distance (Fig. 5.32d, *AP initiation*), because the local input resistance varies inversely with dendrite diameter, and is higher for thin (distal) dendritic segments. On the other hand, the chance that a dendritic AP propagates down to the soma, and leads to soma/axon APs, is higher for proximal sites and gradually decreased with distance (Fig. 5.32d, *AP propagation*). Remarkably, these two effects compensate such that the probability of evoking a soma/axon AP (the product of these two probabilities) is approximately independent on the distance to soma (Fig. 5.32d, *somatic response*). This effect is typically observed only in the presence of conductance-based background activity and is not present in quiescent conditions or by using current-based models of synapses. Thus, these results show that the location-independent impact of synaptic events under in vivo-like conditions is due to a compensation between an opposite distance dependence of the probabilities of AP initiation and propagation.

It was shown in Rudolph and Destexhe (2003b) that the same dynamics is present in various pyramidal cells (Fig. 5.33) suggesting that this principle may apply to a large variety of dendritic morphologies. It was also found to be robust to variations in ion channel densities and kinetics, such as NMDA conductances (Fig. 5.34a), passive properties (Fig. 5.34b), and different types of ion channels (Fig. 5.34c), including high distal densities of leak and hyperpolarization-activated I_h conductances (Fig. 5.34c, *gray line*). In the latter case, the presence of I_h affects EPSPs in the perisomatic region, in which there is a significant contribution of passive signaling, but synaptic efficacy is still remarkably location independent for the remaining part of the dendrites where the I_h density was highest. Location independence is also robust to changes in membrane excitability (Fig. 5.35a, b) and shifts in the Na^+ current inactivation (Fig. 5.35c). Most of these variations change the absolute probability of evoking spikes, but do not affect the location independence induced by background activity. The location-independent synaptic efficacy is, however, lost when the dendrites have too strong K^+ conductances, either with high I_{KA} in distal dendrites (Fig. 5.34c, *dotted line*), or with a high ratio between K^+ and Na^+ conductances (Fig. 5.35b). In other cases, synaptic efficacy is larger for distal dendrites (see Fig. 5.35a, *high excitability*, and Fig. 5.35c, *inactivation shift* $= 0$).

5.7.2.4 Activity-Dependent Modulation of Synaptic Efficacy

To determine how the efficacy of individual synapses varies as a function of the intensity of synaptic background activity, the same stimulation paradigms as used in Fig. 5.32 can be repeated, but by varying individually the release rates of excitatory

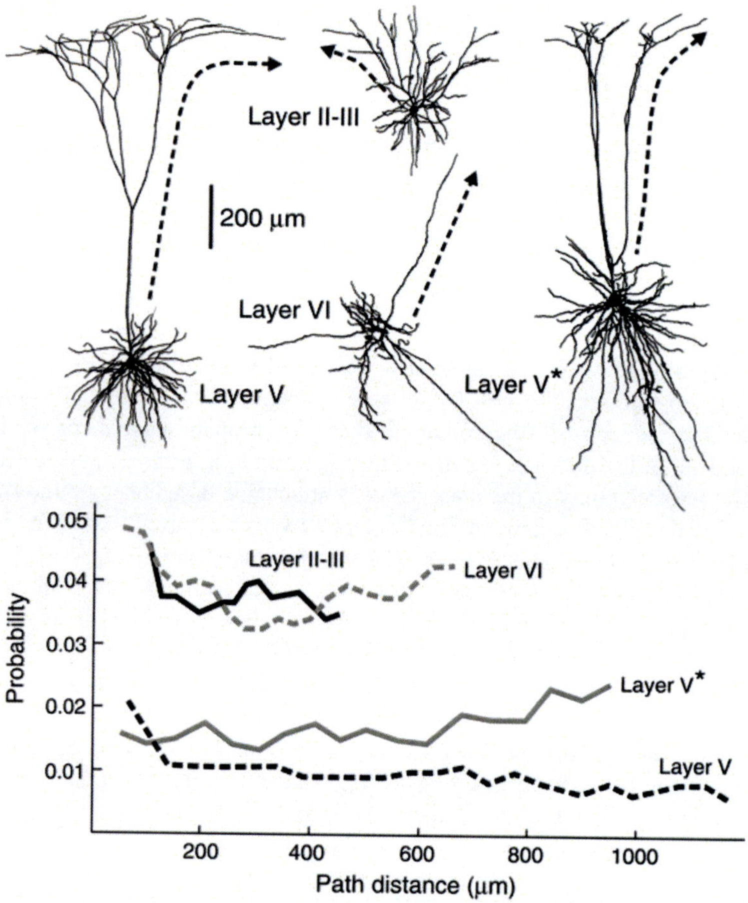

Fig. 5.33 Location-independent impact of synaptic inputs for different cellular morphologies. The somatic response to AMPA stimulation (12 nS amplitude) is indicated for different dendritic sites (corresponding branches are indicated by *dashed arrows*; equivalent electrophysiological parameters and procedures as in Fig. 5.32) for four different cells (one layer II-III, two layer V and one layer VI), based on cellular reconstructions from cat cortex (Douglas et al. 1991; Contreras et al. 1997). Somatic responses (integrated PSTH) are represented against the path distance of the stimulation sites. In all cases, the integrated PSTH shows location independence, but the averaged synaptic efficacy was different for each cell type. Modified from Rudolph and Destexhe (2003b)

(Fig. 5.36a) or inhibitory (Fig. 5.36b) inputs of the background, by varying both (Fig. 5.36c) or by varying the correlation with fixed release rates (Fig. 5.36d). In all cases, the synaptic efficacy (integrated PSTH for stimuli which are subthreshold under quiescent conditions) depends on the particular properties of background activity, but remains location independent. Moreover, in the case of "balanced" excitatory and inhibitory inputs (Fig. 5.36c), background activity can be changed continuously from quiescent to in vivo-like conditions. In this case, the probability

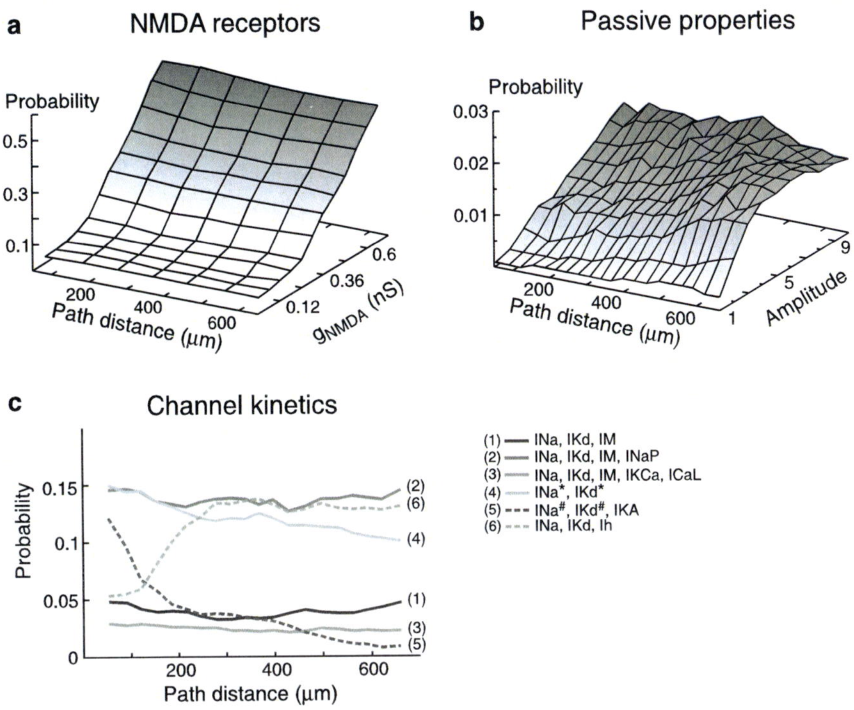

Fig. 5.34 Location independence for various passive and active properties. (**a**) Synaptic efficacy as a function of path distance and conductance of NMDA receptors. The quantal conductance (g_{NMDA}) was varied between 0 and 0.7 nS, which corresponds to a fraction of 0 to about 60% of the conductance of AMPA channels (Zhang and Trussell 1994; Spruston et al. 1995). NMDA receptors were colocalized with AMPA receptors (release frequency of 1 Hz) and stimulation amplitude was 12 nS. (**b**) Synaptic efficacy as a function of path distance and stimulation amplitude for a nonuniform passive model (Stuart and Spruston 1998). (**c**) Synaptic efficacy as a function of path distance for different ion channel models or different kinetic models of the same ion channels (stimulation: 12 nS). Simulations were done using the Layer VI cell in which AMPA-mediated synaptic stimuli were applied at different sites along the dendritic branch indicated by a *dashed arrow* in Fig. 5.33. Modified from Rudolph and Destexhe (2003b)

steadily rises from zero (Fig. 5.36c, *clear region*), showing that subthreshold stimuli can evoke detectable responses in the presence of background activity, and reaches a "plateau" where synaptic efficacy is independent of both synapse location and background intensity (Fig. 5.36c, *dark region*). This region corresponds to estimates of background activity based on intracellular recordings in vivo (Destexhe and Paré 1999). Thus, it seems that synaptic inputs are location independent for a wide range of background activities and intensities. Modulating the correlation, or the respective weight of excitation and inhibition, allows the network to globally modulate the efficacy of all synaptic sites at once.

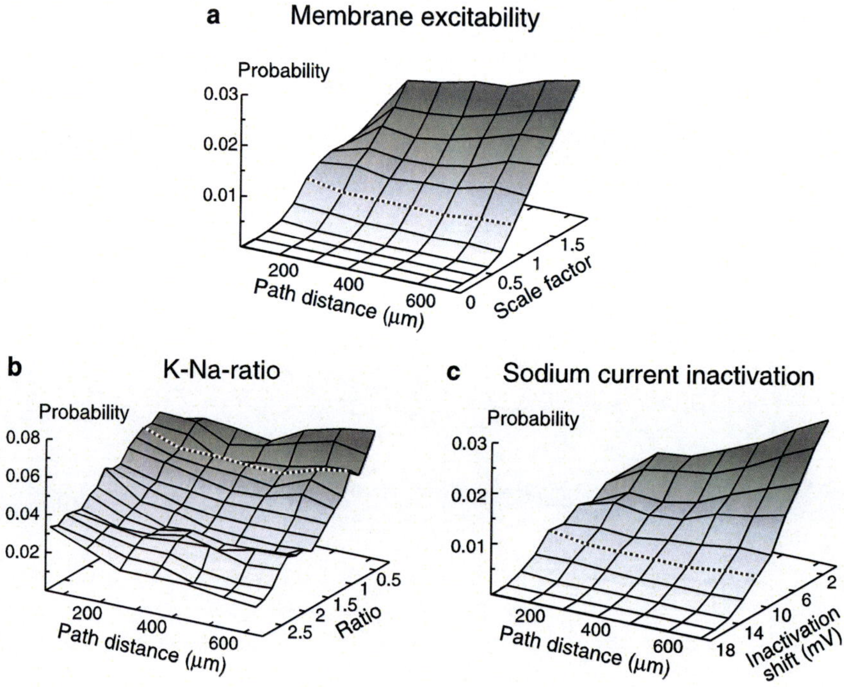

Fig. 5.35 Location-independence for various active properties. (**a**) Synaptic efficacy as a function of path distance and membrane excitability. Both Na^+ and K^+ conductance densities were changed by a common multiplicative scaling factor. The *dotted line* indicates a dendritic conductance density of 8.4 mS/cm^2 for the Na^+ current, and 7 mS/cm^2 for the delayed rectifier K^+ current. The stimulation amplitude was in all cases 12 nS. (**b**) Synaptic efficacy obtained by changing the ratio between Na^+ and K^+ conductances responsible for action potentials (conductance density of 8.4 mS/cm^2 for I_{Na}; the *dotted line* indicates 7 mS/cm^2 for I_{Kd}). (**c**) Synaptic efficacy as a function of path distance obtained by varying the steady-state inactivation of the fast Na^+ current. The inactivation curve was shifted with respect to the original model (Traub and Miles 1991) toward hyperpolarized values (stimulation amplitude: 12 nS). The *dotted line* indicates a 10 mV shift, which approximately matches the voltage clamp data of cortical pyramidal cells (Huguenard et al. 1988). All simulations were done using the layer VI cell in which AMPA-mediated synaptic stimuli were applied at different sites along the dendritic branch indicated by a *dashed arrow* in Fig. 5.33. Modified from Rudolph and Destexhe (2003b)

5.7.2.5 Location Dependence of the Timing of Synaptic Events

Another aspect of location dependence is the timing aspects, because it is well known that significant delays can result from dendritic filtering. Figure 5.37a illustrates the somatic membrane potential following synaptic stimuli at different locations. In quiescent conditions, as predicted by cable theory (Segev et al. 1995; Koch 1999), proximal synaptic events lead to fast rising and fast decaying somatic EPSPs, whereas distal events are attenuated in amplitude and slowed in duration (Fig. 5.37a, *Quiescent*). The time-to-peak of EPSPs increases monotonically with

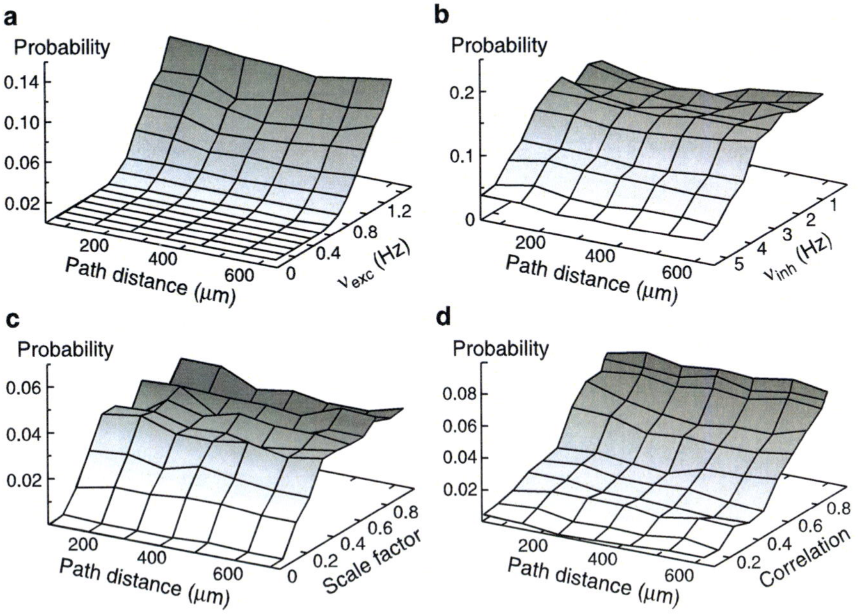

Fig. 5.36 Modulation of synaptic efficacy by background activity. (**a**) Integrated PSTH obtained for different intensities of background activity obtained by varying the release rates at glutamatergic synapses (ν_{exc}) while keeping the release rates fixed at GABAergic synapses ($\nu_{inh} = 5.5$ Hz). (**b**) Integrated PSTH obtained by varying the release rates at inhibitory synapses (ν_{inh}) with fixed excitatory release rates ($\nu_{exc} = 1$ Hz). (**c**) Integrated PSTH obtained by varying both excitatory and inhibitory release rates, using the same scaling factor. The plateau region (*dark*) shows that the global efficacy of synapses, and their location independence, are robust to changes in the intensity of network activity. (**d**) Integrated PSTH obtained for fixed release rates but different background correlations. In all cases, the integrated PSTHs represent the probability that a spike was specifically evoked by synaptic stimuli (12 nS, AMPA-mediated), as in Fig. 5.32d. Modified from Rudolph and Destexhe (2003b)

distance (Fig. 5.37b, *Quiescent*). In the presence of background activity, the average amplitude of these voltage deflections is much less dependent on location (Fig. 5.37a, *In vivo-like*), consistent with the PSTHs in Fig. 5.32b, and the time-to-peak of these events is only weakly dependent on the location of the synapses in dendrites (Fig. 5.37b, *In vivo-like*). These observations suggest that *in vivo-like* conditions set the dendrites into a fast-conducting mode, in which the timing of synaptic inputs shows little dependence on their distance to soma.

The basis of this fast-conducting mode can be investigated by simulating the same paradigm while varying a number of parameters. First, to check if this effect is attributable to the decreased membrane time constant due to the high-conductance imposed by synaptic background activity, the latter can be replaced by an equivalent static conductance. This leads to an intermediate location-dependent relation (Fig. 5.37c, *Quiescent, static conductance*), in between the quiescent and

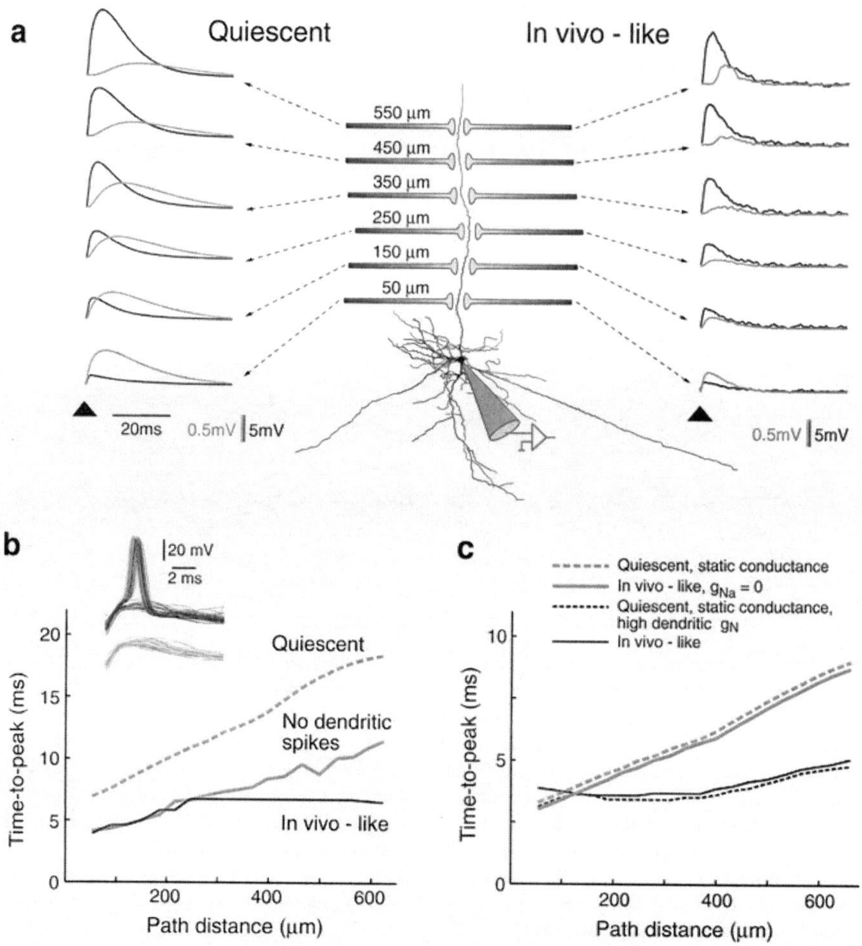

Fig. 5.37 Fast conduction of dendrites under in-vivo–like conditions. (**a**) Somatic (*black*) and dendritic (*gray*) voltage deflections following stimuli at different locations (somatic responses are shown with a magnification of 10x). There was a reduction of the location dependence at the soma under in vivo-like conditions (averages over 1,200 traces) compared to the quiescent state (all stimuli were 1.2 nS, AMPA-mediated). (**b**) Location-dependence of the timing of EPSPs. In the quiescent state, the time-to-peak of EPSPs increased approximately linearly with the distance to soma (*Quiescent*). This dependence on location was markedly reduced under in vivo-like conditions (*In vivo-like*), defining a fast-conducting state of the dendrites. This location dependence was affected by removing dendritic APs (*no dendritic spikes*). *Inset*: examples of dendritic EPSPs at the site of the synaptic stimulation (50 traces, stimulation with 8.4 nS at 300 μm from soma) are shown under in vivo-like conditions (*black*) and after dendritic APs were removed (*gray*). (**c**) Mechanism underlying fast dendritic conduction. Replacing background activity by an equivalent static conductance (*Quiescent static conductance*), or suppressing dendritic Na^+ channels (*In vivo-like*, $g_{Na} = 0$) led to an intermediate location dependence of EPSP time-to-peak. On the other hand, using high dendritic excitability together with strong synaptic stimuli (12 nS) evoked reliable dendritic APs and yielded a reduced location dependence of the time-to-peak in quiescent conditions (*Quiescent, static conductance, high dendritic* g_{Na}), comparable to in vivo-like conditions. The fast-conducting mode is, therefore, due to forward-propagating dendritic APs in dendrites of fast time constant. Modified from Rudolph and Destexhe (2003b)

in vivo-like cases. The reduced time constant, therefore, can account for some, but not all, of the diminished location dependence of the timing. Second, to check for contributions of dendritic Na^+ channels, the same stimulation protocol can be used under in vivo-like conditions, but by selectively removing Na^+ channels from dendrites. This also leads to an intermediate location dependence (Fig. 5.37c, *In vivo-like*, $g_{Na} = 0$), suggesting that Na^+-dependent mechanisms underlie the further reduction of timing beyond the high-conductance effect. Finally, to show that this further reduction is due to dendritic APs, a quiescent state with equivalent static conductance, but higher dendritic excitability (twice larger Na^+ and K^+ conductances), can be used, such that strong synaptic stimuli are able to evoke reliable forward-propagating dendritic APs. In this case only, the reduced location dependence of the timing can be fully reconstructed (Fig. 5.37c, *Quiescent, static conductance, high dendritic* g_{Na}). The dependence on dendritic APs is also confirmed by the intermediate location dependence obtained when EPSPs are constructed from trials devoid of dendritic APs (Fig. 5.37b, *no dendritic spikes*). This analysis shows that the fast-conducting mode is due to forward-propagating APs in dendrites of fast time constant.

Finally, it is important to note that the location independence properties described here only apply to cortical neurons endowed with the "classic" dendritic excitability, mediated by Na^+ and K^+ currents. Some specific classes of cortical neurons, such as intrinsically bursting cells, or thick-tufted Layer 5 pyramidal cells, are characterized by a dendritic initiation zone for calcium spikes (Amitai et al. 1993). Such calcium spikes can heavily influence the dendritic integration properties of these cells (Larkum et al. 1999, 2009), as modeled recently (Hay et al. 2001). However, how such dendritic calcium spikes interact with in vivo levels of background activity is presently unknown, and constitutes a possible extension of the present work.

5.8 Consequences on Integration Mode

For a long time, the discussion about the neural code utilized in biological neural systems was dominated by the question of whether individual neurons encode and process information by using precise spike timings, thus, working as coincidence detectors, or spike rates, thus, working as temporal integrators (Softky and Koch 1993; Shadlen and Newsome 1994, 1995; Softky 1995; Shadlen and Newsome 1998; Panzeri et al. 1999; Koch and Laurent 1999; Segundo 2000; Lábos 2000; Panzeri et al. 2001, for a review of original work see deCharms and Zador 2000). It has been argued both based on experimental studies (Smith and Smith 1965; Noda and Adey 1970; Softky and Koch 1993; Holt et al. 1996; Stevens and Zador 1998; Shinomoto et al. 1999) and through numerical investigations (Usher et al. 1994; Tsodyks and Sejnowski 1995; van Vreeswijk and Sompolinsky 1996; Troyer and Miller 1997) that the irregular firing activity of at least cortical neurons is inconsistent with the temporal integration of synaptic inputs, and that coincidence detection is the preferred operating mode of cortical neurons.

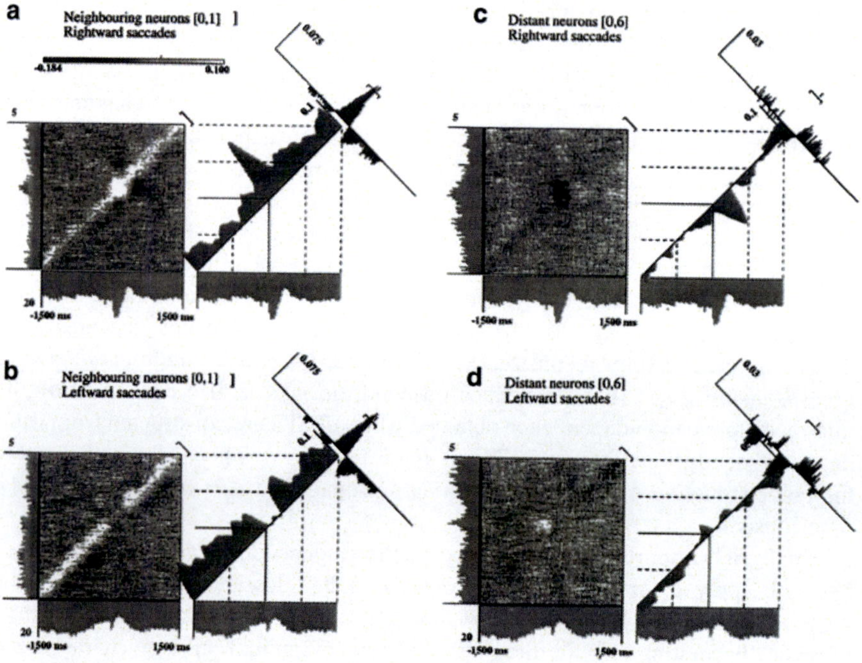

Fig. 5.38 Dynamic modification of correlated firing between neurons in the frontal eye field in relation to the onset of saccadic eye movements. (**a**, **b**), JPSTHs for neurons (0, 1), recorded by one microelectrode ("neighboring" neurons). (**c**, **d**), JPSTHs for neurons (0, 6). Neuron 0 is from the first pair, neuron 6 was recorded by another electrode ("distant" neurons). The following features are apparent. First, the averaged cross-correlograms for each pair are very similar. However, the correlation dynamics are temporally linked to the saccades and depend strongly on their direction, as shown by the matrix and the coincidence-time histograms ((**a**) compared with (**b**) and (**c**) compared with (**d**)). Second, the time-averaged correlation between neighboring neurons (**a**, **b**) is positive (i.e., the probability that either of the neurons will fire a spike is higher around the times the other neuron fires), whereas the correlation between distant neurons (**c**, **d**) is negative. Third, the temporal changes of correlation could not be predicted from the firing rates of the two neurons. The correlation either increased (**a**) or decreased (**c**) near the time of saccade initiation, whereas the firing rates of both neurons increased around the onset of saccades, regardless of its direction. The normalization and format of the JPSTHs are the same as in Fig. 2. Bin size, 30 ms. The JPSTHs around onsets of rightward saccades (**a**, **c**) were constructed from 776 saccade, 33,882 spikes of neuron 0 (**a**, **c**), 4,299 spikes of neuron 1 (**a**) and 6,927 spikes of neuron 6 (**c**). The JPSTHs in (**b**) and (**d**) were constructed from 734 saccades, 32,621 spikes of neuron 0 (**b**, **d**), 4,167 spikes of neuron 1 (**b**) and 5,992 spikes of neuron 6 (**d**). Modified from Vaadia et al. (1995)

Experimental evidence for the functional role of cortical neurons as coincidence detectors was provided by a number of researchers through studies of the cat visual cortex (Gray 1994; König et al. 1995) and the monkey frontal cortex (Vaadia et al. 1995). The latter study, for instance, showed that the discharge activity of cortical neurons recorded simultaneously exhibits rapid correlations linked to behavioral events (Fig. 5.38a). Such correlations occurred on timescales as low as a few tens of millisecond, and were not coupled to changes of the average firing rate of the individual neurons. Based on these results, it was suggested that neurons can simultaneously participate in different computations by rapidly changing their coupling to other neurons, i.e., temporally correlate their responses, without associated changes in their firing rate, and that these rapid transients of coinciding activity do give rise to behavioral changes.

Such experimental observations (for early studies see also McClurkin et al. 1991; Engel et al. 1992 for visual cortex; Reinagel and Reid 2000 for cat LGN; Bell et al. 1997; Han et al. 2000 for mormyrid electric fish; Panzeri and Schultz 2001 for rat somatosensory cortex; Bi and Poo 1998 for cultured neurons; Bair and Koch 1996 for cortical area MT neurons in monkey; Prut et al. 1998 for behaving monkey) emphasize the importance of the exact timing of spikes, a view which found support in a number of modeling studies (e.g., Abeles 1982; Bernander et al. 1991; Softky and Koch 1993; Murthy and Fetz 1994; Theunissen and Miller 1995; König et al. 1996). Using a morphologically reconstructed layer V pyramidal neuron, Bernander and colleagues could demonstrate (Bernander et al. 1991) that distributed conductance-based synaptic activity not just alters the electrotonic structure of neurons due to dramatic changes in the membranes input resistance and, thus, time constant, but that these changes also lead to significant differences in the cellular response to the timed release of a selected group of synapses. Specifically, whereas the number of synapses necessary to evoke a response generally increased for increasing synaptic noise due to the decrease in input resistance, much less synapses were needed if their activity was temporally synchronized (Fig. 5.39a). Moreover, a reliable cellular response to periodic stimulation of temporally correlated group of synaptic inputs could be observed even in the presence of a strong synaptic background, whereas temporally desynchronized synaptic inputs could not evoke a reliable response (Fig. 5.39b), thus stressing that neurons subjected to sustained network activity act as reliable coincidence detectors.

The mechanism for coincidence detection outlined above rests solely on changes of the electrotonic distance of the dendritic arborization due to synaptic noise. Another mechanism was proposed in theoretical studies by Softky and Koch (Softky and Koch 1993; Softky 1994, 1995). Here, the presence of active currents for spike generation might endow the cell with the ability to detect coinciding weak synaptic inputs in its far distal region at timescales significantly faster, up to 100 times, than the membrane time constant. The presence of such active currents in thin distal dendrites was found to either result in fast and strong local depolarizations, or evoke fast voltage deflections of several millivolt amplitude in the soma. If two or more synaptic inputs coincide, the generated membrane potentials may generate a dendritic spike, or directly generate a somatic response to the inputs. On the other hand, active currents, in particular the delayed-rectifier currents, prevent at the same time the soma from temporally summating dispersed synaptic stimuli (Fig. 5.40), hence rendering cortical neurons with active dendrites an efficient detector for temporally coinciding synaptic inputs.

While the case for coincidence detection, hence an integrative mode based on the precise timing of spikes, enjoyed a clear experimental and theoretical support,

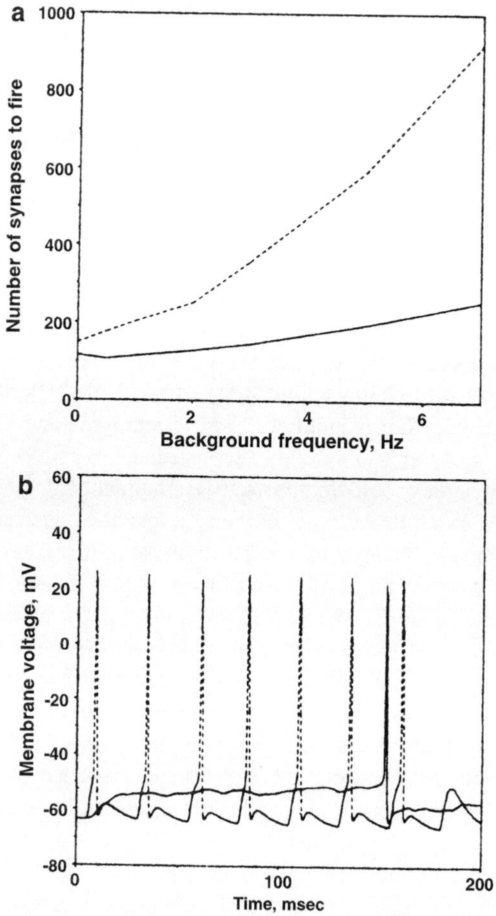

Fig. 5.39 Coincidence detection in a detailed biophysical model of cortical neurons. (**a**) A group of excitatory synapses superimposed onto a synaptic background activity were distributed across the dendritic tree and released either simultaneously (*solid*) or temporally desynchronized (*dashed*). For desynchronized inputs, the minimum number of synapses required to evoke a cellular response was higher and increased much faster as a function of the background activity compared to the case where the selected group of synapses released simultaneously. (**b**) An example of the somatic membrane potential in response to the period activity of a selected group of 150 synapses in the presence of background activity of 1 Hz. Synchronized inputs lead to a reliable response reflecting the periodic stimulus, while the same group of synapses activated in a temporally dispersed manner over the first 12.5 ms of each cycle led only to one response. Modified from Bernander et al. (1991)

evidence for the other extreme in the spectrum of possible operating modes, namely rate-based coding, remains more sparse. Through the use of principal component analysis, Tovée and colleagues observed in the visual cortex of rhesus monkeys that not only the most information in responses of single neurons was contained in the

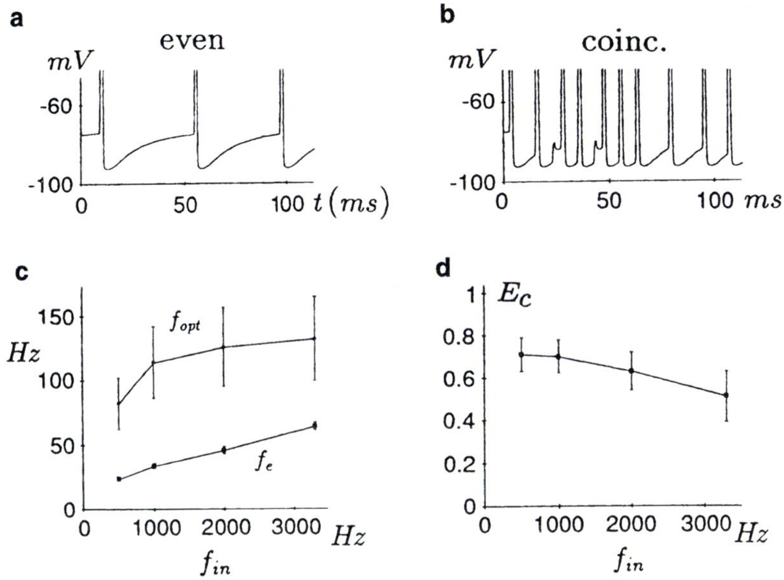

Fig. 5.40 Coincidence detection in a detailed biophysical model of cortical neurons with active dendrites. Evenly timed synaptic inputs which are to weak to initiate dendritic spikes lead to low-frequency somatic responses (*top left*; f_c, *bottom left*), whereas the same synaptic inputs occur at the same rate in coincident pairs inside the same dendrite (*top right*). In the latter case, dendritic spikes are evoked which propagate to the soma and lead to a higher-frequency response (f_{opt}, *bottom left*). This preference for coincident EPSPs can be quantified by values of "effectiveness" $E_c = 1 - f_c/f_{opt} > 0$ (*bottom right*). Modified from Softky (1994)

first principal component but also that the latter was strongly correlated with the mean firing rate of the studied neurons (Tovée et al. 1993). Modeling support for a rate-based coding paradigm, however, only indirectly proved their point by arguing that the high irregularity in the spiking pattern of cortical neurons, in fact, hinders the resolution of precise temporal pattern in its inputs (Barlow 1995; Bugmann et al. 1997).

Shadlen and Newsome (1998) pointed out that in a simple IAF model with balanced inhibition and excitation operating in a "high-input regime," the cellular response naturally displays a high variability similar to that observed experimentally in cortical neurons. However, it was argued that due to this variability (Fig. 5.41a), detailed information about temporal patterns in the synaptic inputs cannot be recovered from the cellular response alone (Fig. 5.41b), and that instead only the information contained in average rates of an ensemble of up to 100 neurons is represented on the network level down to a temporal resolution of a typical ISI.

A more intermediate position was taken later, with a series of experimental (e.g., see Krüger and Becker 1991) and theoretical (e.g., Kretzberg et al. 2001) studies proposing the view that cortical neurons could operate according to both of these modes, or even in a continuum between temporal integration and coincidence

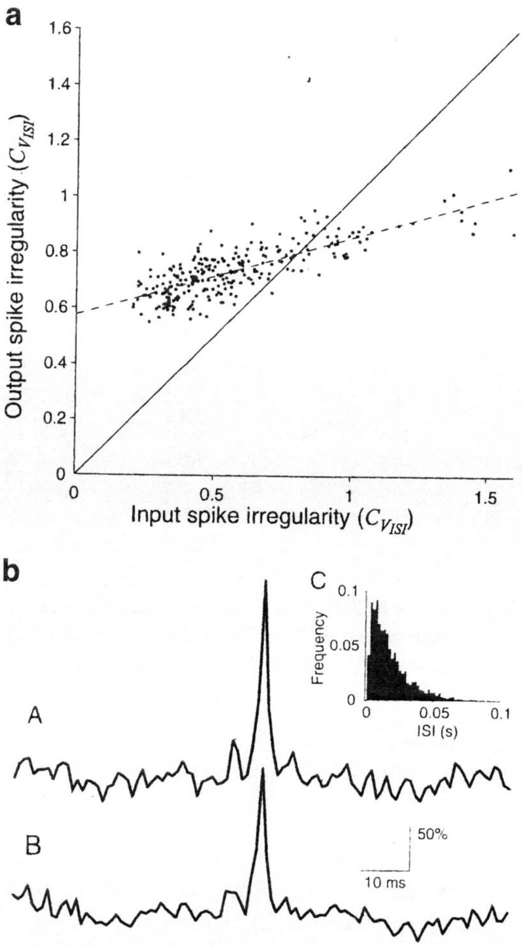

Fig. 5.41 (**a**) The relation between the irregularity in input and output of a simple integrate-and-fire neuron model operating in a "high-input regime" with balanced inhibition and excitation. The input irregularity was obtained by using interspike intervals following a gamma distribution. Interestingly, the degree of input irregularity had only little impact on the distribution of output-spike intervals, with the latter remaining high even for more regular inputs. This suggests that precise temporal input pattern cannot be preserved. (**b**) Homogeneity of synchrony among input and output ensembles of neurons. The *upper trace* shows the normalized cross-correlogram from a pair of neurons operating in a balanced high-input regime sharing 40% of their inhibitory and excitatory inputs. The *lower trace* shows the average cross-correlogram of neurons serving as inputs. Although both correlograms show a clear peak, suggesting the detection of the synchronous inputs by the receiving neurons, this synchrony does not lead to a detectable structure in the output-spike train (*inset*). Modified from Shadlen and Newsome (1998)

detection (e.g., Maršálek et al. 1997; Kisley and Gerstein 1999). In a detailed modeling study using both IAF neurons and biophysically more realistic models of cortical pyramidal cells with anatomically detailed passive dendritic arborization,

Maršálek and colleagues found that, under physiological conditions, the output jitter is linearly related to the input jitter (similar to the relation of Shadlen and Newsome mentioned above) but with a constant of less than one (Fig. 5.42; Maršálek et al. 1997). This finding suggests that not only the response irregularity could converge to smaller values in successive layers of neurons in a network but also that the temporal characteristics of the input can serve as a factor determining the operating mode of the cell. When inputs are broadly distributed in time, neurons tend to respond to the average firing rate of afferent inputs, whereas the same neurons can also respond precisely to a large number of synchronous synaptic events, therefore acting as coincidence detectors.

These aforementioned studies were later complemented by an investigation of the response of morphologically reconstructed biophysical models with active dendrites (Rudolph and Destexhe 2003c). Using Gaussian-shaped volleys of synaptic inputs spatially distributed in the dendritic structure and superimposed with a Poisson-distributed background activity, the input synchronization, or temporal dispersion, could be controlled and the ability of the cell to respond as a function of the synaptic noise fully investigated (Fig. 5.43). In this study, the relation between input and output synchrony was assessed by using the ratio

$$\xi = \frac{\sigma_{\rm in}}{\sigma_{\rm out}} \tag{5.12}$$

with $\sigma_{\rm out}$ obtained from Gaussian fits of the PSTH. Similarly, the number of synaptic activations in the Gaussian input event $N_{\rm Gauss}$ and the number of responses $N_{\rm resp}$ for a fixed number of trials served as a measure of the reliability of the cellular response, defined as

$$R = \frac{N_{\rm resp}}{N_{\rm Gauss}} . \tag{5.13}$$

It was found that in quiescent conditions, the cell shows a reliable response ($R = 1$) to Gaussian events of nearly all widths (Fig. 5.44a1) in agreement with earlier studies (Segundo et al. 1966; Kisley and Gerstein 1999), suggesting that the cell is capable of acting as both a coincidence detector (for small $\sigma_{\rm in}$) or a temporal integrator (for large $\sigma_{\rm in}$). The minimal number of synaptic inputs N required to evoke a response (as indicated by the boundary of the $R = 1$ region in Fig. 5.44a1) is lower for more synchronized input events (smaller $\sigma_{\rm in}$). In agreement with other studies (Abeles 1982; Bernander et al. 1991; Softky 1995; Aertsen et al. 1996; König et al. 1996), this result indicates that coincidence detection is the more "efficient" operating mode. However, the flat boundary for $R = 1$ also shows that, in quiescent conditions, temporal integration needs only a small increase in the strength N of the temporally dispersed synaptic input in order to be effective.

However, this picture changes quantitatively in the presence of synaptic background activity. Here, coincidence detection is still the most efficient operating mode, but the higher slope of the boundary for $R = 1$ (see Figs. 5.44a2, a3) indicates

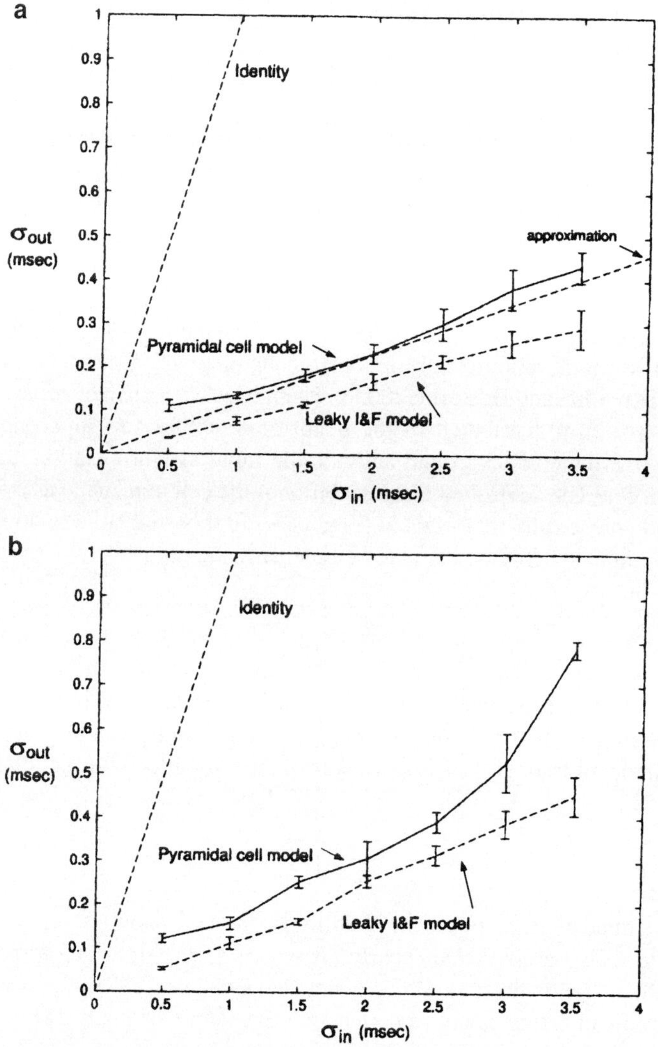

Fig. 5.42 Relationship between input and output jitter for excitatory input only (**a**) and with both excitation and inhibition present (**b**). An approximation for the nonleaky integrate-and-fire neuron according to $\sigma_{\text{out}} \sim \sigma_{\text{in}} 2\sqrt{3}\frac{1}{n}\sqrt{n_{\text{th}}}$, where n_{th} denotes the number of inputs needed to reach spiking threshold and n is the number of synaptic inputs in a volley of activity is indicated in (**a**), with slope of 0.116. The numerical simulations always fall below the identity line, as does the output jitter associated with our anatomical and biophysical accurate model of a pyramidal cell with a passive dendritic tree. Error bars correspond to sample standard deviation from 5 runs from 50 threshold passages. These results show that in a cascade of such neurons in a multilayered network and in the absence of large timing uncertainty in synaptic transmission and inhomogeneous spike-propagation times, the timing jitter in spiking times will converge to zero. Modified from Maršálek et al. (1997)

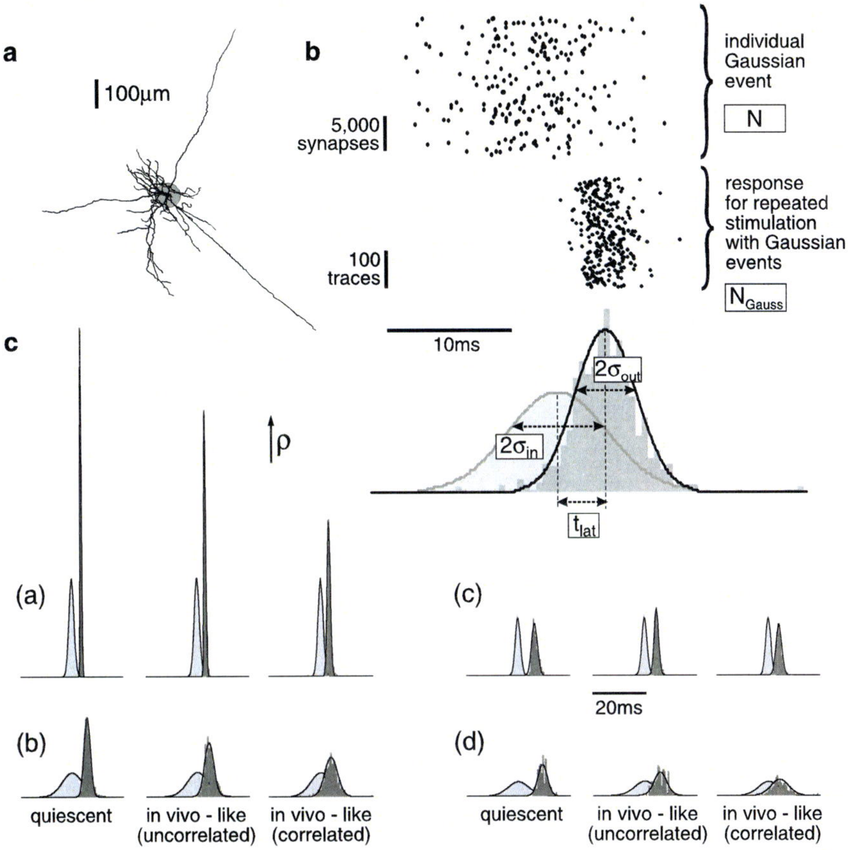

Fig. 5.43 (**a**) Morphologically reconstructed neocortical pyramidal layer VI neuron of a cat used in the modeling studies. The *shaded area* indicates the proximal region (radius $\leq 40\,\mu$m). Inside that region there were no excitatory synapses, whereas inhibitory synapses were spread over the whole dendritic tree. (**b**) Scheme of the simulation protocol. Individual Gaussian events (*top panel*) were obtained by distributing N synaptic inputs randomly in time according to a Gaussian distribution of standard deviation σ_{in} (*light gray curve, bottom panel*). The cellular response was recorded for repeated stimulation with N_{Gauss} individual Gaussian events (*middle panel*), yielding a Gaussian shaped PSTH of width σ_{out} and a mean shifted by the latency against the mean of the input events (*dark gray curve, bottom panel*). (**c**) Representative examples of Gaussian input events (*light gray*) and corresponding cumulated responses (*dark gray*) for quiescent conditions, and under (correlated and uncorrelated) in vivo-like activity. Characteristics of Gaussian input events: (**a**) $N = 220$, $\sigma_{\text{in}} = 1$ ms (**b**) $N = 220$, $\sigma_{\text{in}} = 4$ ms (**c**) $N = 130$, $\sigma_{\text{in}} = 1$ ms and (**d**) $N = 130$, $\sigma_{\text{in}} = 4$ ms. The relative probability ρ is defined as $\rho = (\text{number of spikes in time interval} T)/(N_{\text{resp}} \times T)$. Modified from Rudolph and Destexhe (2003c)

that an effective temporal integration can be obtained only for a marked increase in the strength of the temporally dispersed input signal (see also Bernander et al. 1991). For fixed N, the cell is less capable of responding reliably to Gaussian events of higher widths compared to quiescent conditions, and for correlated background

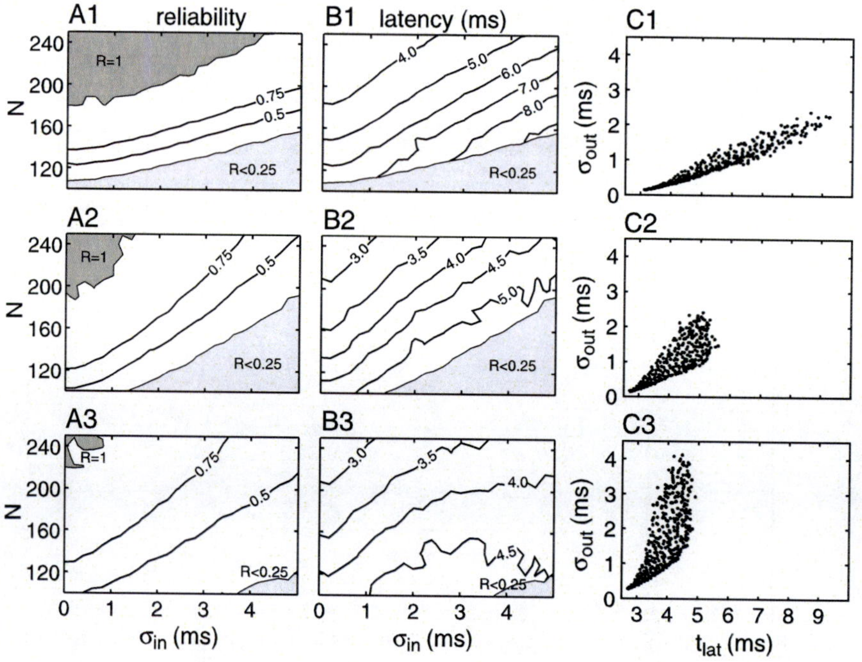

Fig. 5.44 (**a**) Reliability R with which the Gaussian events drive the postsynaptic response. In the quiescent case (**a**1), the cell is capable of responding reliable to events of nearly all widths, whereas the region with $R = 1$ decreased in the presence of uncorrelated (**a**2) and correlated (**a**3) background activity. Moreover, under quiescent conditions, the dependence of R on the strength of the Gaussian signal N becomes weaker (smaller slope of iso-reliability lines). (**b**) The mean latency as a function of Gaussian stimuli characteristics in the quiescent case (**b**1) and under uncorrelated (**b**2) and correlated (**b**3) in vivo-like conditions. (**c**) Relation between latency and the output jitter. In the quiescent case (**c**1), there is a nearly linear relation, allowing to determine the time of the occurrence of the input events by measuring the jitter of the output, whereas no corresponding relation can be evidenced under in vivo-like conditions ((**c**2) uncorrelated and (**c**3) correlated). Modified from Rudolph and Destexhe (2003c)

activity a reliability of $R = 1$ is only obtained for very strong input signals (large N) with a very narrow temporal distribution ($\sigma_{in} < 1$ ms). This overall increase in the strength of the synaptic input necessary to evoke a response as well as the reduction in reliability for stronger input events (small σ_{in}, larger N; compare Figs. 5.44a1, a2) is a direct result of the smaller input resistance (or increase in the effective membrane conductance shunting the dendrites) caused by the intense synaptic background activity impinging on the cell.

On the other hand, the spontaneous discharge activity shifts the parameter region with no or only low reliable responses ($R < 0.5$) toward lower input strength N. The latter effect, which can be interpreted as *enhanced responsiveness* (see, e.g., Hô and Destexhe 2000), is pronounced for correlated background activity (compare Figs. 5.44a2, a3). Interestingly, both effects together are found to yield a weaker dependence of the reliability on the strength of the Gaussian events in the presence of background activity (as indicated by the broader "band" between $R = 1$ and $R < 0.25$ in Fig. 5.44a), suggesting a more "variable" response to discriminate input settings compared to the quiescent case.

The mean latency of the output with respect to the center of the input events shows that, in general, lower σ_{in} and large N lead to responses with shorter latencies (Fig. 5.44b). In the quiescent case (Fig. 5.44b1), a minimal time of about 3 to 4 ms corresponds to the average time dendritic spikes evoked by strong synaptic inputs need to propagate along the spatially extended dendritic tree and impact on the soma (Rudolph and Destexhe 2001b). Weaker signals or signals with a broader temporal distribution need more time to be integrated, leading to a corresponding increase in latency. A clear increase of the latency with σ_{in} further suggests that somatic spikes generated by coincidence detection occur with a shorter delay than those caused by temporal integration (König et al. 1996; Shadlen and Newsome 1995). However, the presence of background activity markedly decreases the latency (between 25% for strong Gaussian events and 50% for weaker Gaussian events), especially for higher σ_{in} values, yielding a latency which is much less dependent on σ_{in} and N and, thus, the operating mode for correlated background activity.

Finally, in the quiescent state, tightly synchronized input distributions (small σ_{in}) cause less jitter in the timing of output spikes (Fig. 5.45a1) with a σ_{out} considerably smaller than σ_{in} (Fig. 5.45c1), in agreement with various models (Kisley and Gerstein 1999; Maršálek et al. 1997). However, σ_{out} depends much more on N than in passive models. In the presence of synaptic background activity, in addition to the overall decrease of the latency, also the range of σ_{out} increases with the background correlation. Here, there is no longer a linear relation between latency and σ_{out}. The output jitter σ_{out} depends more on σ_{in}, whereas the dependence on N becomes weaker, as indicated by the higher slope of the σ_{out}.

Concerning the impact of the input synchronization on the output jitter of the cell, it was found (Rudolph and Destexhe 2003c) that the input parameter range for which $\xi \geq 1$ is large under quiescent conditions (Fig. 5.45b1), and only minimally changes in the presence of background activity (Fig. 5.45b2, b3). However, background activity and correlation markedly lowers ξ, as indicated by the increasing light gray region in Fig. 5.45b2, b3. In all cases, σ_{out} is roughly proportional to σ_{in} with a slope smaller than one in the quiescent state or in the presence of uncorrelated background activity (Fig. 5.45c1, c2). Only with correlated background activity, the slope increases and is close to one for a broad range of input settings (Fig. 5.45c3), suggesting that under these conditions the cell nearly conserves the synchronization of the input signal. The output jitter for fixed σ_{in} stays nearly constant in all cases for all N, with σ_{out}, suggesting a much higher variability when synaptic background activity was present. Interestingly, there is a relation between output jitter and reliability of the response in the presence of background activity: a higher jitter in the output, hence less precision, is accompanied by a decrease in reliability. No such relation can be evidenced in the quiescent state.

In summary, in a number of computational studies of detailed biophysical models of cortical neurons, synaptic background activity was found to modulate and

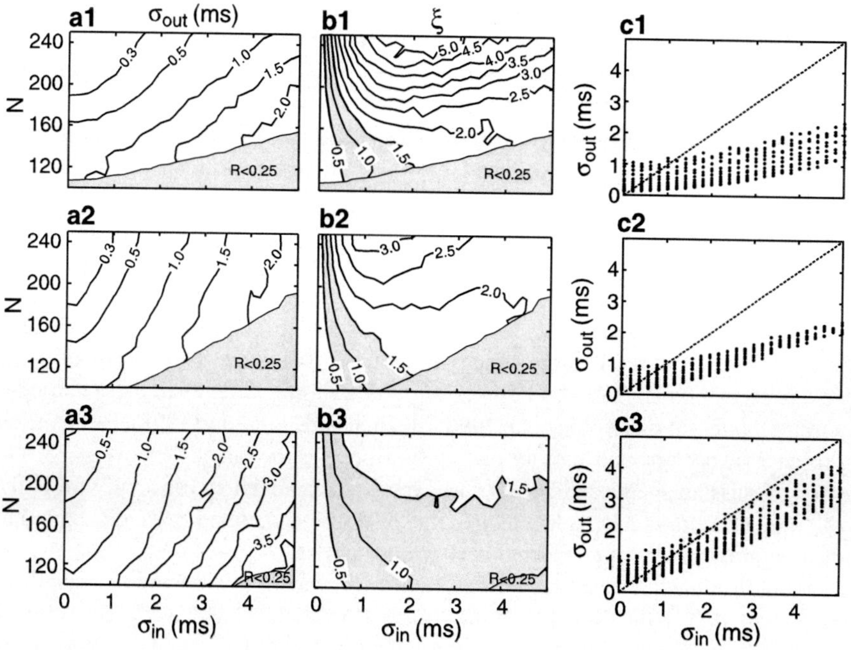

Fig. 5.45 (**a**) Output jitter σ_{out} as a function of the input settings. In all three cases, σ_{out} is lower than σ_{in} for most of the parameter range. Moreover, the dependence of σ_{out} on N is higher in the quiescent case (**a**1), whereas the output jitter becomes nearly independent on N in the presence of correlated background activity (**a**3; indicated by higher slope of iso-σ_{out} lines). (**a**2) shows the corresponding results for uncorrelated background activity. (**b**) $\xi = \sigma_{\text{in}}/\sigma_{\text{out}}$ as a function of the input settings. In all three cases, the input parameter range covers settings for which $\xi \sim 1$ (increasing *light gray region*). This range markedly increases for the correlated case (**b**3). **b**1, **b**2 show the results for quiescent and uncorrelated in vivo-like conditions, respectively. (**c**) Relation between output jitter and the synchrony in the input. In all three cases ((**c**1) quiescent, (**c**2) in vivo-like uncorrelated and (**c**3) in vivo-like correlated), σ_{out} is nearly proportional to σ_{in}, but with a slope smaller than one. Only in the correlated case the slope is close to unity, suggesting a neuronal response which preserves the synchrony in the input signal. Modified from Rudolph and Destexhe (2003c)

control the integrative mode of individual cells by allowing to tune the (temporal) characteristics of their outputs. Changes in the correlation of the presynaptic activity alter the reliability of the response as well as the ability of the cell to discriminate between synaptic inputs arriving simultaneously or dispersed in time. Furthermore, the impact of background activity on the latency of the response suggests that the time needed for integrating synaptic inputs and, thus, the "speed" of propagation of information across cortical layers is not fixed but can be tuned by changes in the background activity. The temporal dispersion of the response was found to depend on the characteristics of background activity, which could provide a mechanism through which the cell can focus on specific temporal patterns in its synaptic inputs.

Taken together, these results suggests that "information contents" is not only a property of the signal itself, but that it also depends on the background activity, which this way, although still widely viewed as noise, will loose its detrimental attributes. Both signal and background activity must, therefore, be considered together when investigating information processing paradigms of neocortical neurons.

5.9 Spike-Time Precision and Reliability

In the previous section, the question whether cortical neurons act as detectors for coinciding or synchronous inputs instead of mere integrators of average input rates was addressed, with recent experimental data and theoretical studies supporting the idea that both integration modes might be simultaneous at work, and that the temporal pattern of synaptic noise itself can tune, hence alter, the encoding paradigm. An important question playing a decisive role in both coding paradigms, however, is under which conditions neurons can and do show a reliable and precise response to a given temporal pattern of synaptic inputs.

Earliest theories of neural encoding, in particular in sensory systems, considered the mean firing rate as the relevant quantity (e.g., Adrian and Zotterman 1926; Barlow 1995). However, it has been long recognized that sensory information may also be encoded by the temporal pattern of the neural activity (MacKay and McCulloch 1952). As mentioned in the previous section, a number of experimental and theoretical results suggested that simple coding by firing rate alone, as classically considered (e.g., Barlow 1995; see also Tovée et al. 1993; Bugmann et al. 1997; Shadlen and Newsome 1998), may be at odds with observed data. Experimental studies, primarily performed in the visual system, have found a significant role for precise timing of individual spikes in coding as well as precisely reproducible spike time patterns or *relational codes* (see for instance Krüger and Becker 1991; McClurkin et al. 1991; Engel et al. 1992; Bair and Koch 1996; Thorpe et al. 1996; Reinagel and Reid 2000; Panzeri and Schultz 2001). This *temporal coding* hypothesis states that the precise timing of spikes, in addition to the firing rate, carries information (e.g., Bialek and Rieke 1992; Gray 1994; Theunissen and Miller 1995; Prut et al. 1998; for a comprehensive overview see deCharms and Zador 2000).

A prerequisite for the spike-time code to work is that spikes must be evoked precisely and reliably by a given stimulus. Over the past years, several converging lines of research have found that spike generation in cortical neurons can indeed be precise and reliable, depending on the nature of the inputs. In one of the first experimental studies aimed at investigating mode locking and spike timing precision, Bryant and Segundo (1976) elegantly demonstrated that repeated injections of white noise in Aplysia neurons led to a remarkable invariance in the firing times accompanied by a high degree of reliability in the response (Fig. 5.46). In vivo recordings have shown that the neural responses were robust and reproducible when the stimulus leads to fast fluctuations in the firing rates (e.g., in

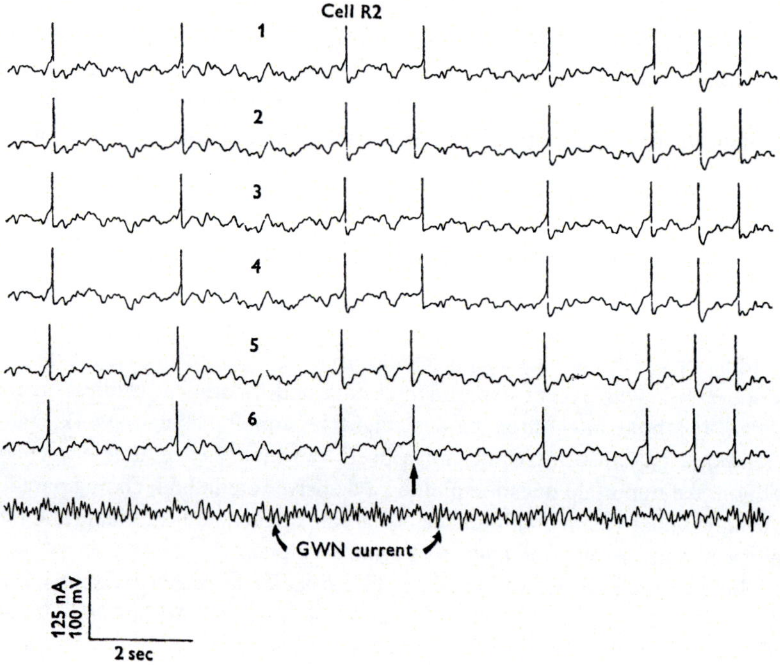

Fig. 5.46 Response of a cell to a repeated segment of Gaussian white noise. The *bottom trace* is a 15 s segment of Gaussian current (±25 mA) which was repeatedly injected into the cell at 3 to 5 min intervals. *Traces 1–6* show the response to six such identical stimulus sequences and reveal a high degree of time invariance of the current-to-spike triggering system. This result also suggests a high degree of reliability in the response of individual neurons to the presentation of an identical stimulus. Modified from Bryant and Segundo (1976)

monkey MT: Britten et al. 1993; in the LGN: Reinagel and Reid 2000) and under stimuli with statistics of natural scenes (e.g., for motion-sensitive H1 neurons in the fly's visual system: de Ruyter van Steveninck et al. 1997. In vitro experimental work has examined more closely the conditions for precise spike-timing (e.g., Calvin and Stevens 1968; Mainen and Sejnowski 1995; Nowak et al. 1997; Tang 1997; Hunter et al. 1998)). The main finding of these studies is that spike-timing is rather imprecise for constant (Fig. 5.47; Mainen and Sejnowski 1995) and low frequency (as in Nowak et al. 1997) driving currents, but relatively precise for stimuli with pronounced temporal structure.

Precision of spike-timing has also received considerable attention in the computational literature, utilizing a variety of modeling approaches (e.g., see Howeling et al. 2001; Needleman et al. 2001; Kretzberg et al. 2001; van Rossum 2001) and analytical methods (e.g., Brunel et al. 2001). Using a simplified canonical model for neurons of type I excitability, the stochastic θ-neuron (Ermentrout and Kopell 1984, 1986; Hoppensteadt and Izhikevich 1997; Gutkin and Ermentrout 1998), it was shown that the spike- timing precision and reliability results reported previously

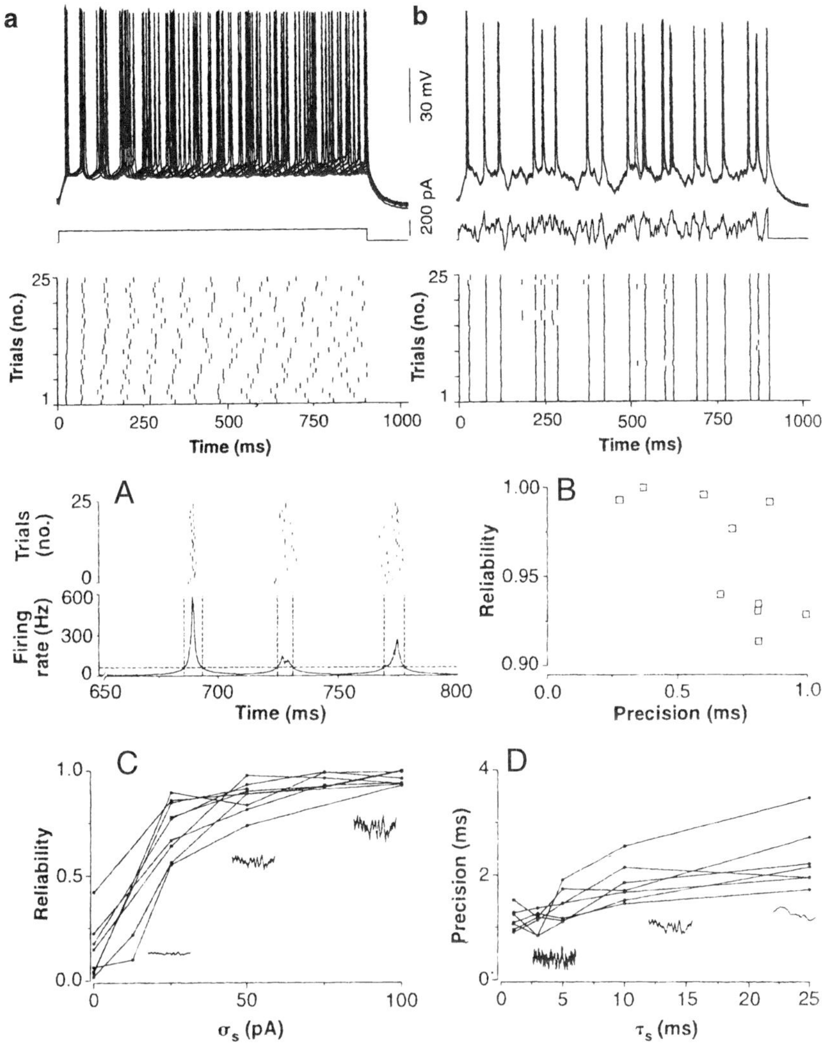

Fig. 5.47 (**a**) Reliability of firing pattern of cortical neurons evoked by constant and fluctuating current. A superthreshold DC current pulse (150 pA) evoked a train of action potentials (*left, top*) which became increasingly unreliable in their temporal pattern when subsequent trials are considered (*rasterplot*; *bottom left*). In contrast, the same cell exposed to a fluctuating current in the form of Gaussian white noise (average 150 pA; standard deviation 100 pA, time constant 3 ms) showed a highly reliable firing pattern across subsequent trials (*right*). (**b**) Dependence of the reliability and precision of spike-timing on stimulus current statistics. Using a PSTH of spikes collected over a number of trials (*top left*), blocks of responses (events) can be distinguished which allow to define "reliability" (fraction of total spikes occurring in an event) and "precision" (standard deviation of spike times within an event). The most reliable event in each of the recorded cells shows that reliability increased with the fluctuation amplitude (σ_s) of the stimulus (*bottom left*), whereas the "precision" of these events was smaller (hence the events sharper) for higher noise (*bottom right*). All recorded cells were capable of responding to fluctuating input currents with nearly 100% reliability in clusters with a standard deviation of less than 1 ms (*top right*). Modified from Mainen and Sejnowski (1995)

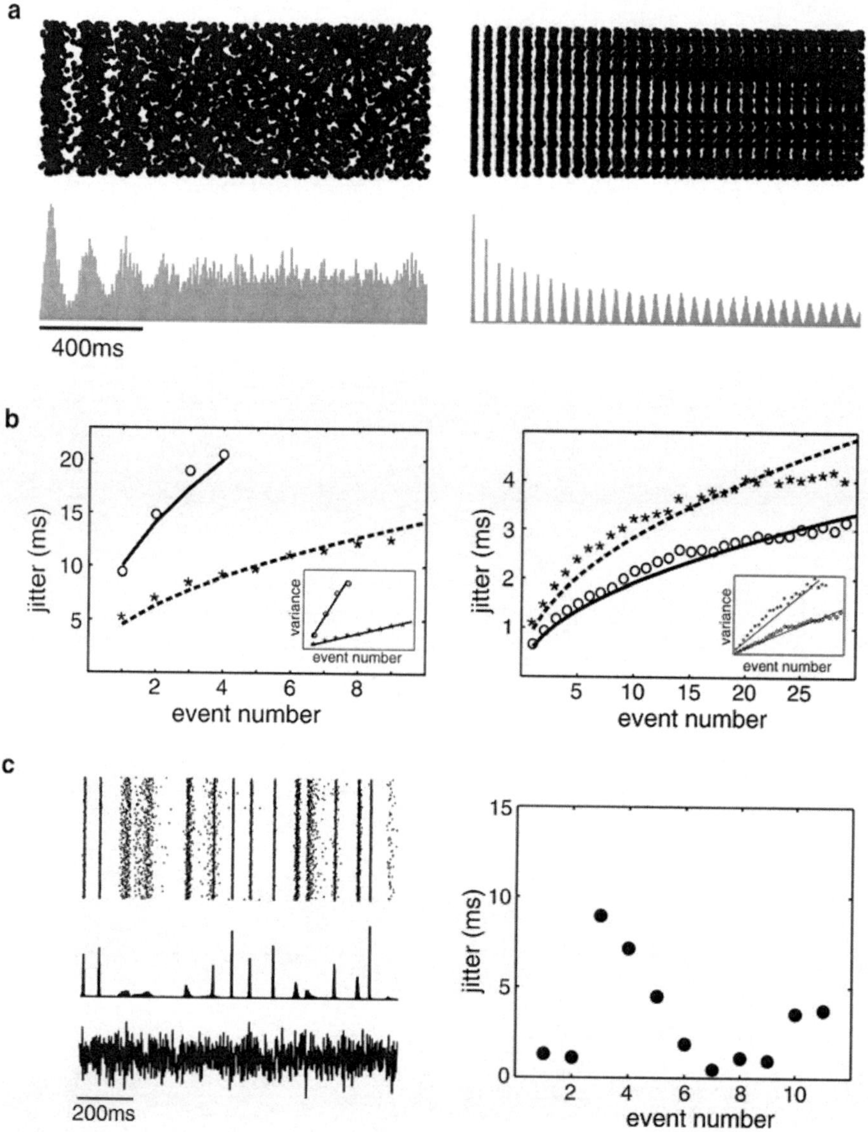

Fig. 5.48 Spike-timing jitter under constant current injection is compatible with a renewal process. The jitter grows as a square root of the event number (variance is proportional to event number), and is reduced for higher firing rates. (**a**) Typical rasterplots (*upper panels*) and PSTHs (*lower panels*) for constant stimuli. The bias was adjusted to give 10 Hz (*left*) and 30 Hz (*right*) average firing rate. The noise strength was fixed to $\sigma = 0.001$. (**b**) Spike-time jitter and spike-time variance (*insets*) as a function of the event number. *Left*: result for mean firing rate of 10 Hz and noise amplitude of $\sigma = 0.001$ (*stars*) and $\sigma = 0.003$ (*dots*). *Right*: results for $\sigma = 0.001$ (*stars*) and $\sigma = 0.003$ but higher bias (average firing rate 30 Hz). (**c**) Example of spike-time jitter in the θ-neuron under an aperiodic stimulus. *Left*: rasterplots (*upper panel*) and PSTH (*middle*) for repeated injection of aperiodic current (*lower panel*; mean firing rate 17 Hz, amplitude of aperiodic

from in vitro experiments can be reproduced (Gutkin et al. 2003). Indeed, the spike trains evoked by constant current injections (oscillating regime, Fig 5.48a, b), the overall spike-time jitter is high and increases with successive spikes. Here, the response is caused by regenerative activity of the intrinsic currents and the jitter in each spike depends on the jitter in the preceding spikes. On the other hand, under aperiodic stimuli (Fig 5.48c), the response is caused by the rapid fluctuations in the stimulus itself, and the cell acts as an excitable threshold element. In this case, a given spike is largely statistically independent from the previous spike and its jitter is stationary and relatively low compared to the constant case with constant stimulation.

Results obtained in the aforementioned studies were reproduced by using also a single-compartment model with Hodgkin–Huxley spike generation and conductance noise modeled as effective stochastic OU process (Rudolph and Destexhe 2003e). There it was found that in the presence of conductance noise, the cell is able to resolve stimulating frequencies in the whole tested range (up to 150 Hz and beyond) with high precision and reliability, while keeping the average firing frequency nearly constant (Fig. 5.49). This corresponds to a coding scheme built on the precise timing of spikes. On the other hand, the strength of the input translates into a change of the firing rate, with the input frequency showing only minimal impact. This suggests that both coding dimensions are not exclusive but simultaneously present to encode different aspects of the input: conductance noise shows a modulating effect on the firing rate and reliability, while keeping the precision nearly unaffected.

5.10 Summary

In this chapter, we have overviewed the consequences of synaptic noise on many aspects of the integrative properties of neurons. These consequences were obtained from detailed biophysical models directly constrained by the quantitative measurements of synaptic background activity (see Chap. 3 for measurements and Chap. 4 for details of the models).

These models allow to derive a series of interesting properties and behavioral characteristics, which are only valid in the presence of high amounts of synaptic noise comparable to that found in vivo.

- *Enhanced attenuation* (Sect. 5.2). As found by early studies (Barrett 1975; Holmes and Woody 1989), the presence of background activity has a marked effect on the effectiveness of synaptic inputs.

Fig. 5.48 (continued) stimulus 0.05, noise strength $\sigma = 0.003$). *Right*: jitter as a function of even number. Note that under such strong aperiodic signal the event jitter does not depend on event number. Modified from Gutkin et al. (2003)

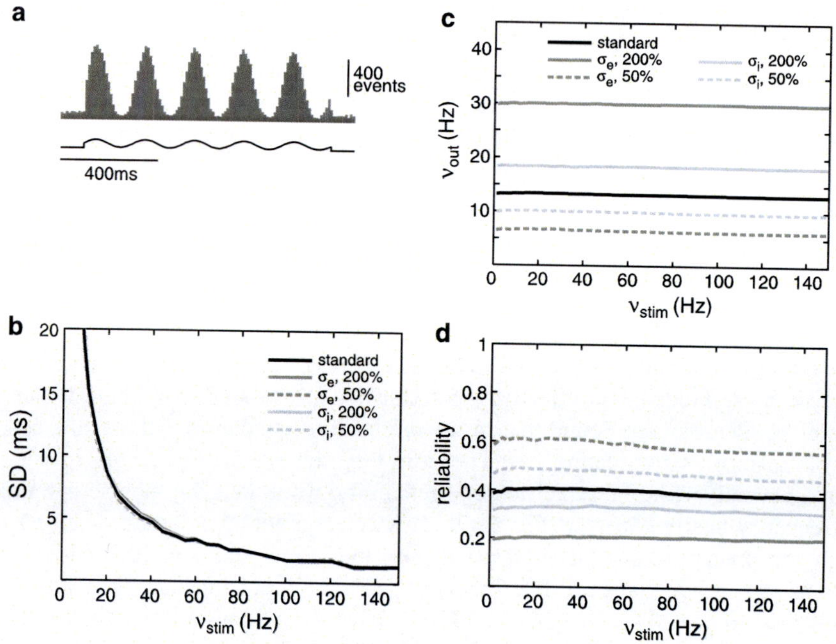

Fig. 5.49 (**a**) PSTH obtained from injection of sinusoidal excitatory conductance waveforms superimposed on Ornstein–Uhlenbeck conductance background noise in a single-compartment model of a cortical neuron. (**b**) Average SD of the response as function of stimulation frequency $\nu_{\rm in}$ for fixed stimulation amplitude ($g_{\rm amp} = 3$ nS, (**b**)) for different background parameters. The standard deviation, or "precision" follows the frequency dependency of the width of the sinusoidal stimulus even at high background noise levels. (**c**), (**d**) ν_{out} (**c**) and reliability (**d**) as a function of $\nu_{\rm in}$ for fixed stimulation amplitude ($g_{\rm amp} = 3$ nS). Both the output rate and reliability of response depend solely on the background noise level, but are unaffected by the stimulus frequency. Modified from Rudolph and Destexhe (2003e)

- *Enhanced responsiveness* (Sect. 5.3). The presence of background activity was found to markedly change the cell's excitability and produce a detectable response to inputs that are normally subthreshold (Hô and Destexhe 2000). This prediction was verified in dynamic-clamp experiments (see Chap. 6).
- *High discharge variability* (Sect. 5.4). A high discharge variability is paralleled with a high-conductance state, and remains mainly insensitive to active and passive cellular properties or the balance between excitation and inhibition. This suggests that the fluctuating high-conductance state caused by the ongoing activity in the cortical network in vivo can be viewed as a natural determinant of the highly variable discharges of these neurons.
- *Stochastic resonance* (Sect. 5.5). The mechanism of enhanced responsiveness is very similar to the well-studied phenomenon of stochastic resonance, which is a peak of the SNR for a nonzero level of noise. In neurons, a very similar

phenomenon is present, although more complicated due to the multiplicative character of conductances.

- *Correlation detection* (Sect. 5.6). As a consequence of their exquisite sensitivity to correlations, pyramidal neurons constitute a very powerful device for detecting correlations (Rudolph and Destexhe 2001a). Even small and brief changes of correlations can be detected by the neuron in high-conductance states.
- *Location independence* (Sect. 5.7). The effectiveness of synaptic inputs becomes much less dependent on their position in dendrites, as found in cerebellar (De Schutter and Bower 1994) and cortical neurons (Rudolph and Destexhe 2003b), although based on very different mechanisms.
- *Different integrative mode* (Sect. 5.8). As initially predicted by Bernander et al. (1991), synaptic noise can have profound effects on the integrative mode of cortical pyramidal neurons. This important property was indeed confirmed with models constrained by experimental measurements (Rudolph and Destexhe 2003b).
- *Enhanced temporal processing* (Sect. 5.9). As a direct consequence of the "high-conductance state" of the neurons under background activity, the faster membrane time constant allows the neuron to perform finer discrimination, which is essential for coincidence detection (Softky 1994; Rudolph and Destexhe 2003b; Destexhe et al. 2003a) or detecting brief changes of correlation (Rudolph and Destexhe 2001a). The latter prediction was also verified experimentally (Fellous et al. 2003), as detailed in the next chapter.
- *Modulation of intrinsic properties*. It was found that in the presence of synaptic background activity, the responsiveness of bursting neurons is strongly affected (Wolfart et al. 2005). This aspect was not covered here, but will be considered in detail in the next chapter (see Sect. 6.4), devoted to dynamic-clamp experiments.

Chapter 6
Recreating Synaptic Noise Using Dynamic-Clamp

This chapter will cover one of the most promising and elegant approach for studying the effect of synaptic noise on neurons: the dynamic-clamp injection of artificial conductance-based synaptic noise. We start by an introduction to the dynamic-clamp technique, and we next describe the "re-creation" of in vivo-like activity states in neurons maintained in vitro. We then overview the consequences of synaptic noise on the integrative properties of neurons, as found by dynamic-clamp experiments.

6.1 The Dynamic Clamp

6.1.1 Introduction to the Dynamic-Clamp Technique

The technique of *dynamic clamp* is a special electrophysiological recording configuration to enable injecting conductances into a neuron. Most of the time, the injected conductance is calculated using a computational model (as we will show below), thus rendering the dynamic clamp a remarkably close interaction between living cells and models.

As reviewed in different articles (Prinz et al. 2004; Prinz 2004; Goaillard and Marder 2006; Piwkowska et al. 2009; Economo et al. 2010), books (Destexhe and Bal 2009) or theses (Piwkowska 2007), the dynamic clamp was and is used by a large number of researchers to study a variety of physiological questions at the level of single cells, as well as tissues of interacting cells. The earliest type of dynamic-clamp experiment was called the *Ersatz Nexus*, and was introduced in cardiac physiology as early as 1979, in a PhD thesis studying the impact of electrical synapses (gap junctions) on the synchronization of clusters of cardiomyocytes in the chicken (Scott 1979). Later studies of cardiac tissue reintroduced a technique named *coupling clamp* (Tan and Joyner 1990) for bidirectionally connecting two isolated myocytes by a virtual gap junction of chosen conductance. The injected current flowing through the virtual gap junction is calculated according to a driving force

A. Destexhe and M. Rudolph-Lilith, *Neuronal Noise*, Springer Series in Computational Neuroscience 8, DOI 10.1007/978-0-387-79020-6_6,

determined in real time and equal to the difference of membrane potential between the two cells (see Verheijck et al. 1998 for an example of an application which explored the synchronization between two spontaneously active rabbit cardiac cells). An extension of this technique, named the *model clamp* by the authors, consists in coupling, through such a virtual gap junction, a real myocyte and a model myocyte simulated in real time (Wilders et al. 1996).

The first application of the dynamic-clamp technique in neuroscience was introduced independently by Hugh Robinson (Robinson and Kawai 1993) and by a team led by Eve Marder and Larry Abbott, based on a collaboration with Gwendal Le Masson (Le Masson et al. 1992; Sharp et al. 1993a,b; see also Le Masson et al. 1995). Based on the same principle of injecting a membrane-potential-dependent current into a neuron, different implementations and applications were explored by the different groups: using digital systems, Robinson and Kawai injected synaptic inputs into cultured hippocampal neurons of the vertebrate CNS, while Sharp and colleagues studied various conductances and artificial networks in the stomatogastric ganglion (STG) of decapod crustaceans (lobsters and crabs) nervous system. Le Masson and colleagues (Le Masson et al. 1995) developed an analog and a digital approach simultaneously for studies of the invertebrate preparation (and subsequently combined both approaches in a single study of mammalian thalamus networks, see Le Masson et al. 2002). Since then, dynamic-clamp has been widely used in both vertebrate and invertebrate preparations (for a comprehensive overview, see Destexhe and Bal 2009).

In the context of the present monograph, we will review a particular, and popular, type of application of the dynamic-clamp technique, to "re-create" in vivo-like synaptic inputs into neurons recorded in vitro. Here, one injects time-varying conductance waveforms mimicking thousands of synaptic inputs that converge in vivo onto single neurons.

This paradigm was applied to various brain structures such as the cortex, the thalamus, the hippocampus, the cerebellum, or the basal ganglia. The conductance waveforms are generated either by the convolution of presynaptic spike trains with unitary synaptic conductances (Reyes et al. 1996; Jaeger and Bower 1999; Harsch and Robinson 2000; Gauck and Jaeger 2000; Chance et al. 2002; Hanson and Jaeger 2002; Gauck and Jaeger 2003; Kreiner and Jaeger 2004; Suter and Jaeger 2004; de Polavieja et al. 2005; Zsiros and Hestrin 2005; Dorval and White 2006; Tateno and Robinson 2006; Morita et al. 2008) or by effective stochastic models of synaptic bombardment or synaptic noise, without explicit representation of the presynaptic spike trains (Destexhe et al. 2001; Shu et al. 2003a,b; McCormick et al. 2003; Fellous and Sejnowski 2003; Fellous et al. 2003; Rudolph et al. 2004; Wolfart et al. 2005; Hasenstaub et al. 2005; Shu et al. 2006; Desai and Walcott 2006; Prescott et al. 2006; Pospischil et al. 2007; Piwkowska et al. 2008; Pospischil et al. 2009; for a more comprehensive overview, see Destexhe and Bal 2009). In the present chapter, the latter technique is reviewed.

6.1.2 Principle of the Dynamic-Clamp Technique

The dynamic-clamp technique is implemented in current-clamp mode, in which the membrane potential (V) is recorded, while a current $I_{\rm inj}$ is injected in the membrane. This situation corresponds to the following membrane equation:

$$C_{\rm m}\frac{{\rm d}V}{{\rm d}t} = -g(V - E_{\rm rev}) + I_{\rm inj}\,, \qquad (6.1)$$

where $C_{\rm m}$ is the membrane capacitance, g is the "natural" membrane conductance, and $E_{\rm rev}$ is its reversal potential. Note that the injected current $I_{\rm inj}$ is of opposite sign than this ionic current because, by convention, positive charges injected into the cell are considered positive current, while outward is positive for ionic currents. From this, it is easy to see that, if this injected current is made to depend, in a controlled way, on the recorded membrane potential V of the cell, according to this equation, this current is strictly equivalent to adding an additional open ion channel to the membrane:

$$I_{\rm inj} = -g_{\rm add}(V - E_{\rm add}) \qquad (6.2)$$

and

$$C_{\rm m}\frac{{\rm d}V}{{\rm d}t} = -g(V - E_{\rm rev}) - g_{\rm add}(V - E_{\rm add}), \qquad (6.3)$$

where $g_{\rm add}$ is the added conductance and $E_{\rm add}$ is its reversal potential. This is exactly what happens in dynamic clamp: a loop between recorded and injected current is implemented, calculated according to (6.2) from preestablished $g_{\rm add}$ and $E_{\rm add}$. This calculation can be done either with an analog device, the dependency between $I_{\rm inj}$ and V being really instantaneous in this case, or digitally using a computer, in which case the time-step required for the calculation, i.e. the delay between V measure and $I_{\rm inj}$ injection, has to be as small as possible.

In this way, any conductance waveform can, indeed, be inserted in the membrane since $g_{\rm add}$ can be time-dependent. $g_{\rm add}$ can also be negative, which provides a way for subtracting existing channels from the membrane, through the injection of a negative image of the current flowing through the biological channels, i.e. the two currents canceling each other out. In principle, any type of channel can be added (or subtracted) by this procedure: to insert a voltage-dependent channel, $g_{\rm add}$ itself has to be calculated in real time, using the recorded V and any necessary equations describing the voltage-dependency; to insert a virtual chemical synapse, the synaptic conductance change has to be triggered by some presynaptic signal. This signal can be set in advance, but it can also correspond to an AP detected in a biological cell, in which case a virtual connection can be created between two previously unconnected cells. It can also originate from a model of a cell or network simulated in real time, i.e. with the time-step used to numerically integrate the model equations equal to the time-step of the dynamic-clamp system. Moreover, such a model cell can also receive inputs triggered by APs detected in the recorded cell, which establishes a bidirectional connection between the biological cell and the model cell, creating a small hybrid network (Le Masson et al. 1995).

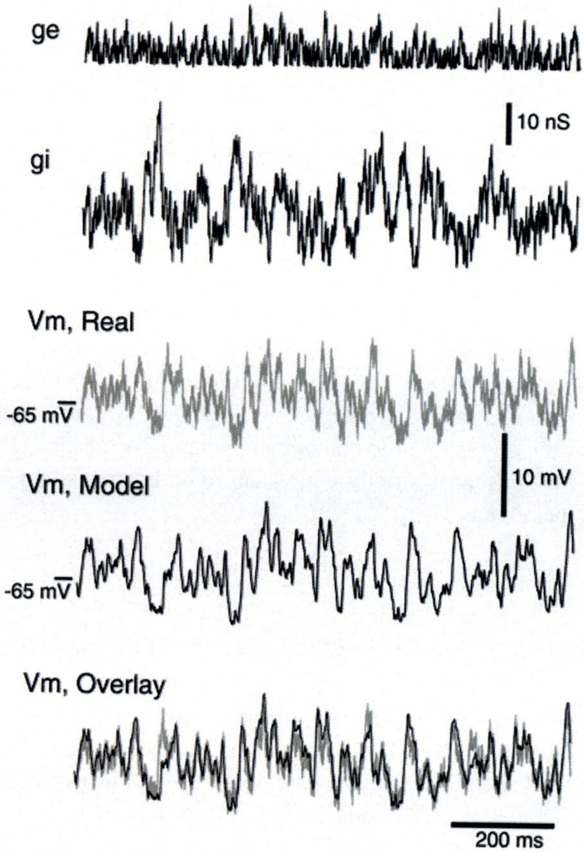

Fig. 6.1 Test of the dynamic-clamp method by comparing real and model V_m fluctuations. From *top* to *bottom*: g_e, g_i: excitatory and inhibitory fluctuating conductances injected in a thalamic relay neuron in dynamic-clamp (DCC mode). *Vm, Real*: V_m activity of this neuron. *Vm, Model*: same injection of conductance in a passive single-compartment model. The leak conductance was adjusted to match the input resistance (47 MΩ) and resting V_m (−63.5 mV) of that particular neuron. V_m, Overlay: both V_m overlayed (*gray* = real V_m, *black* = model V_m), showing that most of the fluctuations are due to the conductance injection. Modified from Wolfart et al. (2005)

From the analysis of the equations above, it is clear that the dynamic-clamp loop is, indeed, equivalent to the insertion of a chosen ion channel, or combination of ion channels since an equivalent conductance and an equivalent E_{rev} can be used in the same manner, at the site of the recording. In a recent review (Piwkowska et al. 2009), it was clarified to which extend such an equivalence can, or is, reached, if the dynamic clamp is able to shunt the membrane, and how the dynamic clamp allows change of the effective conductance and, thus, time constant of the cellular membrane of real neurons.

A test of the dynamic-clamp technique is illustrated in Fig. 6.1. The stochastic conductances of the point-conductance model were injected in a thalamic relay neuron, and the V_m activity resulting from this injection was compared to a passive single-compartment model. The very good match of both V_m activities shows that most of the fluctuations are due to the conductance injection, while other factors (channel noise, instrumental noise, capacitive artifacts, and intrinsic conductances) are of much smaller amplitude at this membrane potential level (Wolfart et al. 2005; for dynamic-clamp experiments on thalamic neurons, see Sect. 6.4).

6.2 Recreating Stochastic Synaptic Conductances in Cortical Neurons

In the remaining sections of this chapter, we will outline various applications of the dynamic-clamp protocol, with focus on experimental models of synaptic noise and their impact on the integrative properties of neurons. We start with a dynamic-clamp application which demonstrates the recreation of in vivo type activity in neurons in vitro.

6.2.1 *Recreating High-Conductance States In Vitro*

One of the most efficient way to recreate, and study, high-conductance states in vitro is to make use of the point-conductance model introduced in Sect. 4.4. This model offers the possibility of independently controlling the mean and variance of conductances, which, as we will see below, turns out to be a critical feature to fully understand the effect of synaptic noise on neurons.

This approach was initiated by Destexhe and colleagues, in which the point-conductance model was injected into cortical pyramidal cells in rat prefrontal cortex slices (Destexhe et al. 2001). When injecting the fluctuating conductances $g_{\mathrm{e}}(t)$ and $g_{\mathrm{i}}(t)$ given by this model, one first has to estimate and adjust the specific model parameters to account for the different size and input resistance of each cell. Table 4.2 shows examples of the so estimated optimal parameters of the point-conductance model obtained from biophysical models with the same densities of all conductances, but various different cellular morphologies. As can be inferred from this example, the average conductances ($g_{\mathrm{e}0}$, $g_{\mathrm{i}0}$) clearly depend on cellular morphology, which is expected because larger cells have larger number of synapses. Consequently, $g_{\mathrm{e}0}$ and $g_{\mathrm{i}0}$ vary approximately as the inverse of the input resistance of the cell (Table 4.2). Interestingly, although the absolute values of $g_{\mathrm{e}0}$ and $g_{\mathrm{i}0}$ clearly depended on the particular cell morphology, their ratio is approximately constant ($g_{\mathrm{i}0} \simeq 5\, g_{\mathrm{e}0}$; see Table 4.2). Table 4.2 also shows that the time constants τ_{e} and τ_{i} seem approximately independent of cellular morphology, with the exception of the layer III pyramidal cell. The latter effect is presumably due to the fact that these cells are electrotonically more compact and have fewer excitatory synapses than large cells from deep layers.

Taking advantage of the relation between the parameter values and the input resistance of a given cell (Table 4.2), these parameters can be obtained heuristically using the following criteria. The parameters $g_{\mathrm{e}0}$ and $g_{\mathrm{i}0}$ are estimated such as to provide a $\sim$15 mV depolarizing drive and 4–5 times reduction of input resistance, as found experimentally (Paré et al. 1998b; Destexhe and Paré 1999). The SDs of the conductance σ_{e} and σ_{i} (or D_{e} and D_{i}) can then be adjusted to yield voltage fluctuations with a standard deviation around $\sigma_V = 4$ mV. These conditions are illustrated in Fig. 6.2a for the voltage effect and Fig. 6.2b for the input resistance.

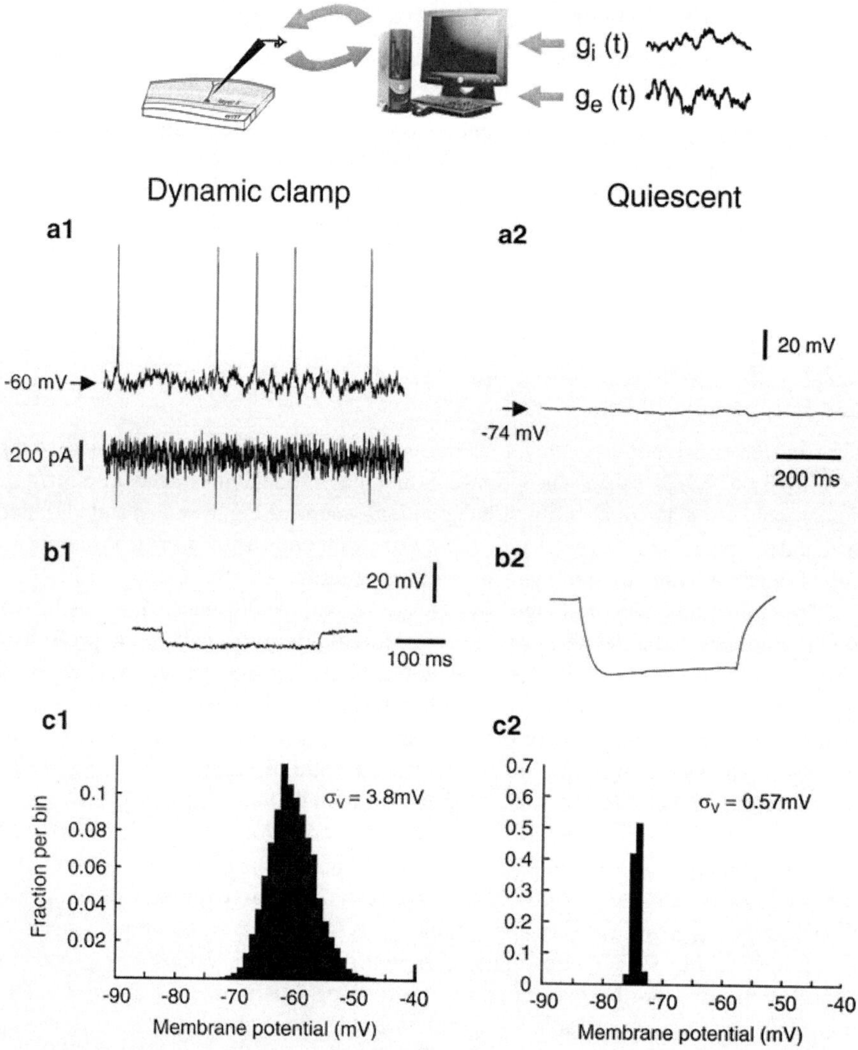

Fig. 6.2 Dynamic-clamp injection of the point-conductance model in neurons from rat prefrontal cortex in vitro. *Top*: scheme of the dynamic-clamp, the point-conductance model is simulated and the excitatory and inhibitory conductances are injected in a living neuron from rat prefrontal cortex using dynamic-clamp. (**a**) Intracellular recording of a prefrontal cortex layer V pyramidal cell in control condition (**a**2), and injected with the point-conductance model (**a**1; $g_{e0} = 0.014\,\mu$S, $g_{i0} = 0.05\,\mu$S, $\sigma_e = 0.0058\,\mu$S, $\sigma_i = 0.0145\,\mu$S, $\tau_e = 2.7$ ms and $\tau_i = 10.7$ ms). The current computed by the point-conductance model, and injected in real time is depicted in (**a**1), lower trace. The point-conductance clamp depolarized the cell by about 15 mV, and introduced membrane potential fluctuations. (**b**) Average response of a different cell to a 200 pA hyperpolarizing pulse in control (**b**2) and in point-conductance clamp (**b**1) conditions (10 trials; $g_{e0} = 0.02\,\mu$S, $g_{i0} = 0.1\,\mu$S, $\sigma_e = 0.005\,\mu$S, $\sigma_i = 0.012\,\mu$S, $\tau_e = 2.7$ ms and $\tau_i = 10.7$ ms). The input resistance has been decreased by 4.3 fold. (**c**) Distribution of membrane potential before (**c**2) and after (**c**1) the activation of the point-conductance clamp [same cell as in (**a**)]. The standard deviation of the membrane fluctuations (σ_V) has been increased to about 4 mV. Modified from Destexhe et al. (2001)

The V_m distributions are depicted in Fig. 6.2c. In this study, similar results were obtained for all the neurons tested. Here, again, the absolute values of the parameters g_{e0}, g_{i0}, σ_e, σ_i were found to depend on the cell, but their ratio was kept constant: $g_{e0}/g_{i0} = 0.2$, $\sigma_e/\sigma_i = 0.4$). Thus, the point-conductance representation, when applied in conjunction with the dynamic-clamp protocol, does indeed recreate intracellular conditions similar to that found in in vivo recordings (see Chap. 3).

6.2.2 High-Discharge Variability

To test whether the timing of fluctuations, and therefore the spike discharges, are also similar to in vivo, the statistics of the spontaneous discharge recreated by the point-conductance model can be analyzed (see Sect. 5.4). A quantification of the irregularity of an observed neuronal spike trains is provided in terms of the coefficient of variation C_V (5.1). In an experiment by Destexhe and colleagues (Destexhe et al. 2001), the C_V was computed for spike trains from various models and corresponding dynamic-clamp experiments mimicking the spontaneous activity observed in vivo. It was found that already for 20 s of spontaneous activity, the C_V converges towards a constant value. Here, high C_V values are observed in the detailed biophysical model (Fig. 6.3a), in the point-conductance single-compartment representation (Fig. 6.3b) and in the point-conductance clamp in vitro (Fig. 6.3c). In the latter case, however, the C_V-ISI curve seems to be shifted towards higher ISI values compared to the models. In all cases in these experiments, the C_V values obtained for large ISI (>200 ms) were comparable ($C_V = 0.93 \pm 0.21, 0.94 \pm 0.14$, and 0.72 ± 0.13 for Fig. 6.3a,b,c, respectively). These values are consistent with the averaged C_V values measured in awake monkeys (Softky and Koch 1993; Shadlen and Newsome 1998).

By comparing the conductance noise and current noise injected during a dynamic-clamp experiment, it is possible to determine to what extend the high-conductance state of the neuron is necessary to reproduce the high variability of discharges. As shown above, the neuronal discharge at rates typical for spontaneous activity in the cortex (Fig. 6.4a, top) displays a high C_V (C_V >0.7) and is characterized by a gamma-distributed ISIH and flat autocorrelogram (Fig. 6.4b). In order to investigate now to which extend the irregular discharge statistics in high-conductance states can be reproduced by injecting current noise into the cell, the statistics of spontaneous responses was studied for current noise with variable time constant τ_I (Destexhe et al. 2001). For small τ_I (<20 ms), the C_V was found to be low around 0.6 (Fig. 6.5a), and the ISIH is, in this case, well described by a gamma distribution (Fig. 6.5b, top). Increasing τ leads to an increase in the ISI variability, and for large noise time constants (>30 ms), the C_V is around 0.8. However, the distribution of ISI and autocorrelograms shows a clear deviation from what is expected from a gamma process (Fig. 6.5b, middle and bottom). The strong peak at low ISIs indicates the preference to produce "bursts" in response to

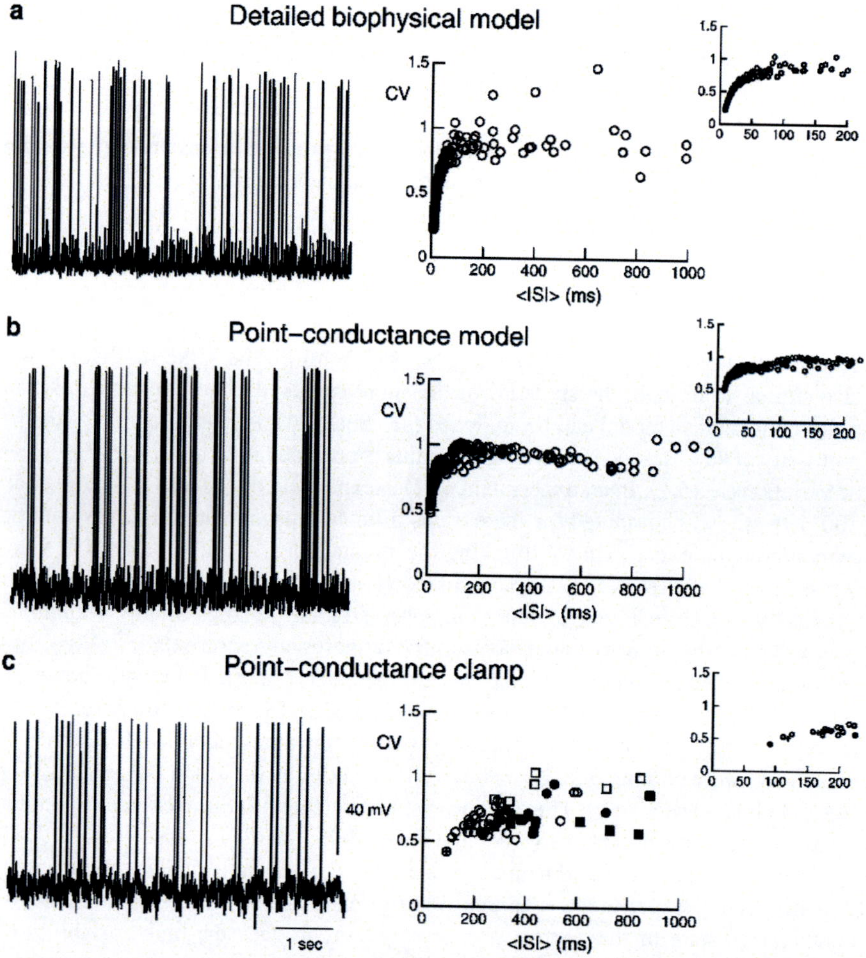

Fig. 6.3 High variability of spontaneous discharges in models and dynamic-clamp experiments. The *left panels* show a six-second trace of spontaneous activity to illustrate the variability of discharges; the *right panels* show the coefficient of variation [C_V, (5.1)] calculated for periods of > 33 s of activity in different conditions and represented against the mean interspike interval (<*ISI*>). (**a**) Detailed biophysical model (same parameter settings as in Fig. 4.10). The different C_V values on the left were obtained by varying excitatory and inhibitory release frequencies (range of 0.5–2.9 Hz and of 4.0–8.0 Hz, respectively). (**b**) Point-conductance model (identical parameters as in Fig. 4.17; g_{e0} and g_{i0} were varied with the range of 0.003–0.035 μS and 0.017–0.145 μS, respectively). (**c**) Dynamic-clamp injection of the point-conductance model in vitro (same parameters as in Fig. 6.2). The different symbols in the right panel indicate different cells; the different points correspond to variations of the parameters (g_{e0} = 0.005–0.0375 μS; g_{i0} = 0.025–0.05 μS; σ_e = 0.00025–0.009 μS and σ_i = 0.00025–0.033 μS). All three models gave high C_V values. Modified from Destexhe et al. (2001)

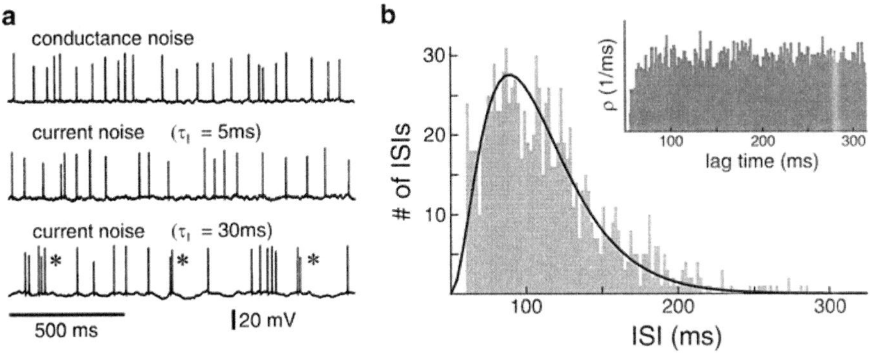

Fig. 6.4 Discharge activity in cortical neurons under current and conductance noise. In the presence of conductance noise, dynamic-clamp experiments show that neurons fire highly irregular ((**a**), *top*; $C_V = 0.7$) in a gamma distributed fashion, characterized by a gamma distributed ISIH ((**b**), *black solid*; fit: $q = 2$, $a = 1{,}996.5$, $r = 0.051\,\text{ms}^{-1}$) and flat autocorrelogram ((**b**), *inset*). In contrast, when current noise was injected in the same cells, the discharge was either more regular ((**a**), *middle*; $C_V = 0.65$), or irregular due to "burst-like" clustering of spikes ((**a**), *bottom*; *stars*; $C_V = 0.78$). Parameters: $I_0 = 0.225\,\text{nA}$, $\sigma_I = 0.1\,\text{nA}$, $\tau_I = 2\,\text{ms}$ ((**a**), *middle*) and $\tau_I = 30\,\text{ms}$ ((**a**), *bottom*); $g_{e0} = g_{i0} = 6\,\text{nS}$, $\sigma_e = 1\,\text{nS}$, $\sigma_i = 2\,\text{nS}$, $\tau_e = 2.72\,\text{ms}$ and $\tau_i = 10.49\,\text{ms}$. Modified from Badoual et al. (2005)

the driving current, a behavior which can also be seen in the autocorrelograms. Here, a peak for small lag times suggests that the spikes are statistically not independent. Indeed, similar results are also reproduced by models (Badoual et al. 2005).

These results show that high C_V values can be obtained by injection of current noise with a larger time constant, but in this case the gamma statistics of the discharge is lost. It appears that the high-conductance state caused conductance-based synaptic background activity is a more natural determinant for the highly variable discharges of cortical neurons (Badoual et al. 2005).

6.3 Integrative Properties of Cortical Neurons with Synaptic Noise

After having briefly outlined in the previous section, the potential of the dynamic-clamp technique to recreate in experiments in a controlled way in vivo type activity, we will now show that this potential can be applied to investigate various consequences of synaptic noise on the dynamics and response behavior of cortical neurons. The results presented here can be viewed as an experimental complement to a number of modeling studies presented in Chap. 5.

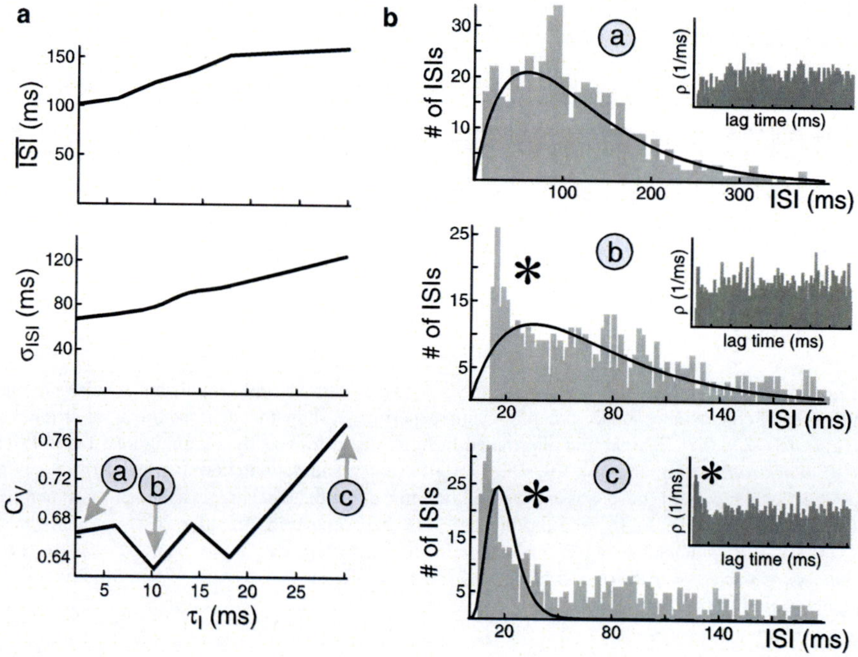

Fig. 6.5 Discharge statistics for neurons in vitro subject to current noise of variable time constant τ_I. (**a**) Mean ISI, ISI standard deviation, and C_V as functions of τ_I. An increase in τ_I leads to an increase in the C_V (**a**), but the spike distribution does no longer follow gamma statistics. (**b**) ISI histograms and autocorrelograms (*insets*) for different τ_I ((**a**) 2 ms; (**b**) 10 ms; (**c**) 30 ms). The ISIHs follow gamma distributions (*black solid*) only for small τ_I values (low C_V; top), but deviate for larger τ_I values from gamma statistics (*middle* and *bottom*). Fits: $q = 1$, $a = 3{,}786.7$, $r = 0.018\,\text{ms}^{-1}$ (**a**); $q = 1$, $a = 1{,}108.9$, $r = 0.028\,\text{ms}^{-1}$ (**b**); $q = 3$, $a = 538.3$, $r = 0.188\,\text{ms}^{-1}$ (**c**). Modified from Badoual et al. (2005)

6.3.1 Enhanced Responsiveness and Gain Modulation

In Sect. 5.3, we saw how synaptic noise might lead to a significant enhancement of the responsiveness of model cells. The same paradigm can be tested experimentally by injecting noise into real neurons in vitro using the dynamic-clamp protocol. In one such experimental implementation, Fellous and colleagues assessed the spiking probability following somatically injected current pulses of fixed duration (20 ms) and varying amplitude (Fellous et al. 2003). Here it was found that, if the cells does not receive simulated background synaptic activity, their responses are all-or-none (Fig. 6.6, *dashed curve*), marking the presence of a current threshold below which signals are not detected, and above which signals are always detected. In the same experiment, the stimulation protocol was then repeated in the presence of simulated synaptic (conductance) noise ($g_{e0} = 7\,\text{nS}$, $g_{i0} = 26\,\text{nS}$, $\sigma_e = 2.5\,\text{nS}$ and $\sigma_i = 7.5\,\text{nS}$). In this case, the slope of the input–output response function

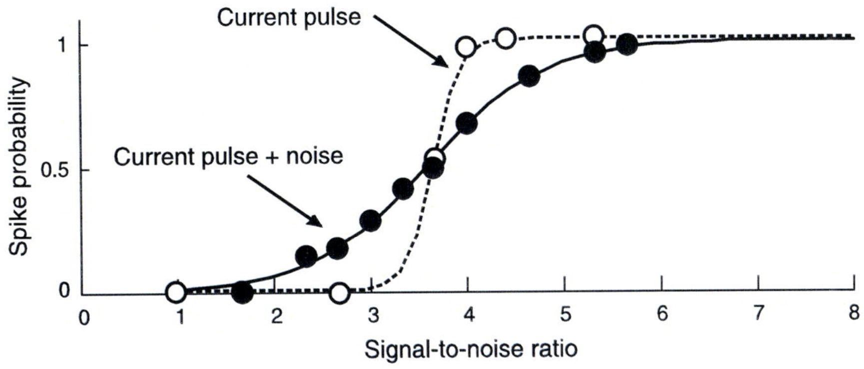

Fig. 6.6 Enhanced responsiveness of neurons in vitro following dynamic-clamp injection of synaptic background activity. The continuous curves show the sigmoid fits to the data points (*circles*) representing the spiking probability of a cell undergoing simulated synaptic noise ($g_{e0} = 7\,\text{nS}$, $g_{i0} = 26\,\text{nS}$, $\sigma_e = 2.5\,\text{nS}$, $\sigma_i = 7.5\,\text{nS}$, spontaneous firing 0.5 Hz) in response to 20 ms current pulses of increasing amplitudes. The cell was able to detect amplitudes as small as 2.8 times the standard deviation of the current resulting from the injection of the synaptic noise. The firing probability was, however, smaller than 0.5. The dashed curves were obtained without noise but equivalent conductance and V_m level ($g_{e0} = 7\,\text{nS}$, $g_{i0} = 26\,\text{nS}$, $\sigma_e = \sigma_i = 0$ nS; no spontaneous firing; *open circles*). Current pulses smaller than 4.5 times the standard deviation of the current resulting from the noise injected previously rarely succeeded in eliciting spiking. Above this value, the probability of spiking rapidly became one. Each data point was obtained from 200 trials. Each curve was established on the basis of at least ten data points. The experiment was repeated 4 times in each condition to assess the robustness of the data acquisition and analysis procedures. Modified from Fellous et al. (2003)

(see definition in Sect. 5.3.2) changed (Fig. 6.6, *continuous line*), indicating that the cell is able to partially detect signals that are below the "classical" threshold. However, the detection probability remains smaller than 0.5. At a probability of 0.5, the ratio of the slopes in the noise case to the no-noise case is 0.51 ± 0.25 (average over 6 cells, 19 curves with pulse widths of 10, 20 or 30 ms; the absolute values for the mean and SD of excitatory and inhibitory conductance differed slightly from cell to cell, due to their difference in input resistance and threshold).

Enhanced responsiveness was also investigated using dynamic-clamp in Layer 5 pyramidal neurons from ferret visual cortex (Shu et al. 2003b). This study examined the contribution of changes in membrane potential, conductance, and variance on the response probability and spike variability measures. This was done by independently varying each of these parameters in the point-conductance model. Changing the membrane potential of the cortical neuron through the intracellular injection of current resulted in a shift in the input–output response function but no effect on the slope (Fig. 6.7a). Conversely, increasing the conductance (by augmenting only the parameters g_{e0} and g_{i0}) resulted in a shift of the input–output response function in the opposite direction (to the right), so that larger synaptic conductances were required to reach the AP threshold (Fig. 6.7b). Here again, the slope of the response function was not significantly affected. Finally, increasing the amount of "noise"

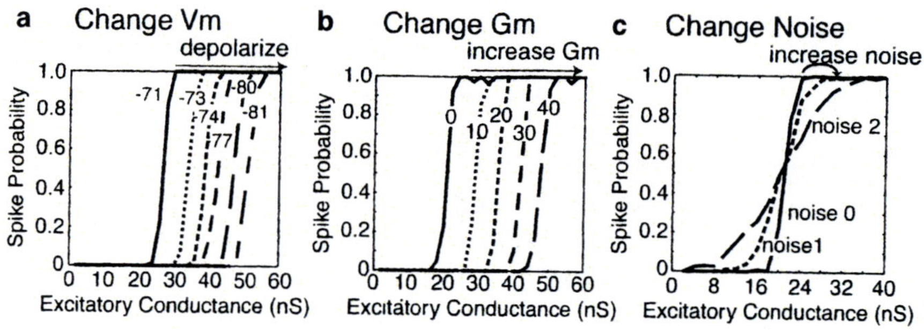

Fig. 6.7 Effects of increase in membrane potential, membrane conductance, and noise on neuronal responsiveness. (**a**) Depolarization of the membrane potential with the intracellular injection of current results in a leftward shift in the input–output response function. (**b**) Increasing the background membrane conductance (with the dynamic clamp system) from 0 to 40 nS results in a shift of the input–output response function to the right. (**c**) Increases in the variance of the membrane potential (noise) result in a change in the slope and a "smoothing" of the response function. Three conditions are compared, *noise 0*: no additional noise (although the baseline conductance for the noise is active); *noise 1*: increases in the standard deviation of both g_e and g_i to 1 nS; *noise 2*: increase of these standard deviations to 2 nS. Modified from Shu et al. (2003b)

was tested by keeping the same average conductance (g_{e0}, g_{i0} fixed) but varying their variances (σ_e, σ_i). This resulted in an increase in the SD of the membrane potential to 2.87 $\pm$ 0.34 mV (n = 7 cells), which is similar to the SD of the membrane potential during natural Up-states. Varying the "noise" resulted here in a smoothing of the input–output response function and a decrease in slope, such that the probability of responding to smaller inputs was enhanced, whereas the response to medium and larger inputs was decreased (Fig. 6.7c). These effects on the input–output response function pivoted on the 50% spike probability point (Fig. 6.7c), as expected from computational models (Destexhe et al. 2001).

The same study (Shu et al. 2003b) also investigated the responsiveness obtained with natural background activity by comparing the above results to the responsiveness during the natural up-states occurring spontaneously in the ferret visual slice. The main finding here was that the spontaneously occurring up-states result in an increase in responsiveness as well as a change in slope of the input–output response function, very similar to the results obtained with artificial background activity (Fig. 6.8). This phenomenon was very reproducible and robust from cell to cell, as the normalized input–output response functions were superimposable (Fig. 6.8b). The comparison between natural and artificial up-states also resulted in superimposable response functions (Fig. 6.9), further indicating their equivalence in terms of responsiveness.

As reviewed in Chap. 5, the conductance fluctuations were found to change the shape of the response function (Hô and Destexhe 2000), an effect which was later called *gain modulation* (Chance et al. 2002). The property of gain modulation by synaptic noise was heavily investigated by dynamic-clamp studies (Chance et al. 2002; Fellous et al. 2003; Mitchell and Silver 2003; Prescott and De Koninck 2003;

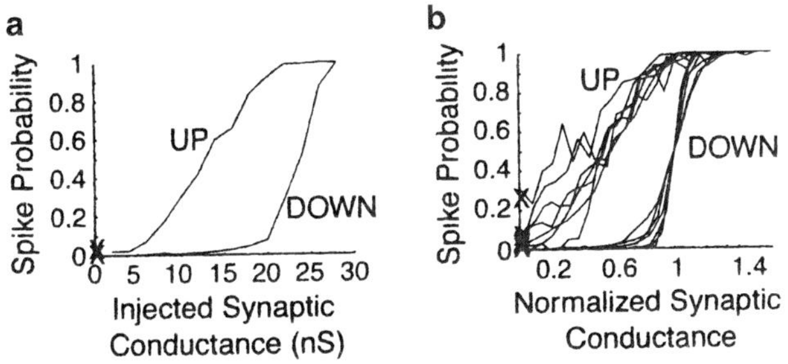

Fig. 6.8 Spontaneously occurring up-states result in an increase in responsiveness to mimicked synaptic conductances as well as a change in slope of the input–output response function. (**a**) Probability of different-amplitude mimicked synaptic conductances to evoke an action potential during the down- and up-states. The up-state is associated with an increase in responsiveness to inputs and a decrease in slope of the input–output response. (**b**) Normalized data from seven cells illustrating the reproducibility of these effects. The synaptic conductance for each cell was normalized such that the conductance that resulted in a 0.5 probability of action potential discharge during the down-state was given a value of 1. The amplitude of each synaptic conductance was randomly chosen from a distribution of 2 to 60 nS in steps of 2 nS. Extrapolation of the graphs to 0 conductance yields an estimate of the probability of spontaneous activity (denoted by an X). Modified from Shu et al. (2003b)

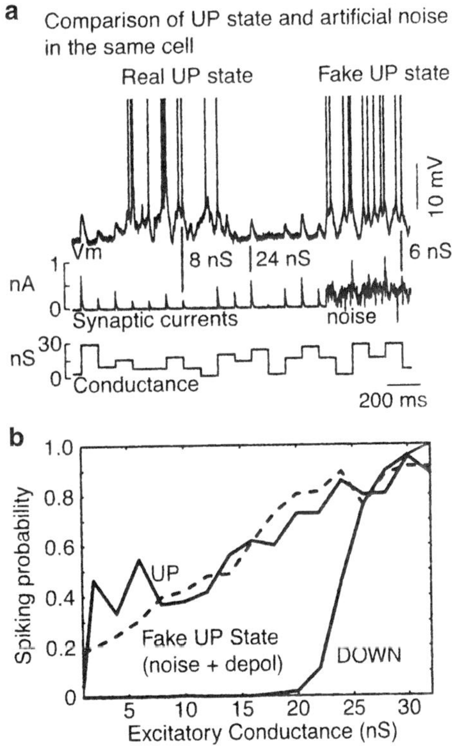

Fig. 6.9 Comparison of up-state and artificial noise in the same cell. (**a**) Mimicking the up-state with combined depolarization (6.6 mV depolarization), an increase in conductance (11 nS combined excitatory and inhibitory conductance), and an increase in noise (4.2 mV standard deviation) results in effects similar to the real up-state. (**b**) The natural up-state resulted in a large shift to the left and a change in the slope of the response function in this cell. Mimicking the up-state with a similar level of depolarization (*depol*) and membrane potential variance resulted in a similar effect. Modified from Shu et al. (2003b)

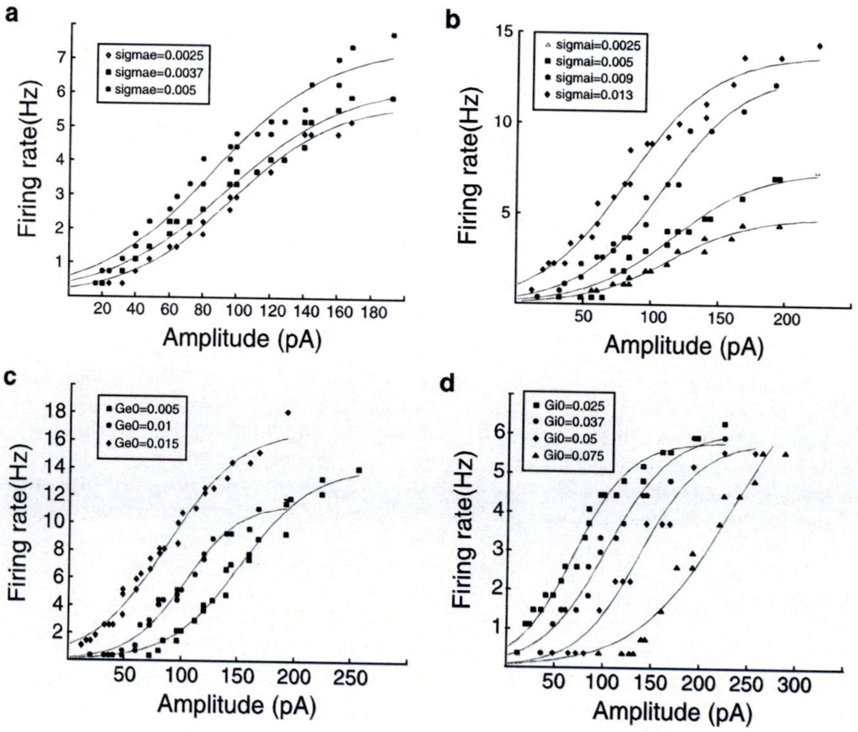

Fig. 6.10 Influence of the point conductance parameters on the F-I curves of prefrontal cortical cells undergoing simulated synaptic background activity. (**a**) Increases in the standard deviation of excitatory inputs slightly increased the slope of the response functions of this cell. (**b**) An increase in the standard deviation of inhibitory inputs increased the slope of the response functions (gain of the cell), and increased its maximum firing rate. (**c**) An increase in the mean excitatory inputs shifted the response functions leftward, keeping their slope constant and increasing its maximal value only slightly. (**d**) An increase in the mean inhibitory conductance drive shifted the response functions toward the right, while their slope (gain) and maximal value remain constant. Panels (**b**) and (**c**) are from the same cell. Panels (**a**) and (**d**) are from two other cells. Modified from Fellous et al. (2003)

Shu et al. 2003a,b; Tateno and Robinson 2006). In one of these studies (Fellous et al. 2003), the gain was calculated by computing the input–output response function of the neuron (from rat prefrontal cortex) using 3 s long current pulses injected at the soma. Figure 6.10 shows the firing rate of a cell when the four parameters of the point conductance model were systematically varied. An increase in mean excitatory or inhibitory conductances resulted in a leftward (7.5 pA/nS) or rightward (2.8 pA/nS) shift of the F-I curve without any significant change to the gain of the cell (Fig. 6.10c,d). The maximal firing rate allowed by the cell given its adaptation currents (saturation) remained almost unaffected by changes in mean conductances. Increases in the SD of the simulated excitatory inputs resulted in a slight shift of the F-I curve upwards (0.6 Hz/nS), and an increase in the slope of the sigmoid fit (in

Fig. 6.10a, with a 100 pA input, the gain of the cell increased by 3.2 Hz/pA per nS increase in σ_e).

Increases in the SD of inhibitory inputs had two effects on the cell's F-I curve. The first was to increase its maximal firing rate for a given current pulse amplitude. The second was to increase the mid-height slope of the curve (in Fig. 6.10b, with a 100 pA input, this slope increased by 6.1 Hz/pA per nS increase in σ_i) compatible with other recent studies performed in constrained excitatory and inhibitory balanced conditions (Chance et al. 2002). The slope (also called *gain*) of the F-I curve taken at mid-height between the spontaneous firing rate, and the maximal firing rate is a measure of the sensitivity of the cell to its inputs. A low gain (slope) indicates that large inputs will be required to induce noticeable changes in firing rate; at high gain, small variations in the inputs will results in large variations in the cell's output firing rate. Note that for this specific cell in the aforementioned study, the increase in gain varied nonlinearly with σ_i: a doubling in σ_i with σ_i = 2.5 nS results in a smaller slope increase than a doubling of σ_i with σ_i = 9 nS. Increases in SD of either the excitatory or inhibitory inputs had the same general effects on the maximal firing rate and slope.

In general, long experiments are required to obtain the response curves needed to investigate gain modulation and cellular responsiveness. For instance, in the experiment displayed in Fig. 6.10, it was not possible to collect data for more than three or four values for each of the four parameters g_{e0}, g_{i0}, σ_e, and σ_i of the stochastic model. In order to better assess the effects of these parameters on the gain of the cell, computational model placed in the same condition as in the experiments can be used. Such simulations show that the working point of the cell is mainly determined by the balance of mean inhibition and excitation, and the SDs of excitatory and inhibitory inputs can individually modulate the gain. In the computational study accompanying the experiment shown in Fig 6.10, the slope range due to σ_e variations was 75 to 89 Hz/nA and was 72 to 92 Hz/nA for σ_i.

Simulations performed with the same model but using stimuli consisting of AMPA conductance changes (instead of current transients; Fellous et al. 2003) yielded here qualitatively similar results for the impact of the various parameters g_{e0}, g_{i0}, σ_e, and σ_i. In fact, these simulation results were in excellent qualitative agreement with the experimental findings of Fig. 6.10: the mean excitation and mean inhibition modulated the working point, and the excitatory and inhibitory variances modulated primarily the gain. Three currents (I_{Na}, I_{kd}, and I_M) were, therefore, sufficient to capture the influence of synaptic background noise on the F-I curve observed experimentally.

6.3.2 Variance Detection

Cortical pyramidal neurons can detect changes in the correlation of background inputs. Previous computational models have shown that in high-conductance states, cortical neurons can, indeed, detect correlation changes as short as 2 ms duration

(Rudolph and Destexhe 2001a). As shown earlier (see Fig. 4.15), the correlation of synaptic inputs translates into the variance of synaptic conductances. Together, these results predict that cortical neurons should be able to detect brief changes in the variance of synaptic conductances.

To test this prediction, the ability of cells recorded in vitro to detect transient changes in the variance of their background synaptic conductances can be assessed. Figure 6.11 shows an example of a cell that received continuous simulated noise, firing spontaneously at less than 1 Hz, with a membrane potential fluctuating around -68 mV $\pm$ 3.6 mV. At predetermined times, the SDs of the noise (both σ_e and σ_i) are doubled for 30 ms every 330 ms, mimicking a 3 Hz signal consisting of synchronous inhibitory and excitatory inputs.

Figure 6.11a (*inset*) shows the average membrane potential and SD around such a pulse, across all trials: the former increases during the signal, but remains smaller than the SD before or after the signal (*horizontal dashed lines*). The cell, however, fires preferentially during these 30 ms transients, as indicated by the firing histogram across about 100 trials (Fig. 6.11a). This shows that the cell is able to detect events that are as short as 10 ms (Fig. 6.11b, left, *dashed curve*), a timescale much shorter than the cells' typical membrane time constant (about 30 ms). Qualitatively similar results were obtained experimentally in several cells, and reproduced in the point-conductance model (Fig. 6.11b right; Fellous et al. 2003).

6.3.3 *Spike-Triggering Conductance Configurations*

Another application of dynamic-clamp experiments with synaptic noise is to investigate the optimal conductance patterns leading to spikes, or the STA of conductances (Pospischil et al. 2007; Rudolph et al. 2007; Piwkowska et al. 2008; Destexhe 2010, 2011).

6.3.3.1 STA of Synaptic Conductances in the Point-Conductance Model

Figure 6.12 shows simulations of the point-conductance model in two different conductance states leading to similar V_m dynamics. With equal conductances (Fig. 6.12, left panels, *Equal conductances*), g_e and g_i are both of low amplitude and comparable magnitude, and the V_m is fluctuating around -60 mV. With inhibition-dominated conductances (Fig. 6.12, right panels, *Inhibition-dominated*), g_e is larger, but g_i needs to be several-fold stronger to yield V_m fluctuations comparable to the "Equal-conductance" model. Such conductance values are more typical of what is usually measured in vivo (Rudolph et al. 2005, 2007; Monier et al. 2008). Both conductances are larger than the resting conductance, a situation which corresponds to a high-conductance state. In fact, a continuum of values of g_e and g_i are possible to yield V_m fluctuations around -60 mV, and these two cases represent the extreme values of this continuum.

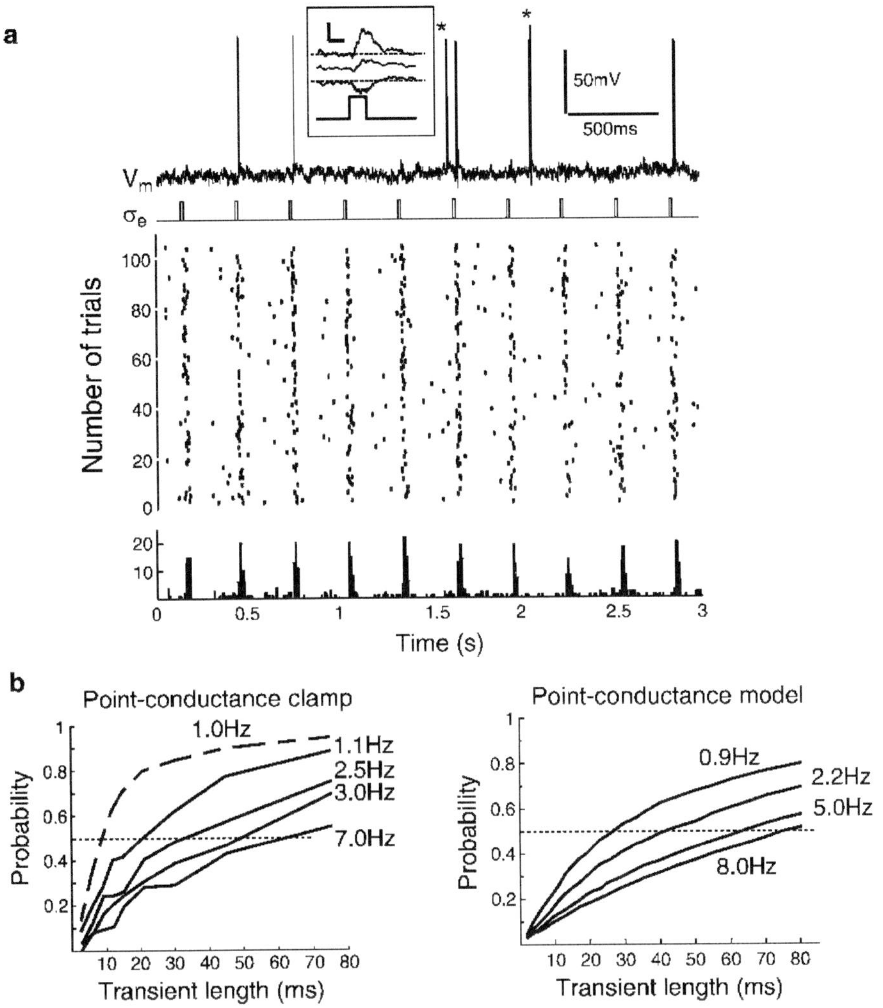

Fig. 6.11 Detection of transient changes in the variance of synaptic inputs. (**a**) A pyramidal cell was injected with background synaptic noise ($g_{e0} = 5$ nS, $g_{i0} = 25$ nS, and $\sigma_e = 5$ nS, $\sigma_i = 12.5$ nS). The standard deviation of the excitatory and inhibitory stochastic variables was doubled for a duration of 30 ms (only σ_e is represented), mimicking the arrival of correlated synaptic inputs. The cell was able to detect this transient by emitting a spike time locked to the signal onset. *Top*: sample trace showing two spontaneous spikes (*) and four evoked spikes. *Middle*: spike rastergram with about 100 trials shown. *Inset*: the *thin curve* shows the average membrane potential computed around all transients, in all trials. The *thick curves* represent the standard deviation of the membrane potential around the transient. Note that the average membrane potential during the transient stayed within the standard deviation of the membrane noise (*horizontal dashed lines*). Inset scale bars = 4 mV, 30 ms. *Bottom*: spike histogram (10 ms bins) of the rastergram above. (**b**) *Left*: signal detection capability (probability that an action potential indicated the presence of a transient input) for varying transient lengths. The *dashed curve* corresponds to the cell shown in panel (**a**). Note that this cell is able to detect about 50% of 10 ms long stimuli. The cell had a spontaneous firing rate of about 1 Hz. The four other curves are from a different cell. Four different levels of spontaneous firing (1.1, 2.5, 3, and 7 Hz corresponding to g_{e0} values of 10, 13, 17, 24 nS, g_{i0} fixed at 60 nS) are represented. *Right*: the point conductance model reproduces qualitatively the experimental data. Note that for low spontaneous firing rates, the detection capabilities of the cell depended nonlinearly on transient lengths (model and experiments). Modified from Fellous et al. (2003)

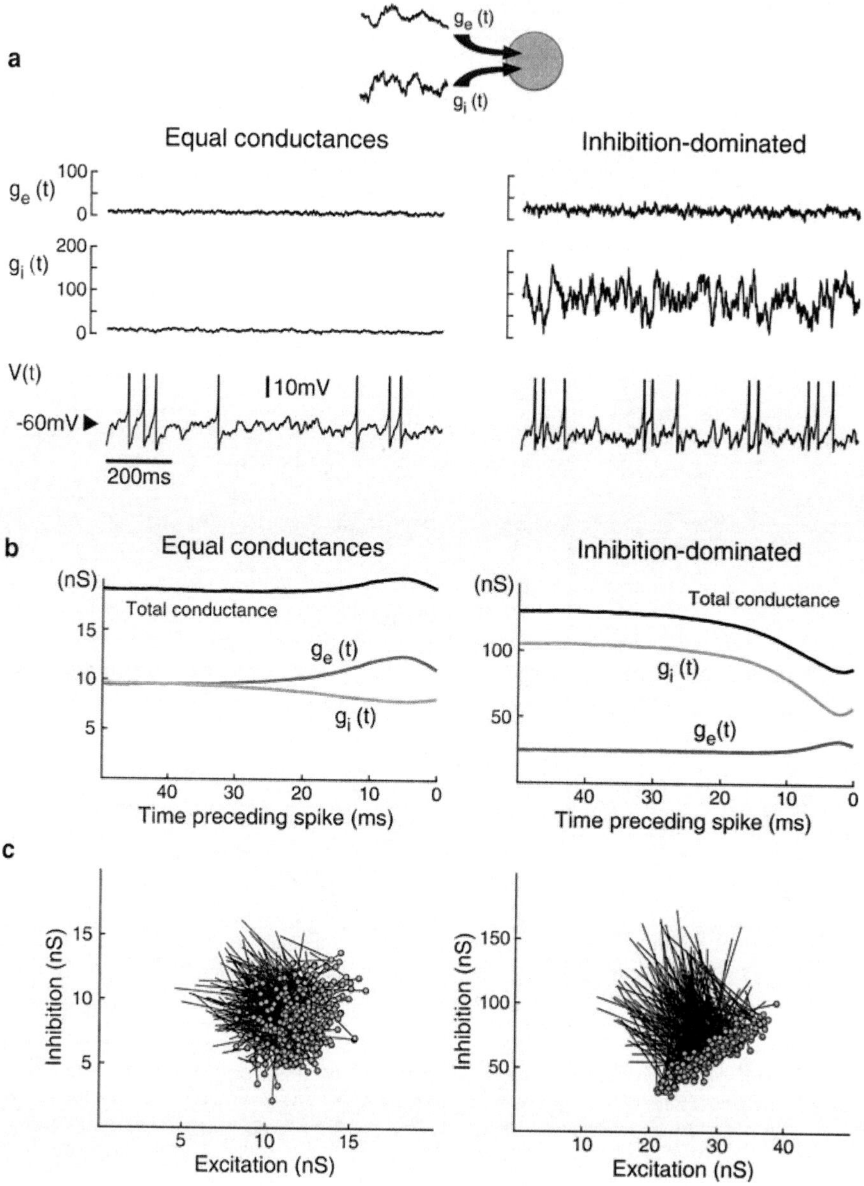

Fig. 6.12 Comparison between equal conductances and inhibition-dominated states in a computational model. (**a**) Equal conductance (*left*; $g_{e0} = g_{i0} = 10\,\text{nS}$, and $\sigma_e = \sigma_i = 2.5\,\text{nS}$) and inhibition-dominated states (*right*; $g_{e0} = 25\,\text{nS}$, $g_{i0} = 100\,\text{nS}$, $\sigma_e = 7\,\text{nS}$ and $\sigma_i = 28\,\text{nS}$) in the point-conductance model. Excitatory and inhibitory conductances, and the membrane potential, are shown from *top* to *bottom*. Action potentials (truncated here) were described by Hodgkin–Huxley type models (Destexhe et al. 2001). (**b**) Average conductance patterns triggering spikes. Spike-triggered averages (STAs) of excitatory, inhibitory, and total conductance were computed in a window of 50 ms before the spike. (**c**) Vector representation showing the variation of synaptic conductances preceding each individual spike. The excitatory and inhibitory conductances were averaged in two windows of 30–40 ms and 0–10 ms (*circle*) before the spike, and a vector was drawn between the obtained values. Modified from Piwkowska et al. (2008)

To determine how these two states differ in their spike selectivity, the spike-triggering conductances are evaluated by averaging the conductance traces collected in 50 ms windows preceding spikes (Piwkowska et al. 2008). This average pattern of conductance variations leading to spikes is shown in Fig. 6.12b. For equal-conductance states, there is an increase of total conductance preceding spikes (*black solid* in Fig. 6.12b, left), as can be expected from the fact that excitation increases (g_e curve in Fig. 6.12b, left). In contrast, for inhibition-dominated states, the total conductance decreases prior to the spike (*black solid* in Fig. 6.12b, right), and this decrease necessarily comes from a similar decrease of inhibitory conductance, which, in this case, is stronger than the increase of excitation (g_i curve in Fig. 6.12b, right). Thus, in such states the spike seems primarily caused by a drop of inhibition.

This pattern is seen not only in the average but also at the level of single spikes. Using a vector representation to display the conductance variation preceding spikes (each vector links the conductance state in a window of 30–40 ms before the spike with that in the 10 ms preceding the spike) shows that the majority of spikes follow the average pattern (Fig. 6.12c). The same features is also present in an IAF model, and thus do not seem to depend on the spike-generating mechanisms.

6.3.3.2 STA of Synaptic Conductances in Dynamic-Clamp Experiments

These patterns of conductance variations preceding spikes were also investigated in real neurons by using dynamic-clamp experiments to inject fluctuating conductances in vitro (Piwkowska et al. 2008). In this case, performing the same analysis as above reveals similar features: STAs of the injected conductances display either increase or decrease in total conductance, depending on the conductance parameters used (Fig. 6.13a), and the vector representations are also similar (Fig. 6.13b). It suggested that these features are independent of the spike generating mechanism but rather are caused by subthreshold V_m dynamics.

6.3.3.3 A Geometrical Interpretation of Conductance STA

In a previous article (Piwkowska et al. 2008), a qualitative explanation was provided to account for the configuration of synaptic conductances just before spikes. The idea is to assume that the neuron behaves as an IAF, in which case one can consider the condition in which the total current is positive at spike time:

$$g_e(E_e - V_t) + g_i(E_i - V_t) + G_L(E_L - V_t) > 0,$$

where V_t is the spike threshold. This inequality defines a half-plane in which (g_e, g_i) must lie at spike time. Figure 6.14a shows graphically how this inequality affects the synaptic conductances. The variable (g_e, g_i) is normally distributed, so that isoprobability curves are ellipses in the plane (plotted in light gray). In that plane, the line $\{g_e + g_i = g_{e0} + g_{i0}\}$ going through the center of the ellipses defines

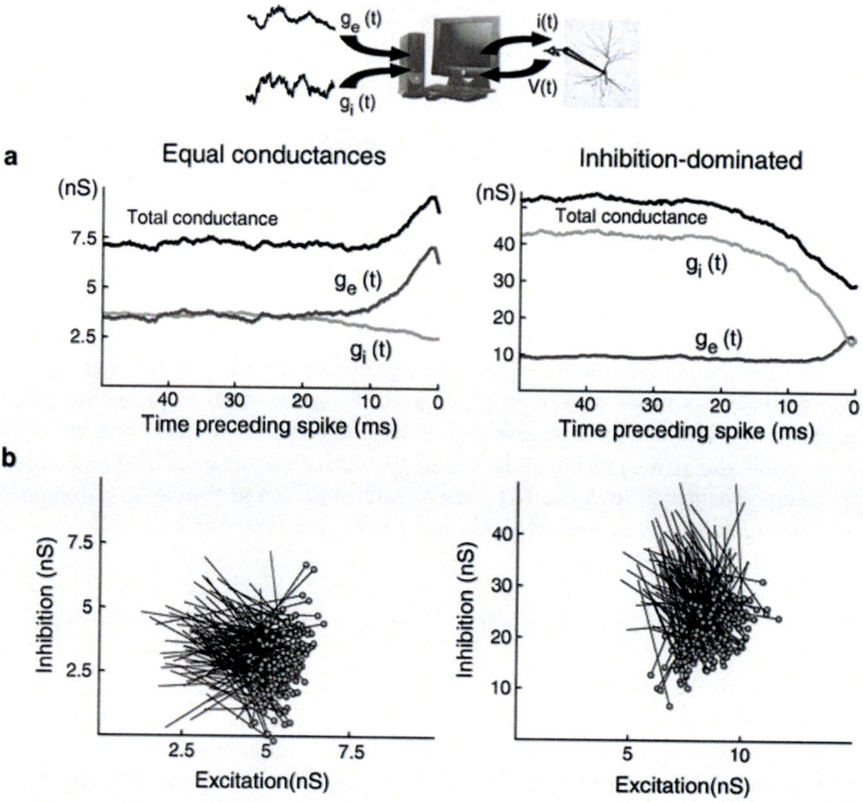

Fig. 6.13 Average conductance patterns triggering spikes in dynamic-clamp experiments. (**a**) Spike-triggered averages of excitatory, inhibitory, and total conductance in a window of 50 ms before the spike in a cortical neuron subject to fluctuating conductance injection. The two states, equal conductances (*left*) and inhibition-dominated (*right*), were recreated similar to the model of Fig. 6.12. Conductance STAs showed qualitatively similar patterns. (**b**) Vector representation showing the variation of synaptic conductances preceding each individual spike (as in Fig. 6.12c). Modified from Piwkowska et al. (2008)

the points for which the total conductance equals the mean conductance, and the line $\{g_e(E_e - V_t) + g_i(E_i - V_t) + G_L(E_L - V_t) = 0\}$ defines the border of the half-plane in which conductances lie at spike time. In the equal conductances regime (Fig. 6.14a, left), synaptic conductances are small and have similar variances, so that isoprobability curves are circular; the intersection of the half-plane with those circles is mostly above the mean total conductance line, so that the total conductance is higher than average at spike time.

In the inhibition-dominated regime (Fig. 6.14a, right), synaptic conductances are large and the variance of g_i is larger than the variance of g_e, so that isoprobability curves are vertically elongated ellipses; the intersection of the half-plane with those ellipses is essentially below the mean total conductance line, so that the total conductance is lower than average at spike time. More precisely, when

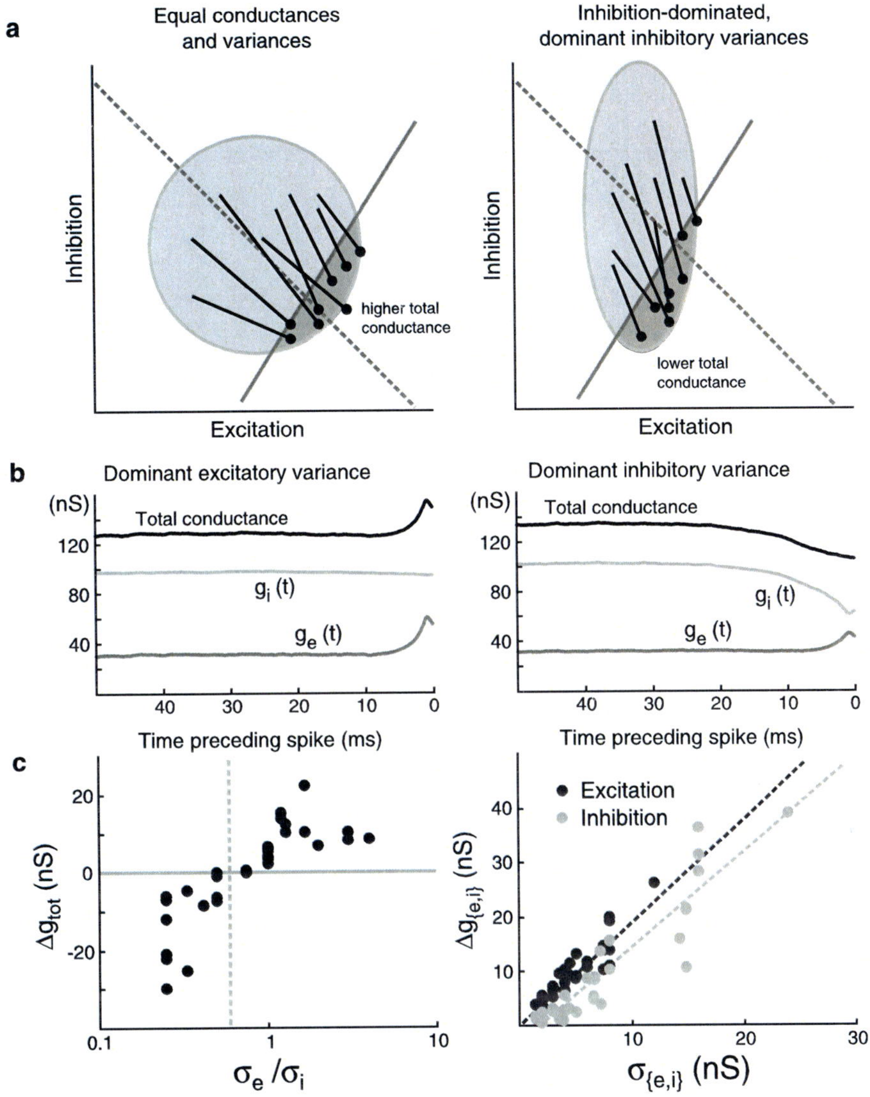

Fig. 6.14 Geometrical interpretation of the average conductance patterns preceding spikes and test in dynamic-clamp. (**a**) *Light-gray ellipses*: isoprobable conductance configurations. *Dark-gray area*: conductance configurations for which the total synaptic current is positive at spike threshold. The vector representations of Figs. 6.12 and 6.13 are schematized here, and compared to the lines defined by $\{g_e(E_e - V_t) + g_i(E_i - V_t) + G_l(E_L - V_t) = 0\}$ (*solid gray*) and $\{g_e + g_i = g_{e0} + g_{i0}\}$ (*dashed gray*). The angle between the two lines and the aspect ratio of the ellipse determine whether spikes are preceded, on average, by total conductance increase (*left*) or decrease (*right*). (**b**) Spike-triggered average conductances obtained in dynamic-clamp, illustrating that for the same average conductances, the variances determine whether spikes are preceded by total conductance increase (*left*) or decrease (*right*). (**c**) Geometrical prediction tested in dynamic-clamp (*left*): grouped data showing total conductance change preceding spikes as a function of the ratio σ_e/σ_i. The *dashed line* ($\sigma_e/\sigma_i = 0.6$) visualizes the predicted value separating total conductance increase cases from total conductance decrease cases. In addition (*right*), dynamic-clamp data indicate that the amplitude of change of each of the conductances before a spike is linearly correlated with the standard deviation parameter used for this conductance. Modified from Piwkowska et al. (2008)

isoprobability curves are circular (equal variances), then the expected total conductance is unchanged at spike time when the lines

$$\{g_e(E_e - V_t) + g_i(E_i - V_t) + G_L(E_L - V_t) = 0\}$$

and

$$\{g_e + g_i = g_{e0} + g_{i0}\}$$

are orthogonal, i.e., when $E_e - V_t + E_i - V_t = 0$. Spikes are associated with increases in conductance when the first line has a higher slope, i.e., when $E_e - V_t > V_t - E_i$, which is typically the case.

When isoprobability curves are not circular, one can look at the graph in the space $(\frac{g_e}{\sigma_e}, \frac{g_i}{\sigma_i})$ where isoprobability curves are circular. Then the orthogonality condition between the lines

$$\left\{\frac{g_e}{\sigma_e}\sigma_e(E_e - V_t) + \frac{g_i}{\sigma_i}\sigma_i(E_i - V_t) + G_L(E_L - V_t) = 0\right\}$$

and

$$\left\{\frac{g_e}{\sigma_e}\sigma_e + \frac{g_i}{\sigma_i}\sigma_i = g_{e0} + g_{i0}\right\}$$

reads

$$\sigma_e^2(E_e - V_t) + \sigma_i^2(E_i - V_t) = 0\,.$$

It follows that spikes are associated with increases in total conductance when the following condition is met:

$$\frac{\sigma_e}{\sigma_i} > \sqrt{\frac{V_t - E_i}{E_e - V_t}}\,.$$

One can also recover this result by calculating the expectation of the conductance change conditionally to the fact that the current at spike threshold is positive (implicitly, this neglects the correlation time constants of the synaptic conductances). Using typical values ($V_t = -55$ mV, $E_e = 0$ mV, $E_i = -75$ mV), it was concluded that spikes are associated with increases in total conductance when $\sigma_e > 0.6\ \sigma_i$. This inequality is indeed satisfied in the equal conductances regime and not in the inhibition-dominated regime investigated above (see details in Piwkowska et al. 2008).

6.3.3.4 Testing the Geometrical Prediction with Dynamic-Clamp

The geometrical reasoning presented in the previous section predicts that the sign of the total synaptic conductance change triggering spikes depends only on the ratio of synaptic variances, and not on the average conductances. In a study by Piwkowska and colleagues, this prediction was systematically tested using dynamic-clamp injection of fluctuating conductances in vitro (Piwkowska et al. 2008). There, in eight regular spiking cortical neurons, different parameter regimes in a total of 36 fluctuating conductance injections were scanned. Figure 6.14b shows two examples from the same cell: both correspond to an average "high conductance" regime, dominated by inhibition, but in one case it is the variance of excitation, in the other case the variance of inhibition, that is higher. One can see that the total conductance before the spike increases in the first case, but decreases in the second. Figure 6.14c (left) further shows the average total conductance change preceding spikes as a function of σ_e/σ_i, for all the 36 injections: the vertical dashed line represents the predicted value of $\sigma_e/\sigma_i = 0.6$, which indeed separates all the "conductance drop" configurations from the "conductance increase" configurations. Even though the prediction is based on a simple IAF extension of the point-conductance model, the ratio of synaptic variances can predict the sign of the total conductance change triggering spikes in biological cortical neurons subjected to fluctuating excitatory and inhibitory conductances.

In addition (Fig. 6.14c, right), the dynamic-clamp data show that the average amplitude of change ($\Delta g_e = g_{e0}k_e$) of each synaptic conductance preceding a spike is related, in a linear way, to the SD of this conductance. It was found that for a fixed value of SD, there is no significant influence of the average conductance. This observation is consistent with the idea that in all the cases studied in Piwkowska et al. (2008), the firing of the cell was driven by fluctuations in the V_m, rather than by a high mean V_m value.

Taken together, these theoretical and experimental analyses indicate that the average total conductance drop preceding spikes, as seen in the "high conductance" case initially considered (Fig. 6.12), is not a direct consequence of the high-conductance state of the membrane itself, but is in fact related to the high inhibitory variance, which is indeed to be expected especially when the mean inhibitory conductance is also high (as confirmed by conductance measurements, see Chap. 9).

This result stresses the importance of evaluating the variances of synaptic conductances, in addition to their averages, when analyzing V_m fluctuations recorded in vivo. The VmD method (Rudolph et al. 2004; see Sect. 8.2) is one of the methods available so far that provides an estimate of the variance of conductances. Other methods have been proposed (reviewed in Monier et al. 2008), but these methods estimate the "trial-to-trial variability" of conductances, which is different than the variance which is defined in term of the temporal evolution of the conductance. In Chap. 9, we will see that this pattern of conductance STA is indeed identified in awake and naturally sleeping cats (Rudolph et al. 2007), and leads to the observation of both types of firing regimes described above, i.e., average total conductance increase and average total conductance drop, with a majority of cases displaying the inhibition-dominated, conductance-drop pattern.

6.4 Integrative Properties of Thalamic Neurons with Synaptic Noise

The previous section focused on the characterization of the integrative properties of cortical neurons in noisy states, generated experimentally through dynamic-clamp injection of stochastic conductances and currents. In this section, we will explore the integrative properties of thalamic neurons in the presence of synaptic noise, using dynamic-clamp techniques (Wolfart et al. 2005). The interest of this study is that like cortical neurons, thalamic neurons are also characterized by high-amplitude V_m fluctuations and synaptic noise in vivo. But most importantly, these neurons display prominent intrinsically generated bursting properties, which gives the opportunity to study the effect of noise on burst-generating neurons.

6.4.1 Thalamic Noise

Like the cortex, the membrane potential of thalamic neurons highly fluctuating during nonanesthetized states (Fig. 6.15), such as SWS or REM sleep, as well as in the awake condition (not shown). However, how these neurons function with such a massive amount of "noise" is unknown. According to the classic view, thalamocortical cells function in two intrinsically generated firing modes (Llinás and Jahnsen 1982). At depolarized membrane potentials, these neurons fire single APs, faithfully transmitting synaptic inputs. This relay or *single-spike mode* is mainly found during the awake state (McCormick and Bal 1997; Sherman and Guillery 2001). At hyperpolarized potentials, the activation of low-threshold calcium (T-type) channels triggers high-frequency bursts of APs (Jahnsen and Llinás 1984). The *burst mode* is mostly found during SWS and epileptic absence seizures, when thalamocortical cells participate in synchronous bursting of the thalamic network, functionally uncoupling the cortex from visual input (McCormick and Feeser 1990).

As described in much detail in Chaps. 3 and 5, neurons in vivo generally experience a noisy high-conductance state which is likely to interact with their built-in integrative properties (Steriade 2001; Destexhe et al. 2003a). Thalamocortical cells recorded in vivo display as well a highly irregular intracellular activity (see Fig. 6.15) and are in a high-conductance state, in particular during corticothalamic barrages (Contreras et al. 1996; see Fig. 6.16). A given thalamocortical cell receives 4,000 to 8,000 synapses (Wilson et al. 1984; Liu et al. 1995), of which about 30% have direct cortical origin (Liu et al. 1995; Erisir et al. 1997a,b; Van Horn et al. 2000; Sherman and Guillery 2002). In addition, they are innervated by intrathalamic inhibitory neurons (interneurons and reticular thalamic neurons), which also receive direct cortical inputs and account for about 30% of synapses onto thalamocortical cells (Liu et al. 1995; Erisir et al. 1997; Van Horn et al. 2000). Thus, about 60% of synapses of thalamocortical neurons are directly or indirectly related to the activity of corticothalamic axons (in addition to other afferents), but it remains uncertain

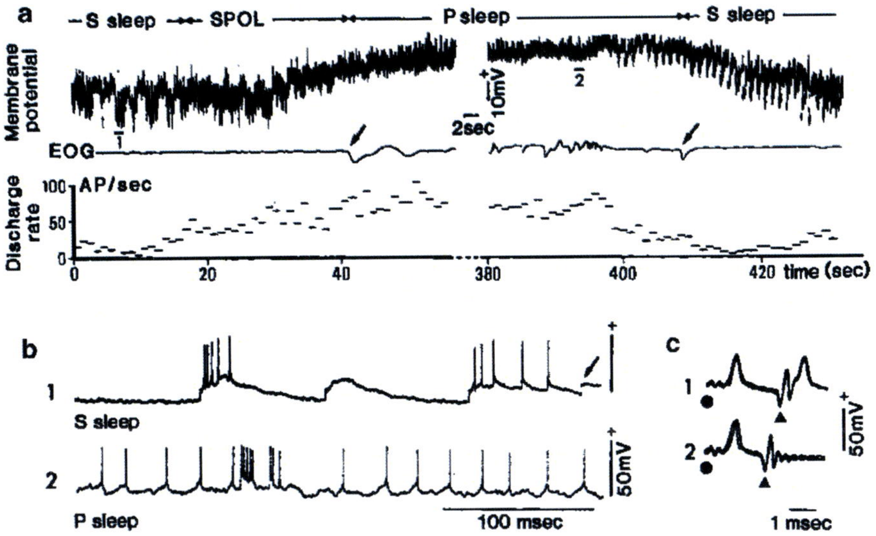

Fig. 6.15 Tonic depolarization of an LGB relay neuron during REM sleep. (**a**) Simultaneous display of behavioral state, DC cell recording, eye movements (EOG), and discharge rate (1 s bin width). *Left*: neuron depolarized during SPOL and had already depolarized by 8 mV when the animal entered REM sleep (P sleep; *arrow*: first eye movement). Depolarization was maintained throughout REM sleep. *Right*: upon last eye movement (*arrow*) of REM sleep, the neuron repolarized as the animal went back to slow-wave sleep (S sleep). (**b**) Enlarged segments of spontaneous activity. (**b**1) slow-wave sleep activity was composed of isolated excitatory postsynaptic potentials (EPSPs) (*arrow*) that did not always fire the cell, and of large depolarizations with or without a burst of spikes. (**b**2) During REM sleep, action potentials arose from small depolarizations. The rapid discharge was disrupted by short depolarizations which gave birth to a cluster of spikes of decreasing amplitude. (**c**) Identification of neuron. (**c**1) orthodromic response to optic tract stimulation (*circle*), antidromic action potential to visual radiation stimulation (*triangle*). (**c**2) positive collision test. Modified from Hirsch et al. (1983)

whether overall this feedback is excitatory or inhibitory for thalamocortical cells (Koch 1987). It has been proposed that corticothalamic feedback could switch thalamocortical neurons between burst and single-spike modes (Destexhe et al. 1998a; Sillito and Jones 2002), but this was never investigated directly.

To assess the effect of synaptic noise on thalamic neurons, Wolfart and colleagues recorded thalamocortical neurons in dorsolateral geniculate nucleus (LGNd) slices of guinea pigs using intracellular electrodes and the dynamic-clamp technique (Wolfart et al. 2005). In this study, 52 neurons from 36 animals were analyzed. They had a resting potential of -63 ± 0.4 mV, an input resistance of $61 \pm 4\,\mathrm{M}\Omega$, and exhibited rebound burst discharges accompanied by low-threshold calcium spikes (LTS) upon repolarization following hyperpolarization (Fig. 6.17a, left inset). These properties, as well as morphological reconstructions of biocytin-filled cells unambiguously identify the neurons as thalamocortical relay neurons.

In Wolfart et al. (2005), the standard protocol used to assess the integration properties of thalamocortical cells consisted of a recording at resting potential,

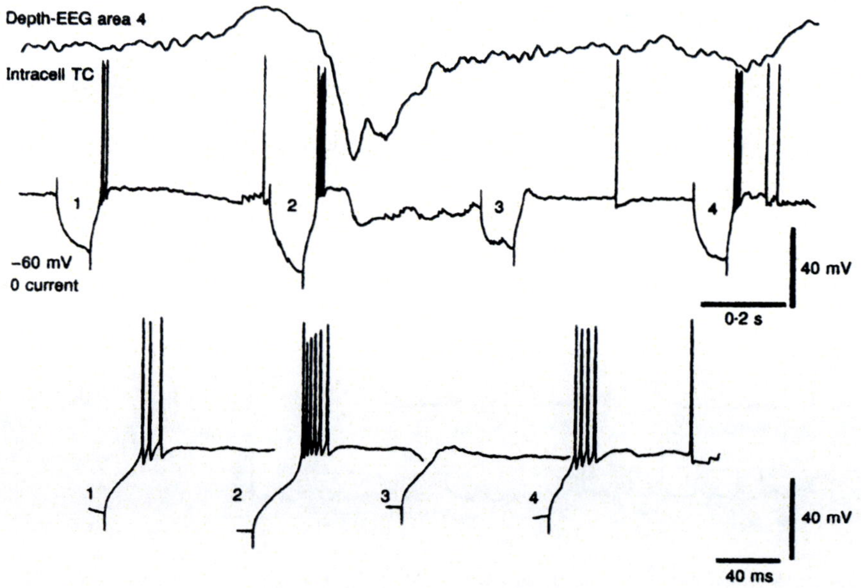

Fig. 6.16 Changes of input resistance in thalamic relay cells during periods of slow oscillatory behavior. The responsiveness of a TC cell from VL (*Intracell TC*), recorded simultaneously with the corresponding motor cortical depth-EEG, was tested with hyperpolarizing pulses (1-4). Maximum voltage deflection and rebound burst response were obtained during the depth-positive electroencephalogram (EEG) wave ("Down-state"; detail of the burst expanded below in 2). The smallest voltage deflection which did not trigger a burst, but a subthreshold low-threshold calcium spikes (LTS), was obtained during the depth-negative phase in the EEG (being of up-state; detailed below in 3). The response to pulse 1 was intermediate between pulses 3 and 4. Modified from Contreras et al. (1996)

where excitatory signal input conductances were injected at 5 Hz without additional background conductances (Fig. 6.17a, *Quiescent*). Subsequently, excitatory and inhibitory background conductances were added without fluctuations (*Static*). Furthermore, the total background conductance was adjusted such that the cells input resistance was approximately 50% of its initial value (Fig. 6.17a, compare insets of quiescent and static traces), similar to the conductance estimates in thalamocortical cells in vivo (see Fig. 6.16). The same background conductances were then injected with stochastic fluctuations (*Noise*). In order to separate the effect of noise from simple depolarization or hyperpolarization, when necessary, the effect of background conductances was compensated by injecting a small DC current such that the mean potential was similar under these conditions (Fig. 6.17a). Finally, it was found (Wolfart et al. 2005) that the variance of the membrane potential was low in the quiescent (0.71 $\pm$ 0.03 mV) and static conditions, and was increased with synaptic noise (3.65 $\pm$ 0.13 mV for 24 cells recorded in Wolfart et al. 2005) to an amplitude consistent with voltage fluctuations in vivo (Paré et al. 1998b).

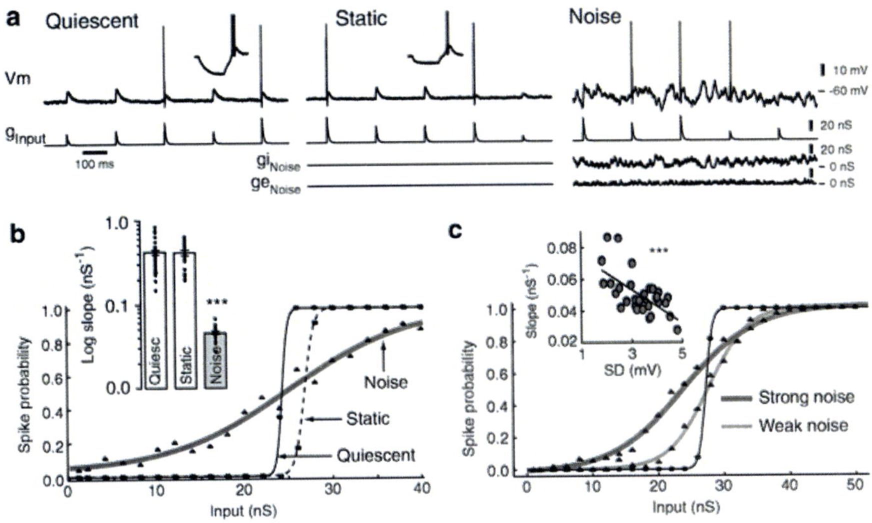

Fig. 6.17 The influence of conductance (g) noise changes the input–output response function of thalamocortical cells recorded in vitro. (**a**) Voltage during injection of input g alone (g_{input}, *Quiescent*) and with additional inhibitory plus excitatory background g's (gi_{Noise}, ge_{Noise}) that were either nonfluctuating (*static*) or stochastically fluctuating (*noise*). Combined noise conductances reduced the input resistance to ~50% (*insets*: *quiescent*, *static*). (**b**) Probabilities of input g strengths to evoke ≥1 spike, fitted to sigmoid functions. Noise but not static induced a gain (slope) reduction of the response function (*inset*). (**c**) Decreasing the variance of noise g's (*strong*, *weak*) increased the slope of the input–output response function. The response gain was correlated with the noise-induced voltage variance (SD). Error bars: s.e.m. Modified from Wolfart et al. (2005)

6.4.2 *Synaptic Noise Affects the Gain of Thalamocortical Neurons*

To assess whether inputs are transmitted and how this transfer is affected by synaptic background noise, Wolfart and colleagues evaluated the probability of a given input magnitude to evoke at least one AP within a 20 ms delay under quiescent, static, and noise conditions (Fig. 6.17b; Wolfart et al. 2005). In general, the slope, or gain, of the input–output response function can be determined by fitting a sigmoid function to input–output cell's response functions and extracting its slope at the 0.5 probability. While a step-like response function, typically, characterizes the quiescent and static conditions, under the influence of synaptic noise, the response probability is linearized, adopting intermediate values between 0 and 1 over a larger input range (Fig. 6.17b). Indeed, in Wolfart et al. (2005) it was found that the gain was not significantly different between quiescent and static conductance injection, but was strongly reduced with synaptic noise (Fig. 6.17b, insets, ANOVA, $P < 0.001$, quiescent $0.413 \pm 0.029\,\text{nS}^{-1}$, $n = 41$, vs. noise $0.046 \pm 0.002\,\text{nS}^{-1}$, $n = 27$, $P < 0.001$; quiescent vs. static $0.409 \pm 0.029\,\text{nS}^{-1}$, $n = 24$, $P = 0.72$). Similar results were obtained using a measure of the total spike output.

Thus, consistent with previous results from the cortex (see Chap. 5 and Sect. 6.3), the variance of background conductance, indeed, reduces the input–output response gain of thalamocortical cells and increases the sensitivity to small inputs.

Because the effect of corticothalamic feedback is expected to vary considerably with the state of the animal and the signals that are processed (Sherman and Guillery 2001), different variances of the fluctuating conductances were explored. For example, in Wolfart et al. (2005), it was found that in the "strong noise" condition, the default for the results described above, the conductance variance for excitatory and inhibitory noise were 3 and 12 nS, respectively. A decrease in noise conductance variances (to 1 and 4 nS respectively, "weak noise" voltage variance 2.61 $\pm$ 0.14 mV, n = 5) increased the input–output gain (Fig. 6.17c, 0.070 $\pm$ 0.007 nS^{-1}, n = 5, n = 27, $P < 0.01$). The gain in the various noise conditions correlated with the actual degree of voltage variance induced by synaptic noise (Fig. 6.17c, *inset*, n = 33, $r = 0.63$, $P < 0.01$). Thus, these results clearly demonstrate that physiologically plausible levels of background synaptic activity are able to modulate the response function of thalamocortical neurons in a multiplicative manner.

6.4.3 Thalamic Gain Depends on Membrane Potential and Input Frequency

The responsiveness of thalamic neurons at depolarized membrane potentials is very different from that at hyperpolarized potentials. At hyperpolarized levels, incoming EPSPs can be amplified by T-type channels producing LTS burst discharges. It was found (Wolfart et al. 2005) that, unexpectedly and unlike cortical cells, the gain of thalamocortical cells is markedly reduced at hyperpolarized compared to resting potentials (Fig. 6.18a, ANOVA, $P < 0.001$, hyperpolarized 0.146 $\pm$ 0.028 nS^{-1}, n = 12, vs. resting, 0.413 $\pm$ 0.029 nS^{-1}, n = 41, $P < 0.001$).

Wolfart and colleagues also investigated the origin of the voltage dependence of the gain. T-type channel activation at hyperpolarization is not expected to reduce the gain but rather to increase the all-or-none character of thalamocortical cell responses (Sherman and Guillery 2002). Yet, since deinactivation time constants of T-type channels are in the range of hundreds of milliseconds (Jahnsen and Llinás 1984), variable degrees of T-type channel recruitment are to be expected at 5 Hz input rate with randomized EPSP amplitudes. This could account for the gain reduction described above. Further analysis (Wolfart et al. 2005) showed that much of the LTS variability at hyperpolarization could indeed be attributed to the recent LTS activation history (see Wolfart et al. 2005; supplementary Fig. 1).

These results suggest that the voltage dependence of gain in the quiescent thalamic neuron is itself dependent on the input frequency. To explore this further, Wolfart et al. (2005) tested four input frequencies (1, 5, 10, and 20 Hz) at all voltage conditions. They found that, at resting and depolarized potentials, there was no

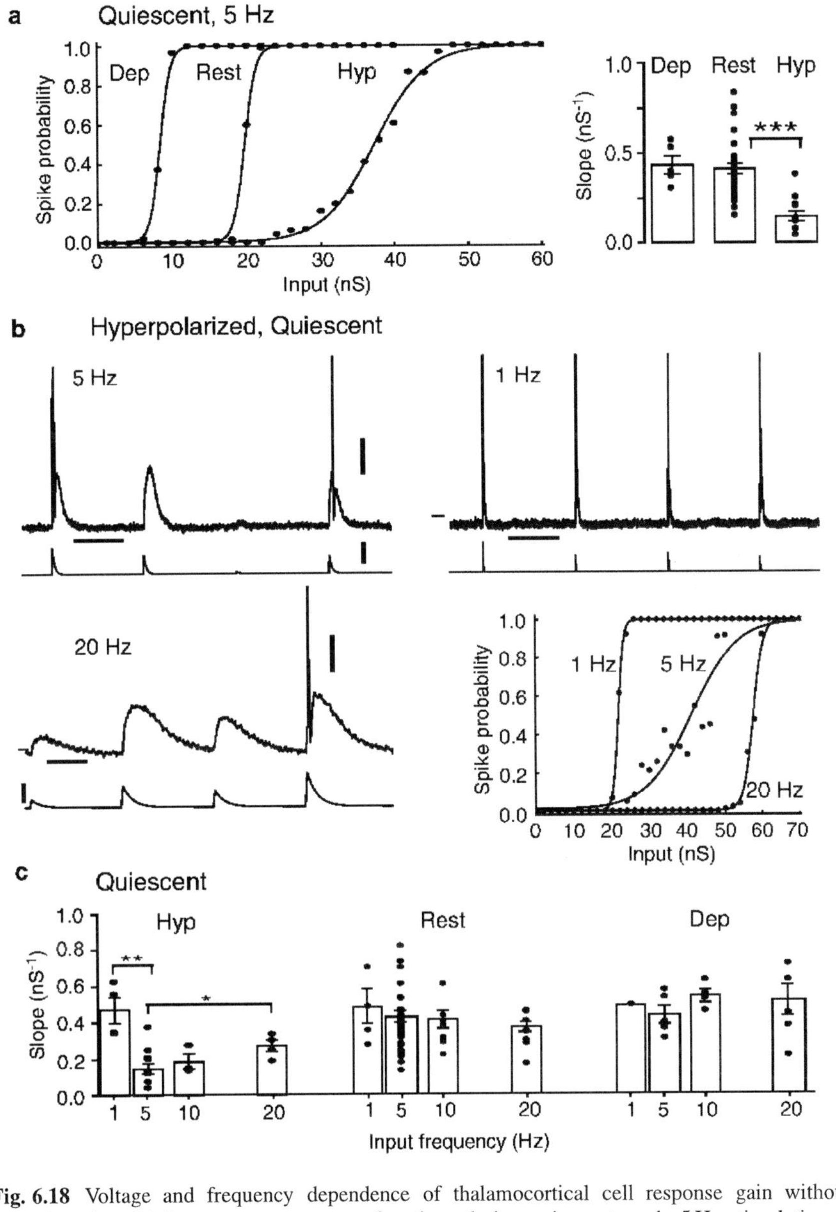

Fig. 6.18 Voltage and frequency dependence of thalamocortical cell response gain without synaptic noise. (**a**) Input–output response functions during quiescent mode 5 Hz stimulation at depolarized (*Dep*), resting (*Rest*), and hyperpolarized potentials (*Hyp*). The response gain (slope) was equal at Dep and Rest but reduced at Hyp (*inset*). (**b**) Different input frequencies (5 Hz, 1 Hz, 20 Hz) in the Hyp, quiescent condition. Response functions showed lower gain only in the 5 Hz condition. (**c**) Summary of experiments as in (**a**) and (**b**). Response gains were all-or-none at *Rest* and *Dep* at all input frequencies and at 1 Hz at *Hyp*, but reduced at Hyp with 5–10 Hz input. Scale bars in (**b**); horizontal below traces, 5 Hz: 0.1 s, 1 Hz: 0.5 s, 20 Hz: 20 ms; between traces: −70 mV; vertical (*upper*, *lower*): 10 mV, 20 nS. Error bars: s.e.m. Modified from Wolfart et al. (2005)

frequency dependence of gain (ANOVA, $P = 0.29$). However, at hyperpolarization a frequency-dependence of gain was clearly visible, as shown in Fig. 6.18b,c (ANOVA, $P < 0.001$). This is in agreement with the idea that voltage and frequency dependence of gain are due to the T-type channel gating behavior: the low gain at hyperpolarization is reverted to an all-or-none gain when the input frequency is lowered to 1 Hz such that T-type channels can recover from inactivation between stimuli (Fig. 6.18b,c; 1 Hz, $0.468 \pm 0.072\,nS^{-1}$, n = 4, vs. 5 Hz, $0.146 \pm 0.028\,nS^{-1}$, n = 12, $P = 0.008$). On the other hand, increasing the input frequency to 10 and 20 Hz gradually increases the gain at hyperpolarization (Fig. 6.18b,c; 20 Hz, $0.270 \pm 0.033\,nS^{-1}$, n = 4, vs. 5 Hz, $P = 0.039$). This effect can be explained by a cumulative inactivation of T-type channels at higher frequencies: the inability of T-type channels to follow high frequencies endows thalamocortical cells with low-pass filter properties when they are hyperpolarized. Thus, these results suggest the following scenario: at hyperpolarization and input frequencies around 5 Hz, T-type channels favor low gain-response functions, while at more depolarized potentials thalamocortical cells are expected to show all-or-none gain regardless of the input frequency.

6.4.4 *Synaptic Noise Renders Gain Independent of Voltage and Frequency*

To test how conductance noise does interfere with the voltage- and frequency dependence of gain, the characterization shown in Fig. 6.18 can be performed in the presence of synaptic noise (Fig. 6.19). Here, it was found (Wolfart et al. 2005) that the injection of synaptic noise reduces the gain at all voltage conditions (Fig. 6.19a). Unlike in the quiescent condition, the gain is very similar at all membrane potentials although still slightly reduced at hyperpolarization with 5 Hz input (Fig. 6.19a, *inset*, y-axis scaling different from Fig. 6.18a; see also Fig. 6.19c, *inset*, ANOVA voltage dependence, $P = 0.007$, across all frequencies $P = 0.076$, with frequencies pooled $P = 0.145$). This shows that, indeed, the voltage dependence of gain is strongly reduced with synaptic noise.

Remarkably, and in agreement with the hypothesis that voltage and frequency dependence are due to the same mechanism, synaptic noise nearly abolishes the frequency dependence of the gain. In Fig. 6.19b, three input frequencies are compared at hyperpolarized potentials in the presence of synaptic noise (compare with Fig. 6.18b). The marked distinction in gain that is seen without noise at hyperpolarization between 1 and 5 Hz or 5 and 20 Hz (Fig. 6.18c) is much reduced with synaptic noise (Fig. 6.19c, left panel; compare y-axis scaling of Figs. 6.18c and 6.19c). At all other frequencies and membrane potentials, the gain is equally low when synaptic noise is present (Fig. 6.19c, ANOVA across all conditions, frequency dependence, $P = 0.049$, with potentials pooled, $P = 0.087$). Further analysis of the underlying events showed (Wolfart et al. 2005) that synaptic noise

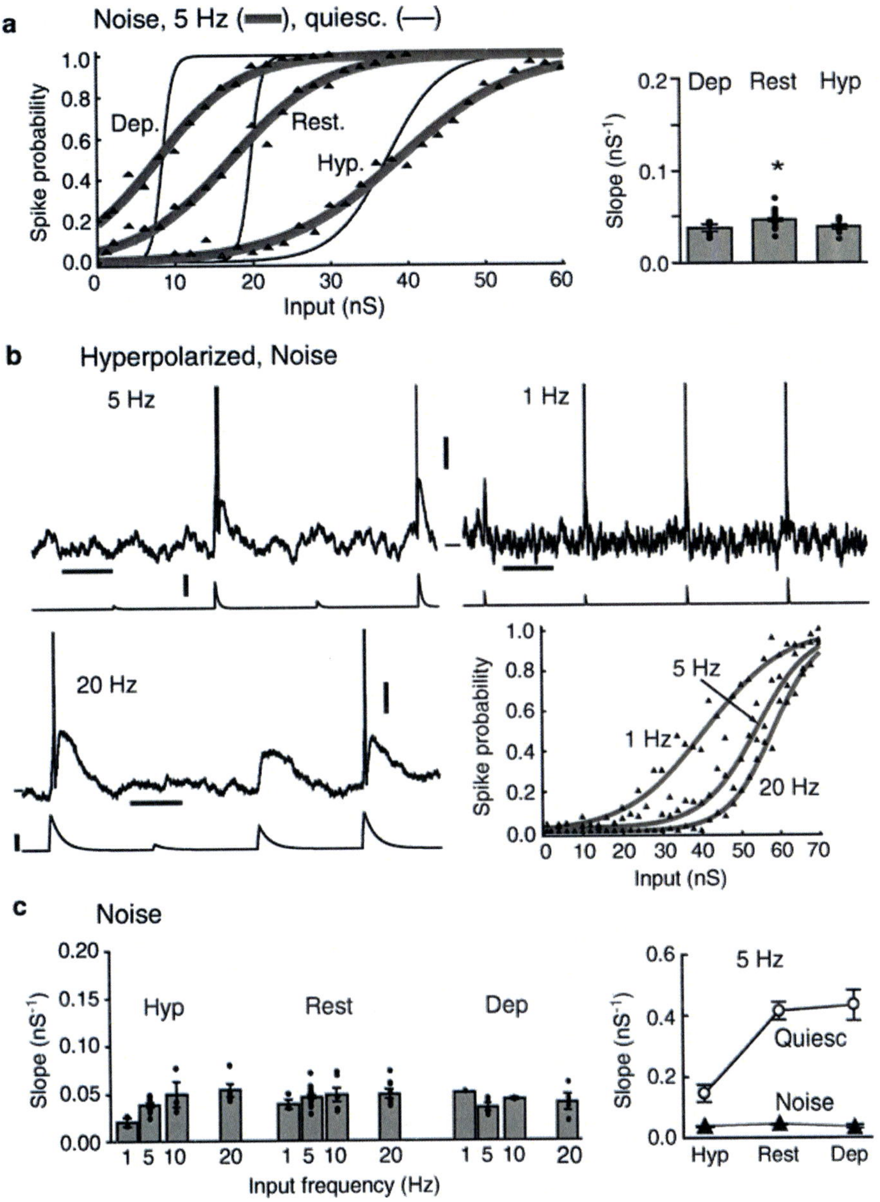

Fig. 6.19 Influence of synaptic noise on voltage and frequency dependence of gain. (**a**) *Thick gray curves*: response probabilities during synaptic noise with 5 Hz stimulation at depolarized (*Dep*), resting (*Rest*), and hyperpolarized potentials (*Hyp*). *Thin black lines*: quiescent curves from Fig. 6.18a (same cell). The response gain (*slope*) during synaptic noise was almost equally low at all potentials (*inset*). (**b**) Different input frequencies (5 Hz, 1 Hz, and 20 Hz) in the Hyp, noise condition. Response functions showed almost equally low gain at all input frequencies. (**c**) Summary of experiments as in (**a**) and (**b**). Response gains were low at all input frequencies and membrane potentials. Compare to Fig. 6.18c (note: y-axis scaling different). *Inset* shows loss of voltage dependence of gain with synaptic noise compared to the quiescent condition (at 5 Hz input). Scale bars in (**b**); horizontal below traces, 5 Hz: 0.1 s, 1 Hz: 0.5 s, 20 Hz: 20 ms; between traces: −70 mV; vertical (*upper*, *lower*): 10 mV, 20 nS. Error bars: s.e.m. Modified from Wolfart et al. (2005)

increases the variability of subthreshold responses to an overall high level, overwhelming the variability due to LTS, thereby reducing the response gain of thalamocortical cells to an overall low level, equal across different membrane potentials and input frequencies (see Wolfart et al. 2005; supplementary Fig. 1). Thus, one can conclude that synaptic noise effectively masks the intrinsic, nonlinear response behavior of thalamocortical cells and equips them with a robust, voltage-independent transfer function.

6.4.5 Stimulation with Physiologically Realistic Inputs

The next step in the assessment of the behavior of thalamic neurons under noisy conditions is not only to consider realistic background conditions but also realistic input. In the visual thalamus, the relay neurons receive direct input from retinal ganglion cells, whose mean firing frequencies in vivo are in the range of 5 to 50 Hz and gamma or Poisson distributed (Troy and Robson 1992). Even if the magnitude of a single-retinogeniculate EPSP may vary little, the effective postsynaptic retinogeniculate EPSPs depend, among other factors, on variable degrees of temporal summation, such that the effective input has a larger magnitude range (Turner et al. 1994).

In the experiments described above, the variability of effective input was achieved by randomizing the input conductances while fixing the input frequency, thereby separating magnitude from frequency allowing better control and comparison with cortical neurons (Shu et al. 2003a,b) (see also Fellous et al. 2003). To check whether the voltage- and frequency dependence of the spike probability occurs under more physiological input conditions, one can compare the response properties of thalamocortical cells during stimulation with Poisson-distributed retinal input. This was done in the study by Wolfart et al. (2005) by using a mean frequency of 10 Hz for the retinal input. Even though the retinogeniculate input conductance in this study was fixed, Poisson-rate stimulation led to variable effective input EPSP magnitudes due to summation (Fig. 6.20a). Moreover, at the beginning of each experiment, the conductance magnitude was adjusted such that evoked subthreshold EPSPs at resting potential were in a physiological range (5–15 mV; Turner et al. 1994). Since the degree of input summation is dependent on the frequency, the interstimulus interval (sISI) immediately preceding the response was used to measure input strength (Fig. 6.20b).

Wolfart et al. (2005) found that at resting potential with subthreshold input magnitude, spikes are only evoked by summed inputs at smaller sISIs (Fig. 6.20a,b): sISIs in the range of 50 to 600 ms are related to spike probabilities in the range of 0.00 to 0.021 ± 0.016 (n = 4), whereas sISIs shorter than 50 ms are associated with a spike probability of 0.530 ± 0.123 (n = 4). In contrast, it was observed that at hyperpolarization, spike response can be evoked not only by input summation but also by long sISIs (Fig. 6.20a,b, ANOVA, $P < 0.001$): sISIs in the range of 300 to 600 ms are related to spike probabilities in the range of 0.253 ± 0.099

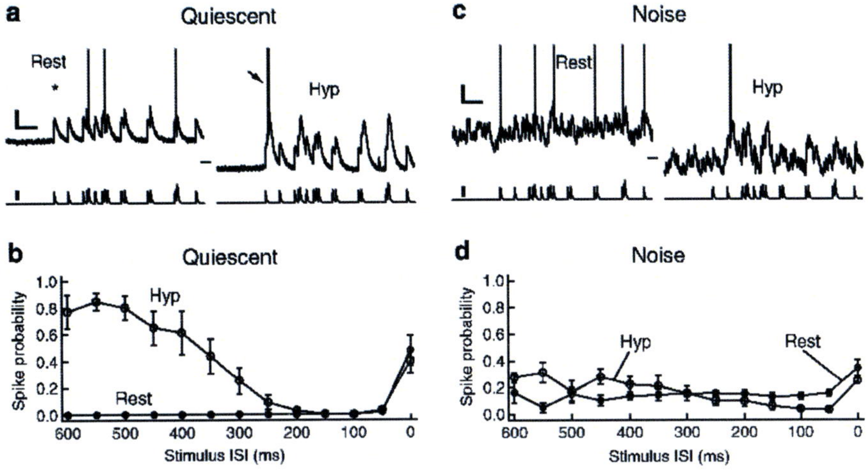

Fig. 6.20 Physiologically realistic Poisson-distributed inputs. (**a**) "Retinogeniculate" input conductances (lower traces) were injected at a fixed magnitude with Poisson-distributed rate (mean frequency 10 Hz). At resting potential without synaptic noise, action potentials were only evoked by summation of inputs occurring at short intervals while inputs with long pre-stimulus intervals (Stimulus ISIs) did not trigger action potentials (*left panel, asterisk*). At hyperpolarized potential, subthreshold inputs led to variable degrees of EPSP summation occasionally accompanied by LTS activation. Large sISIs led to the activation of LTS-driven bursts (*right panel, arrow*). (**b**) The probability to evoke at least one spike was plotted against the sISIs. While at resting potential spiking probability was increased only with high- frequency inputs, at hyperpolarized potentials the spiking probability increased strongly with low- frequency inputs. (**c**), (**d**) During the injection of synaptic noise, the difference in frequency-dependent response behavior of thalamocortical cells was strongly reduced. Spiking probabilities were approximately equal at all input frequencies in the presence of synaptic noise. Scale bars in (**a**) and (**c**), 100 ms, 10 mV, lower trace, 10 nS; between traces: −80 mV. Error bars: s.e.m. Modified from Wolfart et al. (2005)

to 0.847 ± 0.061 ($n = 4$), even higher than those evoked by sISIs shorter than 50 ms (0.398 ± 0.089, $n = 4$). Thus, consistent with the experiments using fixed input frequencies and randomized input magnitudes, spike probabilities induced by Poisson rate input have an all-or-none character at resting potential, but adopt intermediate values depending on input frequency at hyperpolarized potentials.

The Poisson rate experiment described above can be repeated in the presence of synaptic noise (Fig. 6.20c,d; Wolfart et al. 2005). With synaptic noise, input summation also increases the spike probability at resting and hyperpolarized potentials. However, unlike the quiescent condition, here larger sISIs do not lead to different spiking probabilities at hyperpolarized compared to resting potential (ANOVA, $P = 0.43$). Although no direct comparison between fixed rate and Poisson rate experiments (such as comparing the gain) is feasible, these results match: in both experimental conditions voltage and frequency dependencies are abolished with synaptic background noise.

6.4.6 Synaptic Noise Increases Burst Firing

Action potential burst firing in thalamocortical cells is classically considered as the result of LTS activation following hyperpolarization (Lu et al. 1992; Ramcharan et al. 2000; Weyand et al. 2001). However, with synaptic noise, even at resting and depolarized potentials, high frequency burst responses often occur (Fig. 6.21a, left panel). Since noisy voltage fluctuations hamper LTS identification, a burst detection algorithm was proposed (Ramcharan et al. 2000) and used by Wolfart et al. (2005) to assess the "burstiness" (percent burst responses per spike-evoking input). The latter study found that, indeed, not only the voltage but also the presence of synaptic noise does have an influence on burstiness (ANOVA across all membrane potentials and frequencies, $P < 0.001$). At resting and depolarized potentials there is an increased burstiness with synaptic noise (in 24 of 27 cases in Wolfart et al. 2005; see Fig. 6.21b, inset, ANOVA, $P = 0.002$, resting quiescent 5 Hz, $5.7 \pm 1.9\%$, $n = 40$, vs. noise $21.2 \pm 3.4\%$, $n = 27$, $P < 0.001$; depolarized quiescent 5 Hz, 0.00%, $n = 7$, vs. noise $27.6 \pm 6.1\%$, $n = 7$, $P = 0.004$). As expected, burst firing was found more pronounced at hyperpolarization (5 Hz quiescent $44.5 \pm 9.6\%$, $n = 12$) and without clear change of burstiness in the presence of synaptic noise (Fig. 6.21c, left panel).

Another approach to quantify burstiness is to determine the "burst threshold," i.e., the input level at which bursts first occur. This analysis again showed that at resting and depolarized potential, bursting was more likely with synaptic noise since the bursting threshold was much lower than in the quiescent condition (Wolfart et al. 2005; Fig. 6.21c, resting 5 Hz quiescent 58.5 ± 3.8 nS, $n = 16$ vs. noise 33.3 ± 3.4 nS, $n = 25$, $P < 0.001$; depolarized 5 Hz quiescent no bursting, vs. noise 13.5 ± 8.0 nS, $n = 6$).

Further dynamic-clamp experiments were realized to investigate the reasons for the increased burstiness with synaptic noise. In principle, decrease in the membrane time constant resulting from conductance increase could play a role, but no increased burstiness was so far detected in the static condition (e.g., in the study by Wolfart et al. 2005; burstiness: resting 5 Hz quiescent $6.2 \pm 3.3\%$ vs. static $6.5 \pm 3.3\%$, $n = 22$, paired test, $P = 0.44$). It was also observed that the mean membrane potentials of noise and quiescent conditions are not significantly different (resting 5 Hz quiescent -65.3 ± 0.8 mV vs. noise -65.7 ± 0.8 mV, $n = 24$, paired test, $P = 0.80$ in Wolfart et al. 2005) arguing against a role of T-type channels. Indeed, the gating parameters of T-type channels suggest little involvement in these conditions (Perez-Reyes 2003), although native T-type currents of thalamic neurons can be available at resting potential (Leresche et al. 2004).

Finally, comparing STAs of single-spike and burst responses during noise at resting potential revealed that there was a small but significant difference in the pre-response voltage (see Fig. 6.22). This suggests that even if overall, noise is neutral to the membrane potential, short hyperpolarizations preceding inputs statistically recruit more T-type channels compared to the quiescent state. In addition, the occasional co-occurrence of noise-induced depolarizations with retinal inputs clearly

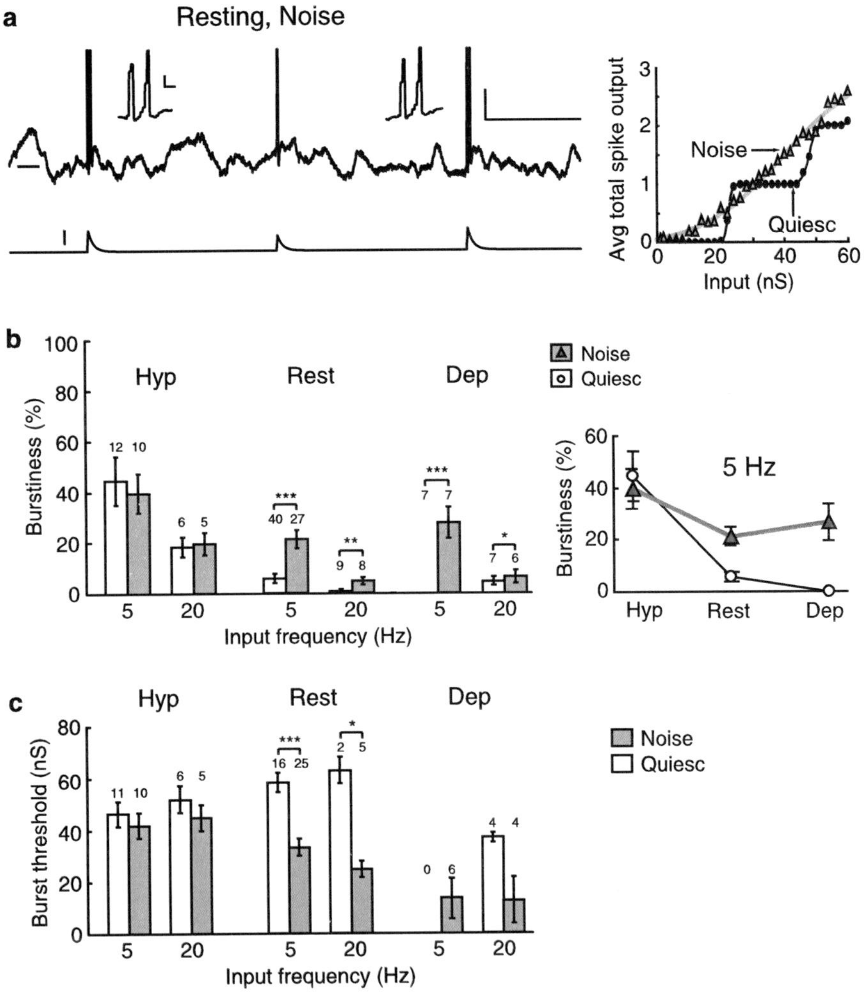

Fig. 6.21 Synaptic noise increases the occurrence of bursts at resting potential. (**a**) With synaptic noise, high-frequency bursts of action potentials occurred at resting potential (*left panel*). Plotting the average total number of spikes per burst response against the input shows that synaptic noise linearized the staircase-like response function across the whole input range (*right panel*). (**b**) Percent bursts per spike-evoking stimulation with different input frequencies and membrane potentials. Noise increased burstiness at resting and depolarized potentials (e.g., at 5 Hz, *inset*). This effect was smaller with higher input frequencies. Noise did not increase the burstiness at hyperpolarization. (**c**) The input value at which bursting occurred (*Burst threshold*) was decreased with synaptic noise at resting and depolarized but not at hyperpolarization. Scale bars in (**a**), 100 ms, 10 mV; lower trace, 10 nS; before trace −80 mV; inset, 10 mV, 1 ms. Error bars: s.e.m. Modified from Wolfart et al. (2005)

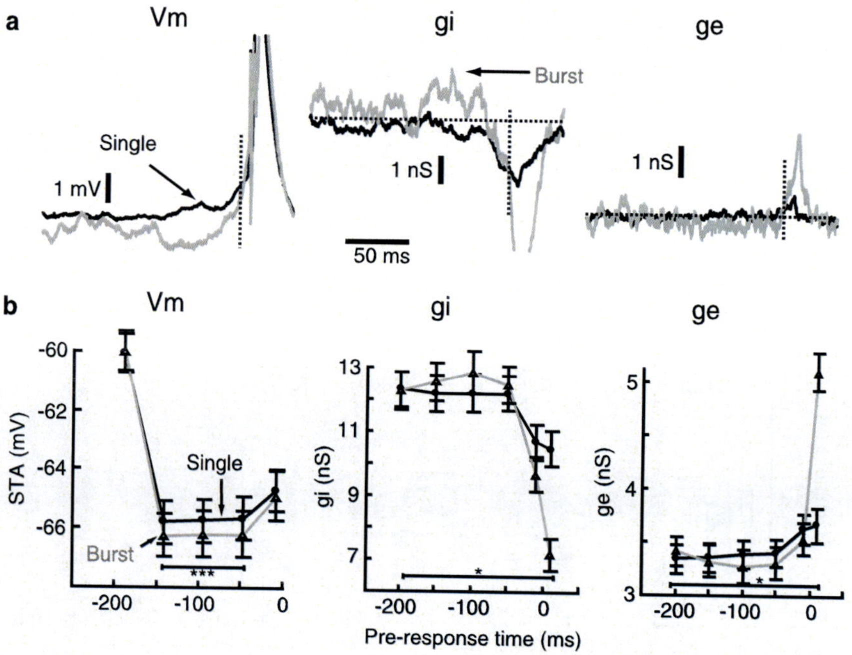

Fig. 6.22 Comparison of preresponse potential and background conductances preceding bursts and single- spike responses during noise recordings at resting potential. (**a**) Spike-triggered average (*STA*) of 857 single spikes (*black traces*) and 134 burst responses (*gray traces*) of membrane potential (V_m, *left panel*), inhibitory (g_i, *middle*), and excitatory background conductances (g_e, *right*). 50–150 ms before burst responses, g_i was increased and V_m was hyperpolarized compared to single-spike responses, while ge was similar. Several milliseconds before spike, g_e increased and g_i decreased, as in cortical neurons (compare with Fig. 9.21). (**b**) Quantification and time course of observations in (**a**). Time-averaged STAs were extracted at different time windows before the spike (0 to −10 ms; −40 to −50 ms; −90 to −100 ms; −140 to −150 ms; −190 to −200 ms; and 0 to +10 ms). This analysis reveals two mechanisms of bursting during noise: (**a**) small hyperpolarizations due to increased g_i preceding the response; (**b**) simultaneous g_e increase and g_i decrease during (or immediately prior to) the response. Modified from Wolfart et al. (2005)

facilitated bursts, as seen from STAs of excitatory and inhibitory conductances (see Fig. 6.22). These results show that noise increases the occurrence of burst responses at resting and depolarized but not at hyperpolarized potentials.

6.4.7 Synaptic Noise Mixes Single-Spike and Burst Responses

If the number of spikes in the response grows proportionally to the strength of the input, a spike count can be used to reliably encode sensory information. Without synaptic background, such a reliable response function does not exist at resting and depolarized states, the cell behaving as a high-pass filter detecting only strong inputs with no discrimination of strength until the latter passes a threshold (see the step-like response in Fig. 6.21a, right panel). In contrast, noise does make cells fire on average and, with the number of spikes generated being proportional to the input strength (Fig. 6.21a, right panel), this provides a more linear response function at all potentials. Thus, in the presence of synaptic background activity, probabilistic "mixing" of single-spike and burst responses potentially provides better encoding capabilities.

To quantify this mixing, single-spike responses can be separated from responses with 2-spike and 3-spike bursts, as done in the study by Wolfart et al. (2005). There, the different responses at the respective input levels were measured and their distribution was plotted (Fig. 6.23a,b). The relative overlap of the 1-spike, 2-spike, and 3-spike response functions was calculated by integration and the "percent mixing" (*overlap*) for the different conditions was compared (Fig. 6.23c,d). It was observed that with noise, at resting potential, the overlap of 1-spike and 2-spike response functions is increased (Fig. 6.23a,c, ANOVA, $P < 0.001$, 5 Hz: quiescent $4.3 \pm 0.52\%$, $n = 16$, vs. noise $14.6 \pm 1.4\%$, $n = 25$, $P < 0.001$). In the quiescent condition, in the aforementioned study, only 16 of 40 cells showed 2-spike bursts and one cell showed 3-spike bursts with very strong inputs, whereas with noise 25 of 27 cells displayed 2-spike bursts and 13 displayed 3-spike bursts, the overlap of the latter being zero in the quiescent condition but notable in the noise condition (Fig. 6.23d, $4.6 \pm 1.4\%$). Moreover, Wolfart et al. (2005) found that the same difference in the mixing of single-spike and burst responses is true for depolarized and hyperpolarized potentials although it is less marked for the latter (Fig. 6.23c,d, inset in d). At 20 Hz input, the difference in percent mixing was generally reduced (Fig. 6.23c,d). Thus, while thalamocortical cells recorded in vitro are usually either in single-spike or in burst mode, the present results suggest that thalamocortical cells under the influence of synaptic background activity may show both burst firing and single-spike responses.

6.4.8 Summing Up: Effect of Synaptic Noise on Thalamic Neurons

In summary, the dynamic-clamp investigation of synaptic noise in thalamic neurons by Wolfart et al. (2005), as reviewed here, has demonstrated that physiologically plausible levels of background synaptic activity can modulate the response function of thalamocortical neurons in many ways. The presence of bursting properties (T-type current) in thalamic relay neurons confers complex response properties with respect to the gain of the neurons, which is dependent on membrane potential and input frequency in quiescent neurons. Interestingly, this complex dependence disappears in the presence of synaptic noise. This suggests that synaptic noise effectively masks the intrinsic, nonlinear response behavior of thalamocortical cells and equips them with a robust, voltage-independent transfer function.

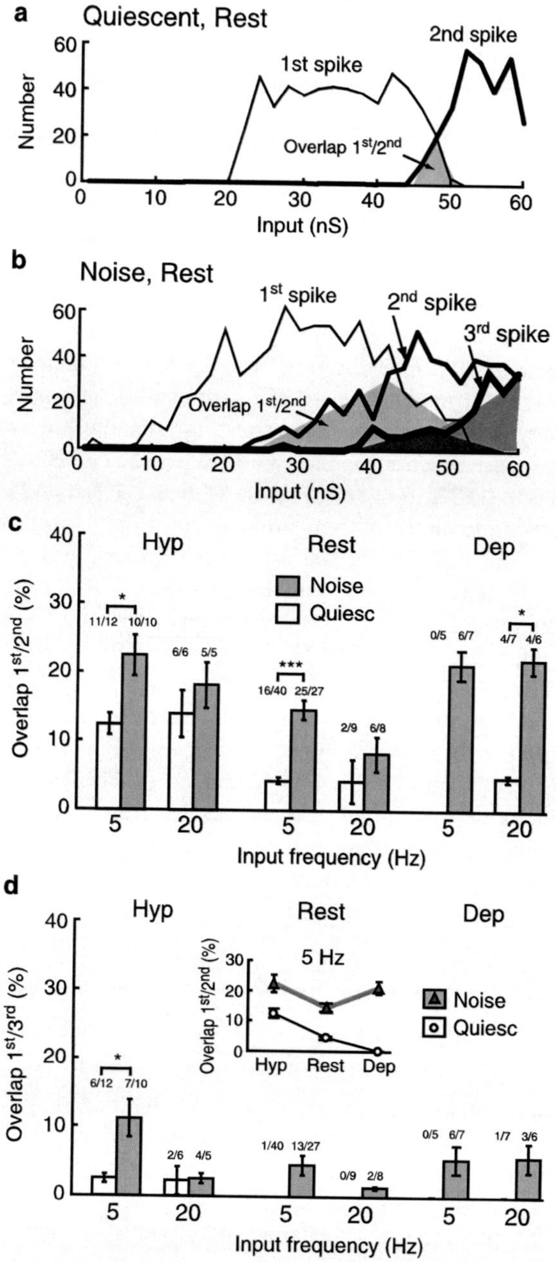

Fig. 6.23 Noise mixes single-spike and burst responses of thalamocortical neurons. (**a**) Example of input strength distributions evoking single-spike (*first spike*) and burst responses (*second spike*, with interspike intervals ≤5 ms). Without noise, only very strong inputs could evoke spike doublets. (**b**) With noise, most input levels could evoke single-spike and burst responses (e.g., at 5 Hz, *inset*). Different shades of gray mark areas of overlap between first and second spike (*light*), second and third spike (*middle*), first and third spike (*dark*). (**c**), (**d**) Mixing (*overlap*) of single-spike and multiple-spike responses ((**c**), *first / second*; (**d**), *first / third*). Noise increased the mixing of single-spike and burst responses at most potentials and input frequencies. Error bars: s.e.m. Modified from Wolfart et al. (2005)

Investigating this issue further revealed that the intrinsic properties of thalamic neurons, responsible for a duality of firing modes, are expressed differently in the presence of synaptic noise. The firing modes of thalamic neurons are no longer distinct, but are mixed at all potentials. Interestingly, for realistic inputs, the response curve of thalamocortical cells may become independent of the V_{m} level (Fig. 6.20). This shows that in order to fully characterize the transfer properties of thalamic neurons, we need not only know the intrinsic properties of the cells but also the exact amount of synaptic noise. The response function of neurons is a function of both properties, and is therefore predicted to be very different during active states in vivo compared to quiescent states in vitro.

The fact that the response function of thalamic neurons becomes voltage independent and that this holds for physiologically realistic inputs, suggests that the "relay" of information by the thalamus takes a very different character in the presence of synaptic noise. It is possible that the T-type current is present in thalamic neurons precisely for the purpose of making this "relay" particularly robust and voltage-independent. Because most of the synaptic noise originates from the corticothalamic synapses, we can say that the cortex has the ability to finely tune the relay of information, and possibly amplify the transmission of information, by controlling the amount of "noise" sent to the thalamus. Again, this shows that the presence of noise in the system is not detrimental, but confers many advantages and flexibility. Like cortical neurons, the modulation of the amount of synaptic noise could also be linked to attentional mechanisms.

6.5 Dynamic-Clamp Using Active Electrode Compensation

The intracellular recording of the neuronal membrane potential is currently the only method for studying the integration of excitatory synaptic inputs, inhibitory inputs, and intrinsic membrane currents underlying the spiking response. In vivo, these recordings are done by using either single high-resistance sharp microelectrodes with low capacitance (Steriade et al. 2001; Wilent and Contreras 2005a; Crochet et al. 2006; Higley and Contreras 2007; Haider et al. 2007; Paz et al. 2007) that are also used in some adult in vitro preparations (Thomson and Deuchars 1997; Shu et al. 2003a,b), or by using single patch electrodes that can display a whole range of resistances and capacitances depending on the age and species of the animal. Here, rather low resistance values could be obtained in rats up to a certain age (see Margrie et al. 2002), as well as in cat (Borg-Graham et al. 1998; Monier et al. 2008), whereas resistances closer to those of sharp electrodes were used in other studies (see, e.g., Pei et al. 1991; Hirsch et al. 1998; Anderson et al. 2000; Wehr and Zador 2003; Mokeichev et al. 2007).

The problem inherent to such single-electrode recordings is that the injected current biases the measurement because of the voltage drop through the electrode. This electrode bias imposes restrictions on the use of advanced electrophysiological techniques like voltage clamp or dynamic clamp (Robinson and Kawai 1993; Sharp et al. 1993a,b; Prinz et al. 2004), which require injection of a current dependent on the simultaneously recorded V_{m}. These techniques allow the dissection of the importance of specific intrinsic and synaptic channels, and their interactions, for neuronal function, as shown in the previous sections of this chapter.

In this section, we review a recent method called *AEC*, which was introduced to compensate for the electrode bias (Brette et al. 2007a,b, 2008, 2009). This method is of unprecedented accuracy and is based on a model of the electrode interfaced in real time with the electrophysiological setup. AEC opens the way for the improved and simplified use of these advanced techniques in many preparations requiring high resistance and/or capacitance electrodes, for example, in vivo recordings of cortical neurons. Moreover, the AEC turns out to be particularly relevant in the case of injecting synaptic noise, as we will see below.

6.5.1 Active Electrode Compensation

Electrode compensation circuits implemented in intracellular amplifiers usually reflect the assumption that the electrode is equivalent to a simple RC (resistor and capacitor) circuit (Thomas 1977), but this simplification does not account for distributed capacitance and the resulting compensation produces artifactual voltage transients (Fig. 6.24a, *bridge compensation*, middle). In situations where the injected current depends on the V_m, the artifacts are injected back and can be amplified by the control loop, which leads to oscillatory instabilities (Fig. 6.24a, *bridge compensation during dynamic-clamp*, right). An option for voltage clamp, and the only one for dynamic clamp when electrode resistance is high, is to use a discontinuous mode, alternatively injecting current and recording the V_m (Brennecke and Lindemann 1971, 1974a,b; Finkel and Redman 1984) with a frequency set by the electrode time constant (typically 1.5–3 kHz with sharp electrodes in the present experiments in cortical neurons in vitro and in vivo). Unfortunately, the alternation method is valid only when the electrode response is at least two orders of magnitude faster than the recorded phenomena (Finkel and Redman 1984), because the membrane response must be quasi-linear in the sampling interval. Moreover, recordings in discontinuous modes are very noisy and sampling frequency is limited, which makes the precise recording of fast phenomena like spikes impossible (Fig. 6.24b).

The AEC method allows the sampling of the V_m during current injection with a frequency only limited by the speed of the computer used for the digital convolution at the core of the technique (Fig. 6.24c). In the following, we briefly describe how this method is tested in demanding experimental situations: high-resistance sharp microelectrode recordings for both current-clamp with fast current injection and dynamic clamp protocols. Such recordings were performed by Brette et al. (2008) in vitro and in vivo in two experimental preparations widely used for the study of mammalian cortical function: in slices of visual cortex from adult animals (in this case, guinea pigs and ferrets) and in the primary visual cortex of the anesthetized and paralyzed cat. These experiments demonstrated that with AEC it is now possible, for the first time, in cases when single high-resistance electrodes are used, to inject white noise in a cell at a high sampling frequency, and to accurately inject complex conductance stimuli with the goal of precisely analyzing high-frequency features of

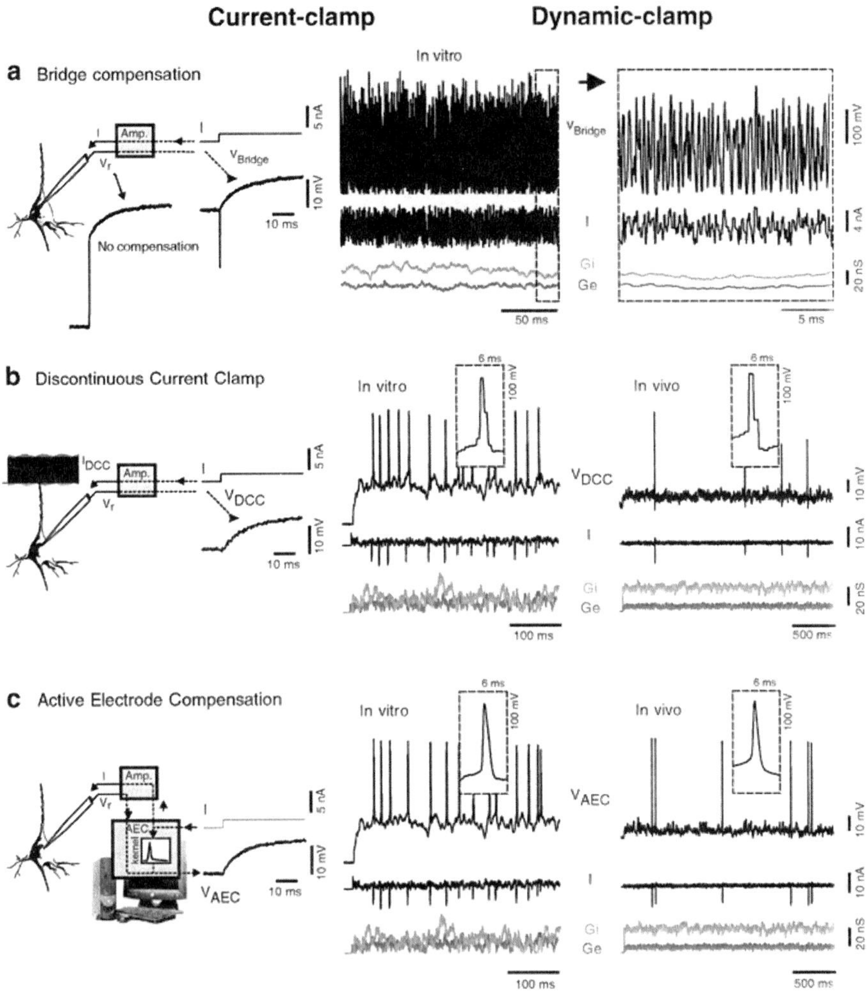

Fig. 6.24 Compensating for the electrode response during simultaneous current injection and recording of neuronal membrane potential (V_m). (**a**) "Bridge" performed by the amplifier (*left*). The capacitive properties of the electrode lead to a capacitive transient at the onset of the response to a current step (*middle*). A loop is established between V_m recording and current injection when inserting virtual conductances G in dynamic-clamp (*right*). When fast fluctuating conductances are inserted, the transients lead to a strong "ringing" oscillation in the recorded potential V_{Bridge} and the injected current I ($R_e = 93\,\mathrm{M\Omega}$, IB cell). (**b**) Discontinuous current clamp (DCC) in vitro and in vivo. There is no capacitive transient at the onset of the response to a current step (*middle*). Conductance injection using dynamic-clamp can be performed (*right*) without oscillations, but the sampling resolution of the V_m is low, as seen when zooming in on single spikes (*insets*). (**c**) Active electrode compensation (AEC) in vitro and in vivo, a new method for high-resolution V_m recording during simultaneous current injection. This digital compensation is performed in real-time by a computer (*left*). No capacitive transient is seen at the onset of the response to a current step (*middle*; $R_e = 87\,\mathrm{M\Omega}$, RS cell, for all the current step examples shown). Conductance injection using dynamic-clamp (*right*; $R_e = 63\,\mathrm{M\Omega}$, RS cell in vitro; $R_e = 103\,\mathrm{M\Omega}$, RS cell in vivo) is performed with a high V_m sampling frequency (10 kHz), so that the shape of single spikes can be resolved (*inset*). Modified from Brette et al. (2008)

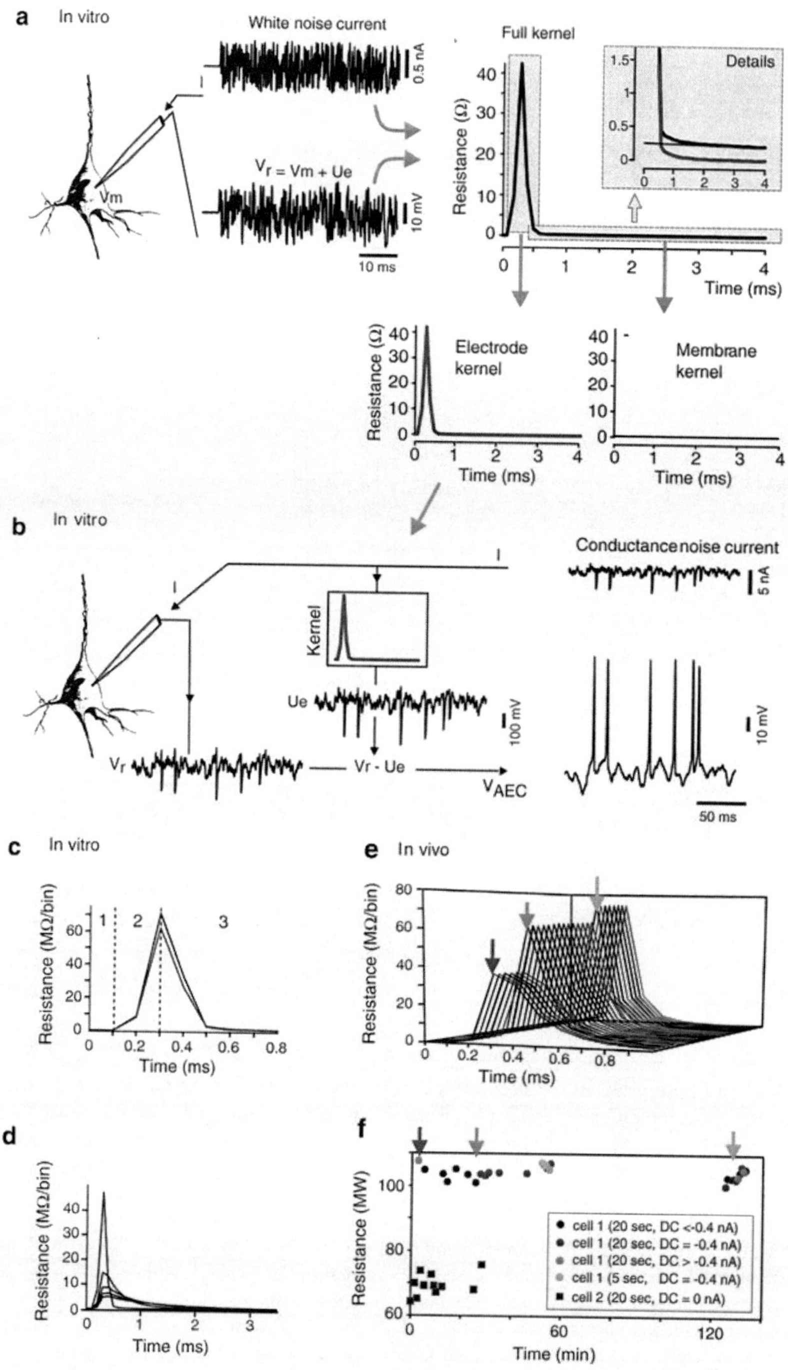
a In vitro
White noise current
0.5 nA
I
Vr = Vm + Ue
Vm
10 mV
10 ms
Full kernel
Details
Resistance (Ω)
Time (ms)
Electrode kernel
Membrane kernel
b In vitro
Conductance noise current
5 nA
Kernel
Ue
100 mV
Vr
Vr - Ue
VAEC
10 mV
50 ms
c In vitro
Resistance (MΩ/bin)
Time (ms)
e In vivo
d
f
Resistance (MW)
Time (min)
cell 1 (20 sec, DC <-0.4 nA)
cell 1 (20 sec, DC = -0.4 nA)
cell 1 (20 sec, DC > -0.4 nA)
cell 1 (5 sec, DC = -0.4 nA)
cell 2 (20 sec, DC = 0 nA)

the cell's response like the onset of spikes. More generally, AEC will be especially useful for the study of electrical neuronal phenomena happening at time-scales of the order of the electrode time constant, and it widens in an important way the repertoire of electrophysiological protocols easily applicable to the study of central mammalian neurons in vivo and of other preparations where very low resistance electrodes cannot be routinely employed.

6.5.2 *A Method Based on a General Model of the Electrode*

The essential idea behind the AEC method is to represent the electrode by an arbitrarily complex linear circuit, extract the properties of this circuit for each particular recording (Fig. 6.25a), and actively compensate for the effect of the electrode by subtracting the voltage drop through this circuit from the recording (Fig. 6.25b). The classic compensation methods, "bridge" compensation and capacitance neutralization, are equivalent to modeling the electrode as a resistor plus a capacitor (RC circuit), but this model has proven too simple in many practical situations. A more general model of the electrode should be used instead, consisting of an unknown linear circuit. A particular case of such a linear circuit is given by two resistors and two capacitors, as hypothesized by Roelfsema et al. (2001). It could also be much more complicated: in fact, Brette et al. (2008) showed that elements of the amplifier (e.g., filters) should be included in the recording circuit.

Fig. 6.25 The two stages of the AEC method. (**a**) Electrode properties are estimated first: white noise current is injected into the neuron, and the total response V_r, corresponding to the sum of the membrane potential V_m and the voltage drop across the electrode U_e, is recorded (*left*; R_e = 76 MΩ, RS cell recorded in vitro). The cross-correlation between the input current and the output voltage gives the kernel (or impulse response) of the neuronal membrane + electrode system (*full kernel*). The neuronal membrane kernel has a total resistance of 31 MΩ here, however the response is spread over a long time: the vertical axis gives the resistance of each 0.1 ms time bin, so that the resistance value is very small for each such bin (though not 0: see detail of *black* and *gray traces* in the *inset*) but adds up to 31 MΩ when all the bins are added (not all the bins are shown). This full kernel is separated into the electrode kernel and the membrane kernel. (**b**) The electrode kernel is then used in real time for electrode compensation: the injected current is convolved with the electrode kernel to provide the electrode response U_e to this current. U_e is then subtracted from the total recorded voltage V_r to yield the V_m (V_{AEC}). (**c**) Kernel of the same electrode estimated in the slice before the impalement of a neuron (*black*, R_e = 118 MΩ), and after impalement, in a cell (*gray*, R_e = 108 MΩ, IB cell). The numbers above the graph indicate the three phases of a typical electrode kernel. (**d**) Electrode kernels obtained in the slice for different levels of capacitance neutralization (the sharpest kernel corresponds to the highest level of capacitance neutralization; R_e = 87–89 MΩ). (**e**) Electrode kernel obtained in vivo after impalement in a cell (*dark gray*, R_e = 103 MΩ, RS cell). (**f**) Temporal stability of the electrode properties in vivo (cell 1 and 2 are RS cells). The kernel was estimated using 5 or 20 s white noise current injections. In addition, different constant current levels (DC) were injected, preventing spiking activity during the estimation. In all cases but for the light gray points, moderate spiking did not perturb the kernel estimation. Modified from Brette et al. (2008)

The voltage across the electrode U_e is modeled as the convolution of the injected current I_e and a kernel K_e which characterizes the electrode:

$$U_e(t) = (K_e * I_e)(t) = \int_0^{+\infty} K_e(s) I_e(t-s) \mathrm{d}s. \tag{6.4}$$

Thus the voltage across the electrode depends linearly on all past values of the injected current: it is the sum of all these past values weighted by the coefficients of the kernel K_e. This formulation encompasses any time-invariant linear model, e.g., a circuit with a resistor and a capacitor (the kernel K_e is then an exponential function). For any linear model, the kernel completely characterizes the system, i.e., it allows the calculation of the system's response to any input. For digitally sampled signals, the formula reads

$$U_e(n) = \sum_0^{+\infty} K_e(p) I_e(n-p), \tag{6.5}$$

with the remark that, in general, the discrete and continuous kernels are not identical (see details in Brette et al. 2008).

The procedure consists in two passes: first, measure the electrode kernel, i.e., the values of $K_e(p)$ (Fig. 6.25a, right) and, second, inject and record at the same time in continuous mode, with the true V_m recording obtained by subtracting the voltage across the electrode U_e from the raw recording. The compensation involves the digital convolution of I with K_e to obtain U_e, which is performed in real time by a computer (Fig. 6.25b).

6.5.3 *Measuring Electrode Properties in the Cell*

The next step is to measure and extract the electrode kernel in the cell. For small injected currents, the response of the cell can also be considered as linear, so that the recorded potential can be expressed as the linear convolution

$$V_r = V_{rest} + K * I,$$

where V_{rest} is the resting potential and K is the total kernel comprising both the electrode kernel K_e and the membrane kernel K_m and fully characterizing the system electrode + cell. K is derived from the recorded response to a known injected current, then the contributions of the electrode and the membrane are separated (Fig. 6.25a). Electrode properties before and after cell impalement can be quite different (Fig. 6.25c), so it is essential to estimate the electrode kernel in the neuron by means of this separation.

To this end, Brette et al. (2008) injected 5–20 s of noisy current consisting of a sequence of independent random current steps (white noise) at a sampling resolution of 0.1 ms, with an amplitude uniformly distributed between −0.5 nA and 0.5 nA. In principle, here any current can be used, provided it is small enough to prevent spiking and nonlinear effects from the cell (see supplementary methods in Brette et al. 2008 for details about how to set the current intensity), but this choice was found not to be arbitrary: a uniform amplitude distribution makes the best use of the D/A converters in the acquisition board, provided their range is adjusted accordingly, while using a fast-varying current with minimum autocorrelation enhances the electrode contribution in the recording relatively to the membrane contribution, because the electrode response is at least one order of magnitude faster than the membrane response. Therefore, in vivo, it is also possible to inject in addition a constant negative current (Fig. 6.25f) in order to prevent spiking. The kernel K is then derived mathematically from the autocorrelation of the current and the correlation between the current and the recorded potential.

The separation of the total kernel K into the electrode kernel K_e and the membrane kernel K_m is based on two facts: first, the electrode kernel is very short compared to the membrane kernel, so that after a couple of milliseconds ($p > 50$, i.e. 5 ms at 10 kHz) $K_e(p)$ vanishes and $K(p) \approx K_m(p)$; second, if the injected current is small enough, the membrane response is mostly linear and can be approximated on short timescales by a decaying exponential

$$K_m(t) = \frac{R}{\tau} e^{-t/\tau}$$

($t = p\Delta t$, where Δt is the sampling step; R and τ are passive membrane parameters). Thus, the membrane kernel K_m is estimated from the tail of the total kernel K (least square fitting to an exponential) and deduce the electrode kernel K_e (Fig. 6.25a; see also supplementary Fig. 1 in Brette et al. 2008). The sensitivity of the method to these two hypotheses as tested in numerical simulations (Brette et al. 2008): first, the estimated electrode kernel degrades continuously as the ratio of electrode and membrane time constants (τ_e/τ_m) increases (see supplementary Fig. 1D in Brette et al. 2008), and remains acceptable for ratios better than 1/10 (the error in electrode resistance is about 10% and AEC can still be used for compensating the voltage in dynamic-clamp protocols, see supplementary Fig. 3 in Brette et al. 2008). Secondly, the membrane kernel deviates from a single exponential function in the presence of a dendritic tree and fast passive dendritic contributions can be confused with the electrode response, but the impact on the estimated electrode kernel remains small (see supplementary Fig. 4 in Brette et al. 2008). In practice, nonlinear membrane properties (even APs if the firing rate is not too large) have a small impact on the estimation of the electrode kernel (see Fig. 6.25f, *light gray points*, and supplementary Fig. 4 in Brette et al. 2008), the important requirement being the linearity of the electrode (see below).

In fact, the electrode kernel K_e captures not only the characteristics of the electrode but also of the recording device, i.e., the whole circuit between the digital output of the computer and the tip of the electrode, including all circuits in the amplifier (e.g., capacitance neutralization) and acquisition filters, but not subsequent digital signal processing. All measured electrode kernels consisted of three phases (Fig 6.25c): first, a short phase where the kernel vanishes, second, a sharp, but noninstantaneous increase for about 0.2 ms, and third, a decrease. The first phase corresponds to the feedback delay of the acquisition system and always lasts two sampling steps (0.2 ms). The second, noninstantaneous phase also appears when the amplifier is plugged into an electronic circuit, named the model cell, consisting of a resistor modeling the electrode and a resistor plus a capacitor modeling the neuron membrane: its nonzero rise-time is likely due to the electronics of the acquisition system rather than to the electrode properties. The third phase varies between experiments and, when fitted by an exponential, displays a time constant around 0.1 ms (0.11 ± 0.09 ms, Brette et al. 2008, $n = 67$ cells in vitro, for maximal levels of capacitance neutralization by the amplifier). Lowering the level of capacitance neutralization increases this time constant (Fig. 6.25d). In vivo, the time constant estimated in four cells in Brette et al. (2008) was of the same order of magnitude (for example, 0.1 ms for the cell Fig. 6.25e).

6.5.4 Estimating the Electrode Resistance

As a first test of the AEC method, Brette et al. (2008) assessed the quality of the estimation of the electrode resistance derived from the measurement of the electrode kernel. The electrode resistance R_e is defined as the ratio of the stationary voltage across the electrode over the amplitude of a constant injected current. It equals the integral of the electrode kernel (or the sum in the digital formulation). In 67 cortical neurons in vitro, the R_e estimated from the kernel was compared to the R_e estimated by manually adjusting the "bridge" on the amplifier, which revealed a difference of only $1.4 \pm 4.2\%$ (AEC-bridge; $n = 67$ cells). Bridge compensation only provides an estimate of the electrode resistance, but AEC also deals with the effects of the residual capacitance of the electrode (after maximal neutralization, residual electrode capacitance was 1.3 ± 0.001 pF as assessed from the decay phase of the electrode kernel; $n = 67$ cells).

The linearity of the electrode is an important assumption of the AEC method, since one considers that the electrode response can by fully characterized by the kernel K_e. It has been described that for sharp electrodes, R_e can change with the amplitude of injected current, presumably due to differences in ion concentrations inside and around the pipette tip (Purves 1981). This might constitute a problem since during AEC, the same kernel, with a fixed R_e, is used for predicting the electrode response to all levels of injected current. This effect was systematically tested in 23 intracellular recordings in vitro (using 20 different electrodes; Brette et al. 2008), by estimating the electrode kernel in the recorded cell while simultaneously injecting different levels of constant current in addition to the white noise current (see supplementary Fig. 2 in Brette et al. 2008). The degree of nonlinearity λ can be expressed in MΩ/nA as a current-dependent change in resistance, and the resulting voltage measurement error for a given depolarization ΔV (induced by constant current injection) is then $(\lambda / R_m^2) * V$, where R_m is the membrane resistance. Across all tested electrodes in the study by Brette and colleagues, this slope averaged −2.2 (±5.0) MΩ/nA, giving, for a cell with $R_{\mathrm{m}} = 60\,\mathrm{M}\Omega$, a voltage error −0.0006 mV/mV (e.g., −0.55 mV for 30 mV depolarization). In 9 out of 23 cases, it was found that the correlation between R_{e} and the level of constant injected current was not significant (see supplementary Fig. 2 in Brette et al. 2008), meaning that the current-dependent variability of the electrode resistance was comparable to the intrinsic variability of the electrode resistance (for these electrodes, the measured nonlinearity was −0.5 ± 1.1 MΩ/nA).

Thus the kernel estimation procedure provides a fast and automated way to measure electrode nonlinearities in the cell, and possibly discard some recordings if the change of R_{e} during given current injections leads to unacceptable errors in the recorded V_{m}. Electrode nonlinearities are generally described as slow processes (Purves 1981), which was, in the aforementioned study, the present working hypothesis, but it was also checked in the dynamic clamp protocols that large transiently injected currents did not impair the electrode compensation, as could result from fast nonlinearities (see below). In addition, in four intracellular recordings (using three electrodes), the electrode kernels obtained by injecting 0.5 or 1 nA of white noise current were compared, and a difference was found on average of only −0.1 MΩ (range −0.9 to 0.5; see supplementary Fig. 2B in Brette et al. 2008), which indicates that possible electrode nonlinearities (i.e., changes in R_{e}) do not manifest themselves for very fast current injections.

In vivo, it was found that electrode properties could remain stable for up to two hours, as assessed with kernel estimations obtained repetitively, also when using different constant current injections and different durations of white noise injection (Fig. 6.25e,f).

6.5.5 White Noise Current Injection

A first application of the AEC method is the possibility to accurately record the response of a neuron to an injection of white noise current sampled at a high frequency (e.g., at 10 kHz). This type of stimulus has been used to characterize neuronal response properties (Bryant and Segundo 1976). It has the advantage of allowing the comparison of the recorded V_{m} with a theoretical prediction for a passive neuronal membrane. It was confirmed (n = 18 injections, in three cortical cells in vitro and n = 8, in two cortical cells in vivo in Brette et al. 2008) that the subthreshold response of neurons to such an injection corresponds to the theoretical prediction based on the passive parameters of the cell (Fig. 6.26a). The recorded V_{m} distributions closely matched the predicted distributions (Fig. 6.26b,d) (8.9 ± 9.9% relative error on the SD of the distributions for in vitro recordings), and power spectra of the response matched the theoretical power spectra up to a frequency of 1 kHz (Fig. 6.26c). In addition, spikes occurring during the white noise injection could be recorded with good temporal resolution (Fig. 6.26c; a higher sampling frequency might be required for fast spiking cells). Attempts to inject a white noise current sampled at 10 kHz with discontinuous current clamp (DCC) at

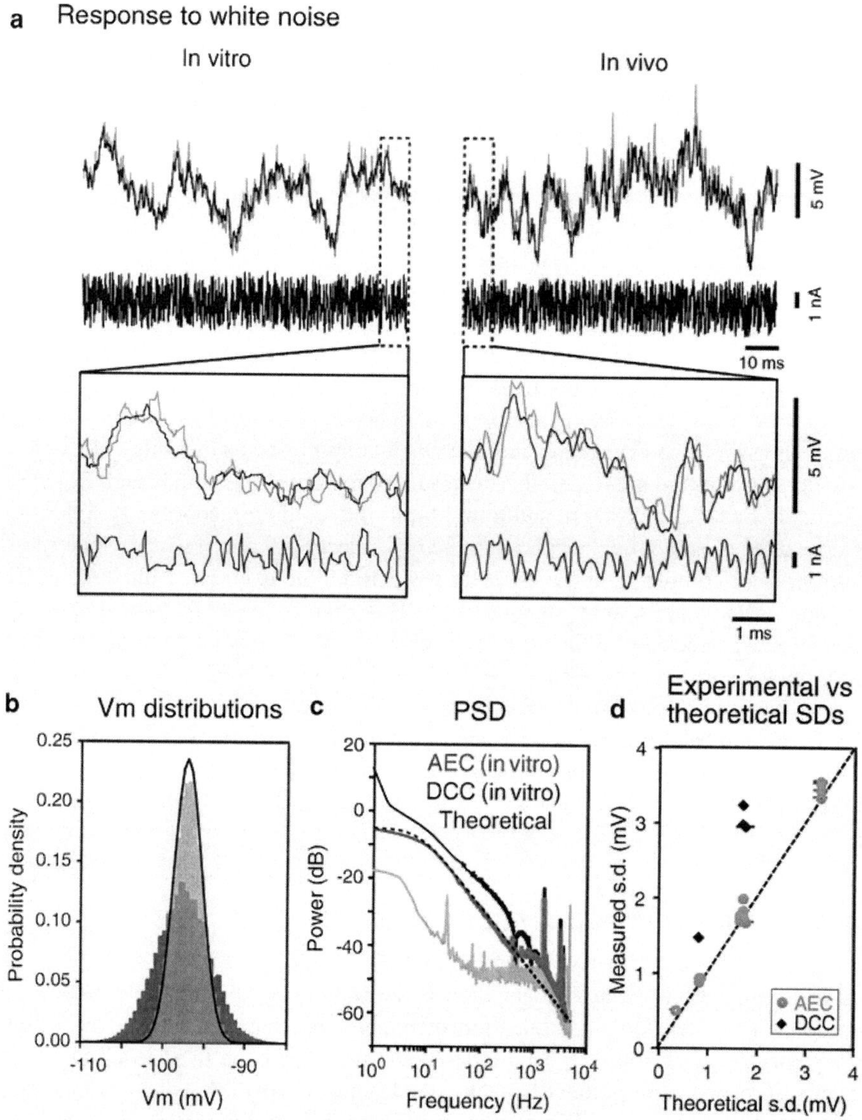

Fig. 6.26 White noise current injection using AEC. (**a**) Example V_m responses of one neuron recorded in vitro (*left*) and another one in vivo (*right*) using AEC (*gray*), and V_m responses obtained by simulating the same noise injection in a point model neuron using the leak parameters of the recorded cell (*black*). The injected current is displayed below. (**b**) V_m distribution recorded using AEC (*light gray*) and theoretical distribution in response to the same injected current (*black solid*). (**c**) Power spectral density (PSD) of the V_m recorded using AEC (*dark gray*) and theoretical PSD (*black dashed*) in response to the same injected current. The PSD of the baseline V_m (*light gray*) shows that the bump in the higher frequencies is not due to AEC, but rather to the power of recording noise reaching the level of the signal. All examples shown (**a–d**) were obtained in a RS cell in vitro, with $R_e = 63\,\text{M}\Omega$, excepted for the in vivo trace, obtained in the same RS cell with $R_e = 103\,\text{M}\Omega$. (**d**) Pooled data: standard deviation of the V_m distributions obtained using AEC (*gray*) or DCC (*black*), vs. theoretical standard deviation based on leak parameters of the recorded neuron (error bars represent the range of theoretical standard deviations obtained for different estimates of passive cell parameters; *dashed line*: $y = x$). Modified from Brette et al. (2008)

1–2 kHz switching frequency, however, failed to match the prediction (Fig. 6.26d), which is expected since the input current is changing significantly faster than the sampling clock of the utilized DCC. Generally, white noise inputs are an interesting choice for probing the input–output functions of systems due to their lack of autocorrelation and they could now with AEC be more widely applied in single-cell neurophysiology.

6.5.6 Dynamic-Clamp Experiments

The initial motivation for developing the AEC method was to improve the quality of dynamic-clamp performed with single high-resistance electrodes. As outlined in Sect. 6.1, the dynamic-clamp can be advantageously employed in a variety of contexts, and in particular to re-create in vivo conditions in neurons recorded in vitro by injecting noisy synaptic conductances. The problem is that the technique relies on a loop in which the current injected into the cell is a function of the recorded V_m. It is, thus, crucial to use the real V_m of the cell, uncontaminated by electrode artifacts which can lead to oscillatory instabilities or, simply, inaccurate results. With bridge compensation, dynamic clamp injections are unstable if the conductance is too large; theoretical analysis (see Brette et al. 2008) shows that this critical conductance is determined by the electrode resistance: $g_c = 1/R_e$ (same magnitude as the typical membrane leak conductance). With complete digital compensation of the electrode, the critical conductance is theoretically limited by the sampling frequency f: $g_c = \tau_m f/(4R_m)$, where τ_m is the membrane time constant and R_m is the membrane resistance. A typical values here is $g_c = 25$ times the leak conductance.

The performance of the AEC method in dynamic-clamp experiments was tested by Brette et al. (2008) in cortical neurons in vitro and in vivo with three different dynamic-clamp protocols of increasing complexity: square conductance pulses and fluctuating synaptic conductance input without and with additional discrete AMPA synaptic inputs. Here, the AEC was compared to the only alternative method, DCC, and, whenever possible, to theoretical predictions of the response.

6.5.6.1 Square Conductance Pulses

In Brette et al. (2008), a simple conductance stimulus was considered for a case where the cell's response can be computed analytically, assuming the membrane is in the passive regime and the leak parameters of the cell are known (from the response to small current pulses). This choice enabled comparison of the responses obtained in AEC and in DCC with theoretical predictions, but this time with dynamic clamp rather than current clamp. The stimulus was a square wave of alternating excitatory (G_e) and inhibitory (G_i) conductance pulses (Fig. 6.27a). Different conductance amplitudes (range 10–100 nS) and frequencies (range 10–1,000 Hz) were scanned (n = 57 AEC-DCC pairs in total, in eight cells in vitro).

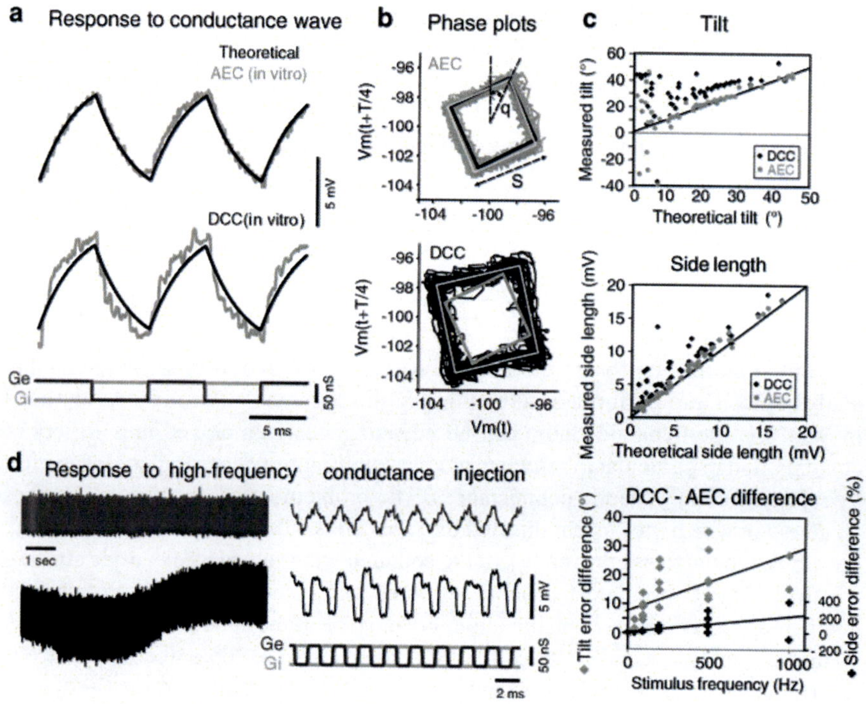

Fig. 6.27 Conductance square wave injection in dynamic-clamp. (**a**) Example V_m responses (RS cell) to a square wave (50 nS amplitude) of alternating excitatory and inhibitory conductances, using AEC (*top*, *gray*) or DCC (*middle*, *gray*). The response obtained by simulating the same conductance injection in a point model neuron using the passive parameters of the recorded cell is shown in *black*. (**b**) Phase plots of the V_m responses shown in (**a**): instead of plotting V_m vs. time, here V_m at time $t+T/4$ is plotted against V_m at time t (T: period of injected wave). *Black* (*top*) and *dark gray* (*bottom*): theoretical phase plot calculated using the stimulus parameters and the cell's leak parameters. *Light gray*: square providing the best fit to the experimental phase plot. Each square is characterized by its side length S and its tilt relative to the vertical, θ. (**c**) Pooled data from all cells and injection parameters used (AEC: *gray*; DCC: *black*): tilt of the recorded phase plot vs. the theoretical tilt (*top*); side length of the recorded phase plot vs. theoretical side (*middle*); and difference (DCC–AEC) between tilt and side errors (relative to the theoretical prediction) in AEC and in DCC vs. square wave frequency. (**d**) V_m response ($R_e = 89\,\text{M}\Omega$, RS cell) to high-frequency (500 Hz) conductance wave injection (50 nS amplitude), in AEC (*gray*, *top*) and DCC (*black*, *bottom*). Low- frequency oscillations appear in DCC as a result of asynchronous sampling when the frequency of the conductance wave is close to the frequency of the DCC sampling clock (the phase of sampling points with respect to the input slowly drifts). Modified from Brette et al. (2008)

To avoid alignment problems and to separate the quality of the response shapes from the amount of recording noise, the responses are represented in a phase space where they appear as noisy squares and can be compared to theoretical predictions (see Fig. 6.27b). Moreover, three error measures can be derived from the comparison of the actual response to the theoretical prediction: the "side" measure quantifies the amplitude of the V_m response, the "tilt" measure quantifies the shape of the response (between a triangular wave and a square wave), and the "distance" measure quantifies the amplitude of the noise around the best-fit response. All those error measures are significantly lower for AEC (tilt error: 5.7 ± 9.2; side error: 21.2 ± 34%; distance: 0.2 ± 0.16 mV) than for DCC (tilt error: 14.6 ± 11.8; side error: 66.6 ± 86%; distance: 0.43 ± 0.49 mV) ($P < 0.0001$ for all three error measures, n = 56). Figure 6.27c (top and middle) shows measured vs. theoretical values for tilt and side. In addition, for all three measures investigated in Brette et al. (2008), the advantage of AEC over DCC grows with the waveform's frequency (Fig. 6.27c, bottom; linear regression analyses: $P < 0.0001$ for tilt and side error difference, $P = 0.0066$ for distance difference), indicating that AEC allows partly overcoming the limitations due to the low sampling frequency in DCC. Low frequency aliasing artifacts in DCC at very high stimulus frequencies are the most striking examples of these limitations (Fig. 6.27d). In conclusion, AEC allows better quality dynamic-clamp injection of conductance and recording of the V_m response than DCC, especially for high-frequency stimuli.

In addition, Brette et al. (2008) used also large currents, injected transiently at times of switching between excitation and inhibition: up to ±2 nA for the highest conductance values. If, in these experiments, there were important changes of R_e with such strong current injection, they would have resulted in an asymmetry between positive and negative parts of the response, which was not observed, confirming that electrode nonlinearities are probably slow processes that do not develop when strong but transient current passes though the electrode.

6.5.6.2 Colored Conductance Noise

A second application, and test, of the AEC method is provided by the injection of fluctuating synaptic conductances according to the point-conductance model of synaptic background activity (Destexhe et al. 2001). This paradigm was realized by Brette et al. (2008) both in vitro and in vivo (Fig. 6.28a). It must be noted that this stimulus was applied here for the first time in an in vivo experiment. In both cases, the AEC or DCC injections appeared of comparable accuracy. More quantitative methods are needed to compare them in such a complex paradigm, as we will see below.

During fluctuating synaptic conductance injection, excellent approximations are available for the V_m distribution (Rudolph and Destexhe 2003d; Rudolph et al. 2004) (Sect. 8.2) and for the V_m power spectrum (Destexhe and Rudolph 2004) (Sect. 8.3). These expressions were used to compare responses recorded in both AEC and DCC with theoretical predictions (see Brette et al. 2008). Here, a very good match was observed between the predicted average V_m and the average V_m measured both in AEC (0.5% average relative error, range 0.003–1.2%) and in DCC (0.5% average relative error, range 0.1–1.8%, no significant difference when compared with AEC, $P = 0.87$) (Fig. 6.28b,c left). The measured SDs are slightly higher than the prediction (Fig. 6.28b,c right), both in AEC (14.7% average relative error,

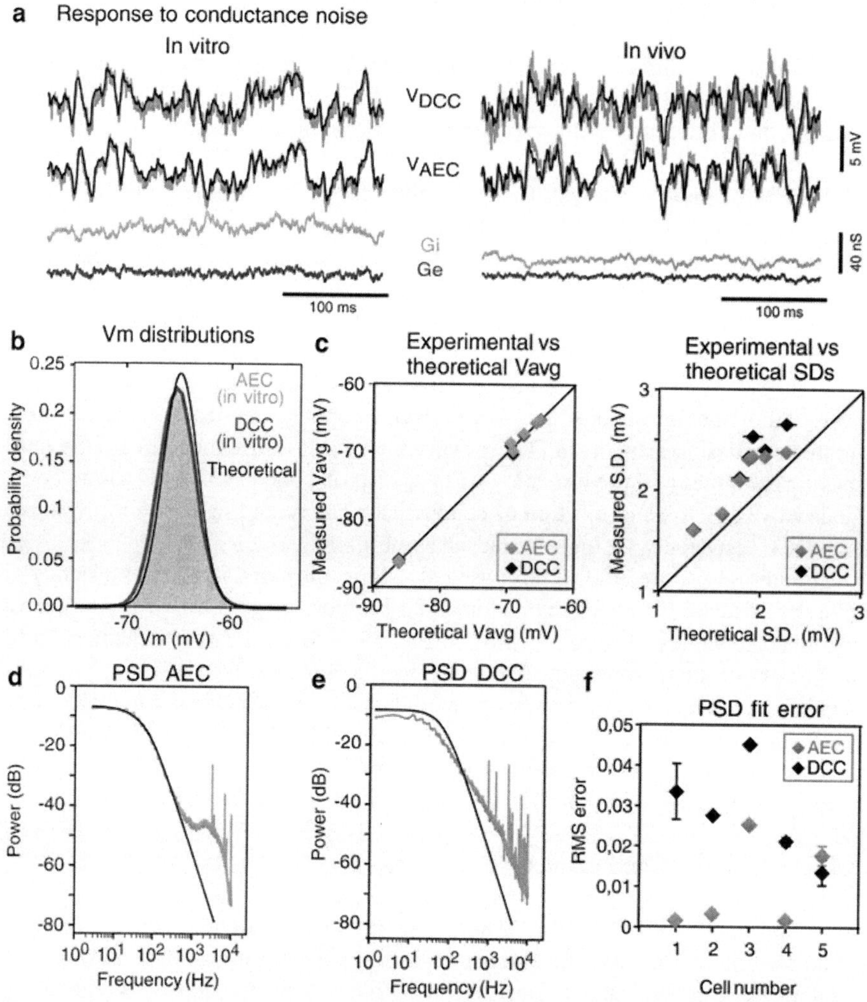

Fig. 6.28 Colored conductance noise injection in dynamic-clamp. (**a**) Example V_m responses to fluctuating excitatory and inhibitory conductances , using AEC or DCC (*gray*) in in vitro (*left*) and in vivo (*right*) neurons. The response obtained by simulating the same conductance injection in a point model neuron using the passive parameters of the recorded cell is shown in *black*. (**b**) V_m distribution using AEC (*light gray*), DCC (*dark gray*), and theoretical distribution in response to the same injected conductance noise (*black*). (**c**) Pooled data: mean of the V_m distributions obtained using AEC (*gray*) or DCC (*black*) vs. theoretical mean based on leak parameters of the recorded neuron (*left*); standard deviation of the V_m distributions obtained using AEC or DCC, vs. theoretical standard deviation based on leak parameters of the recorded neuron (*right*). (Error bars represent the range of theoretical values obtained when varying estimates of leak parameters by $\pm$ 10%; *black solid*: $y = x$). (**d**) Power spectral density (PSD) of the V_m recorded using AEC (*gray*) and best fit with the theoretical template for the PSD (*black*). (**e**) PSD of the V_m recorded using DCC (*gray*) and best fit with the theoretical template for the PSD (*black*). All examples shown (**a**, **b**, **d**, **e**) were obtained in the same IB cell in vitro, with $R_e = 89\,\mathrm{M}\Omega$, excepted for the in vivo trace, obtained in an RS cell with $R_e = 103\,\mathrm{M}\Omega$. (**f**) Pooled data: root-mean-square (RMS) error of the best fit to the experimental V_m PSDs obtained when using the theoretical template (AEC: *gray*; DCC: *black*; each point is an average for PSDs obtained from fragments of conductance injections done in the same cell, error bars are standard deviations). Modified from Brette et al. (2008)

range 5–21.6%) and in DCC (18.8% average relative error, range 7.7–31%), and the error difference between the two methods, albeit small, is significant ($P = 0.028$). The most striking difference between the two methods, however, appeared in the frequency content of the V_m fluctuations. In four out of five cells in the study by Brette et al. (2008), the PSD of the V_m in AEC could be fitted very well with the theoretical template, which provided a good match up to the frequencies where recording noise becomes important (Fig. 6.28d): the PSD scaled as f^{-4} in the high frequencies as predicted. In DCC, the f^{-4} scaling was never observed, the exponent was always smaller (Fig. 6.28e), showing that the correct frequency scaling could only be obtained in AEC (Fig. 6.28f; except for one cell in which all methods yielded erroneous scaling). Again, this result demonstrates that AEC is superior to DCC in allowing the accurate investigation of high- frequency components of the neuronal response, and stresses that this advantage is present not only with simple stimuli like conductance pulses but also when more realistic stimuli are applied.

6.5.7 Analysis of Recorded Spikes

A final application of the AEC method which we will present here is its use for the analysis of individual spikes. In Brette et al. (2008), spikes were recorded in both DCC and AEC in vitro and compared with recordings obtained from a more complex dynamic-clamp protocol: AMPA excitatory synaptic inputs (Destexhe et al. 1994b) of five different amplitudes occurred in a randomized fashion, at 10 Hz, in a synaptic background of fluctuating excitation and inhibition modeled as above. Such a protocol has the advantage of distinguishing two types of inputs: driving AMPA inputs and modulatory background inputs (Shu et al. 2003b; Fellous et al. 2003; Wolfart et al. 2005). To illustrate the new possibilities offered by high-resolution recording of spikes using AEC (Fig. 6.29a, *AEC*), Brette and colleagues performed an analysis on spike threshold variability. In cortical neurons, the V_m value at spike threshold has been shown to correlate with the slope of the preceding depolarization in a way such that faster depolarizations evoke spikes from a more negative threshold (Azouz and Gray 2000; de Polavieja et al. 2005; Wilent and Contreras 2005b). Such a significant negative correlation was, indeed, found in the aforementioned study in all AEC recordings (average slope of linear regression: -1.4 ms, range -3.6 to -0.5 ms; average coefficient of regression: 0.407, range 0.137 to 0.621), similar to the example shown in the top row of Fig. 6.29b (*AEC*).

Such an analysis cannot be done with DCC recordings because spike threshold is not detected with enough accuracy (Fig. 6.29a, *DCC*): the time of spike onset is not locked to the (low) DCC sampling frequency, and thus the first V_m value to be effectively recorded after spike onset is only weakly correlated with the actual V_m value at spike threshold. Instead of detecting the threshold, one detects here an approximately random value at the beginning of the spike, and so the slope-threshold correlation (average slope of linear regression: 0.5 ms, range 0.07 to 1.43 ms; average coefficient of regression: 0.177, range 0.028 to 0.475) is either

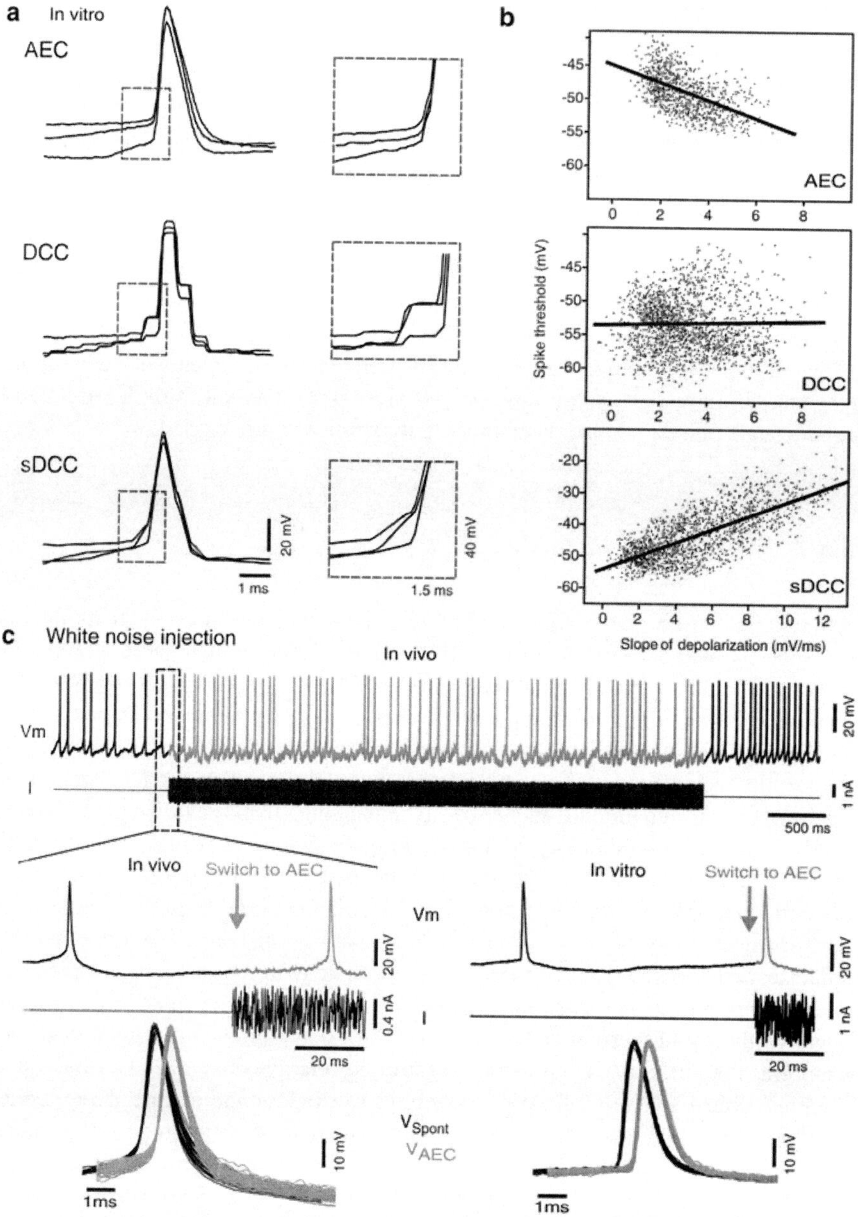

Fig. 6.29 Spikes evoked by a complex dynamic-clamp conductance injection and spikes recorded during white noise current injection. (**a**) Examples of single spikes recorded in AEC (*top*) or DCC (*middle*), for different slopes of depolarization preceding the spike. Spikes obtained when the DCC recording is smoothed are shown at the bottom (*sDCC*). *Insets*: zoom on the onset of spikes (R_e = 64 MΩ, RS cell). (**b**) Spike threshold vs. slope of depolarization preceding spikes, from injections corresponding to the examples shown in (**a**). Lines are linear regressions to the clouds of points. (**c**) Spontaneous spikes (*black*, V_{spont}) are compared to spikes recorded in the same cell during white noise current injection using AEC (*gray*, V_{AEC}), in vivo (central trace and zoom, *left*; RS cell, R_e = 104 MΩ) and in vitro (*right*; IB cell, R_e = 91 MΩ). Modified from Brette et al. (2008)

nonsignificant or positive. This relation (Fig. 6.29b, *DCC*) reflects the fact that V_m and V_m slope are positively correlated along the rising phase of a spike. Smoothing the DCC trace can make the spikes look better by eye (Fig. 6.29a, *sDCC*), but cannot retrieve high-frequency information like spike threshold, and so the results of slope-threshold analysis are similar to the ones obtained from a raw DCC trace (significant positive correlations were detected in Brette et al. 2008 in all cases; average slope of linear regression: 2.8 ms, range 2.1 to 4.1 ms; average coefficient of regression: 0.787, range 0.686 to 0.895; Fig. 6.29b, *sDCC*).

Such spike recordings have proven to also be a good opportunity for testing the presence of fast electrode nonlinearities, because large negative currents are usually injected at spike times (up to −5 nA). It was found (Brette et al. 2008) that the correlation between the injected current and the voltage at spike peaks is very small (between −0.1 and −0.6 MΩ), thus confirming the absence of significant fast nonlinearities that would manifest themselves as over- or under-compensation of R_e during spikes.

Finally, one can test if the shape of fast cellular events like spikes is correctly recorded with AEC. To that end, in Brette et al. (2008), 36 spontaneous spikes recorded without any current injection in one cell in vitro were compared to 36 spikes recorded in the same cell using AEC during white noise current injection with 0 nA mean (Fig. 6.29c, right bottom). Under these conditions, the average V_m and the firing frequency were found to be the same, so that spike shape is not expected to change due to adaptation. In this study, no significant difference were found between spikes recorded with and without current injection in terms of spike height (54.0 ± 1.5 mV vs. 53.9 ± 1.1 mV, respectively; $P = 0.8413$), spike width at half-amplitude (0.91 ± 0.04 ms vs. 0.90 ± 0.03 ms, $P = 0.2859$) nor spike threshold (−50.4 ± 1.3 mV vs. −50.9 ± 1.0 mV, $P = 0.0967$). Spikes recorded using AEC in vivo are also very similar to spontaneous spikes (Fig. 6.29c, top trace and left bottom). The possibility to precisely analyze spike onset and spike shape during finely controlled current- and dynamic-clamp stimuli, demonstrated with AEC but not feasible before with high-resistance electrodes due to the limited sampling frequency of the DCC, appears crucial especially in the light of recent reports indicating that in cortical networks, spikes are more than all-or-none signals (de Polavieja et al. 2005) and that their shape might influence synaptic transmission and thus the functioning of the network (Shu et al. 2006).

6.5.8 *Concluding Remarks on the AEC Method*

In conclusion, this section has overviewed the AEC method to obtain high-resolution intracellular recordings, and dynamic-clamp experiments in vivo and in vitro (Brette et al. 2008). Because synaptic noise is precisely the most difficult signal to use in dynamic-clamp experiments (due to the presence of very fast frequencies), it is essential to use an appropriate technique to avoid artifacts. It was shown here, and in the corresponding publications (Brette et al. 2007a,b, 2008,

2009) that the AEC method provides an efficient "cleaning" of electrode artifacts, and enables reliable dynamic-clamp investigations with conductance-based synaptic noise.

As we noted above, the dynamic-clamp is a nice example of intimate cooperation between modeling and experimental techniques, interacting in real-time. The AEC method is another example of such a tight association, the recording of the living neurons is made by a direct, online and real time interaction with a computational model of the electrode. While doing dynamic-clamp with the AEC, not only computational models are used to calculate the injected conductances, as in standard dynamic-clamp, but models are also used in the recording technique itself. As illustrated by such a paradigm, theory and experiments have never been so close in neuroscience.

6.6 Summary

In this chapter, we have overviewed the study of synaptic noise in neurons using the dynamic-clamp technique. This technique combines computational models (models of synaptic noise in the present case) with living neurons. The computer-controlled conductances are injected in real time in the neuron. The bases of this technique are reviewed in Sect. 6.1.

The "recreation" of in vivo-like high-conductance states in cortical neurons using the dynamic-clamp is overviewed in Sects. 6.2 and 6.3. These experiments used the point-conductance model reviewed in detail in Chap. 4. The comparison to the in vivo measurements reviewed in Chap. 3 shows that, under dynamic-clamp, cortical neurons in vitro display all the essential features of neurons recorded in vivo. As predicted by models (Hô and Destexhe 2000; reviewed in Chap. 5), these dynamic-clamp experiments revealed an important effect of the stochastic synaptic activity on neuronal responsiveness (Destexhe et al. 2001; Chance et al. 2002; Fellous et al. 2003; Mitchell and Silver 2003; Prescott and De Koninck 2003; Shu et al. 2003a,b; Higgs et al. 2006). These investigations were greatly facilitated by models in which the fluctuations and the mean conductances can be controlled independently (Destexhe et al. 2001).

Perhaps the most unexpected property of synaptic noise was found when investigating the effect of noise on thalamic neurons (Wolfart et al. 2005; Sect. 6.4). These neurons are classically known to display two distinct firing modes a single-spike (tonic) mode, and a burst mode at more hyperpolarized levels (Llinás and Jahnsen 1982). However, thalamic neurons are also known to receive large amounts of synaptic noise through their numerous direct and indirect synaptic connections from descending corticothalamic fibers, and this activity accounts for about half of the input resistance of thalamic neurons in vivo (Contreras et al. 1996). Based on these measurements, the effect of synaptic noise was simulated using dynamic-clamp on thalamic neurons in slices, and remarkably, it was found that under such in vivo-like conditions, the transfer properties of the neurons are radically changed, as covered in detail in Sect. 6.4. These experiments show that both neuronal intrinsic properties and synaptic noise are necessary to understand the transfer function of central neurons in vivo.

Finally, in Sect. 6.5, we have reviewed an important technique to perform injection of synaptic noise in neurons. This so-called "Active Electrode Compensation" (Brette et al. 2008) uses a computational model of the electrode to obtain high-resolution recordings and dynamic-clamp. This technique is demonstrated in cortical neurons, both in vivo and in vitro. It is interesting to note that like the dynamic-clamp, the AEC involves computational models interacting in real time with living neurons. The whole chapter illustrates the power of this approach in revealing the integrative properties of neurons under in vivo-like conditions and noise levels.

Chapter 7
The Mathematics of Synaptic Noise

The previous chapters of this book have focused mostly on studies assessing and characterizing synaptic noise under a variety of experimental conditions, and on evaluating its role in shaping neural dynamics through computational models. Although detailed biophysical models of neurons in vivo (see Sect. 4.2) remain, so far, out of reach for a mathematically more rigorous approach, the introduced simplified models (see Sects. 4.3 and 4.4), at least partially, allow for an analytical treatment. The latter can be used to complement experimental and computational studies and, therefore, provide a deeper understanding of neuronal dynamics under noisy conditions. Moreover, a mathematical treatment can also be used to provide unprecedented characterization of synaptic noise and how it affects spiking activity. This will be the subject of this and the forthcoming chapters.

7.1 A Brief History of Mathematical Models of Synaptic Noise

Mathematical models describing and characterizing the consequences of synaptic noise in cortical networks on the neuronal dynamics and cellular response cover a broad range of complexity, ranging from (leaky) integrate-and-fire (IF) neuron models (Lapicque 1907) with Gaussian white noise (Lánský and Lánská 1987; Bindman et al. 1988; Tuckwell 1988; Lánský and Rospars 1995; Doiron et al. 2000; van Rossum 2001; Brunel et al. 2001), stochastic membrane equations (Tuckwell and Walsh 1983; Tuckwell et al. 1984; Manwani and Koch 1999a,b; Tuckwell et al. 2002), up to biophysically faithful models of single neurons in which synaptic background activity is incorporated by the random release at individual synaptic terminals according to Poisson processes (Bernander et al. 1991; Rapp et al. 1992; Lánský and Rodriguez 1999; Manwani and Koch 1999a,b; Tiesinga et al. 2000; Rudolph and Destexhe 2001a; Rudolph et al. 2001; Tuckwell et al. 2002; Rudolph and Destexhe 2003b). Ricciardi and Sacerdote (1979) showed that under point-like

A. Destexhe and M. Rudolph-Lilith, *Neuronal Noise*, Springer Series in Computational Neuroscience 8, DOI 10.1007/978-0-387-79020-6_7,

excitatory and inhibitory synaptic inputs Poisson distributed in time, the neuron's membrane potential undergoes a continuous random walk. The latter is described by a temporally homogeneous Markov process obeying the governing Fokker–Planck equation for a dynamical random process of the OU type. Hence, the OU process, which describes low-pass filtered Gaussian white noise and was originally introduced as a model of Brownian motion (Uhlenbeck and Ornstein 1930; Wang and Uhlenbeck 1945), belongs to the most prominent models for synaptic noise (Ricciardi and Sacerdote 1979; Hänggi and Jung 1994).

As outlined in Sect. 4.4, using detailed biophysical models of cortical neurons subject to stochastic synaptic inputs, the total conductance resulting from a sum of thousands of synaptic inputs has a power spectrum which approximates a Lorentzian whose high-frequency limit follows a $1/f^2$ behavior, where f denotes the frequency. The Gaussian nature of the OU process and its Lorentzian spectrum qualitatively match the behavior of the conductances underlying synaptic noise for higher frequencies, thus providing an effective stochastic representation that captures the amplitude of the conductances, their SD and spectral structure. Moreover, the use of this effective representation in neocortical slices (Destexhe et al. 2001; Fellous et al. 2003; Rudolph et al. 2004) successfully recreated the cellular properties typically found in the intact and activated brain.

However, the complexity of the resulting stochastic equations does allow to address only specific problems analytically (e.g., statistical characteristics like mean and variance of the membrane potential; see Manwani and Koch 1999a,b; Tuckwell et al. 2002), whereas numerical methods (for a review see Werner and Drummond 1997) or approximations (e.g., mean-field approximation, see Van den Broeck et al. 1994; Ibañes et al. 1999; Genovese and Muñoz 1999) remain the standard tools. In addition, the particular mathematical form of noise terms and their incorporation into stochastic neuron models, i.e., the nature of the coupling of the noise to the neural system, are, surprisingly, still subject of controversy. Experimental results show that synaptic transmission underlies a voltage dependent and, thus, state-dependent kinetics (Regehr and Stevens 2001). This suggests that the synaptic current has to be described, according to Ohm's law, by a (possibly voltage-dependent) conductance term coupling multiplicatively to the state variable. In the case of random synaptic inputs, this would translate into a noisy conductance term with multiplicative coupling to the membrane potential (*multiplicative noise*). At leading order, however, the resulting noisy synaptic current can directly be described by a stochastic variable. In this case, the noise term due to synaptic background activity enters the equations governing the membrane potential time course in an additive way (*additive noise*, see, e.g., Kohn 1997).

Representing synaptic noise using an additive term yields a simpler mathematical description of stochastic neural systems and, thus, is the most commonly used model of synaptic noise dating back to the sixties (Stein 1967; for a recent study of additive noise described by an OU process, see Brunel et al. 2001). On the other hand, multiplicative noise appears to be closer linked to the biophysical dynamics observed in nature. For a comparison between both noise couplings, see, e.g., Tiesinga et al. (2000). In this chapter, we will briefly outline the most prominent mathematical models of additive synaptic noise in passive and IF neuronal models (Sect. 7.2) as well as multiplicative, i.e., conductance-based, synaptic noise models (Sects. 7.3 and 7.4) along with a numerical comparison of their ability to predict the statistical properties of membrane potential fluctuations (Sect. 7.5).

7.2 Additive Synaptic Noise in Integrate-and-Fire Models

In the following three sections, we will briefly outline the mathematical approach to assess the response of simple IF neuronal models to additive, i.e., current, noise. We will start with the simplest case, namely Gaussian white noise (Sect. 7.2.1) before moving on to the more realistic case of filtered, i.e., colored, Gaussian noise (Sect. 7.2.2) and correlated synaptic noise (Sect. 7.2.3).

7.2.1 IF Neurons with Gaussian White Noise

One of the simplest and earliest models to describe neuronal dynamics is the integrate-and-fire (IF) neuronal model, introduced by Lapicque more than one hundred years ago (Lapicque 1907). Its dynamics is described by the membrane equation

$$C\frac{\mathrm{d}V(t)}{\mathrm{d}t} = I_{\mathrm{L}}(V) + I_{\mathrm{syn}}(V,t) + I_{\mathrm{ext}}(t), \tag{7.1}$$

where $I_{\mathrm{L}}(V)$, $I_{\mathrm{syn}}(V,t)$ and $I_{\mathrm{ext}}(t)$ denote the leak, synaptic, and external current, respectively, and C is the membrane capacitance. A spike response is generated whenever the membrane potential reaches a fixed threshold V_{th}, after which V is reset to a fixed value V_{reset}.

Several variants of this model are known, two of which will be considered below. In its simplest form, the *simple integrate-and-fire* (SIF) model, the IF neuron has no leak current, i.e.,

$$I_{\mathrm{L}}(V) = 0.$$

Introducing the leak conductance g_{L} and resting (or leak reversal) potential E_{L}, (7.1) describes the *leaky integrate-and-fire* (LIF) model, for which the leak current takes the form

$$I_{\mathrm{L}}(V) = -g_{\mathrm{L}}(V(t) - E_{\mathrm{L}}).$$

For simplicity reasons, it is common to set $E_{\mathrm{L}} \equiv 0$ in mathematical calculations, in which case (7.1) takes the general form

$$\tau_m \frac{dV(t)}{dt} = f(V) + I_{syn}(V,t) + I_{ext}(t),$$
$$f(V) = \begin{cases} 0 & \text{(SIF)} \\ -V(t) & \text{(LIF)} \end{cases} \tag{7.2}$$

where $\tau_m = C/g_L$ denotes the membrane time constant. It is important to note that, here, $I_{syn}(V,t)$ and $I_{ext}(t)$ are expressed in units of voltage.

A first simple model of synaptic transmission is given by an instantaneous rise and fall of a synaptic current at arrival of a presynaptic spike. In this case, the PSC can is described by a delta function with amplitude, or efficacy, J. In this case, the total synaptic current in (7.1) stemming from N_{syn} synaptic input channels takes the form (for simplicity we assume equal amplitude for each synaptic channel):

$$I_{syn}(t) = \tau_m J \sum_{i=i}^{N_{syn}} \sum_k \delta(t - t_i^k). \tag{7.3}$$

Here, the first sum runs over all N_{syn} synaptic channels, whereas the second sum runs over all release times t_i^k for synaptic channel i (t_i^k denotes the kth presynaptic spike arriving at synapse i). Other, more realistic, models of synaptic transmission described by current transients are possible and described in the literature (see, e.g., Fourcaud and Brunel 2002). One of these models will be considered in the next section.

Although (7.1), together with (7.3), provides a very simple description of neurons subject to synaptic activity, the discrete nature of the synaptic inputs, described in terms of sums over discrete time events (with frequencies ranging up to several Hz) and multiple synaptic channels (usually of the order of tens or hundreds of thousand for cortical neurons), makes an analytical treatment unpleasant. However, under the reasonable assumptions that the neuron receives a high barrage of Poissonian distributed and uncorrelated synaptic inputs, and that the amplitude J of a single synaptic current is small compared to the distance between spike threshold and resting potential, i.e. $|J| \ll |V_{th} - E_L|$ (the currents in this notation are expressed in units of voltage), one can average over both sums in (7.3) and obtain

$$I_{syn}(t) \sim \mu + \sigma\eta(t). \tag{7.4}$$

Here, $\eta(t)$ denotes a Gaussian random variable with zero mean and unit SD.

This last equation (7.4) is called *diffusion approximation*. It approximates the total synaptic current caused by an intense barrage with synaptic inputs, (7.3) by an effective stochastic Gaussian white noise process with mean μ and SD σ. Moreover, denoting the mean activation rate of each synapse with ν_{syn}, the parameters of the effective description, (7.4), and the statistical properties of the presynaptic activity can be linked:

$$\mu = \tau_m J N_{syn} \nu_{syn},$$
$$\sigma^2 = \tau_m J^2 N_{syn} \nu_{syn}. \tag{7.5}$$

With this, we arrive at an effective description of the IF neuron subject to synaptic noise described by a Gaussian white noise stochastic process:

$$\tau_m \frac{dV}{dt} = f(V) + \mu + \sigma\sqrt{\tau_m}\eta(t) + I_{ext}(t). \tag{7.6}$$

As a first example, we consider a neuron subject to synaptic noise only, i.e., $I_{ext}(t) \equiv 0$. In this case, the LIF model [(7.6) with $f(V) = -V(t)$] resembles mathematically an OU stochastic process (Uhlenbeck and Ornstein 1930), and its Fokker–Planck equation for the probability density function $\rho(V,t)$ takes the form

$$\tau_m \frac{\partial \rho(V,t)}{\partial t} = \frac{\sigma^2}{2}\frac{\partial^2 \rho(V,t)}{\partial V^2} - \frac{\partial}{\partial V}\left[(\mu + f(V))\rho(V,t)\right]. \tag{7.7}$$

The probability flux $j_V(V,t)$ of $\rho(V,t)$ through V at time t, which defines the (stationary) instantaneous firing rate ν of the neuron, is given by

$$j_V(V,t) = \frac{\mu + f(V)}{\tau_m}\rho(V,t) - \frac{\sigma^2}{2\tau_m}\frac{d\rho(V,t)}{dV} \tag{7.8}$$

with the boundary conditions

$$j_V(V_{th},t) = \nu,$$
$$j_V(V_{reset+},t) - j_V(V_{reset-},t) = \nu. \tag{7.9}$$

These boundary conditions arise out of the fact that, whenever a spike is emitted by crossing the spike threshold V_{th}, the membrane is instantaneously reset to its reset value V_{reset}.

The stationary case is defined by the conditions

$$\frac{\partial \rho}{\partial t} \equiv 0,$$
$$\rho(V,t) \equiv \rho(V).$$

Solving (7.7) in this case for the SIF neuron model, yields (Abbott and van Vreeswijk 1993)

$$\rho(V) = \frac{\nu\tau_m}{\mu}\left[1 - e^{\frac{2\mu(V-V_{th})}{\sigma^2}} - \Theta(V_{reset} - V)(1 - e^{\frac{2\mu(V-V_{reset})}{\sigma^2}}\right],$$
$$\nu = \frac{\mu}{\tau_m}\frac{1}{V_{th} - V_{reset}}, \tag{7.10}$$

where $\Theta(x)$ denotes the Heaviside function. Figure 7.1a shows a typical stationary probability density function $\rho(V)$ for the SIF neuron. Moreover, from the last equation, it can be seen that, somewhat surprising, the response rate of the SIF neuron is only dependent on the mean of the synaptic current, but not its amplitude σ.

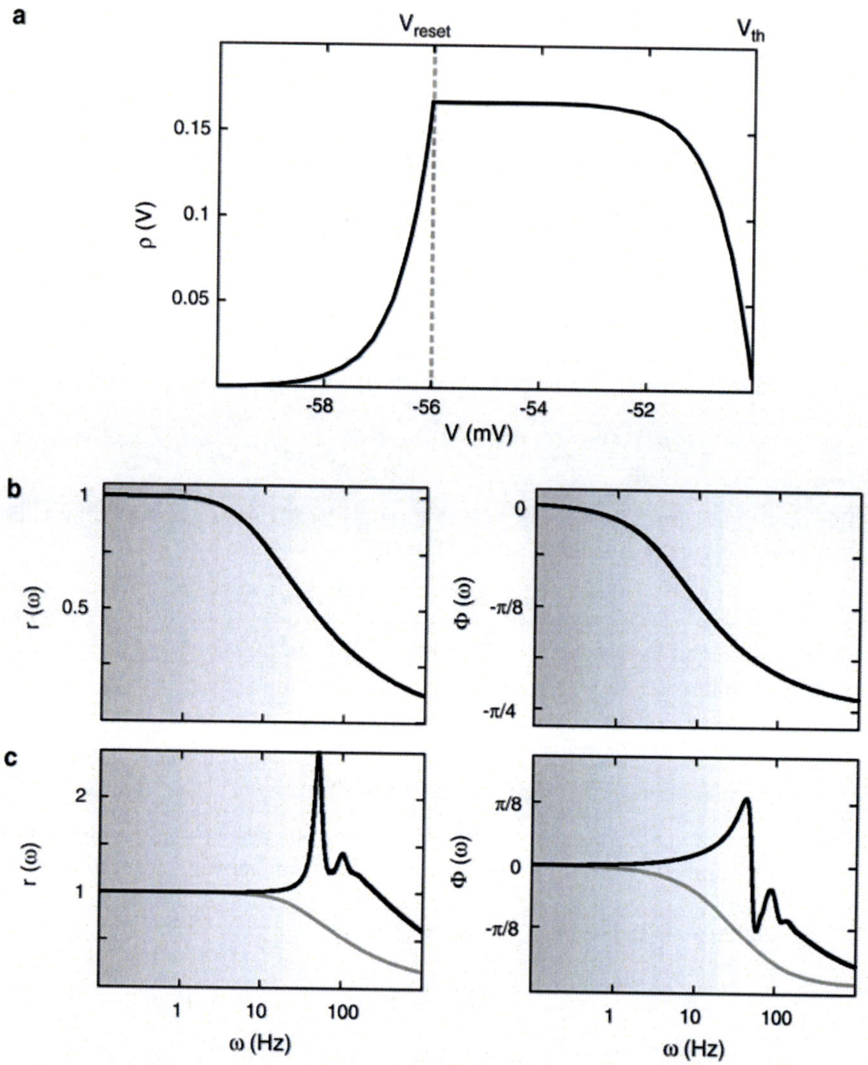

Fig. 7.1 Response of the IF neuron models to Gaussian white noise. (**a**) Stationary probability density function of the SIF neuron, (7.10) for no external inputs. Parameters: $\mu/\tau_m = 1.25\,\text{mV}\cdot\text{ms}^{-1}$, $\sigma^2/\mu = 1.25\,\text{mV}$, $V_{reset} = -56\,\text{mV}$, $V_{th} = -50\,\text{mV}$. (**b**) Numerical simulations of the amplitude and phase of the response for the SIF neuron [(7.2) together with (7.15)]. (**c**) Numerical simulations of the amplitude and phase of the response for the LIF neuron, (7.17), in the case of low ($\sigma = 1\,\text{mV}$, *black solid*) and high ($\sigma = 5\,\text{mV}$, *grey solid*) noise amplitude. In (**b**) and (**c**), the response was normalized to 1 at $\omega = 0.1$ Hz. Response rate was $\nu = 50$ Hz and $\tau_m = 20\,\text{ms}$. Modified from Fourcaud and Brunel (2002)

Similarly, solving (7.7) for the LIF neuron, one obtains (Brunel and Hakim 1999)

$$\rho(V) = \frac{2\nu\tau_m}{\sigma} e^{-\frac{(V-\mu)^2}{\sigma^2}} \times \begin{cases} \int\limits_{(V_{\text{reset}}-\mu)/\sigma}^{(V_{\text{th}}-\mu)/\sigma} ds\, e^{s^2} & \text{if } V < V_{\text{reset}} \\ \int\limits_{(V-\mu)/\sigma}^{(V_{\text{th}}-\mu)/\sigma} ds\, e^{s^2} & \text{if } V > V_{\text{reset}}, \end{cases}$$

$$\frac{1}{\nu} = \tau_m \sqrt{\pi} \int\limits_{(V_{\text{reset}}-\mu)/\sigma}^{(V_{\text{th}}-\mu)/\sigma} ds\, e^{s^2} (1 + \operatorname{erf}(s)), \tag{7.11}$$

where $\operatorname{erf}(x)$ denotes the error function.

In the remainder of this section, we will consider the response of the noisy IF neuron (7.2) subject to oscillatory input of the form

$$I_{\text{ext}}(t) = \varepsilon\mu\cos(\omega t), \tag{7.12}$$

where ε denotes a small parameter which will be used to obtain an approximation of the neuronal response. In this case, the Fokker–Planck equation takes the form

$$\tau_m \frac{\rho(V,t)}{\partial t} = \frac{\sigma^2}{2} \frac{\partial^2 \rho(V,t)}{\partial V^2} - \frac{\partial}{\partial V} \left[f(V) + \mu(1 + \varepsilon e^{i\omega t}) \right] \rho(V,t). \tag{7.13}$$

Note that the probability density function $\rho(V,t)$ is now explicitly time dependent due to the dynamic input. Moreover, to simplify notation, a complex external input is used.

In order to solve (7.13), the probability density $\rho(V,t)$ and instantaneous rate $\nu(t)$ are expanded around the small parameter ε, and only lower order terms are considered:

$$\begin{aligned} \rho(V,t) &= \rho(V) + \varepsilon e^{i\omega t} \hat{\rho}(V,\omega) + O(\varepsilon^2), \\ \nu(t) &= \nu \left[1 + \varepsilon \hat{n}_0(\omega) e^{i\omega t} + O(\varepsilon^2) \right], \end{aligned} \tag{7.14}$$

where

$$\hat{n}_0(\omega) = r_0(\omega) e^{i\Phi(\omega)} \tag{7.15}$$

and $\hat{\rho}(V,\omega)$ denote complex quantities describing the oscillatory component of the instantaneous rate and membrane potential probability density at the input frequency ω, respectively. Here, $r_0(\omega)$ and $\Phi(\omega)$ are called the relative amplitude of the firing rate modulation and the phase lag of the response, respectively. Inserting this expansions into (7.12) yields, in lowest order, for the probability density function and oscillatory component of the response for the SIF neuron (Abbott and van Vreeswijk 1993; Fourcaud and Brunel 2002):

$$\hat{\rho}(V,\omega) = \frac{\nu}{i\omega}\left[e^{\frac{2\mu(V-V_{\rm th})}{\sigma^2}} - e^{r_+(\omega)\frac{2\mu(V-V_{\rm th})}{\sigma^2}} \right.$$
$$\left. -\Theta(V_{\rm reset}-V)\left(e^{\frac{2\mu(V-V_{\rm reset})}{\sigma^2}} - e^{r_+(\omega)\frac{2\mu(V-V_{\rm reset})}{\sigma^2}} \right)\right], \qquad (7.16)$$

where

$$r_+ = \frac{1+\sqrt{1+2i\omega\tau_{\rm eff}}}{2}$$

and

$$\hat{n}_0(\omega) = \frac{\sqrt{1+2i\omega\tau_{\rm eff}}-1}{i\omega\tau_{\rm eff}}.$$

Here, $\tau_{\rm eff} = \sigma^2\tau_{\rm m}/\mu^2$ denotes the effective time constant of the SIF neuron model. For frequencies lower than $1/\tau_{\rm eff}$, the SIF neuron responds with only little attenuation and phase lag to the oscillating input (Fig. 7.1b, *shaded area*). For high input frequency, responses are attenuated by a factor $\sqrt{2/(\omega\tau_{\rm eff})}$ and show a phase lag which approaches $-\pi/4$ (Fig. 7.1b).

In a similar fashion, solving the Fokker–Planck equation (7.12) for the LIF neuron, one obtains (Brunel and Hakim 1999)

$$\hat{n}_0(\omega) = \frac{\mu}{\sigma(1+i\omega\tau_{\rm m})}\frac{\frac{\partial U}{\partial y}(y_{\rm th},\omega) - \frac{\partial U}{\partial y}(y_{\rm reset},\omega)}{U(y_{\rm th},\omega)-U(y_{\rm reset},\omega)}, \qquad (7.17)$$

where $y_{\rm th} = \frac{V_{\rm th}-\mu}{\sigma}$, $y_{\rm reset} = \frac{V_{\rm reset}-\mu}{\sigma}$ and $U(y,\omega)$ is a combination of confluent hypergeometric functions

$$U(y,\omega) = \frac{e^{y^2}}{\Gamma\left(\frac{1+i\omega\tau_{\rm m}}{2}\right)} M\left(\frac{1-i\omega\tau_{\rm m}}{2},\frac{1}{2},-y^2\right) + \frac{2ye^{y^2}}{\Gamma\left(\frac{i\omega\tau_{\rm m}}{2}\right)} M\left(1-\frac{i\omega\tau_{\rm m}}{2},\frac{3}{2},-y^2\right). \qquad (7.18)$$

Examples of the response amplitude and phase lag are shown in Fig. 7.1c. The high input frequency behavior is equivalent to the one observed in the case of the SIF neuron, with

$$\hat{n}_0(\omega) \sim \sqrt{\frac{2}{i\omega\tau_{\rm eff}}} \qquad (7.19)$$

for $\omega \to \infty$. Resonance peaks at frequencies ω corresponding to the input frequency and integer multiples of it occur in both the amplitude and phase of the response (Fig. 7.1c, *black solid*). As expected, these are more pronounced at low noise levels and are even absent at large σ (Fig. 7.1c, compare *black* and *gray solid*). The behavior for $\omega \to \infty$ is equivalent to that observed in the SIF neuron model, with the modulation amplitude $r_0(\omega)$ approaching zero with $\sqrt{1/\omega}$ and the phase lag reaching $-\pi/4$.

In the next section, we will extend this result to a more realistic model of synaptic noise, namely colored Gaussian noise, which can be viewed as an effective model in which synaptic transmission is governed by exponentially decaying synaptic currents.

7.2.2 LIF Neurons with Colored Gaussian Noise

Above we considered the most simple case of synaptic transmission, described by individual synaptic inputs leading to instantaneous finite PSCs. Utilizing the diffusion approximation, an intense barrage with such inputs can mathematically be described by an effective, Gaussian white noise stochastic process (7.4).

As a first extension of this model, we consider now a synaptic current transient described by an instantaneous rise and exponential decay with time constant τ_{syn}. In this case, the defining equation for the synaptic current, (7.3), is replaced by the differential equation

$$\tau_s \frac{\mathrm{d}I_{\text{syn}}(t)}{\mathrm{d}t} = -I_{\text{syn}}(t)\tau_{\text{m}} J \sum_{i=i}^{N_{\text{syn}}} \sum_k \delta(t - t_i^k). \tag{7.20}$$

As in the case of δ-synaptic current transients, an effective stochastic model can be deduced which approximates the total synaptic current caused by the combined effect of many synaptic input channels with activity resembling that of a high-frequency Poissonian process:

$$\tau_s \frac{\mathrm{d}I_{\text{syn}}(t)}{\mathrm{d}t} = -I_{\text{syn}}(t) + \mu + \sigma\sqrt{\tau_{\text{m}}}\eta(t), \tag{7.21}$$

where μ and σ are given by (7.5).

In the following, we will again restrict to the LIF neuron subject to oscillating external input. In this case, (7.2), (7.12), and (7.21) can be rewritten to yield

$$\begin{aligned} \tau_{\text{m}} \frac{\mathrm{d}V(t)}{\mathrm{d}t} &= -V(t) + I_{\text{syn}}(t) + I_{\text{ext}}(t), \\ \tau_s \frac{\mathrm{d}I_{\text{syn}}(t)}{\mathrm{d}t} &= -I_{\text{syn}}(t) + \tilde{\eta}(t), \\ I_{\text{ext}}(t) &= \mu_0 + \mu_1 \cos(\omega t). \end{aligned} \tag{7.22}$$

Here, $\tilde{\eta}(t)$ denotes a Gaussian random variable with zero mean and variance of $\sigma^2\tau_{\text{m}}$.

In order to solve this system of SDEs, the Fokker–Planck calculus is utilized once again (Brunel et al. 2001). However, in contrast to the white noise case ($\tau_s = 0$), an analytic treatment is significantly more difficult. The Fokker–Planck equation takes now the form

$$\tau_s \frac{\partial \rho(y,z,t)}{\partial t}$$
$$= \frac{1}{2}\frac{\partial^2 \rho(y,z,t)}{\partial z^2} + \frac{\partial(z\rho(z,t))}{\partial z} + \frac{\tau_s}{\tau_\mathrm{m}}\frac{\partial(y-y_t)\rho(y,z,t)}{\partial y}$$
$$+\sqrt{\frac{\tau_s}{\tau_\mathrm{m}}}\left[\left(y_t - \frac{\mu_1}{\sigma}(\cos(\omega t) - \omega\tau_s \sin(\omega t))\right)\frac{\partial \rho(y,z,t)}{\partial z} - z\frac{\partial \rho(y,z,t)}{\partial y}\right], \tag{7.23}$$

where

$$y(t) = \frac{V(t)-\mu_0}{\sigma},$$
$$y_\mathrm{th} = \frac{V_\mathrm{th}-\mu}{\sigma},$$
$$z(t) = \sqrt{\frac{\tau_s}{\tau_\mathrm{m}}}\left(\frac{I_\mathrm{syn}(t)+\mu_1\cos(\omega t) - V_\mathrm{th}}{\sigma}\right).$$

Under certain boundary conditions for the probability flux (Brunel et al. 2001), the firing rate is now given by

$$r(t) = \int_{-\infty}^{\infty} \mathrm{d}z\,\rho(z,t), \tag{7.24}$$

where $\rho(z,t)$ corresponds to the probability flux in the direction of y at spiking threshold V_th:

$$\rho(z,t) = \begin{cases} \frac{1}{\sqrt{\tau_s \tau_\mathrm{m}}} z\rho(y_\mathrm{th},z,t) & \text{for } z>0 \\ 0 & \text{for } z<0. \end{cases} \tag{7.25}$$

Unfortunately, (7.23) and, thus, (7.24) defy an analytically exact solution. However, decomposing the response into the average rate r_0 and, similar to (7.15), an oscillating component $\hat{n}_1(\omega)$ with amplitude $r_1(\omega)$ and phase $\Phi(\omega)$, the Fokker–Planck equation can be linearized around its stationary solution $\rho_0(y,z)$, r_0 (Hagan et al. 1989; Brunel and Sergi 1998) if $\mu_1 \ll \mu_0$ and $\tau_s \ll \tau_\mathrm{m}$:

$$\rho(y,z,t) = \rho_0(y,z) + \mathrm{Re}\left[\rho_1(y,z,\omega)\mathrm{e}^{i\omega t}\right]. \tag{7.26}$$

The numerical evaluation of the response amplitude $r_1(\omega)$ and phase $\Phi(\omega)$ are shown in Fig. 7.2. In contrast to the white noise case for corresponding model parameters (Fig. 7.1c), the filtering of the synaptic noise dramatically changes the high-frequency behavior: $r_1(\omega)$ does no longer approach zero but, instead, a finite value for $\omega \to \infty$ (Fig. 7.2, *tick marks*), whereas the phase lag approaches zero. The latter means that the response rate can be modulated by oscillating inputs up to

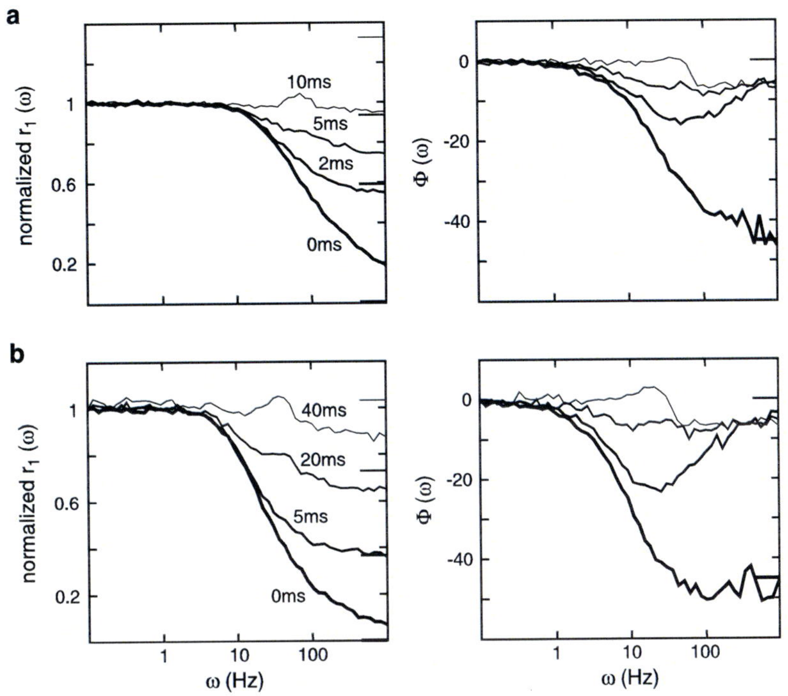

Fig. 7.2 Response amplitude (normalized $r_1(\omega)/r_1(0.1)$ and phase for the LIF neuron model subject to colored Gaussian noise for different values of τ_s between 0 ms and 40 ms. (**a**) Responses with average rate $r_0 = 50$ Hz. (**b**) $r_0 = 10$ Hz. In both cases, $\sigma = 5$ mV. The *tick marks* on the *right side* of the panels indicate the high-frequency limits given by (7.28). Modified from Brunel et al. (2001)

arbitrarily high frequencies without phase lag. Analytically, first-order corrections at high-frequency take the form:

$$\lim_{\omega\to\infty} \rho_1(y,z,\omega) = -\frac{\mu_1}{\sigma}\sqrt{\frac{\tau_s}{\tau_m}}\frac{\partial \rho_0(y,z)}{\partial z},$$

$$\lim_{\omega\to\infty} r_1(\omega)e^{i\Phi(\omega)} = \frac{\mu_1}{\sigma\tau_m}\int_{-\infty}^{\infty} dz\,\rho_0(y,z). \tag{7.27}$$

The latter can be further evaluated (Brunel et al. 2001) for $\tau_{syn} \ll \tau_m$ and $\mu_1/\mu_0 \ll 1$, yielding

$$\lim_{\omega\to\infty} r_1(\omega)e^{i\Phi(\omega)} = \frac{A\mu_1}{\sigma}\sqrt{\frac{\tau_{syn}}{\tau_m}}, \tag{7.28}$$

where $A \approx 1.3238$ is a constant.

7.2.3 IF Neurons with Correlated Synaptic Noise

In the previous two sections, we considered IF neurons with synaptic currents described by instantaneous and exponentially decaying transients. In both cases, we obtained effective stochastic descriptions in terms of Gaussian white and colored noise, respectively. These descriptions were made possible by the diffusion approximation, which links the statistical properties of the combined activity at many independent synaptic channels with low-dimensional effective stochastic processes. However, in reality, the presynaptic activity is not independent but shows a more complicated statistical signature. In this section, we will focus on one of these parameters describing the statistics of presynaptic activity, namely the correlation between individual synaptic input channels.

Again we will consider the LIF neuron model given in (7.2), with the addition of a refractory period τ_{ref} after the occurrence of a spike, in which the neuron does not integrate incoming synaptic inputs and remains at its reset potential V_{reset}. Furthermore, we extend the previous models of the afferent input current by considering N_{e} and N_{i} excitatory and inhibitory synaptic input channels with instantaneous, i.e., δ-like, postsynaptic currents:

$$I(t) = J_{\text{e}} \sum_{i=1}^{N_{\text{e}}} \sum_{k} \delta(t - t_i^k) - J_{\text{i}} \sum_{i=1}^{N_{\text{i}}} \sum_{k} \delta(t - t_j^l). \tag{7.29}$$

Here, $t_{\{i,j\}}^{\{k,l\}}$ denotes the time of the $\{k,l\}$th synaptic release stemming from the ith and jth excitatory and inhibitory presynaptic neuron, respectively. $J_{\{\text{e,i}\}}$ denotes the current amplitude of the respective excitatory and inhibitory inputs.

In what follows, we will restrict to stochastic input trains with exponential autocorrelation with time constant τ_{c} and exponential cross-correlations between individual synaptic channels given by

$$C_p(t,t') = \nu_p \delta(t-t') + \nu_p \left(\frac{F_{\text{p}} - 1}{2\tau_{\text{c}}} \right) \text{e}^{-|t-t'|/\tau_{\text{c}}},$$

$$C_{pq}(t,t') = \sqrt{\nu_p \nu_q} \left(\frac{\rho_{pq} \sqrt{F_{\text{p}} F_q}}{2\tau_{\text{c}}} \right) \text{e}^{-|t-t'|/\tau_{\text{c}}}, \tag{7.30}$$

where $\{p,q\} = \{e,i\}$, $\nu_{\{p,q\}}$ and $F_{\{p,q\}}$ denote the firing rate and Fano factor ($F_{\text{p}} = 1$ for Poisson processes and $F_{\text{p}} > 1$ ($F_{\text{p}} < 1$) for positive (negative) correlated processes) of the synaptic release activity for infinite long time windows, and ρ_{pq} the correlation coefficient determining the magnitude of the cross-correlation.

Summing over all incoming synaptic channels, (7.30), yields the effective correlation of the total synaptic current:

$$C(t,t') = \sigma^2 \delta(t-t') + \frac{\Sigma_2}{2\tau_{\text{c}}} \text{e}^{-|t-t'|/\tau_{\text{c}}}, \tag{7.31}$$

where

$$\sigma^2 = J_e^2 N_e \nu_e + J_i^2 N_i \nu_i \tag{7.32}$$

is the white noise variance and

$$\Sigma_2 = J_e^2 N_e \nu_e [F_e - 1 + f_{ee}(f_{ee}N_e - 1)F_e\rho_{ee}] + J_i^2 N_i \nu_I [F_i - 1 + f_{ii}(f_{ii}N_i - 1)F_i\rho_{ii}]$$
$$-2J_e J_i f_{ei} f_{ie} N_e N_i \sqrt{\nu_e \nu_i}\sqrt{F_e F_i}\rho_{ei} \tag{7.33}$$

is a contribution to the total variance

$$\sigma_{\text{eff}}^2 = \sigma^2 + \Sigma_2.$$

f_{pq} denotes the fraction of correlated synaptic channels, $0 \leq f_{pq} \leq 1$.

In comparing the last expression with the correlation of the effective noise process discussed above, (7.5), one observes that in this more detailed model an additional additive contribution to the effective correlation appears, which describes the effect of correlations between inhibitory and excitatory synaptic inputs (last term in (7.33)). Moreover, the mathematical expressions of the total effective correlation is now linked to higher order statistical properties of the individual synaptic input channels, in particular the Fano factor. This way, the relation between synaptic inputs and the response can be studied in more detail.

In the case of high input rates with individual synaptic inputs causing a membrane depolarization which is small compared to the distance between reset potential and spike threshold, i.e.,

$$\frac{JF}{V_{\text{th}} - V_{\text{reset}}}(1 + fN\rho) \ll 1,$$

the total synaptic current $I(t)$ assumes a Gaussian form and can be characterized in terms of its mean $\mu = J_e N_e \nu_e - J_i N_i \nu_i$, its variance σ^2, a parameter $k = \sqrt{\tau_c/\tau_m}$ and the correlation magnitude $\alpha = \Sigma_2 \sigma^2$:

$$I(t) = \mu + \sigma_w \eta(t) + \sigma_w \frac{\beta}{\sqrt{2\tau_c}} z(t) \tag{7.34}$$

with

$$\dot{z}(t) = -\frac{z}{\tau_c} + \sqrt{\frac{2}{\tau_c}}\eta(t). \tag{7.35}$$

Here, $\eta(t)$ denotes a Gaussian white noise process with unit variance, $\beta = \sqrt{1+\alpha} - 1$ and $z(t)$ is an auxiliary colored random process.

With this description of the synaptic input activity, it can be shown (Moreno-Bote et al. 2002) that the probability density $\rho(x,z)$ to find the neuron in state (x,z) obeys the Fokker–Planck equation

$$\left[L_x + \frac{L_z}{k^2} + \frac{2}{k}\frac{\partial}{\partial x}\left(\frac{\partial}{\partial z} - \frac{\beta z}{2}\right)\right]\rho(x,z) = -\tau\delta(x - \sqrt{2}\hat{H})J(z), \tag{7.36}$$

where

$$
\begin{aligned}
L_u &= \frac{\partial}{\partial u}u + \frac{\partial^2}{\partial^2 u},\\
V &= \mu\tau_{\mathrm{m}} + \sigma_w\sqrt{\tau_{\mathrm{m}}}2x,\\
\hat{H} &= \frac{H-\mu\tau_{\mathrm{m}}}{\sigma_w\sqrt{\tau_{\mathrm{m}}}},\\
\hat{V}_{\mathrm{th}} &= \frac{V_{\mathrm{th}}-\mu\tau_{\mathrm{m}}}{\sigma_w\sqrt{\tau_{\mathrm{m}}}}.
\end{aligned}
$$

In (7.36), $J(z)$ denotes the escape probability current, from which under certain normalization conditions the output firing rate can be determined:

$$
\nu = \int_{-\infty}^{\infty} \mathrm{d}z J(z). \tag{7.37}
$$

For two parameter regimes, analytic expressions for the output rate can be deduced. First, in the case of small correlation time constant, i.e., $\tau_{\mathrm{c}} \ll \tau_{\mathrm{m}}$, the output rate takes the form

$$
\nu = \nu_{\mathrm{eff}} - \alpha\sqrt{\tau_{\mathrm{c}}\tau_{\mathrm{m}}}\nu_0^2 R(\hat{\Theta}), \tag{7.38}
$$

where

$$
R(t) = \sqrt{\frac{\pi}{2}}\mathrm{e}^{t^2}[1+\mathrm{erf}(t)]
$$

and $\mathrm{erf}(t)$ denotes the error function. Moreover,

$$
\begin{aligned}
\nu_{\mathrm{eff}}^{-1} &= \tau_{\mathrm{ref}} + \sqrt{\pi}\tau\int_{\hat{H}_{\mathrm{eff}}}^{\hat{\Theta}_{\mathrm{eff}}} \mathrm{d}t\,\mathrm{e}^{t^2}[1+\mathrm{erf}(t)],\\
\nu_0^{-1} &= \tau_{\mathrm{ref}} + \sqrt{\pi}\tau\int_{\hat{H}}^{\hat{\Theta}} \mathrm{d}t\,\mathrm{e}^{t^2}[1+\mathrm{erf}(t)],
\end{aligned}
$$

with the effective reset and threshold defined as

$$
\begin{aligned}
\hat{\Theta}_{\mathrm{eff}} &= \frac{\Theta-\mu\tau_{\mathrm{m}}}{\sigma_{\mathrm{eff}}\sqrt{\tau_{\mathrm{m}}}},\\
\hat{H}_{\mathrm{eff}} &= \frac{H-\mu\tau_{\mathrm{m}}}{\sigma_{\mathrm{eff}}\sqrt{\tau_{\mathrm{m}}}},
\end{aligned}
$$

respectively. Here, ν_0 denotes the mean firing rate of the IF neuron driven by white noise. This result is exact only for $\tau_{\mathrm{c}} = 0$, but yields a good approximation as long as both k and $\alpha \geq 0$ are small.

In the limit of large correlation time constant, i.e. $\tau_c \gg \tau_m$ and $\tau_{ref} \ll \tau_c$, the Fokker–Planck equation yields for the output rate

$$\nu = \nu_0 + \frac{C}{\tau_c}, \tag{7.39}$$

where

$$C = \alpha \tau_m^2 \nu_0^2 \left[\frac{\tau_m \nu_0 (R(\hat{\Theta}) - R(\hat{H}))^2}{1 - \nu_0 \tau_{ref}} - \frac{\hat{\Theta} R(\hat{\Theta}) - \hat{H} R(\hat{H})}{\sqrt{2}} \right]. \tag{7.40}$$

In this case, ν simply converges to the mean firing rate of IF neurons subjected to white noise, i.e., ν_0.

Examples for the response rate as a function of the correlation time constant and α are shown in Fig. 7.3. In all cases, the theoretical prediction and numerical simulation are in good agreement. As it can be seen, the cellular response is, at fixed α, most sensitive to changes in the input correlation for highly synchronized inputs, i.e., $\tau_c < \tau_m$ (Fig. 7.3, *shaded area*) and decreases (increases) for increasing positive (negative) correlation. Moreover, for fixed τ_c, the rate increases with the correlation magnitude α and is more sensitive to changes in α in the balanced, i.e., fluctuation-dominated (large σ_w, Fig. 7.3, *solid*), state than t is in the unbalanced, i.e., drift-dominated (large μ, Fig. 7.3, *dashed*), state.

7.3 Multiplicative Synaptic Noise in IF Models

In the previous section, we dealt with the most simple cases of neuronal models with synaptic noise, namely IF neurons subjected to stochastic synaptic currents. However, the description of synaptic transmission by current transients must be viewed as only a first approximation of the biophysical dynamics seen in real neurons. In fact, PSCs are a direct consequence of transient changes in the membrane conductance caused by the activation of transmembrane synaptic terminals.

The mathematical characterization of the effect of these synaptic conductances on the cellular behavior is far more challenging, because synaptic conductances couple multiplicatively to the membrane potential, i.e., the state variable. In this section, we will describe two classical approaches to assess the response of IF neuronal models subject to multiplicative, i.e., conductance-based, Gaussian white and colored (Ornstein–Uhlenbeck) synaptic noise.

7.3.1 IF Neurons with Gaussian White Noise

Starting, once more, from the LIF neuronal model defined in (7.1), but describing now the synaptic current as the result of δ-shaped excitatory and inhibitory conductance transients $g_e(t)$ and $g_i(t)$, respectively, we arrive at the following mathematical description of neuronal dynamics incorporating synaptic conductance noise:

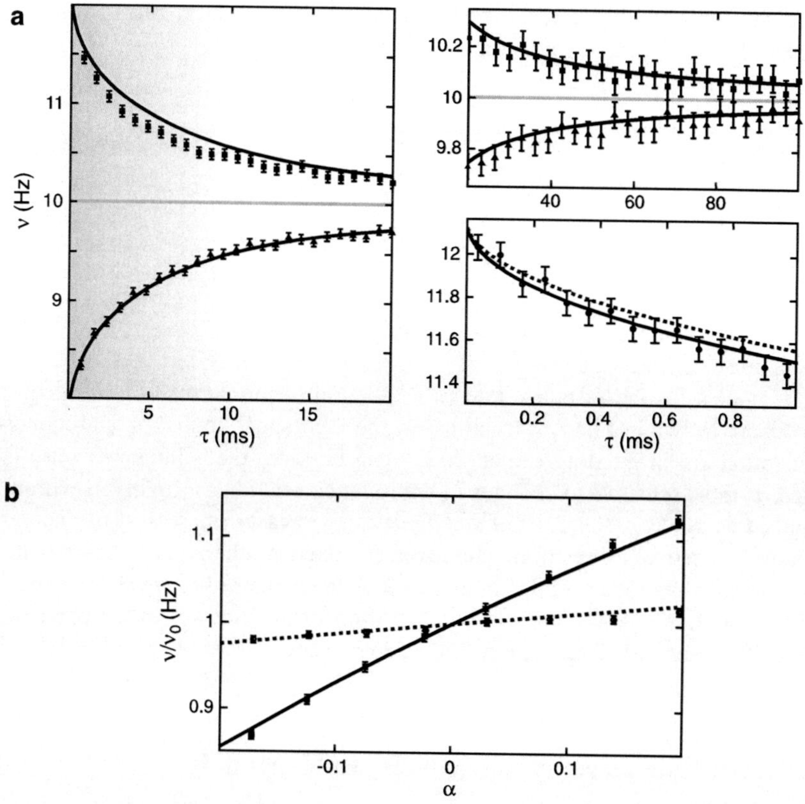

Fig. 7.3 Output rate of LIF neuron models subject to correlated synaptic inputs. (**a**) Comparison between the analytical solution and numerical simulations of the output rate as a function of the correlation time constant τ_c for $\alpha = 0.21$ (*left, upper curve*), $\alpha = -0.19$ (*left, lower curve*), $\alpha = 0.21$ and $\alpha = -0.19$ for large τ_c (*upper right panel*) and very small τ_c (*bottom right*). Interpolations between small and large τ_c from theoretical predictions at interpolation time $\tau_c = 14\,\text{ms}$ (*solid lines*) and results from the analytic solution [(7.38), *dashed lines*] are shown. The *horizontal gray lines* indicate the response rate to white noise. Parameters: $\tau_m = 10\,\text{ms}$, $\tau_{\text{ref}} = 0\,\text{ms}$, $\Theta = 1$, $H = 0$, $\mu = 81.7\,\text{s}^{-1}$, $\sigma^2 = 2.1\,\text{ms}^{-1}$. (**b**) Comparison between theoretical prediction and numerical simulation for the ratio ν/ν_0 as function of α for the fluctuation-dominated (balanced, $\mu = 40\,\text{s}^{-1}$) and drift-dominated (unbalanced, $\mu = 110\,\text{s}^{-1}$) state (*solid* and *dashed lines*, respectively). Parameters: $\sigma^2 = 30\,\text{s}^{-1}$, $\tau_c = 1\,\text{ms}$, other parameters as in (**a**). Modified from Moreno-Bote et al. (2002)

$$
\begin{aligned}
C\frac{\mathrm{d}V(t)}{\mathrm{d}t} &= -g_{\mathrm{L}}(V(t)-E_{\mathrm{L}}) - I_{\mathrm{syn}}(t),\\
I_{\mathrm{syn}}(t) &= g_{\mathrm{e}}(t)(V(t)-E_{\mathrm{e}}) + g_{\mathrm{i}}(t)(V(t)-E_{\mathrm{i}}),\\
g_{\{\mathrm{e,i}\}}(t) &= Ca_{\{\mathrm{e,i}\}} \sum_{i=1}^{N_{\{\mathrm{e,i}\}}} \sum_k \delta(t-t_i^k).
\end{aligned}
\tag{7.41}
$$

Here, $E_{\{e,i\}}$ denote the excitatory and inhibitory reversal potential, respectively, $a_{\{e,i\}}$ a dimensionless parameter quantifying the strength of a single synapse (i.e., their quantal amplitude) and $N_{\{e,i\}}$ the respective number of (uncorrelated) excitatory and inhibitory synaptic channels releasing at Poisson-distributed times $\{t^k\}$ with average rate $\nu_{\{e,i\}}$ for each channel. In (7.41), in contrast to the previous section, the leak reversal potential E_L is no longer considered to be zero.

For small quantal amplitudes $a_{\{e,i\}}$ and high total rates $N_{\{e,i\}}\nu_{\{e,i\}}$, the diffusion approximation provides a mathematical description of the total excitatory and inhibitory synaptic conductances in terms of effective stochastic processes. These take the form

$$g_{\{e,i\}}(t) = Ca_{\{e,i\}}\left[N_{\{e,i\}}\nu_{\{e,i\}} + \sqrt{N_{\{e,i\}}\nu_{\{e,i\}}}\eta(t)\right], \tag{7.42}$$

where $\eta(t)$ denotes a Gaussian white noise process with zero mean and unit SD. With this, the Fokker–Planck equation for the probability density function can be readily deduced (Richardson 2004), yielding

$$\tau_m \frac{\partial \rho(V,t)}{dt} = \frac{1}{\gamma}\frac{\partial^2}{\partial V^2}\left[(V(t)-E_S)^2 + E_D^2\right]\rho(V,t) + \frac{\partial}{\partial V}(V(t)-E)\rho(V,t), \tag{7.43}$$

where τ_m denotes now the total input-dependent average membrane time constant, defined by

$$\frac{1}{\tau_m} = \frac{1}{\tau_L} + N_e\nu_e\tilde{a}_e + N_i\nu_i\tilde{a}_i, \tag{7.44}$$

with the shifted amplitude $\tilde{a}_{\{e,i\}} = a_{\{e,i\}} - \frac{1}{2}a^2_{\{e,i\}}$ and $\tau_L = \frac{C}{g_L}$. As pointed out in Richardson (2004), the latter is a direct result of the *Stratonovich formulation* of the stochastic system in question. In the *Itô formalism*, we have $\tilde{a}_{\{e,i\}} = a_{\{e,i\}}$ (see Sect. 7.4 for an approach utilizing the Itô formalism). Furthermore, in (7.43) we have

$$\begin{aligned}
E &= \tau_m\left(\frac{E_L}{\tau_L} + N_e\nu_e\tilde{a}_eE_e + N_i\nu_i\tilde{a}_iE_i\right),\\
E_S &= (N_e\nu_e a_e^2 E_e + N_i\nu_i a_i^2 E_i)\chi,\\
E_D &= \sqrt{N_eN_i\nu_e\nu_i}a_ea_i(E_e - E_i)\chi,
\end{aligned} \tag{7.45}$$

where $\chi = (N_e\nu_e a_e^2 + N_i\nu_i a_i^2)^{-1}$ and $\gamma = 2\chi/\tau_m$.

In order to simplify notation, a change of variables can be performed in (7.43). This yields, finally, the Fokker–Planck equation for the steady-state membrane potential distribution:

$$-\nu\tau_m\Theta(x - x_{reset}) = \frac{d}{dx}\left(\alpha^2x^2 - 2\alpha\beta x + 1\right)\tilde{\rho}(x) + x\tilde{\rho}(x), \tag{7.46}$$

where ν denotes the firing rate of the cell and

$$
\begin{aligned}
x &= (V-E)\sqrt{\frac{\gamma}{(E-E_S)^2+E_D^2}},\\
x_{\{\mathrm{th,reset}\}} &= (V_{\{\mathrm{th,reset}\}}-E)\sqrt{\frac{\gamma}{(E-E_S)^2+E_D^2}},\\
\alpha &= \frac{1}{\sqrt{\gamma}},\\
\beta &= \frac{E_S-E}{\sqrt{(E_S-E)^2+E_D^2}}
\end{aligned}
$$

and $\Theta(x)$ denoting the Heaviside step function.

A formal solution of the Fokker–Planck equation (7.46) can be obtained, yielding for the steady-state membrane potential distribution and firing rate the following results (Richardson 2004)

$$
\begin{aligned}
\tilde{\rho}(x) &= \frac{r\tau_{\mathrm{m}}\mathrm{e}^{-B(x)}}{(\alpha x-\beta)^2+1-\beta^2}\int_x^{x_{\mathrm{th}}}\mathrm{d}y\,\Theta(y-x_{\mathrm{reset}})\mathrm{e}^{B(y)},\\
\frac{1}{\nu\tau_{\mathrm{m}}} &= \int_{-\infty}^{x_{\mathrm{th}}}\mathrm{d}x\int_x^{x_{\mathrm{th}}}\mathrm{d}y\,\frac{\Theta(y-x_{\mathrm{reset}})\mathrm{e}^{-B(x)+B(y)}}{(\alpha x-\beta)^2+1-\beta^2},
\end{aligned}
\tag{7.47}
$$

with

$$
B(x) = \frac{1}{2\alpha^2}\ln[(\alpha x-\beta)^2+1-\beta^2]+\frac{\beta}{\alpha^2\sqrt{1-\beta^2}}\arctan\left(\frac{\alpha x\sqrt{1-\beta^2}}{1-\alpha\beta x}\right).
$$

In Fig. 7.4, results of the analytical solution for the firing rate from (7.47) are compared with numerical simulations. As expected, the response rate can be effectively modulated by changes in the total synaptic input rates. In particular, an increase in $N_{\mathrm{e}}\nu_{\mathrm{e}}$ (Fig. 7.4b) or decrease in $N_{\mathrm{i}}\nu_{\mathrm{i}}$ will yield an increase in the output rate. Due to (7.45), these changes in the synaptic input rate both result in a depolarization of the cellular membrane through an increase in the equilibrium potential E and, thus, in a decrease of the distance of the average membrane potential to firing threshold (Fig. 7.4a). Differences in the considered cases are due to the different impact on the membrane time constant. For example, an increase in the synaptic drive will yield a smaller τ_{m}, i.e., a faster membrane, due to an increase in the total membrane conductance and, thus, boost the output rate. However, a larger membrane conductance also lowers the amplitude of individual EPSPs and, thus, their impact on driving the output rate of the cell (Fig. 7.4c).

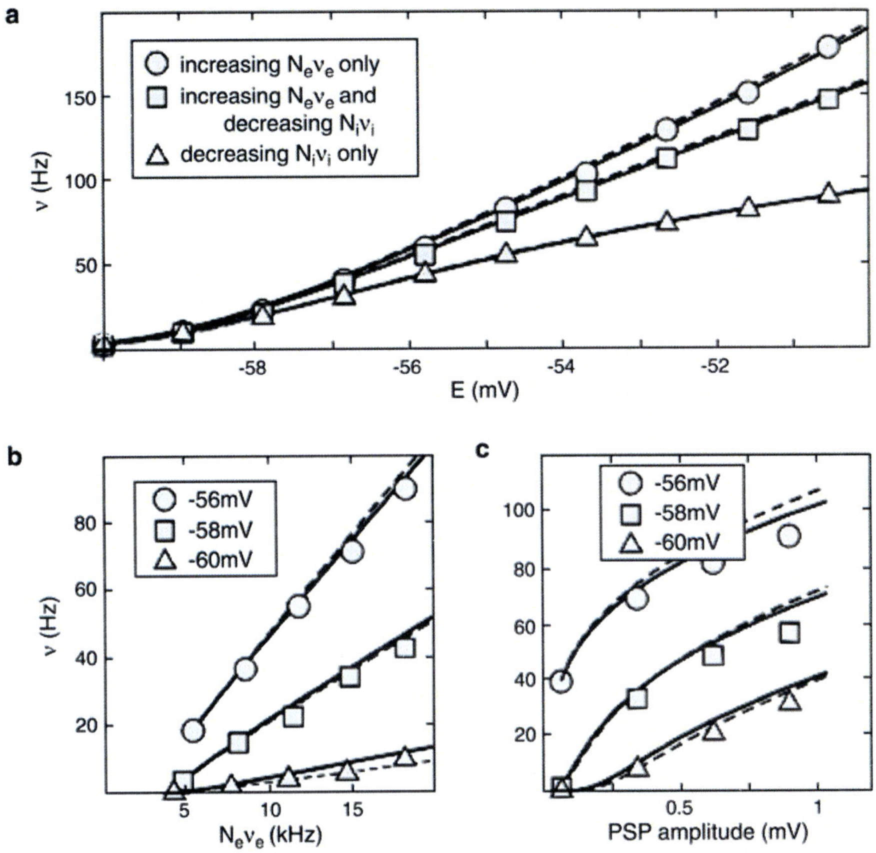

Fig. 7.4 Output rate ν of LIF neuron models subject to excitatory and inhibitory synaptic conductances. (**a**) ν as a function of the equilibrium potential E, (7.45). E was varied by a change in the excitatory drive $N_e \nu_e$ (*circles*), a change in $N_i \nu_i$ and $N_e \nu_e$ which left the total membrane conductance constant (*squares*), and a change in the inhibitory drive $N_i \nu_i$ (*triangles*). Initial values were $N_e \nu_e = 9.17$ kHz and $N_i \nu_i = 3.08$ kHz. Parameters: $\tilde{a}_e = 0.004$, $\tilde{a}_i = 0.026$. (**b**) ν as a function of the excitatory drive under balanced condition $E = -56$ mV (*circles*), -58 mV (*squares*) and -60 mV (*triangles*). Parameters: $\tilde{a}_e = 0.004$, $\tilde{a}_i = 0.026$. (**c**) ν as a function of the PSP amplitude at fixed E (as in B) and $\tau_m = 5$ ms obtained by a change in $\tilde{a}_{\{e,i\}}$ while keeping $N_{\{e,i\}} \nu_{\{e,i\}} \tilde{a}_{\{e,i\}}$ fixed. In all cases, analytic results (*solid*) based on (7.47) are compared with analytic results for current-based approximation (*dashed*) and numerical simulations. Modified from Richardson (2004)

7.3.2 IF Neurons with Colored (Ornstein–Uhlenbeck) Noise

The case of colored (OU) multiplicative noise is considerably more difficult to treat analytically. Before a detailed mathematical deduction of the subthreshold membrane potential distribution utilizing stochastic calculus will be presented in the next section, we will show here an alternative *adiabatic approximation* method (Moreno-Bote and Parga 2005).

The LIF neuron model (7.41) serves once more as basis, with synaptic conductances obeying the differential equation

$$\tau_{\{e,i\}} \frac{g_{\{e,i\}}(t)}{dt} = -g_{\{e,i\}}(t) + g_{\{e,i\}0} + \sigma_{\{e,i\}} \eta_{\{e,i\}}(t), \tag{7.48}$$

where $\eta_{\{e,i\}}(t)$ denote Gaussian white noise processes of zero mean and unit standard deviation, and $\sigma_{\{e,i\}}$ the SDs of the excitatory and inhibitory synaptic conductance, respectively. This system of differential equations can be brought into normalized form by introducing new stochastic processes

$$\tilde{g}_{\{e,i\}}(t) = \frac{1}{\hat{\sigma}_{\{e,i\}}} (g_{\{e,i\}}(t) - g_{\{e,i\}0}) \tag{7.49}$$

with $\hat{\sigma}_{\{e,i\}} = \sigma_{\{e,i\}}/\sqrt{2\tau_{\{e,i\}}}$, which describe only the conductance fluctuation components around zero means. With this, the considered IF neuronal model takes the mathematical form:

$$\tau_{\mathrm{m}}(\tilde{g}_{\mathrm{e}}, \tilde{g}_{\mathrm{i}}) \frac{\mathrm{d}V(t)}{\mathrm{d}t} = -V(t) + \frac{1}{g_{\mathrm{tot}}(\tilde{g}_{\mathrm{e}}, \tilde{g}_{\mathrm{i}})} E_{\mathrm{eff}}(\tilde{g}_{\mathrm{e}}, \tilde{g}_{\mathrm{i}}), \tag{7.50}$$

where

$$\begin{aligned} E_{\mathrm{eff}}(\tilde{g}_{\mathrm{e}}, \tilde{g}_{\mathrm{i}}) &= g_{\mathrm{L}} E_{\mathrm{L}} + (g_{\mathrm{e}0} + \tilde{g}_{\mathrm{e}} \hat{\sigma}_{\mathrm{e}}) E_{\mathrm{e}} + (g_{\mathrm{i}0} + \tilde{g}_{\mathrm{i}} \hat{\sigma}_{\mathrm{i}}) E_{\mathrm{i}}, \\ \tau_{\{e,i\}} \frac{\tilde{g}_{\{e,i\}}(t)}{dt} &= -\tilde{g}_{\{e,i\}}(t) + \sqrt{2\tau_{\{e,i\}}} \eta_{\{e,i\}}(t), \\ g_{\mathrm{tot}}(\tilde{g}_{\mathrm{e}}, \tilde{g}_{\mathrm{i}}) &= g_{\mathrm{L}} + g_{\mathrm{e}0} + g_{\mathrm{i}0} + \tilde{g}_{\mathrm{e}}(t) \hat{\sigma}_{\mathrm{e}} + \tilde{g}_{\mathrm{i}}(t) \hat{\sigma}_{\mathrm{i}}. \end{aligned} \tag{7.51}$$

In this notation, the total membrane conductance $g_{\mathrm{tot}}(\tilde{g}_{\mathrm{e}}, \tilde{g}_{\mathrm{i}})$ and membrane time constant $\tau_{\mathrm{m}}(\tilde{g}_{\mathrm{e}}, \tilde{g}_{\mathrm{i}})$ are dependent on the time-dependent synaptic conductances $\tilde{g}_{\{e,i\}}(t)$, and linked with each other through the membrane capacitance C_{m}:

$$\tau_{\mathrm{m}}(\tilde{g}_{\mathrm{e}}, \tilde{g}_{\mathrm{i}}) = \frac{C_{\mathrm{m}}}{g_{\mathrm{tot}}(\tilde{g}_{\mathrm{e}}, \tilde{g}_{\mathrm{i}})}.$$

Following steps similar to that used in the previous sections, we first write down the Fokker–Planck equation associated with (7.50). However, in contrast to earlier, we interpret now the probability density $\rho(V, \tilde{g}_{\mathrm{e}}, \tilde{g}_{\mathrm{i}})$ as the joint probability density of a large stationary population of independent neurons, each of which receives an independent realization of synaptic activity. With this, the Fokker–Planck equation takes the form

$$-J(\tilde{g}_{\mathrm{e}}, \tilde{g}_{\mathrm{i}}) \delta(V - V_{\mathrm{reset}}) = \left[\frac{\partial}{\partial V} \left(\frac{V - E_{\mathrm{eff}}(\tilde{g}_{\mathrm{e}}, \tilde{g}_{\mathrm{i}})}{\tau_{\mathrm{m}}(\tilde{g}_{\mathrm{e}}, \tilde{g}_{\mathrm{i}})} \right) + \frac{L_{\tilde{g}_{\mathrm{e}}}}{\tau_{\mathrm{e}}} + \frac{L_{\tilde{g}_{\mathrm{i}}}}{\tau_{\mathrm{i}}} \right] \rho(V, \tilde{g}_{\mathrm{e}}, \tilde{g}_{\mathrm{i}}), \tag{7.52}$$

where

$$L_{\tilde{g}_{\{e,i\}}} = \frac{\partial}{\partial \tilde{g}_{\{e,i\}}} \tilde{g}_{\{e,i\}} + \frac{\partial^2}{\partial^2 \tilde{g}_{\{e,i\}}}$$

and

$$J(\tilde{g}_e, \tilde{g}_i) = \frac{1}{C_m} g_{tot}(\tilde{g}_e, \tilde{g}_i) \left(E_{eff}(\tilde{g}_e, \tilde{g}_i) - V_{th}\right) \rho(V_{th}, \tilde{g}_e, \tilde{g}_i) \tag{7.53}$$

denotes the membrane potential component of the probability current. From the latter, the population firing rate ν_{pop} can be assessed under the assumption, that the probability current at spike threshold V_{th} is reinjected at the reset potential V_{reset}:

$$\nu_{pop} = \int_{\Omega} d\tilde{g}_e \, d\tilde{g}_i \, J(\tilde{g}_e, \tilde{g}_i), \tag{7.54}$$

where the integral is taken over the region Ω in which $E_{eff}(\tilde{g}_e, \tilde{g}_i) \geq V_{th}$.

The Fokker–Planck equation (7.52) is, generally, not analytically solvable. However, in the limit of large $\tau_{\{e,i\}}$ and small τ_m, valid in a high-conductance state in which each neuron experiences a huge barrage with synaptic inputs, a good approximation can be obtained (Moreno-Bote and Parga 2005). Analysis shows that the first term on the right-hand side of (7.52) is proportional to $\sqrt{\tau_{\{e,i\}}}$ and can be treated as a constant, whereas the second term is evaluated perturbatively for large $\tau_{\{e,i\}}$. Assuming that the synaptic activity does not change over time, this corresponds to an *adiabatic approximation* of the problem, yielding at leading order for the membrane potential probability distribution below spike threshold

$$\rho_0(V, \tilde{g}_e, \tilde{g}_i) = \frac{C_m J_0(V, \tilde{g}_e, \tilde{g}_i) \Theta(V - V_{reset})}{g_{tot}(\tilde{g}_e, \tilde{g}_i)(E_{eff}(\tilde{g}_e, \tilde{g}_i) - V)} + D_0(\tilde{g}_e, \tilde{g}_i) \delta(V - E_{eff}(\tilde{g}_e, \tilde{g}_i)), \tag{7.55}$$

where $\Theta(x)$ is the Heaviside step function and

$$\begin{aligned} D_0(\tilde{g}_e, \tilde{g}_i) &= \frac{1}{2\pi} e^{-\frac{|g|^2}{2}}, \\ J_0(V, \tilde{g}_e, \tilde{g}_i) &= \frac{1}{2\pi} e^{-\frac{|g|^2}{2}} \nu(\tilde{g}_e, \tilde{g}_i) \end{aligned} \tag{7.56}$$

with $|\tilde{g}|^2 = \tilde{g}_e^2 + \tilde{g}_i^2$ and

$$\begin{aligned} \frac{1}{\nu(\tilde{g}_e, \tilde{g}_i)} &= \tau_m(\tilde{g}_e, \tilde{g}_i) \log\left(\frac{E_{eff}(\tilde{g}_e, \tilde{g}_i) - V_{reset}}{E_{eff}(\tilde{g}_e, \tilde{g}_i)} - V_{th}\right), \\ \nu(\tilde{g}_e, \tilde{g}_i) &= 0 \qquad \text{outside } \Omega \end{aligned} \tag{7.57}$$

denoting the firing rate of the neuron. With this result, the distribution of the membrane potential $\rho_0(V)$ across the considered population of neurons is obtained by integrating $\rho_0(V, \tilde{g}_e, \tilde{g}_i)$ over $\tilde{g}_e$ and $\tilde{g}_i$. A comparison between the analytical

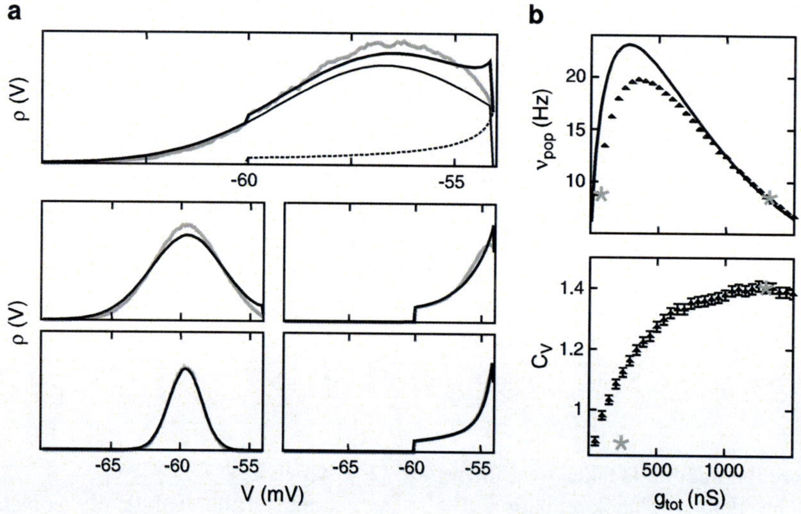

Fig. 7.5 Response of the LIF neuron model subject to OU synaptic noise. (**a**) Comparison between the analytic solution [(7.55), *black thick solid*] and numerical solution (*gray solid*) of the membrane potential probability distribution $\rho_0(V)$. *Top panel*: Parameters: $\tau_{\{e,i\}} = 10\,\text{ms}$, $g_{e0} = 27.5\,\text{nS}$, $\hat{\sigma}_e = 3.95\,\text{nS}$, $\hat{\sigma}_i = 5.59\,\text{nS}$. Contributions of the active (*black thin solid*) and inactive (*black dashed*) subpopulations are shown. *Middle panel*: Same as *top panel*, but with $g_{e0} = 22.5\,\text{nS}$ (*left*) and $g_{e0} = 35\,\text{nS}$ (*right*). *Bottom panel*: Same as *middle panel*, but with $\tau_{\{e,i\}} = 50\,\text{ms}$, $\hat{\sigma}_e = 1.77\,\text{nS}$, $\hat{\sigma}_i = 2.5\,\text{nS}$. Other parameters: $C_m = 0.25\,\text{nF}$, $g_L = 12.5\,\text{nS}$, $g_{i0} = 62.5\,\text{nS}$, $V_{th} = -54\,\text{mV}$, $V_{reset} = -60\,\text{mV}$, $E_L = -65\,\text{mV}$, $E_e = 0\,\text{mV}$, $E_i = -80\,\text{mV}$, $\sigma_e^2 = 0.31\,\text{nS}^2$, $\sigma_i^2 = 0.62\,\text{nS}^2$. (**b**) Comparison of the population firing rate [(7.58), *black solid*] with numerical simulations (*upper panel*) and the coefficient of variation C_V (*bottom panel*) as a function of the total membrane conductance. The input conductance was changed by keeping $\sigma_e^2 = 1.25\times10^{-3} C_m g_{tot}$, $\sigma_i^2 = 25\times10^{-3} C_m g_{tot}$ and $E_{eff} = -55.7\,\text{mV}$ fixed. Other parameters as in (**a**), except $\tau_e = 3\,\text{ms}$, $\tau_i = 10.5\,\text{ms}$, $E_i = -70\,\text{mV}$. The *gray stars* indicate populations with a low conductance (*left*) and high conductance (*right*) with about the same output rate, respectively. Modified from Moreno-Bote and Parga (2005)

result and the result of the numerical simulation of a large population of neurons is shown in Fig. 7.5a.

In a similar fashion, an approximation of the population rate is given by integrating $\nu(\tilde{g}_e, \tilde{g}_i)$ over $\tilde{g}_e$ and $\tilde{g}_i$:

$$\nu_{pop} = \int_{\Omega} d\tilde{g}_e d\tilde{g}_i \frac{1}{2\pi} e^{-\frac{|g|^2}{2}} \nu(\tilde{g}_e, \tilde{g}_i). \tag{7.58}$$

An analytic result of this integration is hard to obtain, but under certain assumptions this equation can be further simplified (Moreno-Bote and Parga 2005). Results of the numerical simulations and analytic approximation are shown in Fig. 7.5b. Interestingly, the population rate reaches a maximum for intermediate values of the total conductance. This is, indeed, a direct consequence of the nature of synaptic conductances, whereas for lower inputs the membrane is effectively in a low-conductance state in which synaptic inputs have a decisive and proportional effect on the spiking response; at higher levels, these inputs effectively shunt the membrane, thus reducing the amplitude of voltage fluctuations and, hence, firing rate of the cell.

7.4 Membrane Equations with Multiplicative Synaptic Noise

In the previous two sections, we exclusively dealt with neurons of the integrate-and-fire type and deduced, utilizing the Fokker–Planck formalism, expressions of the subthreshold and spiking response of neuronal models subject to Gaussian white and colored noise. The latter describe, in an effective stochastic fashion, synaptic currents and conductances. All these approaches are based on the solution of a system of SDEs and utilize what is known in the theory of stochastic systems as the *Stratonovich calculus*.

In this section, we will present an alternative and mathematically more rigorous approach, based on the *Itô formalism*, to analytically assess the subthreshold response of a neuron subjected to colored synaptic noise described by the OU stochastic process. Specifically, this model will be analyzed mathematically with the aim to derive an analytic description of the statistical properties of the membrane potential. Using the Itô–Stratonovich calculus (van Kampen 1981; Gardiner 2002) and the Fokker–Planck approach (Risken 1984), the stochastic *Langevin equation* (Genovese and Muñoz 1999) is solved. This will yield an analytic expression for the steady-state membrane potential distribution in the presence of noise. For a general introduction into this approach, we refer to Gardiner (2002, Chap. 4).

7.4.1 *General Idea and Limitations*

Before detailing the mathematics behind this approach, a few notes about the applicability and limitations need to be made. In the past decades, many attempts were undertaken to solve stochastic differential equations describing systems with multiplicative (and colored) noise (see last section for two examples). Here, the spectrum of mathematical frameworks is broad. On its one end reside *spectral approaches*, which utilize the PSD obtained after Fourier transforming the SDEs in question (see, e.g., Manwani and Koch 1999a). As the spectral density of neuronal membranes subject to various noise sources (such as thermal, synaptic, or channel noise) closely resembles a Lorentzian function (Fig. 7.6d), a first coarse characterization of the membrane potential distribution in terms of lowest order moments, such as the mean and variance, can be obtained in a straightforward manner (Manwani and Koch 1999a). However, such approximations quickly defy a mathematical assessment if higher order moments are considered, thus limiting the advantage of this approach to the characterization of subthreshold neuronal dynamics at lowest order.

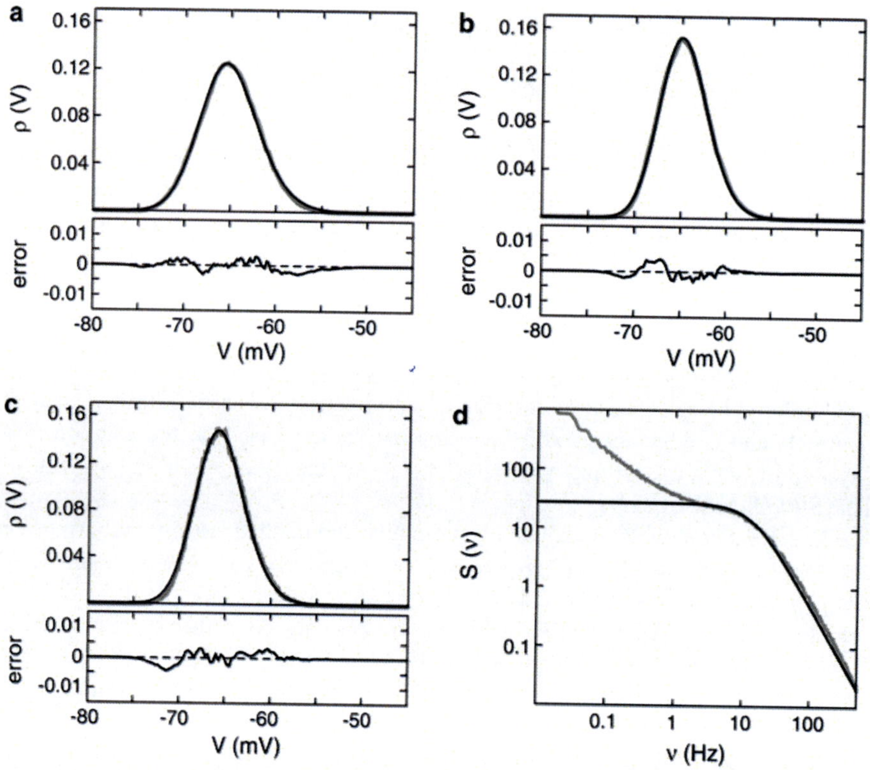

Fig. 7.6 Membrane potential probability distributions $\rho(V)$ for different models of synaptic noise. (**a**) Single-compartment model with two fluctuating synaptic conductances described by OU processes (7.59). (**b**) Single-compartment model (Destexhe et al. 2001), where synaptic activity was simulated by a large number of randomly releasing synapses. 4,472 excitatory and 3,801 inhibitory synapses were described by kinetic models of AMPA and $GABA_A$ receptors (Destexhe et al. 1998b), respectively, which released according to independent Poisson processes. (**c**) Compartmental model of cortical pyramidal neuron (morphologically reconstructed) in which synaptic activity was simulated by randomly releasing synapses distributed in soma and dendrites (see details of this model in Destexhe and Paré 1999; Rudolph and Destexhe 2003d). In all cases shown in (**a–c**), the membrane potential distributions obtained from numerical simulations (*gray solid*) closely matched the analytic solutions $\rho(V)$ for multiplicative, (7.96) (*black solid*). Parameters: $g_{e0} = 0.0121\,\mu S$, $g_{i0} = 0.0573\,\mu S$, $\sigma_e = 0.006\,\mu S$, $\sigma_i = 0.0132\,\mu S$ (**a**); $g_{e0} = 0.0127\,\mu S$, $g_{i0} = 0.0573\,\mu S$, $\sigma_e = 0.0049\,\mu S$, $\sigma_i = 0.0108\,\mu S$ (**b**); $g_{e0} = 0.0148\,\mu S$, $g_{i0} = 0.0702\,\mu S$, $\sigma_e = 0.00696\,\mu S$, $\sigma_i = 0.0153\,\mu S$ (**c**); in all cases: $\tau_e = 2.728$ ms, $\tau_i = 10.49$ ms. The deviation (absolute error) between numerical and analytical distributions is shown in the *bottom panels* of (**a–c**). (**d**) Typical example of the power spectral density $S(\nu)$ of conductances obtained from the single-compartment model of (**b**) (*gray*). The latter yield nearly Gaussian distributions, whose power spectral density (*black*) matched the Lorentzian function $S(\nu) = 2D\tau^2/(1 + (2\pi\tau\nu)^2)$ (D is the diffusion coefficient, τ the noise time constant) expected for OU noise, for frequencies larger than $\sim 1Hz$. Modified from Rudolph and Destexhe (2003d)

The second class of approaches, the *Fokker–Planck approach*, utilizes the Fokker–Planck equation corresponding to the stochastic system in question (see Sect. 7.3). Being a differential equation for the probability distribution of the involved stochastic quantities, such as the membrane potential, the Fokker–Planck approach potentially yields exact expressions for their distributions and temporal characteristics. The advantage of this method is that it utilizes expectation values of the involved quantities, and, thus, is not restricted to systems with Gaussian-distributed noise or the specific coupling of the stochastic sources to the state variable or variables. Moreover, as a differential equation in time, the Fokker–Planck equation preserves the temporal aspects of the distributions of the involved quantities, thus allowing to evaluate transient changes of these probability distributions resulting from transient changes in the system's parameter values. However, even for simple systems, the Fokker–Planck equation usually takes the form of a second or higher order nonlinear differential equation, and analytical approximations or numerical evaluations remain, in almost all cases, the only way to assess it.

Finally, at the other end of the spectrum of mathematical frameworks to solve stochastic systems one finds methods which exclusively utilize expectation values and, hence, the steady-state distributions of the involved stochastic variables. A specific realization of this *expectation value approach* will be presented in the remainder of this section. Whereas the advantage of this method is its more general applicability (independent from the coupling of the stochastic variables, their distributions or the specific form of the underlying stochastic differential equations), it also deprives the system of its spectral signature, as we will show below. To illustrate this issue, consider Gaussian white and colored OU noise, which both have different spectral properties but are described by Gaussian amplitude distributions. That is, both stochastic processes can be characterized indistinguishly by their first and second order nonvanishing moments, irrespective their different spectral signature.

Below, this property of colored OU noise, i.e., the qualitative equivalence of its distribution to Gaussian white noise, will be utilized in order to rewrite the stochastic membrane equation with two multiplicatively coupled colored OU noise sources. By considering that the noise correlation time becomes infinitesimal-small compared to the time needed to obtain a steady-state membrane potential distribution, one can treat the system as being mathematically identical to one governed by Gaussian white noise. Ultimately, this approximation will yield an analytical solution for the steady-state membrane potential distribution. However, in doing so, the spectral signature of the original OU noise sources is altered, in favor of solvability of the stochastic system in question, which results in a mismatch of this solution when compared, for instance, with a numerical assessment. Fortunately, this mismatch is only temporal (spectral) and can be corrected, leading to an analytical approximation of the steady-state solution which achieves unprecedented accuracy over several orders of magnitudes in the corresponding parameter space. The latter expression does form the basis of the "VmD method," which was very successfully tested in dynamic-clamp experiments, and led to remarkable advances in characterizing neuronal noise, as outlined in the next chapters.

7.4.2 The Langevin Equation

We start by incorporating OU stochastic processes for inhibitory and excitatory synaptic conductances into the passive membrane equation

$$C_{\mathrm{m}} \frac{\mathrm{d}V(t)}{\mathrm{d}t} = -g_{\mathrm{L}} \left(V(t) - E_{\mathrm{L}} \right) - \frac{1}{a} I_{\mathrm{syn}}(t). \tag{7.59}$$

Here, we no longer consider the IF neuronal model with spike threshold and reset potential, as in the previous sections, but focus solely on the subthreshold (passive) membrane dynamics. In (7.59), C_{m} denotes the specific membrane capacity, a the membrane area, g_{L} and E_{L} are the leak conductance and reversal potential, respectively. As before, the point-conductance model is used, where the total synaptic current due to synaptic background activity $I_{\mathrm{syn}}(t)$ is decomposed into a sum of two noise terms which describe noisy excitatory and inhibitory conductance components coupling multiplicatively to the membrane potential, $g_{\mathrm{e}}(t)\,(V(t) - E_{\mathrm{e}})$ and $g_{\mathrm{i}}(t)\,(V(t) - E_{\mathrm{i}})$, respectively (Destexhe et al. 2001):

$$I_{\mathrm{syn}}(t) = g_{\mathrm{e}}(t) \left(V(t) - E_{\mathrm{e}} \right) + g_{\mathrm{i}}(t) \left(V(t) - E_{\mathrm{i}} \right). \tag{7.60}$$

Specifically, $g_{\mathrm{e}}(t)$ and $g_{\mathrm{i}}(t)$ denote stochastic variables describing time-dependent excitatory and inhibitory conductances according to the OU stochastic process, and obey

$$\frac{\mathrm{d}g_{\{\mathrm{e,i}\}}(t)}{\mathrm{d}t} = -\frac{1}{\tau_{\{\mathrm{e,i}\}}} \left(g_{\{\mathrm{e,i}\}}(t) - g_{\{\mathrm{e,i}\}0} \right) + \sqrt{D_{\{\mathrm{e,i}\}}}\, \eta_{\{\mathrm{e,i}\}}(t). \tag{7.61}$$

E_{e} and E_{i} are their respective reversal potentials, $g_{\{e,i\}0}$ are the mean excitatory and inhibitory conductances, $\tau_{\{\mathrm{e,i}\}}$ are the respective time constants, and $D_{\{\mathrm{e,i}\}}$ denote the corresponding noise diffusion coefficients. $\eta_{\{\mathrm{e,i}\}}(t)$ are independent Gaussian white noise processes of zero mean and unit SD, i.e.,

$$\begin{aligned} &< \eta_{\{\mathrm{e,i}\}}(t) > = 0 \\ &< \eta_{\{\mathrm{e,i}\}}(t)\eta_{\{\mathrm{e,i}\}}(t') > = \delta(t - t'), \end{aligned}$$

for excitatory and inhibitory conductances, respectively. White noise is obtained for vanishing time constants, i.e., $\tau_{\{\mathrm{e,i}\}} = 0$, whereas a time constant larger than zero yields colored Gaussian noise for the corresponding stochastic process. The noise diffusion coefficients $D_{e,i}$ are related to the SD $\sigma_{\{\mathrm{e,i}\}}$ of the respective stochastic variables by (Gillespie 1996)

$$\sigma^2_{\{\mathrm{e,i}\}} = \frac{1}{2} D_{\{\mathrm{e,i}\}} \tau_{\{\mathrm{e,i}\}}. \tag{7.62}$$

Introducing the new variables

$$\tilde{g}_{\{e,i\}}(t) = g_{\{e,i\}}(t) - g_{\{e,i\}0} \qquad (7.63)$$

yields for (7.59) the one-dimensional *Langevin equation* with two independent multiplicative OU noise terms

$$\frac{dV(t)}{dt} = f(V(t)) + h_e(V(t))\,\tilde{g}_e(t) + h_i(V(t))\,\tilde{g}_i(t), \qquad (7.64)$$

where $\tilde{g}_{\{e,i\}}(t)$ denote now stochastic variables with zero mean for excitatory and inhibitory conductances described by OU processes

$$\frac{d\tilde{g}_{\{e,i\}}(t)}{dt} = -\frac{1}{\tau_{\{e,i\}}}\tilde{g}_{\{e,i\}}(t) + \sqrt{D_{\{e,i\}}}\,\eta_{\{e,i\}}(t). \qquad (7.65)$$

In (7.64), $f(V(t))$ is called the (voltage-dependent) *drift term*

$$f(V(t)) = -\frac{g_L}{C_m}\left(V(t) - E_L\right) - \frac{g_{e0}}{C_m a}\left(V(t) - E_e\right) - \frac{g_{i0}}{C_m a}\left(V(t) - E_i\right) \qquad (7.66)$$

and $h_{\{e,i\}}(V(t))$ the voltage-dependent excitatory and inhibitory conductances noise terms:

$$h_{\{e,i\}}(V(t)) = -\frac{1}{C_m a}\left(V(t) - E_{\{e,i\}}\right). \qquad (7.67)$$

Both $f(V(t))$ and $h_{\{e,i\}}(V(t))$ are nonanticipating functions of the membrane potential $V(t)$.

7.4.3 The Integrated OU Stochastic Process and Itô Rules

The Langevin equation (7.64) describes the subthreshold membrane potential dynamics in the presence of independent multiplicative colored noise sources due to synaptic background activity. Unfortunately, the stochastic terms prevent a direct analytic solution of this differential equation. However, the Itô–Stratonovich stochastic calculus (e.g., van Kampen 1981; Gardiner 2002) allows one to deduce the Fokker–Planck equation corresponding to (7.64), and to describe the steady-state membrane potential probability distribution in the asymptotic limit $t \to \infty$.

In order to solve (7.64), the stochastic variables $\tilde{g}_{\{e,i\}}(t)$ need to be integrated. To that end, one can, formally, define the integrated OU process

$$\tilde{w}(t) = \int_0^t d\tilde{w}(s) = \int_0^t ds\,\tilde{v}(s) \qquad (7.68)$$

of an OU stochastic process $\tilde{v}(t)$. Following a straightforward calculation yields for the cumulants of $\tilde{w}(t)$

$$\ll \tilde{w}^n(t) \gg = \begin{cases} 2\sigma^2\tau t - 2\sigma^2\tau^2\left(1-\mathrm{e}^{-\frac{t}{\tau}}\right) & \text{for } n=2 \\ 0 & \text{otherwise} \end{cases}$$

$$\ll \tilde{w}^{n_0}(t_0)\tilde{w}^{n_1}(t_1) \gg = \begin{cases} 2\sigma^2\tau t_0 - \sigma^2\tau^2\left(1-\mathrm{e}^{-\frac{t_0}{\tau}}-\mathrm{e}^{-\frac{t_1}{\tau}}+\mathrm{e}^{-\frac{\Delta t}{\tau}}\right) \\ \qquad \text{for } t_0 \le t_1,\, \Delta t = t_1 - t_0, n_0 = n_1 = 1 \\ 0 \qquad \text{otherwise}\,. \end{cases} \tag{7.69}$$

From these, one can construct the one-dimensional and multi-dimensional characteristic functions for $\tilde{w}(t)$:

$$\tilde{G}(s,t) = \exp\left[(is)^2\left\{\sigma^2\tau t - \sigma^2\tau^2\left(1-\mathrm{e}^{-\frac{t}{\tau}}\right)\right\}\right] \tag{7.70}$$

and

$$\begin{aligned} &\tilde{G}(s_0,t_0;s_1,t_1) \\ &\quad = \exp\Big[(is_0)(is_1)\left\{2\sigma^2\tau t_0 - \sigma^2\tau^2\left(1-\mathrm{e}^{-\frac{t_0}{\tau}}-\mathrm{e}^{-\frac{t_1}{\tau}}+\mathrm{e}^{-\frac{\Delta t}{\tau}}\right)\right\} \\ &\qquad +(is_0)^2\left\{\sigma^2\tau t_0 - \sigma^2\tau^2\left(1-\mathrm{e}^{-\frac{t_0}{\tau}}\right)\right\} \\ &\qquad +(is_1)^2\left\{\sigma^2\tau t_1 - \sigma^2\tau^2\left(1-\mathrm{e}^{-\frac{t_1}{\tau}}\right)\right\}\Big]. \end{aligned} \tag{7.71}$$

With these equations, the one-dimensional and multi-dimensional moments of the integrated OU stochastic process are given by

$$< \tilde{w}^n(t) > = \begin{cases} \frac{n!}{k!}\left\{\sigma^2\tau t - \sigma^2\tau^2\left(1-\mathrm{e}^{-\frac{t}{\tau}}\right)\right\}^k & \text{for even } n=2k \\ 0 & \text{for odd } n \end{cases} \tag{7.72}$$

and

$$\begin{aligned} < \tilde{w}^{n_0}(t_0)\,\tilde{w}^{n_1}(t_1) > &= n_0!\,n_1! \sum_{m_1,m_2,m_3} \frac{1}{m_1!\,m_2!\,m_3!} \\ &\times\left\{2\sigma^2\tau t_0 - \sigma^2\tau^2\left(1-\mathrm{e}^{-\frac{t_0}{\tau}}-\mathrm{e}^{-\frac{t_1}{\tau}}+\mathrm{e}^{-\frac{\Delta t}{\tau}}\right)\right\}^{m_1} \\ &\times\left\{\sigma^2\tau t_0 - \sigma^2\tau^2\left(1-\mathrm{e}^{-\frac{t_0}{\tau}}\right)\right\}^{m_2} \\ &\times\left\{\sigma^2\tau t_1 - \sigma^2\tau^2\left(1-\mathrm{e}^{-\frac{t_1}{\tau}}\right)\right\}^{m_3}. \end{aligned} \tag{7.73}$$

The sum in the last equation runs over all 3-tuple (m_1, m_2, m_3) obeying the conditions $m_1 + 2m_2 = n_0$ and $m_1 + 2m_3 = n_1$, i.e., $m_1 + m_2 + m_3 = \frac{n_0+n_1}{2}$.

In this condition, the integrated OU stochastic process $\tilde{w}(t)$ becomes a Wiener-process $w(t)$ with one-dimensional and multi-dimensional cumulants:

$$\ll w^n(t) \gg = \begin{cases} 2Dt & \text{for } n = 2 \\ 0 & \text{otherwise} \end{cases} \tag{7.74}$$

and

$$\ll w^{n_0}(t_0)\, w^{n_1}(t_1) \gg = \begin{cases} 2D\min(t_0,t_1) & \text{for } n_0 = n_1 = 1 \\ 0 & \text{otherwise} \end{cases} \tag{7.75}$$

as well as one-dimensional and multi-dimensional moments

$$< w^n(t) >= \begin{cases} \frac{n!}{k!}\,(Dt)^k & \text{for even } n = 2k \\ 0 & \text{for odd } n \end{cases} \tag{7.76}$$

$$< w^{n_0}(t_0)\, w^{n_1}(t_1) >= n_0!\, n_1! \sum_{m_1,m_2,m_3} \frac{2^{m_1}}{m_1!\, m_2!\, m_3!}\, (Dt_0)^{m_1+m_2}\, (Dt_1)^{m_3}, \tag{7.77}$$

where $D = \sigma^2\,\tau$.

In the heart of the mathematical deduction of the Fokker–Planck equation (see next section) from the Langevin equation (7.64) with colored noise sources lies a set of differential rules, called *Itô rules*. It can be proven that for the integrated OU process $\tilde{w}(t)$, the Itô rules read:

$$\begin{aligned} d\tilde{w}_i(t)\, d\tilde{w}_j(t) &= \delta_{ij} \left\{ \sigma^2\,\tau \left(1 - e^{-\frac{t}{\tau}}\right) + \frac{1}{2\tau}\,\tilde{w}_i^2(t) - \sigma^2 t \right\} dt \\ [d\tilde{w}(t)]^N &= 0 \qquad \text{for } N \geq 3 \\ [d\tilde{w}(t)]^N\, dt &= 0 \qquad \text{for } N \geq 1 \\ [dt]^N &= 0 \qquad \text{for } N \geq 2. \end{aligned} \tag{7.78}$$

These rules apply to each of the two stochastic variables $\tilde{w}_{\{e,i\}}(t)$ obtained by integrating the OU processes $\tilde{g}_{\{e,i\}}(t)$. Moreover, the first equality of (7.78) indicates that the independence of $\tilde{g}_{\{e,i\}}(t)$ and $\tilde{I}(t)$ directly translates into the independence between the corresponding integrated stochastic processes.

The rules (7.78) have to be interpreted in the context of integration. Here, the integral

$$S(t) = \int_0^t d\eta(s)\, G(s)$$

over a stochastic variable $\eta(t)$, where $G(t)$ denotes an arbitrary nonanticipating function or stochastic process, is approximated by the sum

$$S^{\alpha}(t) = {}^{\mathrm{ms}}\lim_{n\to\infty} S^{\alpha}_n(t),$$

$$S^{\alpha}_n(t) = \sum_{k=1}^{n} G\big((1-\alpha)t_{k-1} + \alpha t_k\big)\,\big(\eta(t_k) - \eta(t_{k-1})\big) \tag{7.79}$$

which evaluates the integral at n discrete time steps $t_k = k\frac{t}{n}$ in the interval $[0,t]$. The mean square limit ${}^{\mathrm{ms}}\lim_{n\to\infty}$ is defined by the following condition of convergence:

$$S^{\alpha}(t) = {}^{\mathrm{ms}}\lim_{n\to\infty} S^{\alpha}_n(t)$$

if and only if

$$\lim_{n\to\infty} \left\langle \left(S^{\alpha}_n(t) - S^{\alpha}(t)\right)^2 \right\rangle = 0. \tag{7.80}$$

These definitions depend on the parameter α, which allows to choose the position in the interval $[t_{k-1}, t_k]$ where $G(t)$ is evaluated. However, whereas in ordinary calculus the result of this summation becomes independent of α for $n \to \infty$, stochastic integrals do, in general, remain in this limit dependent on α. This is one important difference between ordinary and stochastic calculi, which renders the latter more difficult and less tractable both mathematically and at the level of (physically meaningful) interpretation.

Looking at (7.79), there are two popular choices for the parameter α: $\alpha = 1/2$ defines the *Stratonovich calculus*, which exhibits the same integration rules as ordinary calculus and is a common choice for integrals with stochastic variables describing noise with finite correlation time. However, mathematical rigorous proofs are nearly impossible to perform in the Stratonovich calculus. For instance, the Itô rules listed above can only be derived for $\alpha = 0$, which defines the *Itô calculus*. On the level of SDEs, a transformation between Itô and Stratonovich calculus can be obtained. After applying the Itô rules, which hold in the Itô calculus, we will use this transformation to obtain a physical interpretation and treatment in the context of standard calculus, which, as experience shows, is only meaningful in the framework of the Stratonovich calculus. For more details about both stochastic calculi and their relation, we refer to standard textbooks of stochastic calculus (e.g., Gardiner 2002, Chap. 4).

7.4.4 *The Itô Equation*

Using the Itô rules (7.78), one can now deduce the *Itô equation* corresponding to the Langevin equation (7.64). In order to obtain the steady-state probability distribution of the membrane potential $V(t)$ for the Langevin equation with two independent multiplicative colored noise terms, (7.64), one first deduces *Itô's formula* for the SDE in question. Equation (7.64) together with the definition of the integrated OU stochastic process, (7.68), yields

$$\int_{V(0)}^{V(t)} \mathrm{d}V(s) = \int_0^t \mathrm{d}s\, f(V(s)) + \int_0^t \mathrm{d}\tilde{w}_{\mathrm{e}}(s)\, h_{\mathrm{e}}(V(s)) + \int_0^t \mathrm{d}\tilde{w}_{\mathrm{i}}(s)\, h_{\mathrm{i}}(V(s)). \tag{7.81}$$

The first term on the right-hand side denotes the ordinary Riemannian integral over the drift term $f(V(t))$ given by (7.66), whereas the last two terms are stochastic integrals in the sense of Riemann–Stieltjes. This interpretation, however, does not require the stochastic processes $\tilde{g}_{\{\mathrm{e,i}\}}(t)$ to be a Gaussian white noise processes. Only the mathematically much weaker assumption that the corresponding integrated processes $\tilde{w}_{\{\mathrm{e,i}\}}(t)$ are continuous function of t is required. This condition is fulfilled in the case of OU stochastic processes we do consider here.

The natural choice for an interpretation of stochastic integral equations involving noise with finite correlation time is provided within the Stratonovich calculus (Mortensen 1969; van Kampen 1981; Gardiner 2002). However, in order to solve the integral equation (7.81) in a mathematically satisfying way by applying the Itô rules (7.78), the integrals over stochastic variables in (7.81) have to be written as Itô integrals. For instance, taking the defining (7.79), the stochastic integral

$$S(t) = \int_0^t \mathrm{d}\tilde{w}_{\mathrm{e}}(s)\, h_{\mathrm{e}}(V(s))$$

has to be understood in the Stratonovich interpretation ($\alpha = 1/2$) as

$$\begin{aligned} S(t) &= {}^{\mathrm{ms}}\lim_{n\to\infty} \sum_{k=1}^{n} h_{\mathrm{e}}(V(\tau_k)) \{\tilde{w}_{\mathrm{e}}(t_k) - \tilde{w}_{\mathrm{e}}(t_{k-1})\} \\ &= {}^{\mathrm{ms}}\lim_{n\to\infty} \left[\sum_{k=1}^{n} h_{\mathrm{e}}(V(\tau_k)) \{\tilde{w}_{\mathrm{e}}(t_k) - \tilde{w}_{\mathrm{e}}(\tau_k)\} \right. \\ &\qquad \left. + \sum_{k=1}^{n} h_{\mathrm{e}}(V(\tau_k)) \{\tilde{w}_{\mathrm{e}}(\tau_k) - \tilde{w}_{\mathrm{e}}(t_{k-1})\} \right]. \end{aligned} \tag{7.82}$$

Approximating $h_{\mathrm{e}}(V(\tau_k))$, which is an analytic function of $V(t)$, by power expansion around the left point of the interval $[t_{k-1}, t_k]$, yields in the considered case, due to the linearity of $h_{\mathrm{e}}(V(\tau_k))$ in $V(t)$, the linear function

$$h_{\mathrm{e}}(V(\tau_k)) = h_{\mathrm{e}}(V(t_{k-1})) + \left(\partial_V h_{\mathrm{e}}(V(t_{k-1}))\right) \left(V(\tau_k) - V(t_{k-1})\right). \tag{7.83}$$

Here, $h_{\mathrm{e}}(V(\tau_k))$ does not explicitly depend on t.

To further resolve (7.82), one makes use of the fact that $V(t)$ is a solution of the stochastic Langevin equation (7.64), with an infinitesimal displacement given by

$$\begin{aligned} V(\tau_k) - V(t_{k-1}) = {} & f(V(t_{k-1}))\,(\tau_k - t_{k-1}) \\ & + h_{\mathrm{e}}(V(t_{k-1}))\,(\tilde{w}_{\mathrm{e}}(\tau_k) - \tilde{w}_{\mathrm{e}}(t_{k-1})) \\ & + h_{\mathrm{i}}(V(t_{k-1}))\,(\tilde{w}_{\mathrm{i}}(\tau_k) - \tilde{w}_{\mathrm{i}}(t_{k-1})). \end{aligned}$$

Inserting this equation into (7.83), and the result into the second sum of (7.82), yields after a straightforward calculation

$$\begin{aligned} S(t) = {}^{\mathrm{ms}}\lim_{n\to\infty} \Bigg[& \sum_{k=1}^{n} h_{\mathrm{e}}(V(\tau_k))\,\{\tilde{w}_{\mathrm{e}}(t_k) - \tilde{w}_{\mathrm{e}}(\tau_k)\} \\ & + \sum_{k=1}^{n} \Big(h_{\mathrm{e}}(V(t_{k-1}))\,\{\tilde{w}_{\mathrm{e}}(\tau_k) - \tilde{w}_{\mathrm{e}}(t_{k-1})\} \\ & + 2\,\alpha_{\mathrm{e}}(t_{k-1})\,\{\tau_k - t_{k-1}\}\,h_{\mathrm{e}}(V(t_{k-1}))\,(\partial_V h_{\mathrm{e}}(V(t_{k-1})))\Big)\Bigg], \end{aligned} \tag{7.84}$$

where

$$2\,\alpha_{\mathrm{e}}(t) = \sigma_{\mathrm{e}}^2\,\tau_{\mathrm{e}}\left(1 - \exp\left[-\frac{t}{\tau_{\mathrm{e}}}\right]\right) + \frac{1}{2\,\tau_{\mathrm{e}}}\,\tilde{w}_{\mathrm{e}}^2(t) - \sigma_{\mathrm{e}}^2\,t. \tag{7.85}$$

In order to obtain (7.84), the fact that individual terms of the sum approximate integrals in the Itô calculus [(7.79), $\alpha = 0$] was used, which in turn allows the application of the Itô rules given in (7.78).

For the third term on the right-hand side of (7.81), expressions similar to (7.84) and (7.85) can be obtained. Inserting the corresponding terms into (7.81), yields for an infinitesimal displacement of the state variable $V(t)$:

$$\begin{aligned} dV(t) = {} & f(V(t))\,dt + h_{\mathrm{e}}(V(t))\,d\tilde{w}_{\mathrm{e}}(t) + h_{\mathrm{i}}(V(t))\,d\tilde{w}_{\mathrm{i}}(t) \\ & + \alpha_{\mathrm{e}}(t)\,h_{\mathrm{e}}(V(t))\,\partial_V h_{\mathrm{e}}(V(t))\,dt + \alpha_{\mathrm{i}}(t)\,h_{\mathrm{i}}(V(t))\,\partial_V h_{\mathrm{i}}(V(t))\,dt, \end{aligned}$$

where

$$2\,\alpha_{\{\mathrm{e,i}\}}(t) = \sigma_{\{\mathrm{e,i}\}}^2\,\tau_{\{\mathrm{e,i}\}}\left(1 - \exp\left[-\frac{t}{\tau_{\{\mathrm{e,i}\}}}\right]\right) + \frac{1}{2\,\tau_{\{\mathrm{e,i}\}}}\,\tilde{w}_{\{\mathrm{e,i}\}}^2(t) - \sigma_{\{\mathrm{e,i}\}}^2\,t. \tag{7.86}$$

In deducing (7.86), the fact that $h_{\{\mathrm{e,i}\}}(V(t))$ are linear in $V(t)$ but do not explicitly depend on t [see (7.67)] was used.

Denoting by $F(V(t))$ an arbitrary function of $V(t)$ satisfying (7.86), an infinitesimal change of $F(V(t))$ with respect to $dV(t)$ is given by:

$$dF(V(t)) = F(V(t)+dV(t)) - F(V(t))$$
$$= (\partial_V F(V(t)))\, dV(t) + \frac{1}{2}\left(\partial_V^2 F(V(t))\right) dV^2(t) + O(dV^3(t)), \quad (7.87)$$

where $O(dV^3(t))$ denotes terms of third or higher order in $dV(t)$. Substituting (7.86) back into (7.87) and again applying the Itô rules (7.78), one finally obtains *Itô's formula*

$$\begin{aligned} dF(V(t)) = {} & \partial_V F(V(t))\, f(V(t))\, dt + \partial_V F(V(t))\, h_\mathrm{e}(V(t))\, d\tilde{w}_\mathrm{e}(t) \\ & + \partial_V F(V(t))\, h_\mathrm{i}(V(t))\, d\tilde{w}_\mathrm{i}(t) \\ & + \partial_V F(V(t))\, \alpha_\mathrm{e}(t)\, h_\mathrm{e}(V(t))\, \partial_V h_\mathrm{e}(V(t))\, dt \\ & + \partial_V F(V(t))\, \alpha_\mathrm{i}(t)\, h_\mathrm{i}(V(t))\, \partial_V h_\mathrm{i}(V(t))\, dt \\ & + \partial_V^2 F(V(t))\, \alpha_\mathrm{e}(t)\, h_\mathrm{e}^2(V(t))\, dt + \partial_V^2 F(V(t))\, \alpha_\mathrm{i}(t)\, h_\mathrm{i}^2(V(t))\, dt, \end{aligned} \quad (7.88)$$

which describes an infinitesimal displacement of $F(V(t))$ as a function of infinitesimal changes in its variables. Equation (7.88) shows that, due to the dependence on stochastic variables, this displacement differs from those expected in ordinary calculus.

7.4.5 The Fokker–Planck Equation

Equation (7.88) describes the change of an arbitrary function $F(V(t))$ for infinitesimal changes in its (stochastic) arguments. Averaging over Itô's formula will finally yield the Fokker–Planck equation corresponding to (7.64).

To that end, we take the formal average of Itô's formula (7.88) over time t, which gives after a short calculation

$$\begin{aligned} < dF(V(t)) > = {} & < \partial_V F(V(t))\, f(V(t))\, dt > + < \partial_V F(V(t))\, h_\mathrm{e}(V(t))\, d\tilde{w}_\mathrm{e}(t) > \\ & + < \partial_V F(V(t))\, h_\mathrm{i}(V(t))\, d\tilde{w}_\mathrm{i}(t) > \\ & + < \partial_V F(V(t))\, \alpha_\mathrm{e}(t)\, h_\mathrm{e}(V(t))\, \partial_V h_\mathrm{e}(V(t))\, dt > \\ & + < \partial_V F(V(t))\, \alpha_\mathrm{i}(t)\, h_\mathrm{i}(V(t))\, \partial_V h_\mathrm{i}(V(t))\, dt > \\ & + < \partial_V^2 F(V(t))\, \alpha_\mathrm{e}(t)\, h_\mathrm{e}^2(V(t))\, dt > \\ & + < \partial_V^2 F(V(t))\, \alpha_\mathrm{i}(t)\, h_\mathrm{i}^2(V(t))\, dt >, \end{aligned} \quad (7.89)$$

which gives

$$\left\langle \frac{\mathrm{d}F(V(t))}{\mathrm{d}t} \right\rangle = < \partial_V F(V(t))\, f(V(t)) >$$
$$+ < \partial_V F(V(t))\, \alpha_{\mathrm{e}}(t)\, h_{\mathrm{e}}(V(t))\, \partial_V h_{\mathrm{e}}(V(t)) >$$
$$+ < \partial_V F(V(t))\, \alpha_{\mathrm{i}}(t)\, h_{\mathrm{i}}(V(t))\, \partial_V h_{\mathrm{i}}(V(t)) >$$
$$+ < \partial_V^2 F(V(t))\, \alpha_{\mathrm{e}}(t)\, h_{\mathrm{e}}^2(V(t)) >$$
$$+ < \partial_V^2 F(V(t))\, \alpha_{\mathrm{i}}(t)\, h_{\mathrm{i}}^2(V(t)) > . \tag{7.90}$$

In the last step, the fact that $h_{\{\mathrm{e,i}\}}$ are nonanticipating functions and, thus, are statistically independent of $d\tilde{w}_{\{\mathrm{e,i}\}}$, respectively, was used. Furthermore, the relations

$$< d\tilde{w}(t) > = < \tilde{g}(t)dt > \equiv 0,$$

which are valid for the integrated OU process, were employed.

Defining the average, or expectation value, of the arbitrary function $F(V(t))$

$$< F(V(t)) > = \int \mathrm{d}V(t)\, F(V)\, \rho(V,t), \tag{7.91}$$

where $\rho(V,t)$ denotes the probability density function with finite support in the space of the state variable $V(t)$, one has

$$\frac{\mathrm{d}}{\mathrm{d}t} < F(V(t)) > = \left\langle \frac{\mathrm{d}F(V(t))}{\mathrm{d}t} \right\rangle. \tag{7.92}$$

Performing the time derivative on the right hand side of (7.91) yields, after inserting (7.90) and partial integration, the *Fokker–Planck equation* of the passive membrane equation with multiplicative noise sources:

$$\partial_t \rho(V,t) = -\partial_V \Big(f(V(t))\, \rho(V,t) \Big) + \partial_V \Big(h_{\mathrm{e}}(V(t))\, \partial_V \big(h_{\mathrm{e}}(V(t))\, \alpha_{\mathrm{e}}(t) \rho(V,t) \big) \Big)$$
$$+ \partial_V \Big(h_{\mathrm{i}}(V(t))\, \partial_V \big(h_{\mathrm{i}}(V(t))\, \alpha_{\mathrm{i}}(t) \rho(V,t) \big) \Big), \tag{7.93}$$

where

$$2\,\alpha_{\{\mathrm{e,i}\}}(t) = \sigma_{\{\mathrm{e,i}\}}^2\, \tau_{\{\mathrm{e,i}\}} \left(1 - \exp\left[-\frac{t}{\tau_{\{\mathrm{e,i}\}}} \right] \right) + \frac{1}{2\,\tau_{\{\mathrm{e,i}\}}} < \tilde{w}_{\{\mathrm{e,i}\}}^2(t) > - \sigma_{\{\mathrm{e,i}\}}^2\, t, \tag{7.94}$$

Equation (7.93) describes the time evolution of the probability $\rho(V,t)$ that the stochastic process, determined by the passive membrane equation (7.59), takes the value $V(t)$ at time t. We are interested in the steady-state probability distribution, i.e., $t \to \infty$. In this limit,

$$\partial_t \rho(V,t) \to 0.$$

To obtain explicit expressions for $\alpha_{\{e,i\}}(t)$, defined in (7.94), in the limit $t \to \infty$, one does make use of the fact that, for $t \to \infty$, the ratio

$$\frac{t}{\tau\{e,i\}} \gg 1.$$

This leads to the assumption that in the steady-state limit the variables $\alpha_{\{e,i\}}(t)$ take a form corresponding to a Wiener process. Hence

$$2\,\alpha_{\{e,i\}}(t) \to \sigma^2_{\{e,i\}}\,\tau_{\{e,i\}}\left(1-e^{-\frac{t}{\tau_{\{e,i\}}}}\right)+\frac{1}{\tau_{\{e,i\}}}D_{\{e,i\}}t-\sigma^2_{\{e,i\}}t=\sigma^2_{\{e,i\}}\,\tau_{\{e,i\}} \tag{7.95}$$

for $t \to \infty$. Here, relation (7.76) with $D_{\{e,i\}} = \sigma^2_{\{e,i\}}\,\tau_{\{e,i\}}$ was used. The interpretation of (7.95) is that, in the limit $t \to \infty$, the noise correlation times $\tau_{\{e,i\}}$ become infinitesimal small compared to the time in which the steady-state probability distribution is obtained. It can be shown numerically (see next section) that, indeed, this assumption yields a steady-state probability distribution which closely matches that obtained from numerical simulations for realistic value of the involved parameters for the membrane and synaptic noise properties.

7.4.6 *The Steady-State Membrane Potential Distribution*

With (7.95), the Fokker–Planck equation (7.93) can be solved analytically, yielding the following steady-state probability distribution $\rho(V)$ for the membrane potential $V(t)$, described by the passive membrane equation (7.59) with two independent colored multiplicative noise sources describing excitatory and inhibitory synaptic conductances:

$$\begin{aligned}\rho(V) = N\exp\Bigg[A_1\ \ln\left[\frac{\sigma_e^2\,\tau_e}{(C_m\,a)^2}(V-E_e)^2+\frac{\sigma_i^2\,\tau_i}{(C_m\,a)^2}(V-E_i)^2\right]\\ +A_2\arctan\left[\frac{\sigma_e^2\,\tau_e\,(V-E_e)+\sigma_i^2\,\tau_i\,(V-E_i)}{(E_e-E_i)\sqrt{\sigma_e^2\,\tau_e\,\sigma_i^2\,\tau_i}}\right]\Bigg],\end{aligned} \tag{7.96}$$

where

$$A_1 = -\frac{2aC_m(g_{e0}+g_{i0})+2a^2C_m g_L+\sigma_e^2\tau_e+\sigma_i^2\tau_i}{2(\sigma_e^2\tau_e+\sigma_i^2\tau_i)},$$

$$A_2 = \frac{1}{(E_e-E_i)\sqrt{\sigma_e^2\tau_e\sigma_i^2\tau_i}\,(\sigma_e^2\tau_e+\sigma_i^2\tau_i)} \times\Big[2C_m a\Big(a g_L\big(\sigma_e^2\tau_e(E_L-E_e)+\sigma_i^2\tau_i(E_L-E_i)\big) + (g_{e0}\sigma_i^2\tau_i-g_{i0}\sigma_e^2\tau_e)(E_e-E_i)\Big)\Big]. \tag{7.97}$$

Interestingly, the noise time constants $\tau_{\{e,i\}}$ enter the expressions for the steady-state membrane potential distribution (7.96) only in the combination $\sigma^2_{\{e,i\}}\,\tau_{\{e,i\}}$. This rather surprising result can be heuristically understood by looking at the nature of the effective stochastic processes. As it was earlier shown in the framework of shot-noise processes (see Sect. 4.4.5), correlated activity among multiple synaptic input channels impacts on the variance of the total conductance or current time course, which in the effective model is described by $\sigma^2_{\{e,i\}}$. On the other hand, nonzero noise time constants, which are linked to the finite-time kinetics of synaptic conductances, result in an effective temporal correlation between individual synaptic events. Here, larger time constants yield larger temporal overlap between individual events, which leads to a contribution to the (temporal) correlation of the synaptic inputs. The latter results in an effect which is comparable to that of the correlation in the presynaptic activity pattern. The particular coupling between $\sigma^2_{\{e,i\}}$ and $\tau_{\{e,i\}}$ also indicates that white noise sources with different "effective" variance $\sigma^2_{\{e,i\}}\,\tau_{\{e,i\}}$ will yield equivalent distributions. However, for more complex systems or different couplings (such as in the case of voltage-dependent NMDA currents), this interpretation will no longer hold.

Before assessing the limitations of the presented approach as outlined in Sect. 7.4.1, we will compare the analytically exact solution (7.96) with numerical simulations. Typical examples of membrane potential probability distributions $\rho(V)$ resembling those found in activated states of the cortical network in vivo are shown in Fig. 7.6 (*gray solid* in a–c), along with their corresponding analytic distributions (*black solid*). The chosen parameters in (7.96) matched those in the numerical solution of the passive membrane equation (7.59) (Fig. 7.6a), as well as those obtained from numerical simulations of a passive single-compartment model with thousands of excitatory and inhibitory synapses releasing according to independent Poisson processes (Fig. 7.6b), and of a detailed biophysical model of a morphologically reconstructed cortical neuron (Fig. 7.6c; e.g. Destexhe and Paré 1999; Rudolph et al. 2001; Rudolph and Destexhe 2001a, 2003b). The latter was shown to faithfully reproduce intracellular recordings obtained in vivo (see Paré et al. 1998b; Destexhe and Paré 1999). In all cases, the numerical simulations yield membrane potential distributions which are well captured by the analytic solution in (7.96).

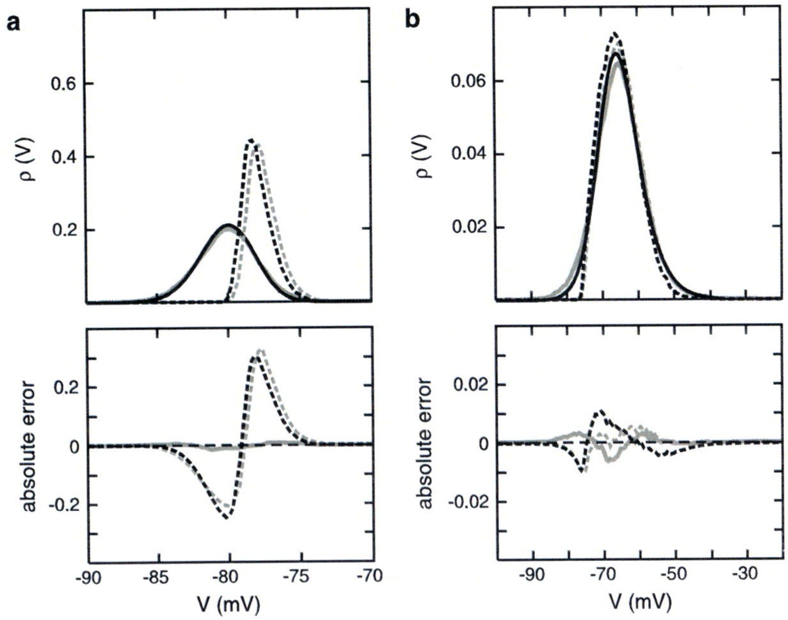

Fig. 7.7 Examples of membrane potential probability distributions for multiplicative synaptic noise (conductance noise) $\rho(V)$. Analytic solutions (*black solid*) are compared to numerical solutions of the passive (*gray solid*: without negative conductance cutoff; *gray dashed*: with negative conductance cutoff) and an active (*black dashed*) model. (**a**) Low-conductance state around the resting potential, similar to in vitro conditions. (**b**) High-conductance state similar to in vivo conditions. The absolute error (*bottom panels*), defined as the difference between the numerical solution and the analytic solution, is markedly reduced in the high conductance. Model parameters were (**a**) $g_{e0} = 0$, $g_{i0} = 0$, $\sigma_e = 0.0012\ \mu S$ and $\sigma_i = 0.00264\ \mu S$; (**b**) $g_{e0} = 0.0121\ \mu S$, $g_{i0} = 0.0573\ \mu S$, $\sigma_e = 0.012\ \mu S$ and $\sigma_i = 0.0264\ \mu S$; for both: $\tau_e = 2.728$ ms and $\tau_i = 10.49$ ms. Modified from Rudolph and Destexhe (2003d)

An apparent limitation of the analytical model, in comparison to numerical simulations, is the presence of (unphysical) negative conductances. Due to the Gaussian nature of the underlying effective synaptic conductances, mean values $g_{\{e,i\}}$ of the order or smaller than $\sigma_{\{e,i\}}$ will yield a marked contribution of the negative tail of the conductance distributions, thus rendering the analytical solution (7.96) less faithful. To evaluate this situation, Fig. 7.7 compares the steady-state membrane potential probability distribution for models with multiplicative noise at a noisy resting state resembling low-conductance in vitro conditions ($g_{\{e,i\}0} = 0$, Fig. 7.7a), and a noisy depolarized state resembling in vivo conditions (Fig. 7.7b). Close to rest, the analytic solution deviated markedly from the numerical solution of the passive membrane equation (with negative conductance cutoff) and the active membrane equation subject to multiplicative noise (Fig. 7.7a, *gray* and *black dashed*, respectively), whereas the error is comparable small for the passive model (Fig. 7.7a, *gray solid*). Due to the smaller fraction of negative conductances in high-conductance states, i.e., for $g_{\{e,i\}} > \sigma_{\{e,i\}}$, the error between the analytic and numerical solution is markedly reduced compared to resting conditions (Fig. 7.7b).

As outlined at the beginning of this section, the derivation of an analytic expression for the membrane potential distribution in the presence of multiplicative colored noise solely utilizes the expectation values, i.e., moments, of the underlying stochastic processes. This *expectation value approach* presented above allows for an easy generalization to more complicated stochastic processes. Stochastic calculus, on the other hand, provides, however, many different ways to assess a specific problem, which might lead, in stark contrast to ordinary calculus, to different results. Thus, the final results have to be checked for their correctness, e.g., by comparing analytic results to numerical simulations of the same problem, and consistency.

It was found that the approach presented above yields a relatively satisfactory agreement with numerical simulations if physiologically relevant parameter regimes are considered (Figs. 7.6 and 7.7; see Rudolph and Destexhe 2003d). However, systematic deviations occur if, for instance, noise time constants much larger or smaller than the total membrane time constant are used (Fig. 7.8a; see Rudolph and Destexhe 2005, 2006b). Indeed, the approach presented here must lead to an alteration of the spectral properties of the noise processes, as the considered OU noise has the same (Gaussian) amplitude distribution as Gaussian white noise. Hence, the same qualitatively indifferent mathematical framework will apply to noise processes which differ qualitatively in their spectral properties.

To illustrate this point in more detail, one can show that solving SDEs solely based on the expectation values of differentials of the underlying stochastic processes does not completely capture the spectral properties of the investigated stochastic system. The core of the problem can be demonstrated by calculating the Fourier transforms of the original Langevin equation (7.59) and that of an infinitesimal displacement of the membrane potential formulated in terms of differentials of the integrated OU stochastic process after application of the stochastic calculus, (7.86). Defining the Fourier transforms $V(\omega)$ and $\tilde{g}_{\{e,i\}}(\omega)$ of $V(t)$ and $\tilde{g}_{\{e,i\}}(t)$, respectively (ω denotes the circular frequency):

$$V(t) = \frac{1}{2\pi} \int_{-\infty}^{\infty} \mathrm{d}\omega V(\omega) \mathrm{e}^{i\omega t}$$

$$\tilde{g}_{\{e,i\}}(t) = \frac{1}{2\pi} \int_{-\infty}^{\infty} \mathrm{d}\omega \, \tilde{g}_{\{e,i\}}(\omega) \mathrm{e}^{i\omega t},$$

the Fourier transform of the membrane equation (7.59) reads

$$\left(i\omega + \frac{1}{\tau_m}\right) \tilde{V}(\omega) = -\frac{1}{2\pi C} \int_{-\infty}^{\infty} \mathrm{d}\omega' \Big\{ \tilde{g}_e(\omega') \big(\tilde{V}(\omega - \omega') - \tilde{E}_e\big) + \tilde{g}_i(\omega') \big(\tilde{V}(\omega - \omega') - \tilde{E}_i\big) \Big\}, \tag{7.98}$$

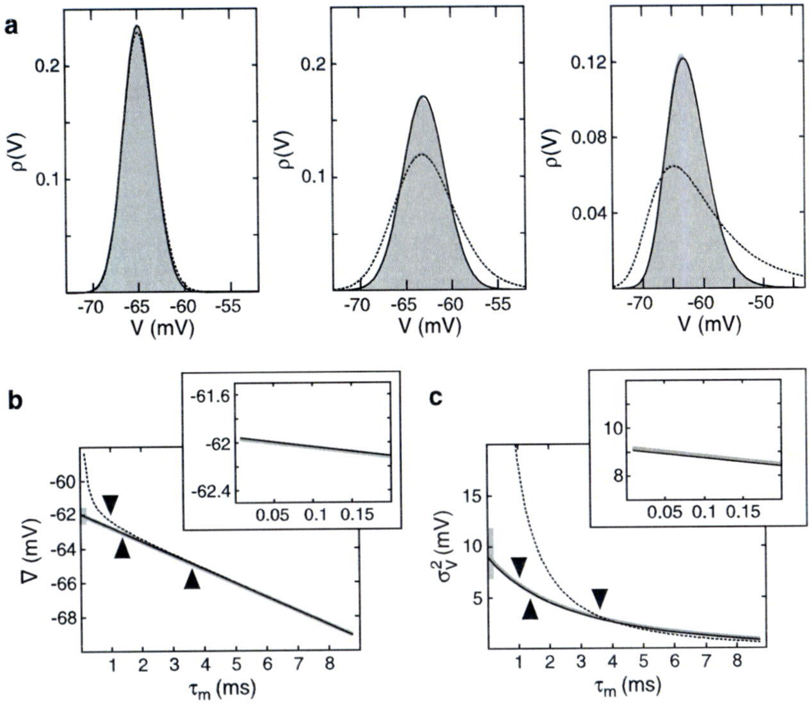

Fig. 7.8 Comparison of the V_{m} distributions obtained numerically and using the extended analytic expression. (**a**) Examples of membrane potential distributions for different membrane time constants τ_{m} (*left*: $\tau_{\mathrm{m}} = 3.63$ ms, *middle*: $\tau_{\mathrm{m}} = 1.36$ ms, *right*: $\tau_{\mathrm{m}} = 1.03$ ms). In all cases, numerical simulations (*gray*) are compared with the analytic solution (*black dashed*; (7.96), see also Rudolph and Destexhe 2003d) and an extended analytic expression (*black solid*) obtained after compensating for the "filtering problem," (7.96) and (7.110). (**b**), (**c**) Mean $\overline{V}$ and variance σ_V^2 of the V_{m} distribution as a function of membrane time constant. Numerical simulations (*gray*) are compared with the mean and variance obtained by numerical integration of the original analytic solution (Rudolph and Destexhe 2003d; *dashed lines*) and the extended analytic expression. The *gray vertical stripes* mark the parameter regimes displayed in the insets. Parameter values: $g_{\mathrm{L}} = G_L/a = 0.0452\,\mathrm{mS\,cm^{-2}}$, $C_{\mathrm{m}} = C/a = 1\,\mu\mathrm{F\,cm^{-2}}$, $E_{\mathrm{L}} = -80$ mV, $g_{\mathrm{e0}} = 12$ nS, $g_{\mathrm{i0}} = 57$ nS, $\sigma_{\mathrm{e}} = 3$ nS, $\sigma_{\mathrm{i}} = 6.6$ nS; A, *right*: $\sigma_{\mathrm{e}} = 3$ nS, $\sigma_{\mathrm{i}} = 15$ nS, $\tau_{\mathrm{e}} = 2.728$ ms, $\tau_{\mathrm{i}} = 10.49$ ms, $E_{\mathrm{e}} = 0$ mV, $E_{\mathrm{i}} = -75$ mV. Membrane area a: $a = 30{,}000\,\mu\mathrm{m}^2$ [(**a**), *left*], $a = 10{,}000\,\mu\mathrm{m}^2$ [(**a**), *middle*], $a = 7{,}500\,\mu\mathrm{m}^2$ [(**a**), *right*], $50\,\mu\mathrm{m}^2 \geq a \geq 100{,}000\,\mu\mathrm{m}^2$ (**b**, **c**). For all simulations, integration time steps of at least one order of magnitude smaller (but at most 0.1 ms) than the smallest time constant (either membrane or noise time constant) in the considered system were used. To ensure that the observed effects were independent on peculiarities of the numerical integration, different values for the integration time step (in all cases, at least one order of magnitude smaller than the smallest time constant in the system) for otherwise fixed noise and membrane parameters were compared. No systematic or significant differences were observed. Moreover, to ensure a valid statistics of the membrane potential time course, the simulated activity covered at least 100 s for each parameter set were used. Modified from Rudolph and Destexhe (2005)

where $\tilde{V}(\omega) = V(\omega) - E_0$ and $\tilde{E}_{\{e,i\}} = E_{\{e,i\}} - E_0$. On the other hand, Fourier transformation of the result (7.86) obtained after application of the stochastic calculus provides another expression for an infinitesimal displacement of the membrane potential $V(t)$:

$$\left(i\omega + \frac{1}{\tau_m}\right)\tilde{V}(\omega) = -\frac{1}{2\pi C}\int_{-\infty}^{\infty} d\omega' \left\{\tilde{g}'_e(\omega')\left(\tilde{V}(\omega-\omega') - \tilde{E}_e\right)\right.$$

$$\left. + \tilde{g}'_i(\omega')\left(\tilde{V}(\omega-\omega') - \tilde{E}_i\right)\right\}. \quad (7.99)$$

In the last equation, $\tilde{g}'_{\{e,i\}}(\omega)$ denotes the Fourier transforms of

$$g'_{\{e,i\}}(t) = \frac{d}{dt}\tilde{w}_{\{e,i\}}(t) - \frac{1}{C}\alpha_{\{e,i\}}(t),$$

where

$$2\alpha_{\{e,i\}}(t) = \sigma^2_{\{e,i\}}\tau'_{\{e,i\}}\left(1 + e^{-t/\tau'_{\{e,i\}}}\right) + \frac{1}{2\tau'_{\{e,i\}}}\tilde{w}^2_{\{e,i\}}(t) - \sigma^2_{\{e,i\}}t. \quad (7.100)$$

Here, $g'_{\{e,i\}}(t)$ is formally well defined but, due to the nontrivial form of $\alpha_{\{e,i\}}(t)$, cannot be calculated explicitly. Furthermore, in (7.99) the noise time constants $\tau_{\{e,i\}}$ from the original equation were replaced by "effective" noise time constants $\tau'_{\{e,i\}}$. As will be demonstrated below, this assumption is justified and provides, although only on a heuristic level, a valid solution of the *filtering problem* at hand. A recent study by Bedard and colleagues, however, suggests that this heuristic approach can be formulated on a mathematically and physically more rigorous basis.

Instead of directly evaluating (7.98) and (7.99), one can follow another path. A comparison of both Fourier transforms shows that the stochastic calculus utilized above introduces a modification of the spectral structure characterizing the original system. This modification is linked to the term $\alpha_{\{e,i\}}(t)$, (7.100), whose appearance is a direct consequence of the use of the integrated OU stochastic process and its differentials. As indicated in (7.99), this translates into an alteration of the spectral "filtering" properties of the stochastic differential equation (7.59). Here, two observations can be made. First, both Fourier transforms, (7.98) and (7.99), show the same functional structure, with $\tilde{g}_{\{e,i\}}(\omega)$ in (7.98) replaced by the Fourier transform $\tilde{g}'_{\{e,i\}}(\omega)$ of a new stochastic variable $g'_{\{e,i\}}(t)$ in (7.99). Secondly, the functional coupling of $\tilde{g}_{\{e,i\}}(\omega)$ and $\tilde{g}'_{\{e,i\}}(\omega)$ to the Fourier transform of the membrane potential $V(\omega)$ is identical in both cases. This, together with the fact that $V(\omega)$ describe the same state variable, provides the basis for deducing an explicit expression for the effective time constants $\tau'_{\{e,i\}}$ and, thus, an extension of (7.96) which will compensate for the loss of the spectral characteristics in the utilized *expectation value approach*.

In order to preserve the spectral signature of $V(t)$, one can assume that the functional form of the Fourier transform of the stochastic process $\tilde{g}'_{\{e,i\}}(t)$ is equivalent to that of the OU stochastic process $\tilde{g}_{\{e,i\}}(t)$. The Fourier transform of the latter is given by

$$\tilde{g}_{\{e,i\}}(\omega) = \sqrt{\frac{2\sigma^2_{\{e,i\}}}{\tau_{\{e,i\}}}}\,\eta_{\{e,i\}}(\omega)\,\frac{1}{i\omega + \frac{1}{\tau_{\{e,i\}}}}. \tag{7.101}$$

Thus, the above assumption can be restated in more mathematical terms as

$$\tilde{g}'_{\{e,i\}}(\omega) = \sqrt{\frac{2\sigma^2_{\{e,i\}}}{\tau'_{\{e,i\}}}}\,\eta_{\{e,i\}}(\omega)\,\frac{1}{i\omega + \frac{1}{\tau'_{\{e,i\}}}}. \tag{7.102}$$

In writing down (7.102), one further assumes that changes in the spectral properties of $\tilde{g}_{\{e,i\}}(\omega)$ and $\tilde{g}'_{\{e,i\}}(\omega)$ reflect in changes of the parameters describing the corresponding Fourier transforms, specifically the noise time constants $\tau_{\{e,i\}}$. This new "effective" time constants $\tau'_{\{e,i\}}$ will later be used to compensate the change in the spectral filtering properties of the analytic solution, (7.96). Importantly, this restriction to changes in the noise time constants is possible, because only the combinations $\sigma^2_{\{e,i\}}\tau'_{\{e,i\}}$ enter (7.96). Thus, each change in $\sigma_{\{e,i\}}$ can be mapped onto a corresponding change of $\tau'_{\{e,i\}}$ only. Finally, due to their definition and (7.99), τ'_e and τ'_i undergo mutually independent modifications.

In order to explicitly calculate the link between the time constants $\tau'_{\{e,i\}}$ and $\tau_{\{e,i\}}$ and, thus, provide a solution with which the above outlined filtering problem can be resolved, one can consider a simpler and dynamically different stochastic system for which the analytic solution is known. As suggested in Rudolph et al. (2005), this approach is possible because the application of stochastic calculus does not impair the qualitative coupling between the stochastic processes $\tilde{g}_{\{e,i\}}(\omega)$, $\tilde{g}'_{\{e,i\}}(\omega)$ and V_m [compare (7.98) and (7.99)]. Thus, a simpler system with the same (conductance) noise processes but different coupling to V_m, such as additive coupling, can be considered.

Solutions to such models were investigated in great detail (e.g., Richardson 2004). Among these solutions, an *effective time constant approximation* describes the effect of colored conductance noise by a constant mean conductance and conductance fluctuations which couple to the mean V_m. The latter leads to a term describing current noise and yields a model equivalent to (7.59), in which $V(t)$ in the noise terms is replaced by its mean E_0, i.e.,

$$\frac{dV(t)}{dt} = -\frac{1}{\tau_m}\left(V(t) - E_0\right) - \frac{1}{C}\tilde{g}_e(t)\left(E_0 - E_e\right) - \frac{1}{C}\tilde{g}_i(t)\left(E_0 - E_i\right), \tag{7.103}$$

where $\tilde{g}_{\{e,i\}}(t)$ are given by (7.61).

In contrast to (7.59), this simplified stochastic system can explicitly be solved by direct integration, which leaves, as required, the spectral characteristics unaltered. The variance of the membrane potential was found to be (Richardson 2004):

$$\sigma_V^2 = \left(\frac{\sigma_e \tau_m}{C}\right)^2 \frac{\tau_e}{\tau_e + \tau_m}(E_0 - E_e)^2 + \left(\frac{\sigma_i \tau_m}{C}\right)^2 \frac{\tau_i}{\tau_i + \tau_m}(E_0 - E_i)^2. \quad (7.104)$$

An equivalent expression for the membrane potential variance can be, more directly, deduced from the PSD of the underlying stochastic processes by approximating the explicit form of σ_V^2 given in Manwani and Koch (1999a).

On the other hand, treating this simplified stochastic system (7.103) within the *expectation value approach* detailed in Sects. 7.4.1–7.4.5, leads to the following Fokker–Planck equation:

$$\begin{aligned}(\partial_t \rho(V,t)) &= -\frac{1}{\tau_m}\rho(V,t) - \frac{V-E_0}{\tau_m}(\partial_V \rho(V,t)) \\ &\quad -\left(\frac{(E_0-E_e)^2}{C^2}\alpha_e(t) + \frac{(E_0-E_i)^2}{C^2}\alpha_i(t)\right)(\partial_V^2 \rho(V,t)). \quad (7.105)\end{aligned}$$

For $t \to \infty$, one obtains the steady-state solution. In this limit,

$$\begin{aligned}\partial_t \rho(V,t) &\to 0, \\ \rho(V,t) &\to \rho(V), \\ 2\alpha_{\{e,i\}}(t) &\to \sigma_{\{e,i\}}^2 \tau'_{\{e,i\}}.\end{aligned}$$

To obtain the latter, one calculates the expectation value of (7.100) and makes use of $\exp[-t/\tau_{\{e,i\}}] \to 0$ for $t \to \infty$, as well as the fact that in this limit the integrated OU stochastic process $\tilde{w}_{\{e,i\}}^2(t)$ yields a Wiener process with 2-dimensional cumulant

$$< \tilde{w}_{\{e,i\}}^2(t) >= 2D_{\{e,i\}}t,$$

where $D_{\{e,i\}} = \sigma_{\{e,i\}}^2 \tau_{\{e,i\}}$. With this, (7.105) takes the form

$$\begin{aligned}0 &= -\frac{1}{\tau_m}\rho(V) - \frac{V-E_0}{\tau_m}(\partial_V \rho(V)) \\ &\quad -\left(\frac{(E_0-E_e)^2}{2C^2}\sigma_e^2 \tau'_e + \frac{(E_0-E_i)^2}{2C^2}\sigma_i^2 \tau'_i\right)(\partial_V^2 \rho(V)). \quad (7.106)\end{aligned}$$

This equation is obtained from (7.105) by performing the limit $t \to \infty$, in which case the ratio $\frac{t}{\tau_{\{e,i\}}} \gg 1$. However, this limit is not equivalent to taking the limit $\tau_{\{e,i\}} \to 0$:

for $t \to \infty$, the noise time constants $\tau_{\{e,i\}}$ become infinitesimally small compared to the time over which the steady-state probability distribution is obtained, hence $\alpha_{\{e,i\}}(t)$ take a form corresponding to that obtained in the case of a Wiener process.

Equation (7.106) can now explicitly be solved, yielding

$$\rho(V) = \mathrm{e}^{-\frac{(V-E_0)^2}{2\sigma_V^2}} \left(C_1\, \mathrm{e}^{\frac{E_0^2}{2\sigma_V^2}} + C_2 \sqrt{\frac{\pi}{2}}\, \sigma_V^2\, \mathrm{Erfi}\left[\frac{V-E_0}{\sqrt{2\sigma_V^2}} \right] \right), \tag{7.107}$$

where Erfi$[z]$ denotes the imaginary error function and

$$\sigma_V^2 = \frac{\tau_\mathrm{m}}{2} \frac{\sigma_\mathrm{e}^2 \tau_\mathrm{e}'}{C^2} (E_0 - E_\mathrm{e})^2 + \frac{\tau_\mathrm{m}}{2} \frac{\sigma_\mathrm{i}^2 \tau_\mathrm{i}'}{C^2} (E_0 - E_\mathrm{i})^2 \tag{7.108}$$

the variance of the membrane potential. With the boundary conditions $\rho(V) \to 0$ for $V \to \pm\infty$ and normalization $\int_{-\infty}^{\infty} \mathrm{d}V\, \rho(V) = 1$, (7.107) simplifies to a Gaussian

$$\rho(V) = \frac{1}{\sqrt{2\pi\sigma_V^2}}\, \mathrm{e}^{-\frac{(V-E_0)^2}{2\sigma_V^2}}. \tag{7.109}$$

This result for the steady-state membrane potential distribution is, formally within the *expectation value approach*, the equivalent of (7.96), obtained from the stochastic system given in (7.59), when considering the stochastic system (7.103) instead.

Comparing now the variance of the membrane potential distribution obtained with two qualitatively different methods, (7.104) and (7.108), respectively, yields the desired link between the time constants

$$\tau'_{\{e,i\}} = \frac{2\tau_{\{e,i\}}\tau_\mathrm{m}}{\tau_{\{e,i\}} + \tau_\mathrm{m}}. \tag{7.110}$$

If the argumentation and assumptions made above are valid, then inserting this relation (7.110) into (7.109) must compensate for the change in the spectral signature introduced by reformulating the original stochastic system, (7.103), within the framework of the *expectation value approach*, i.e., in an approach which utilized solely the expectation values of the differentials of the governing stochastic variables. Moreover, and more crucial, following the above argumentation, (7.110) must also provide this compensation when applied to the original stochastic system (7.59), as the nature of the coupling between the state and stochastic variables was found not to be important. Indeed, this leads to an *extended analytic expression* for the steady-state membrane potential distribution, in which the time constants of the noise are rescaled according to the effective membrane time constant, thus compensating for the filtering effect.

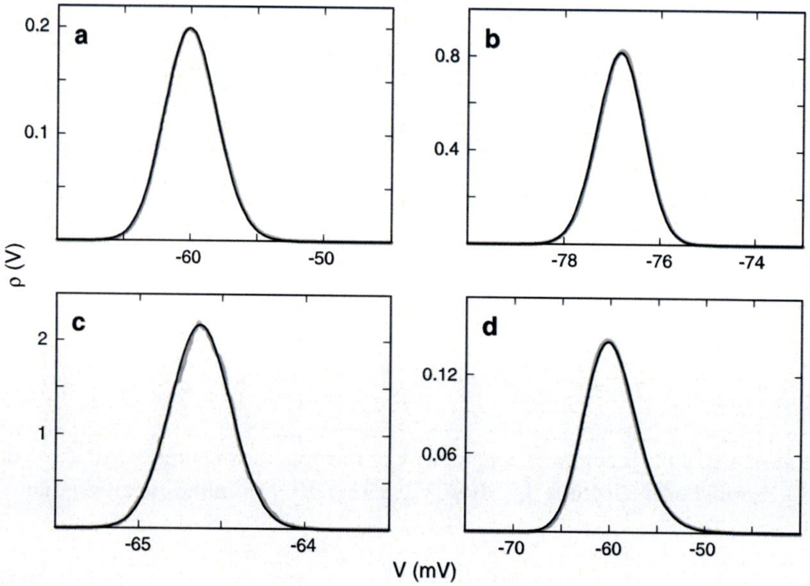

Fig. 7.9 Comparison of the V_m distributions for extreme parameter values, obtained numerically (*gray solid*) and using the extended analytic expression (*black solid*). (**a**) Model with very small effective membrane time constant of 0.005 ms obtained by a small membrane area. Parameters: $a = 38\ \mu m^2$, $G = 75.017176$ nS, $G_L = 0.017176$ nS, $C = 0.00038$ nF, $\tau_m = 22.1$ ms, $\tau_0 = 0.005$ ms, $g_{e0} = 15$ nS, $g_{i0} = 60$ nS, $\sigma_e = 2$ nS, $\sigma_i = 6$ nS, $\tau_e = 2.7$ ms, $\tau_i = 10.5$ ms. (**b**) Model with very small effective membrane time constant of 0.005 ms obtained by a high leak and synaptic conductance. Parameters: $a = 10{,}000\ \mu m^2$, $G = 23{,}728.4$ nS, $G_L = 19{,}978.4$ nS, $C = 0.1$ nF, $\tau_m = 0.05$ ms, $\tau_0 = 0.00421$ ms, $g_{e0} = 750$ nS, $g_{i0} = 3{,}000$ nS, $\sigma_e = 150$ nS, $\sigma_i = 600$ nS, $\tau_e = 2.7$ ms, $\tau_i = 10.5$ ms. (**c**) Model with very large membrane time constants of 5 s obtained by a large membrane area in combination with a small leak and synaptic conductances. Parameters: $a = 100{,}000\ \mu m^2$, $G = 0.1952$ nS, $G_L = 0.0452$ nS, $C = 1$ nF, $\tau_m = 22{,}124$ ms, $\tau_0 = 5{,}123$ ms, $g_{e0} = 3\times10^{-5}$ nS, $g_{i0} = 12\times10^{-5}$ nS, $\sigma_e = 1.5\times10^{-5}$ nS, $\sigma_i = 6\times10^{-5}$ nS, $\tau_e = 2.7$ ms, $\tau_i = 10.5$ ms. (**d**) Model with very large membrane time constants of 50 s obtained by a low leak. Parameters: $a = 20{,}000\ \mu m^2$, $G = 75.004$ nS, $G_L = 0.004$ nS, $C = 0.2$ nF, $\tau_m = 50{,}054$ ms, $\tau_0 = 2.67$ ms, $g_{e0} = 15$ nS, $g_{i0} = 60$ nS, $\sigma_e = 3$ nS, $\sigma_i = 12$ nS, $\tau_e = 2.7$ ms, $\tau_i = 10.5$ ms. In all cases, an excellent agreement between numerical and analytical solution is observed

Numerical simulations show that this extended expression reproduces remarkably well membrane potential distributions in models with a parameter space spanning at least seven orders of magnitude (see Figs. 7.8b, 7.9 and 7.10). However, this extended analytic expression for the membrane potential distribution of the full model with OU stochastic synaptic conductances still does not bypasses two other limitations. First, due to the nature of the distribution of the incorporated conductance noise processes, the presence of unphysical negative conductances cannot be accounted for and will, naturally, lead to a mismatch between numerical

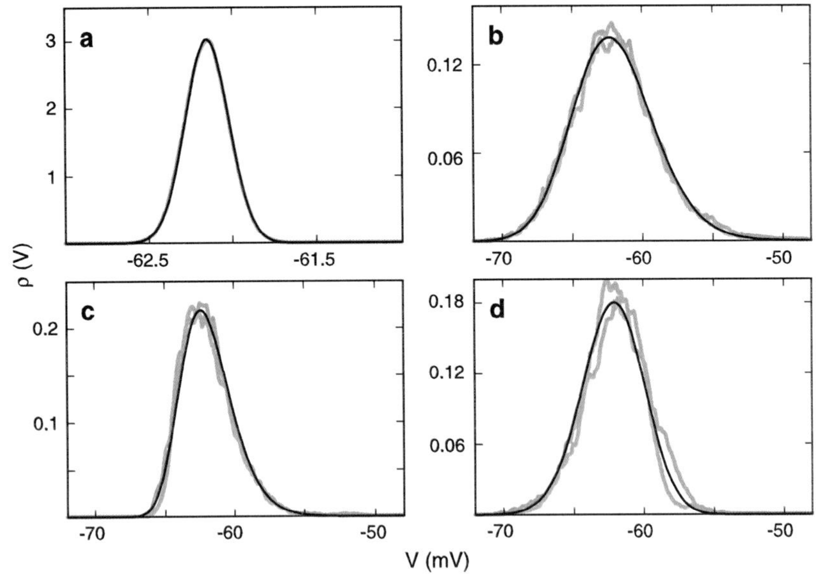

Fig. 7.10 Comparison of the V_m distributions for extreme parameter values, obtained numerically (*gray solid*) and using the extended analytic expression (*black solid*). (**a**) Model with very small excitatory and inhibitory conductance time constants. Parameters: $a = 20{,}000\,\mu\text{m}^2$, $G = 84.04\,\text{nS}$, $G_L = 9.04\,\text{nS}$, $C = 0.2\,\text{nF}$, $\tau_m = 22.12\,\text{ms}$, $\tau_0 = 2.38\,\text{ms}$, $g_{e0} = 15\,\text{nS}$, $g_{i0} = 60\,\text{nS}$, $\sigma_e = 3\,\text{nS}$, $\sigma_i = 12\,\text{nS}$, $\tau_e = 0.005\,\text{ms}$, $\tau_i = 0.005\,\text{ms}$. (**b**) Model with very long excitatory and inhibitory conductance time constants. Parameters: As in (**a**), except $\tau_e = 50{,}000\,\text{ms}$, $\tau_i = 50{,}000\,\text{ms}$. (**c**) Model with a combination of very small and very large excitatory and inhibitory time constants. Parameters: As in (**a**), except $\tau_e = 0.01\,\text{ms}$, $\tau_i = 1{,}000\,\text{ms}$. (**d**) Model with a combination of very small and very large excitatory and inhibitory time constants. Parameters: As in (**a**), except $\tau_e = 1{,}000\,\text{ms}$, $\tau_i = 0.01\,\text{ms}$. In all cases, an excellent agreement between numerical and analytical solution is observed. The multiple traces for numerical simulations (**b–d**) were the result of two identical simulations but with different random seed for the noise generator

simulations and analytic solution. Here, a possible solution is to make use of qualitatively different stochastic processes for conductances, e.g., described by Gamma distributions. Indeed, the approach presented in this section potentially allows to arrive at analytic solutions in cases of more realistic effective models for synaptic noise. Secondly, eventual changes in the sign of the driving force due to crossing of the conductance reversal potentials will lead to a different dynamical behavior of the computational model which, due to the exclusive use of expectation values and averages, can not be captured in the *expectation value approach* described here. The expected deviations are most visible at membrane potentials close to the reversal potentials and large values of the involved conductances. Possible solutions here include the use of different numerical integration methods as well as the use of different boundary conditions (e.g.. $\rho(V) \to 0$ for $V = E_e$ and $V = E_i$).

7.5 Numerical Evaluation of Various Solutions for Multiplicative Synaptic Noise

In Sects. 7.3 and 7.4, different analytic expressions for the membrane potential distribution of passive cellular membranes subject to synaptic conductance noise were presented. These expressions were deduced from the stochastic membrane equation utilizing various mathematical methods. In the last section of this chapter, we will briefly evaluate the different results for $\rho(V)$ found in the literature by comparing the analytic expressions with numerical simulations of the underlying SDE.

As shown in Sect. 4.4, synaptic noise can be faithfully modeled by fluctuating conductances described by OU stochastic processes (Destexhe et al. 2001). This system was later investigated by using stochastic calculus to obtain analytic expressions for the steady-state membrane potential distribution (see Sect. 7.4). Analytic expressions can also be obtained for the moments of the underlying three-dimensional Fokker–Planck equation (Richardson 2004), or by considering this equation under different limit cases (Lindner and Longtin 2006; Hasegawa 2008). One of the greatest promises of such analytic expressions is that they can be used to deduce the characteristics of conductance fluctuations from intracellular recordings in vivo (Rudolph et al. 2004, 2005, 2007). However, a prerequisite of this is to evaluate which analytic expression should, and can, be used in practical situations.

In Rudolph and Destexhe (2005), an extended range of parameters spanning more than seven orders of magnitude was tested and yield an excellent agreement between analytic and numerical results even for extreme, physiologically implausible model parameters. Later, in Rudolph and Destexhe (2006b), the same approach was used to evaluate various analytic expressions for the membrane potential distribution found in the literature by investigating 10,000 models with randomly selected parameter values in an extended parameter space covering a physiologically relevant regime. The results of this study are shown in Fig. 7.11a–c. The smallest error between analytic expressions and numerical simulations is observed for the extended expression of Rudolph and Destexhe (2005), see Sect. 7.4, followed by Gaussian approximations of the same authors and the model of Richardson (2004). The least accurate solution was the static-noise limit by Lindner and Longtin (2006) (see also Hasegawa (2008) for an extensive comparison of these different approaches).

By scanning only within physiologically relevant values based on conductance measurements in cats in vivo (Rudolph et al. 2005), the same ranking is observed (Fig. 7.11d), with even more drastic differences: up to 95% of the cases reveal the smallest error based on the solution proposed in (Rudolph and Destexhe 2005). Manual examination of the different parameter sets, where the extended expression is not the best estimate, further reveals that this happens in cases where both time constants are slow ("slow synapses" with decay time constants $>$ 50 ms). Indeed, performing parameter scans restricted to this region of parameters, Rudolph and Destexhe (2006b) showed that the extended expression, while still providing good fits to the simulations, ranks first for less than 30% of the cases, while the

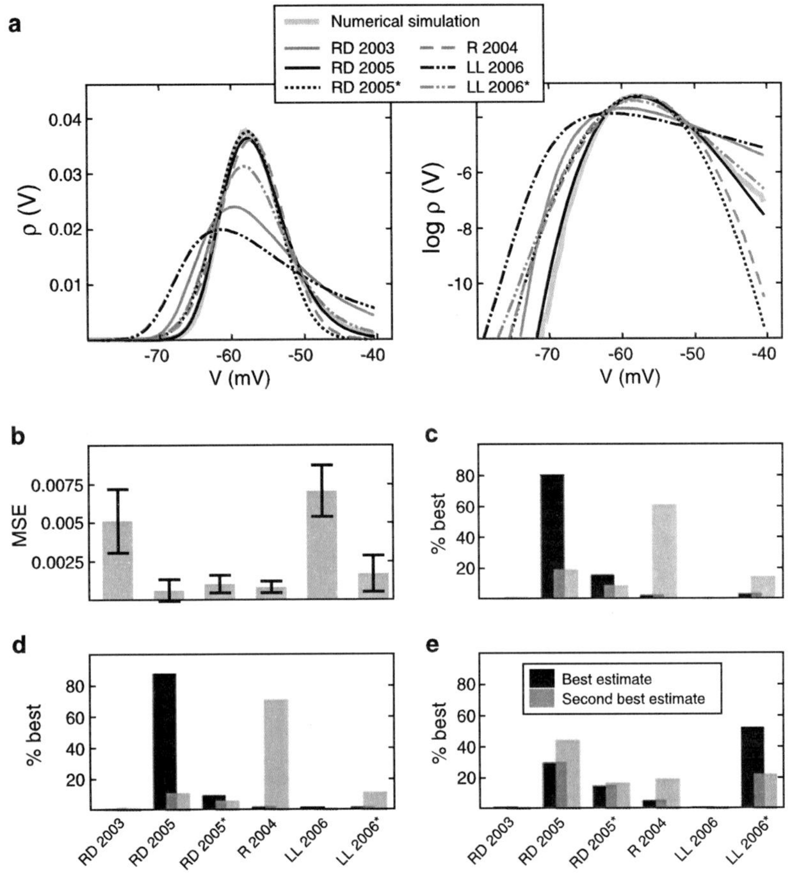

Fig. 7.11 Comparison of the accuracy of different analytical expressions for $\rho(V)$ of membranes subject to colored conductance noise. (**a**) Example of a V_m distribution (*right panel*: log-scale) calculated numerically (*thick gray*; model from Destexhe et al. 2001), compared to different analytical expressions (see legend). Abbreviations: *RD 2003*: Rudolph and Destexhe 2003d; *RD 2005*: Rudolph and Destexhe 2005; *RD 2005**: Gaussian approximation in Rudolph and Destexhe 2005; *R 2004*: Richardson 2004; *LL 2006*: Lindner and Longtin 2006, white noise limit; *LL 2006**: Lindner and Longtin 2006, static noise limit. (**b**) Mean square error (MSE) obtained for each expression by scanning a plausible parameter space spanned by seven parameters (10,000 runs using uniformly distributed parameter values). Varied parameters: $5{,}000 \leq a \leq 50{,}000\ \mu\text{m}^2$, $10 \leq g_{e0} \leq 40\,\text{nS}$, $10 \leq g_{i0} \leq 100\,\text{nS}$, $1 \leq \tau_e \leq 20\,\text{ms}$, $1 \leq \tau_i \leq 50\,\text{ms}$. $\sigma_{\{e,i\}}$ were randomized between 20% and 33% of the mean values to limit the occurrence of negative conductances. Fixed parameters: $g_L = 0.0452\,\text{mS cm}^{-2}$, $E_L = -80\,\text{mV}$, $C_m = 1\ \mu\text{F cm}^{-2}$, $E_e = 0\,\text{mV}$, $E_i = -75\,\text{mV}$. (**c**) Histogram of best estimates (*black*) and second best estimates (*gray*; both expressed in percentage of the 10,000 runs in (**b**). The extended expression (Rudolph and Destexhe 2005) had the smallest mean square error for about 80% of the cases. The expression of Richardson (2004) was the second best estimate, for about 60% of the cases. (**d**) Similar scan of parameters restricted to physiological values (taken from Rudolph et al. 2005: $1 \leq g_{e0} \leq 96\,\text{nS}$, $20 \leq g_{i0} \leq 200\,\text{nS}$, $1 \leq \tau_e \leq 5\,\text{ms}$, $5 \leq \tau_i \leq 20\,\text{ms}$. In this case, Rudolph and Destexhe (2005) was the most performant for about 86% of the cases. (**e**) Scan using large conductances and slow time constants: $50 \leq g_{\{e,i\}0} \leq 400\,\text{nS}$, $20 \leq \tau_{\{e,i\}} \leq 50\,\text{ms}$. In this case, the static noise limit Lindner and Longtin was the most performant for about 50% of the cases. Modified from Rudolph and Destexhe (2006b)

static-noise limit is the best estimate for almost 50% of parameter sets (Fig. 7.11e). Finally, scanning parameters within a wider range of values, including fast/slow synapses and weak/strong conductances, shows that the extended expression is still the best estimate (in about 47% of the cases), followed by the static-noise limit (37%). In conclusion, the approach presented in Sect. 7.4 provides, so far, for practical situations of biophysically plausible conductance values and synaptic time constants the most accurate and generalizable solution.

7.6 Summary

In this chapter, we introduced a number of mathematical models used in the description of synaptic noise. One of the simplest and most widely investigated models is Gaussian white noise entering additively into the neuronal state equation (Sect. 7.2.1). This model is treatable in a mathematically rigorous way, for example within the Fokker–Planck approach. However, additive Gaussian white noise was found to provide only a partially valid description of the stochastic dynamics observed in biological neural systems. The latter is better captured by effective stochastic processes describing colored noise (Sect. 7.2.2), or by considering individual synaptic inputs within the framework of shot noise (Sect. 7.2.3).

Although additive models of synaptic noise provide a good first approximation, only the biophysically more meaningful multiplicative (i.e., conductance) noise allows to capture the dynamical behavior seen in real neurons (Sect. 7.3). The resulting stochastic models, however, are no longer analytically tractable, and approximations, for instance within the *Stratonovich calculus*, remain the only way to mathematically tackle such systems. A novel approach based on the *Itô formalism* was then introduced (Sect. 7.4) to analytically assess the subthreshold response of a neuron subject to colored synaptic noise described by the OU stochastic process. Although this approach allows a mathematically more rigorous exploration, with the possibility of generalization beyond colored (Ornstein-Uhlenbeck) noise processes, it also faces stringent limitations in capturing the spectral properties of the underlying stochastic system.

The chapter ended with a comparative numerical evaluation of the various proposed mathematical models and approximations of the steady-state distribution of the state variable of a passive neuronal system subject to multiplicative synaptic noise (Sect. 7.5). Although none of the known approaches provides an exact solution, huge differences among the individual models do exist, with the extended analytic expression being the most accurate for biophysically realistic parameter regimes. This provides evidence for the usefulness and relevance of the *expectation value approach* utilizing the *Itô formalism* (Sect. 7.4).

This extended expression will form the basis of a new class of methods to analyze synaptic noise, as outlined in the next two chapters.

Chapter 8
Analyzing Synaptic Noise

As we have shown in the previous chapters, specifically in Chaps. 3 and 5, synaptic noise leads to marked changes in the integrative properties and response behavior of individual neurons. Following mathematical formulations of synaptic noise (Chap. 7), we derive in the present chapter a new class of stochastic methods to analyze synaptic noise. These methods consider the membrane potential as a stochastic process. Specific applications of these methods are presented in Chap. 9.

8.1 Introduction

As we have seen in preceding chapters, cortical neurons behave similarly to stochastic processes, as a consequence of their irregularity and dense connectivity. In different cortical structures and in awake animals, cortical neurons display highly irregular spontaneous firing (Evarts 1964; Hobson and McCarley 1971). Together with the dense connectivity of cerebral cortex, with each pyramidal neuron receiving between 5,000 and 60,000 synaptic contacts and a large part of this connectivity originating from the cortex itself (see DeFelipe and Fariñas 1992; Braitenberg and Schüz 1998), one might expect that a large number of synaptic inputs are simultaneously active onto cortical neurons (but see Margrie et al. 2002; Crochet and Petersen 2006; Lee et al. 2006 for reports of sparse firing of cortical neurons in vivo in the awake state). Indeed, intracellular recordings in awake animals reveal that cortical neurons are subject to an intense synaptic bombardment and, as a result, are depolarized and have a low input resistance (Matsumura et al. 1988; Baranyi et al. 1993; Paré et al. 1998b; Steriade et al. 2001) compared to brain slices kept in vitro. This activity is also responsible for a considerable amount of subthreshold fluctuations, called *synaptic noise*. This noise level and its associated *high-conductance state* greatly affect the integrative properties of neurons (reviewed in Destexhe et al. 2003a; Destexhe 2007; Longtin 2011).

A. Destexhe and M. Rudolph-Lilith, *Neuronal Noise*, Springer Series in Computational Neuroscience 8, DOI 10.1007/978-0-387-79020-6_8,

Besides characterizing the effect of synaptic noise on integrative properties, there is a need for analysis methods that are appropriate to this type of signal. In the present chapter, we review different methods to analyze synaptic noise. These approaches are all based on considering the membrane potential ($V_{\rm m}$) fluctuations as a multidimensional stochastic process.

In a first approach, an analytic expression of the steady-state $V_{\rm m}$ distribution (Rudolph and Destexhe 2003d, 2005) is fit to experimental distributions, yielding estimates of mean and variances of excitatory and inhibitory conductances. This so-called *VmD method* (Rudolph et al. 2004) was tested numerically as well as in real neurons using dynamic-clamp experiments. The originality of the VmD method is that it allows to measure not only the mean conductance level of excitatory and inhibitory inputs but also their level of fluctuations, quantified by the conductance variance. However, this approach, like all traditional methods for extracting conductances, requires at least two levels of $V_{\rm m}$ activity, which prevents applications to single-trial measurements (see review by Monier et al. 2008).

Other methods were proposed which can be applied to single-trial $V_{\rm m}$ measurements, such as power spectral analysis, or the *STA method* to compute STA conductances based on a maximum likelihood procedure (Pospischil et al. 2007). Recently, a new method, called the *VmT method*, was introduced (Pospischil et al. 2009) which is based on the fusion between the concepts behind the VmD and STA methods. This VmT method is analogous to the VmD method and estimates excitatory and inhibitory conductances and their variances, but it does so by using a maximum likelihood estimation, and can thus be applied to single $V_{\rm m}$ traces.

In this chapter, we review different methods which are derived from stochastic processes, such as the VmD method (Sect. 8.2). We also detail two methods that can be applied to single $V_{\rm m}$ traces, based on power spectral density (PSD; Sect. 8.3) and STAs (STA; Sect. 8.4). Finally, in Sect. 8.5, we review a recently introduced method, the VmT method, which is aimed at extracting synaptic conductance parameters from single-trial $V_{\rm m}$ measurements.

8.2 The VmD Method: Extracting Conductances from Membrane Potential Distributions

In Sect. 7.4, we introduced the *expectation value approach* and detailed the mathematical derivation of the steady-state membrane potential distribution $\rho(V)$ of a passive membrane subjected to two independent stochastic noise sources describing inhibitory and excitatory synaptic conductances. In this section, we will demonstrate that, after suitable approximation of $\rho(V)$, this solution can be used to infer from a given membrane potential distribution (obtained, for instance, from in vivo intracellular recordings) various properties of the stochastic excitatory and inhibitory conductances, such as their means and variances (Rudolph and Destexhe 2004). Although this *VmD method* can only be applied in cases where two or more $V_{\rm m}$

recordings in the same conductance state are available, it has been successfully applied in a variety of studies, ranging from the characterization of synaptic noise in activated states in vivo during anesthesia (Rudolph et al. 2005; see Sect. 9.2) to the quantification of synaptic noise from intracellular recordings in awake and naturally sleeping animals (Rudolph et al. 2007; see Sect. 9.3).

8.2.1 The VmD Method

In the point-conductance model (Sect. 4.4), excitatory and inhibitory global synaptic conductances are each described by an OU stochastic process (Destexhe et al. 2001). These stochastic conductances, in turn, determine the V_m fluctuations through their (multiplicative) interaction with the membrane potential at the level of the V_m dynamics. Mathematically, this model is defined by the following set of three differential equations:

$$C\frac{dV}{dt} = -g_L(V-E_L) - g_e(V-E_e) - g_i(V-E_i) + I_{ext}$$

$$\frac{dg_{\{e,i\}}(t)}{dt} = -\frac{1}{\tau_e}\left[g_{\{e,i\}}(t) - g_{\{e,i\}0}\right] + \sqrt{\frac{2\sigma^2_{\{e,i\}}}{\tau_{\{e,i\}}}}\,\xi_{\{e,i\}}(t), \qquad (8.1)$$

where C denotes the membrane capacitance, I_{ext} a stimulation current, g_L the leak conductance and E_L the leak reversal potential. $g_e(t)$ and $g_i(t)$ are stochastic excitatory and inhibitory conductances with their respective reversal potentials E_e and E_i. The excitatory synaptic conductance is characterized by its mean g_{e0} and variance σ_e^2 as well as the excitatory time constant τ_e. In (8.1), $\xi_e(t)$ denotes a Gaussian white noise source with zero mean and unit SD. Similarly, the inhibitory conductance $g_i(t)$ is fully characterized by its parameters g_{i0}, σ_i^2 and τ_i, as well as the noise source $\xi_i(t)$. Here, all conductances are expressed in absolute units (e.g., in nS), but an equivalent formulation in terms of conductance densities is possible as well.

The model described by (8.1) has been thoroughly studied both theoretically and numerically. Indeed, different analytic approximations have been proposed to describe the steady-state distribution of the V_m activity of the point-conductance model (Rudolph and Destexhe 2003d, 2005; Richardson 2004; Lindner and Longtin 2006; for a comparative study, see Rudolph and Destexhe 2006b; see also Sect. 7.5). As demonstrated in Rudolph et al. (2004), one of these expressions (Rudolph and Destexhe 2003d, 2005) can be inverted. In turn, this allows to directly estimate the synaptic conductance parameters, specifically g_{e0}, g_{i0}, σ_e and σ_i, solely from experimentally obtained V_m distributions. The essential idea behind this *VmD method* (Rudolph et al. 2004) is to fit an analytic expression to the steady-state subthreshold V_m distribution obtained experimentally.

In the approach proposed by Rudolph and Destexhe (2003d, 2005), the membrane potential distribution $\rho(V)$ takes the form

$$\rho(V) = N \exp\left[A_1 \ln\left[\frac{u_e(V-E_e)^2}{C^2} + \frac{u_i(V-E_i)^2}{C^2}\right] + A_2 \arctan\left[\frac{u_e(V-E_e)+u_i(V-E_i)}{(E_e-E_i)\sqrt{u_e u_i}}\right]\right], \tag{8.2}$$

where $k_L = 2Cg_L$, $k_e = 2Cg_{e0}$, $k_i = 2Cg_{i0}$, $u_e = \sigma_e^2\tilde{\tau}_e$, $u_i = \sigma_i^2\tilde{\tau}_i$ and

$$A_1 = -\frac{k_L + k_e + k_i + u_e + u_i}{2(u_e + u_i)}$$
$$A_2 = 2C\frac{(g_{e0}u_i - g_{i0}u_e)(E_e - E_i) - g_L u_e(E_e - E_L) - g_L u_i(E_i - E_L) + I_{ext}(u_i + u_e)}{(E_e - E_i)\sqrt{u_e u_i}(u_e + u_i)} \tag{8.3}$$

Here, N denotes a normalization constant such that $\int_{-\infty}^{\infty} dV \rho(V) = 1$. $\tilde{\tau}_{\{e,i\}}$ are effective time constants given by (Rudolph and Destexhe 2005, see also Richardson 2004):

$$\tilde{\tau}_{\{e,i\}} = \frac{2\tau_{\{e,i\}}\tilde{\tau}_m}{\tau_{\{e,i\}} + \tilde{\tau}_m}, \tag{8.4}$$

where $\tilde{\tau}_m = C/(g_L + g_{e0} + g_{i0})$ is the effective membrane time constant.

Due to the multiplicative coupling of the stochastic conductances to the membrane potential, the V_m probability distribution (8.2) takes, in general, an asymmetric form. However, with physiologically relevant parameter values, $\rho(V)$ shows only small deviations from a Gaussian distribution, thus allowing an approximation by a symmetric distribution. To this end, one can express (8.2) by a Taylor series expansion of the exponent of $\rho(V)$ around its maximum $\bar{V}$:

$$\bar{V} = \frac{S_1}{S_0}, \tag{8.5}$$

with $S_0 = k_L + k_e + k_i + u_e + u_i$ and $S_1 = k_L E_L + k_e E_e + k_i E_i + u_e E_e + u_i E_i$.

By considering only the first and second order terms in this expansion, $\rho(V)$ one arrives at a simplified expression which takes the Gaussian form

$$\rho(V) = \frac{1}{\sqrt{2\pi\sigma_V^2}} \exp\left[-\frac{(V-\bar{V})^2}{2\sigma_V^2}\right] \tag{8.6}$$

with the SD given by

$$\sigma_V^2 = \frac{S_0^2(u_e E_e^2 + u_i E_i^2) - 2S_0 S_1 (u_e E_e + u_i E_i) + S_1^2(u_e + u_i)}{S_0^3}. \tag{8.7}$$

This expression provides an excellent approximation of the V_m distributions obtained from models and experiments (Rudolph et al. 2004), because the V_m distributions obtained experimentally show little asymmetry (for both Up-states and activated states as well as awake and natural sleep states; for specific examples, see Rudolph et al. 2004, 2005, 2007; Piwkowska et al. 2008).

The main advantage of this Gaussian form is that it can be inverted, which leads to expressions of the synaptic noise parameters as a function of the V_m measurements, specifically its mean $\bar{V}$ and standard deviation σ_V. By fixing the values of τ_e and τ_i, which are related to the decay time of synaptic currents and can be estimated from voltage-clamp data and/or current clamp by using power spectral analysis (see Sect. 8.3), one remains with four parameters to estimate: the means (g_{e0}, g_{i0}) and SDs (σ_e, σ_i) of excitatory and inhibitory synaptic conductances. The extraction these four conductance parameters from the membrane probability distribution (8.6) is, however, impossible because the latter is characterized by only two parameters ($\bar{V}$, σ_V).

To solve this problem, one considers two V_m distributions obtained at two different constant levels of injected current, I_{ext1} and I_{ext2}, in the same state of synaptic background activity, i.e., in states characterized by the same g_{e0}, g_{i0} and σ_e, σ_i. In this case, the Gaussian approximation of the two distributions provides two mean V_m values, $\bar{V}_1$ and $\bar{V}_2$, and two SD values, σ_{V1} and σ_{V2}. The resulting system of four equations relating V_m parameters with conductance parameters can now be solved for the four unknowns, yielding

$$\begin{aligned}
g_{\{e,i\}0} &= \frac{(I_{ext1} - I_{ext2})\left(\sigma_{V2}^2\left(E_{\{i,e\}} - \bar{V}_1\right)^2 - \sigma_{V1}^2\left(E_{\{i,e\}} - \bar{V}_2\right)^2\right)}{\left((E_e - \bar{V}_1)(E_i - \bar{V}_2) + (E_e - \bar{V}_2)(E_i - \bar{V}_1)\right)\left(E_{\{e,i\}} - E_{\{i,e\}}\right)\left(\bar{V}_1 - \bar{V}_2\right)^2} \\
&\quad - \frac{(I_{ext1} - I_{ext2})(E_{\{i,e\}} - \bar{V}_2) + \left(I_{ext2} - g_L(E_{\{i,e\}} - E_L)\right)\left(\bar{V}_1 - \bar{V}_2\right)}{\left(E_{\{e,i\}} - E_{\{i,e\}}\right)\left(\bar{V}_1 - \bar{V}_2\right)} \\
\sigma_{\{e,i\}}^2 &= \frac{2C(I_{ext1} - I_{ext2})\left(\sigma_{V1}^2\left(E_{\{i,e\}} - \bar{V}_2\right)^2 - \sigma_{V2}^2\left(E_{\{i,e\}} - \bar{V}_1\right)^2\right)}{\tilde{\tau}_{\{e,i\}}\left((E_e - \bar{V}_1)(E_i - \bar{V}_2) + (E_e - \bar{V}_2)(E_i - \bar{V}_1)\right)\left(E_{\{e,i\}} - E_{\{i,e\}}\right)\left(\bar{V}_1 - \bar{V}_2\right)^2}.
\end{aligned} \tag{8.8}$$

These relations allow a quantitative assessment of global characteristics of network activity in terms of mean excitatory (g_{e0}) and inhibitory (g_{i0}) synaptic conductances, as well as their respective variances (σ_e^2, σ_i^2), from the sole knowledge of the V_m distributions obtained at two different levels of injected current. This *VmD method* was tested using computational models and dynamic-clamp experiments (Rudolph et al. 2004), as shown in detail in the two next sections. This method was also used to extract conductances from different experimental conditions in vivo (Rudolph et al. 2005, 2007; Zou et al. 2005; see Chap. 9).

8.2.2 Test of the VmD Method Using Computational Models

The equations for estimating synaptic conductance mean and variance (8.8) are based on the extended analytical solution presented in Sect. 7.4 [see (8.2)]. The latter, in turn, is grounded on a simplified effective stochastic model of synaptic noise, namely, the point-conductance model (Sect. 4.4). Before an experimental application of the VmD method introduced in the last section, one, therefore, has to evaluate its validity in more realistic situations, as real neurons exhibit a spatially extended dendritic arborization and receive synaptic conductances through the transient activation of thousands of spatially distributed synaptic terminals instead of two randomly fluctuating stochastic synaptic channels.

Such an evaluation will be presented in the following by using computational models of increasing levels of complexity. The first model is the point-conductance model itself, in which the validity of the expressions (8.8) has to be assessed. Due to the equivalence of the underlying equations [compare (8.1) with (7.59)–(7.61)], here the closest correspondence between estimated and actual (i.e., calculated numerically) conductance parameters is expected.

Figure 8.1 illustrates the VmD method applied to the V_m activity of that model (Fig. 8.1a). Two different values of steady current injection (I_{ext1} and I_{ext2}) yield two V_m distributions (Fig. 8.1b, *gray*). These distributions are fitted with a Gaussian function in order to obtain the means and SDs of the membrane potential at both current levels. Incorporating the values $\bar{V}_1$, $\bar{V}_2$, σ_{V1}, and σ_{V2} into (8.8) yields values for the mean and SD of the synaptic noise, $g_{\{e,i\}0}$ and $\sigma_{\{e,i\}}$, respectively (Fig. 8.1c, *solid line*, and Fig. 8.1d). These conductance estimates can then be used to reconstruct the full analytic expression of the V_m distribution using (8.2), which is plotted in Fig. 8.1b (*solid lines*). Indeed, there is a very close match between the analytic estimates of $\rho(V)$ and numerical simulations (Fig. 8.1b, compare *gray areas* with *black solid*). This demonstrates not just that Gaussian distributions are an excellent approximation for the membrane potential distribution in the presence of synaptic noise, but also that the proposed method yields an excellent characterization of the synaptic noise and, thus, subthreshold neuronal activity.

This can also be seen by comparing the reconstructed conductance distributions (Fig. 8.1c, *black solid*) with the actual conductances recorded during the numerical simulation (Fig. 8.1c, *gray*). Distributions deduced from the estimated parameters

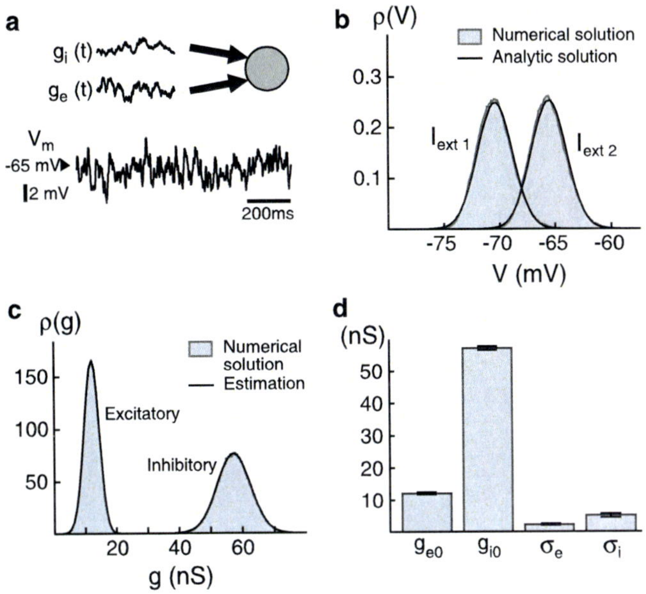

Fig. 8.1 Test of the VmD method for estimating synaptic conductances using the point-conductance model. (**a**) Example of membrane potential (V_m) dynamics of the point-conductance model. (**b**) V_m distributions used to estimate synaptic conductances. Those distributions (*gray*) were obtained at two different current levels I_{ext1} and I_{ext2}. The *solid lines* indicate the analytic solution based on the conductance estimates. (**c**) Comparison between the conductance distributions deduced from the numerical solution of the underlying model (*gray*) with the conductance estimates (*black solid*). (**d**) *Bar plot* showing the mean and standard deviation of conductances estimated from the membrane potential distributions. The error bars indicate the statistical significance of the estimates by using different Gaussian approximations of the membrane potential distribution in (**b**). Modified from Rudolph et al. (2004)

are, again, in excellent agreement with that of the numerical simulations. Thus, this first set of evaluating simulations shows that, at least for the point-conductance model, the proposed approach provides a method which allows to accurately estimate the mean and the variance of synaptic conductances from the sole knowledge of the (subthreshold) membrane potential activity of the cell.

A second test is the application of the VmD method to a more realistic model of synaptic noise, in which synaptic activity is generated by a large number of individual synapses releasing randomly according to Poisson processes. An example is shown in Fig. 8.2. Starting from the V_m activity (Fig. 8.2a), membrane potential distributions are constructed and fitted by Gaussians for two levels of injected current (Fig. 8.2b, *gray*). Estimates of the mean and variance of excitatory and inhibitory conductances are then obtained using (8.8). The analytic solution reconstructed from this estimate (Fig. 8.2b, *black solid*) is in excellent agreement with the numerical simulations of this model, as are the reconstructed conductance distributions based on this estimate (Fig. 8.2c, *black solid*) when compared to the

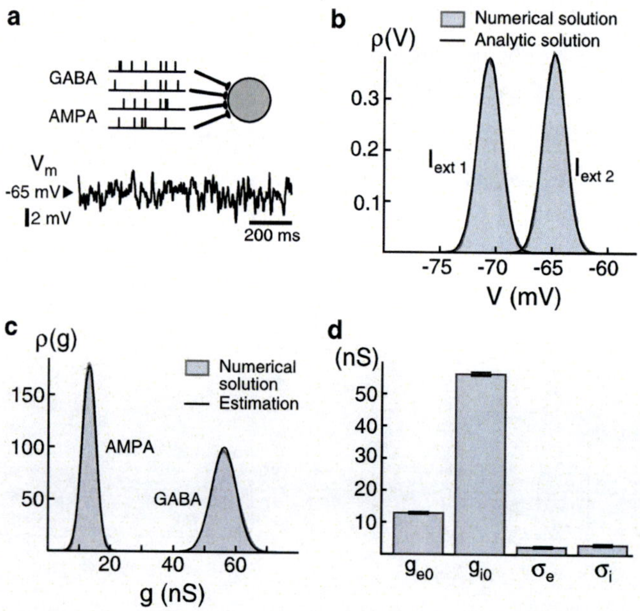

Fig. 8.2 Estimation of synaptic conductances using the VmD method applied to a single-compartment model with realistic synaptic inputs. (**a**) Example of membrane potential (V_m) time course in a single-compartment model receiving thousands of randomly activated synaptic conductances. (**b**) V_m distributions used to estimate conductances. Those distributions are shown at two different current levels I_{ext1}, I_{ext2} (*gray*). The *solid lines* indicate the analytic solution obtained based on the conductance estimates. (**c**) Comparison between the conductance distributions deduced from the numerical solution of the underlying model (*gray*) with those reconstructed from the estimated conductances (*black solid*). (**d**) *Bar plot* showing the mean and standard deviation of conductances estimated from membrane potential distributions. The error bars indicate the statistical significance of the estimates by using different Gaussian approximations of the membrane potential distribution in (**b**). Modified from Rudolph et al. (2004)

total conductance calculated for each type of synapse in the numerical simulations (Fig. 8.2c, *gray*; see Fig. 8.2d for quantitative values and error estimates). Thus, also in case of this more realistic model of synaptic background activity, which markedly differs from the point-conductance model, the estimates of synaptic conductances and their variances from voltage distributions are in excellent agreement with the values obtained numerically. In fact, this agreement can be expected due to the close correspondence between the conductance dynamics in both models (see Sect. 4.4).

A third, more severe test is to apply the VmD method to a compartmental model in which individual (random) synaptic inputs are spatially distributed in soma and dendrites. In a passive model of a cortical pyramidal neuron from layer VI (Fig. 8.3a), the V_m distributions obtained at two steady current levels are approximately symmetric (Fig. 8.3b, left panel, *gray*). Again, application of (8.8) in order to estimate synaptic conductances and their variances, leads to analytic V_m distributions $\rho(V)$ (Fig. 8.3b, left panel, *black solid*) which capture very well

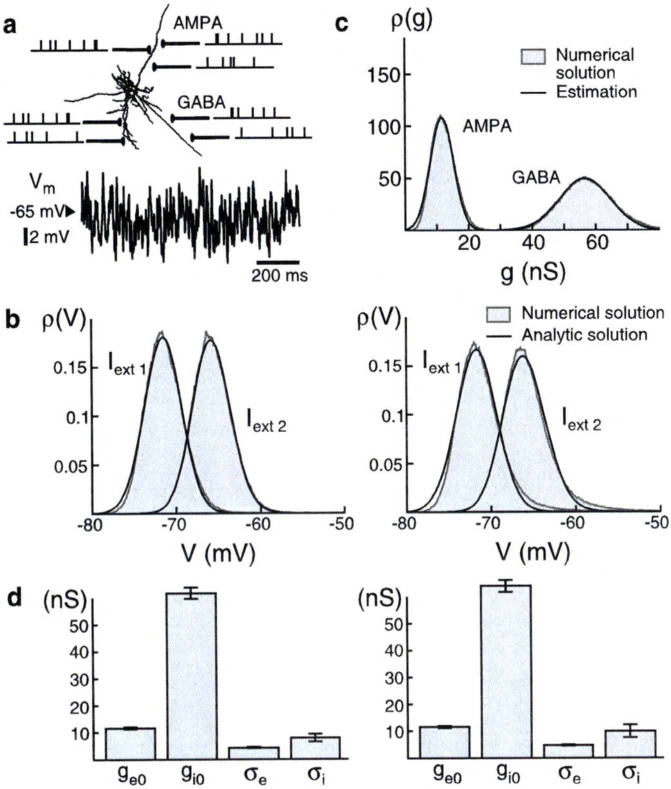

Fig. 8.3 Estimation of synaptic conductances from the membrane potential activity of a detailed biophysical model of synaptic background activity. (**a**) Example of the membrane potential (V_m) activity obtained in a detailed biophysical model of a layer VI cortical pyramidal neuron (scheme on *top*; same model as in Destexhe and Paré 1999). Synaptic background activity was modeled by the random release of 16,563 AMPA-mediated and 3,376 $GABA_A$-mediated synapses distributed in dendrites according to experimental measurements. (**b**) V_m distributions obtained in this model at two different current levels I_{ext1} and I_{ext2}. The *left panel* indicates the distributions obtained in a passive model, while in the *right panel*, these distributions are shown when the model had active dendrites (Na^+ and K^+ currents responsible for action potentials and spike-frequency adaptation, located in soma, dendrites, axon). In both cases, results from the numerical simulations (*gray*) and analytic expression (*black solid*), obtained by using the conductance estimates, are shown. (**c**) Histogram of the total excitatory and inhibitory conductances obtained from the model using an ideal voltage clamp (*gray*), compared to the distributions reconstructed from the conductance estimates based on V_m distributions. (**d**) Bar plot showing the mean and standard deviation of synaptic conductances estimated from V_m distributions. The error bars indicate the statistical significance of the estimates by using different Gaussian approximations of the membrane potential distribution in (**b**). *Left panel*: passive model; *right panel*: model with voltage-dependent conductances. The presence of voltage-dependent conductances had minor ($<10\%$) effects on the estimated conductance values. Modified from Rudolph et al. (2004)

the shape of the V_m distributions obtained numerically, although small deviations are visible at the hyperpolarized and depolarized tail of the distributions. The reconstructed conductance distributions (Fig. 8.3c, *black solid*) are, as well, in excellent agreement with the conductance distributions obtained in this model by using an ideal voltage clamp at the soma (Fig. 8.3c, *gray*; see method in Destexhe et al. 2001). The quantitative comparison of those values (Fig. 8.3d, left panel) shows that the estimation of conductance values from V_m distributions yields estimates comparable to those obtained from an ideal voltage clamp.

This agreement also shows that the dendritic filtering of synaptic inputs, caused by the spatial extension of the dendritic tree, does only have a minor impact on the conductance estimation. Moreover, due to the higher density of GABAergic synapses in the proximal region of cortical neurons, a slight bias in the estimates towards inhibitory conductance is expected. However, also in this case, the results indicate that this effect is small and only minimally impacts on the overall conductance estimates.

Finally, to assess the robustness of the VmD method to the presence of active dendrites, the method can be tested using a biophysical model incorporating voltage-dependent currents (I_{Na}, I_{Kd} for spike generation, and a slow voltage-dependent K^+ current for spike-frequency adaptation, a hyperpolarization-activated current I_h, a low-threshold Ca^{2+} current I_{CaT} and a A-type K^+ current I_{KA} with densities typical for cortical neurons. Here, deviations are expected, because the method is strictly based on passive neuronal dynamics [see (8.1)], which might be strongly altered by the presence of active channels at the site of the recording and the presence of regenerative dendritic spikes. Indeed, as it was shown in (Rudolph et al. 2004), after removing spikes in a broad (10 ms) time window, the subthreshold activity approximates well the passive dynamics, but also shows deviations in the membrane potential distribution at its hyperpolarized and depolarized tails (Fig. 8.3b, *gray*, compare left and right panels). However, these deviations only minimally impact on the mean and variance of the membrane potential obtained by Gaussian fits, which constitute the input for the VmD method. Applying (8.8) leads to estimates (Fig. 8.3b, right panel, *black solid*) which show more significant, albeit still small, deviations from the distributions drawn from the corresponding numerical simulations (Fig. 8.3b, right panel, *gray* and Fig. 8.4, *gray*). In general, the estimated values for synaptic conductances and their variance show larger errors, especially for σ_I (Fig. 8.3d, right panel), and yield V_m distributions that are slightly broader. However, these errors and deviations remain relatively small and the method still provides a good estimate of synaptic conductances, similar to or better than the one provided by ideal voltage clamp (Fig. 8.4).

8.2.3 *Test of the VmD Method Using Dynamic Clamp*

The VmD method can further tested in real neurons during active network activity, such as the recurrent activity (Up-states) occurring spontaneously in

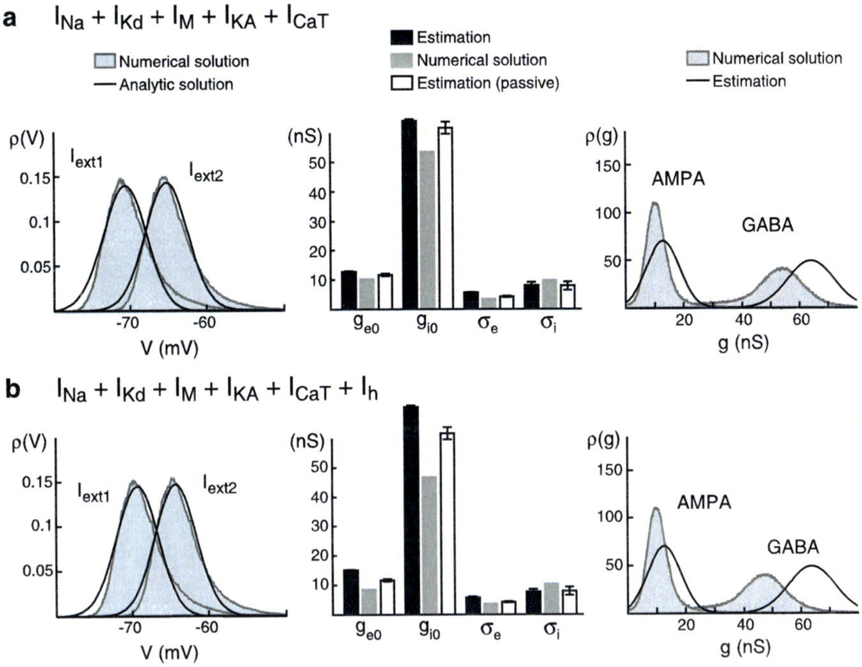

Fig. 8.4 Estimation of synaptic conductances from the membrane potential activity in detailed biophysical models with active dendrites. (**a**) Model with voltage-dependent currents for spike generation (I_{Na}, I_{Kd}), spike-frequency adaptation (I_M), A-type K^+ current (I_{KA}) and low-threshold Ca^{2+} current (I_{CaT}) in soma and dendrites. (**b**) Model with additional hyperpolarization-activated current I_h in soma and dendrites. In both cases, synaptic background activity was modeled by the random release of 16,563 AMPA-mediated and 3,376 $GABA_A$-mediated synapses distributed in dendrites according to experimental measurements. The *left panels* show the V_m distributions at two different current levels (I_{ext1} and I_{ext2}) obtained numerically (*gray*) and analytically using conductance estimates obtained with the VmD method. The comparison of the conductance estimates obtained with the VmD method (*middle panels, black bars*; (**a**) g_{e0} =12.8 nS, g_{i0} =64.0 nS, σ_e =5.7 nS, σ_i =8.1 nS; (**b**) g_{e0} =15.1 nS, g_{i0} =70.4 nS, σ_e =5.8 nS, σ_i =7.7 nS) and ideal somatic voltage clamp (*middle panels, gray bars*; Gaussian fits yield (**a**) g_{e0} =10.3 nS, g_{i0} =53.4 nS, σ_e =3.6 nS, σ_i =9.9 nS; (**b**) g_{e0} =8.3 nS, g_{i0} =46.8 nS, σ_e =3.6 nS, σ_i =10.3 nS) shows that the VmD method gave results which were much closer to the true values of synaptic conductances seen in the corresponding passive model (*middle panels, white bars*; g_{e0} =11.6 nS, g_{i0} =61.7 nS, σ_e =4.3 nS, σ_i =7.9 nS; see also Fig. 8.3). The *right panels* compare the histograms of the total excitatory and inhibitory conductances obtained numerically using ideal somatic voltage clamp (*gray*) and the Gaussian distribution based on the estimates using the VmD method. Modified from Rudolph et al. (2004)

ferret neocortical slices. Intracellularly, this activity consists in a depolarized V_m and relatively large amplitude V_m fluctuations (Fig. 8.5), as described previously (Sanchez-Vives and McCormick 2000). To test the method, Rudolph and colleagues (Rudolph et al. 2004) applied an online protocol (Fig. 8.5a) consisting of estimating synaptic conductances from "natural" Up-states (top traces) and comparing them

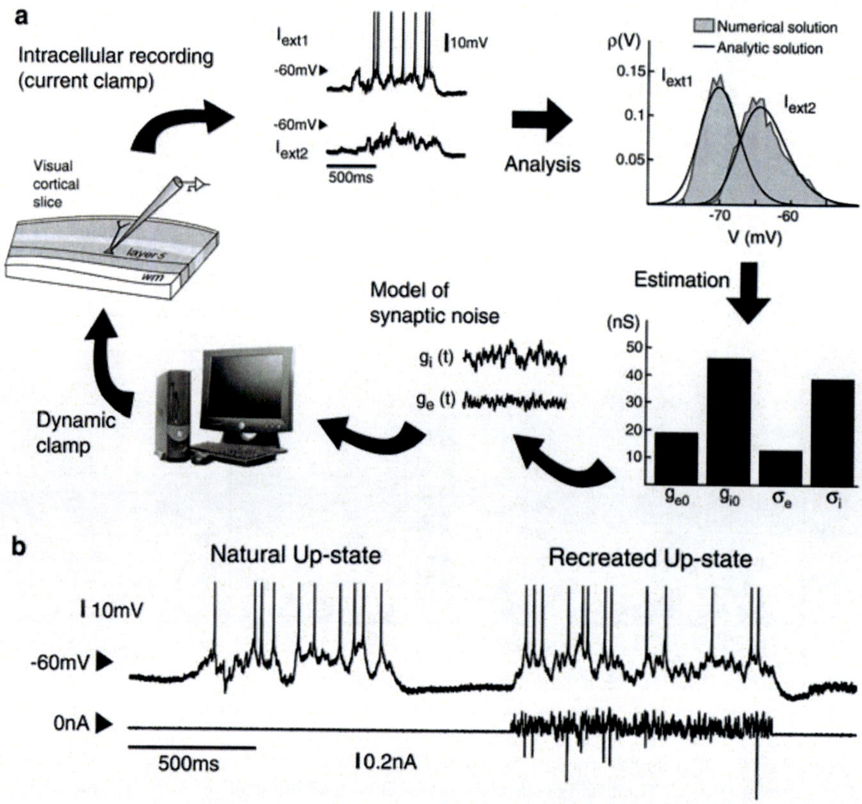

Fig. 8.5 VmD estimation of synaptic conductances from active states in vitro and dynamic-clamp recreation of active states. (**a**) Sketch of the procedure for conductance estimation and test of the estimates. *Top left*: spontaneous active network states (Up-states) were recorded intracellularly in ferret visual cortex slices at two different injected current levels (I_{ext1}, I_{ext2}). *Top right*: the V_{m} distributions (*gray*) were computed from experimental data and used to estimate synaptic conductances. The analytic solution for the V_{m} distribution using those conductance estimates is shown by *solid lines*. *Bottom right*: histogram of the mean and standard deviation of excitatory and inhibitory conductances obtained from the fitting procedure (*gray*). *Bottom left*: a dynamic-clamp protocol was used to inject stochastic conductances consistent with these estimates, therefore recreating artificial Up-states in the same neuron. (**b**) Example of natural and recreated Up-states in the same cell as in (**a**). This procedure recreated V_{m} activity similar to the active state, as shown by the close matching of the V_{m} fluctuations, depolarized level and discharge variability (natural Up-states: $\bar{V} = -67.2$ mV, $\sigma_V = 3.06$ mV, firing rate 14.3 Hz; recreated Up-states: $\bar{V} = -66.96$ mV, $\sigma_V = 2.6$ mV, firing rate 13 Hz). Modified from Rudolph et al. (2004)

to "artificial" Up-states obtained by dynamic-clamp injection of the estimated conductances in the same neuron (bottom traces). Again, these estimates (Fig. 8.5a) are obtained by computing the V_{m} distributions at two different current levels (Fig. 8.5a, top, *gray*) and fitting them using the Gaussian approximation (8.6). For the particular neuron shown in Fig. 8.5a, the estimated inhibitory synaptic conductances parameters are approximately twice as large as those for excitation (see bottom bar plots). These parameters are then used to generate conductances waveforms according to the stochastic process of (8.1) (*Model of synaptic noise* in Fig. 8.5a). These conductance waveforms are then reinjected in the same neuron (*Dynamic clamp* in Fig. 8.5a), leading to recreated Up-states, which are compared to natural Up-states. Using this protocol, Rudolph et al. (2004) tested several cells and showed that the recreated states had similar properties (average V_m, V_m fluctuations, firing behavior) as the natural Up-states (see example in Fig. 8.5b).

This comparison is shown in more detail in Fig. 8.6a. The natural Up-states (Fig. 8.6a, top trace) are used to estimate conductances using the same procedure as in Fig. 8.6a. In the particular cell shown in Fig. 8.6, excitatory and inhibitory conductances are approximately equal, but the variance of inhibitory conductance is relatively high (see bar plot). These estimated values for g_{e0}, g_{i0}, σ_e, and σ_i are then used to generate stochastic conductance waveforms (Fig. 8.6a, bottom traces). Injection of these conductance waveforms during quiescent states into the same cell leads to artificial Up-states whose V_m distributions are in excellent agreement with the natural Up-states (see $\rho(V)$ graph in Fig. 8.6a). This, this dynamic-clamp protocol shows that the network activity as predicted by the V_m distribution method is perfectly consistent with the natural network activity.

An alternative way for testing the VmD method is to analyze artificial Up-states generated by dynamic-clamp injection of known conductances (Fig. 8.6b). In this case, one can inject stochastic conductance waveforms according to a predefined choice of g_{e0}, g_{i0}, σ_e and σ_i parameters. The artificial Up-states obtained are then analyzed using the VmD decomposition method. Also here, this procedure was shown (Rudolph et al. 2004) to lead to estimated values in excellent agreement with the injected values (compare *black* and *white* in bar plot of Fig. 8.6b). Thus, the method provides an acceptable estimate of the injected conductances from the sole knowledge of the V_m activity.

8.2.4 Test of the VmD Method Using Current Clamp In Vitro

Several groups reported the successful application of the VmD method for estimating the synaptic background activity using current-clamp recordings. By comparing the estimated values for inhibitory and excitatory conductances with expected values, these studies further showed the validity of this method.

In a study by Greenhill and Jones (2007), intracellular recordings of Layer III neurons of the rat entorhinal cortex in vitro were used to simultaneously quantify synaptic inhibitory and excitatory synaptic noise as a function of drug-induced changes in the network activity or synaptic excitability. It was observed that blocking the activity dependent release using TTX led to an overall reduced and more balanced background activity less dominated by GABAergic receptors, a result confirmed in whole cell patch-clamp experiments by the same group. In agreement with this observation, application of the VmD method showed not just

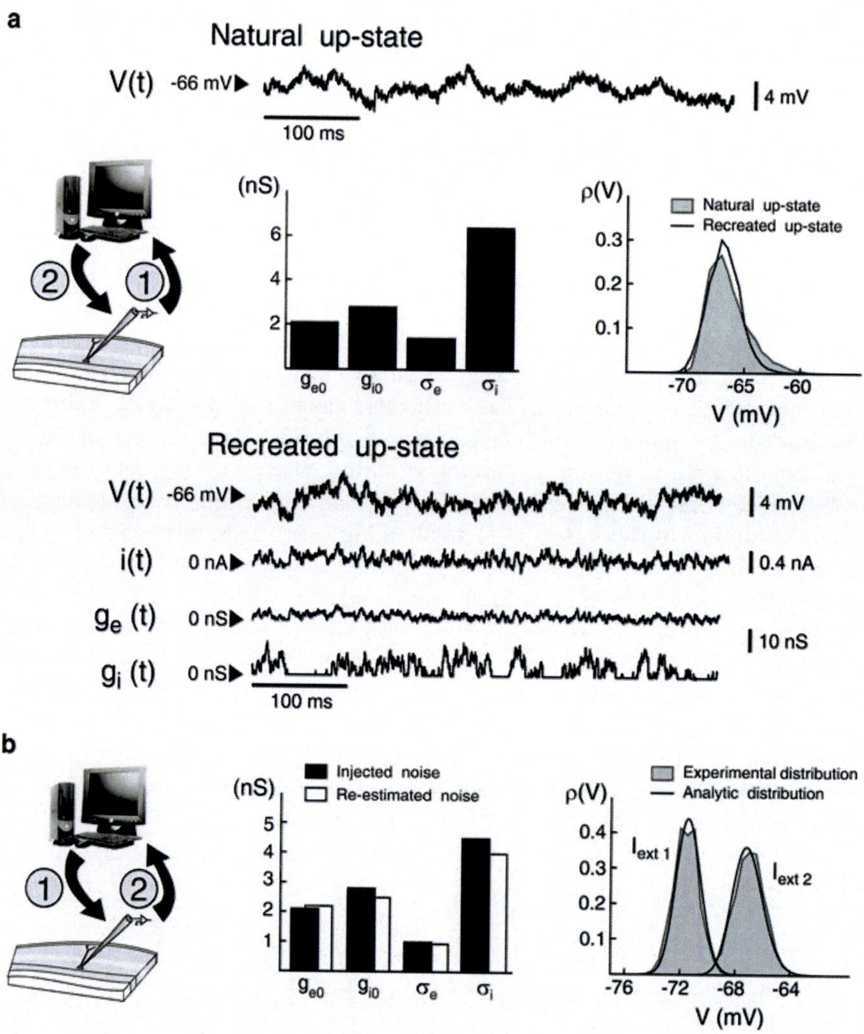

Fig. 8.6 Test of the VmD method using natural and recreated Up-states in vitro under dynamic clamp. (**a**) Reinjection of conductance estimates. The protocol used was similar as in Fig. 8.5 and consisted in first extracting conductances from natural Up-states (*arrow 1* in scheme) and recreating artificial Up-states in the same neuron (*arrow 2*). The natural Up-states (*top trace*) were used to compute V_m distribution and estimate conductances (*middle panels*). These values were then used to generate artificial synaptic noise using stochastically fluctuating conductances (g_e and g_i), which were injected in the same neuron using dynamic clamp (V_m activity shown as $V(t)$ in *bottom traces*). The V_m distribution obtained were in excellent agreement (*gray*: natural Up-states, $\bar{V}$=−66.87 mV, σ_V=1.53 mV; *black solid*: recreated Up-states, $\bar{V}$=−66.81 mV, σ_V=1.36 mV). (**b**) Analysis of artificial Up-states produced by dynamic-clamp injection of known conductances. In this protocol, stochastically varying synaptic conductances were first injected in the neuron (*arrow 1* in scheme). The resulting V_m activity was then used to re-estimate the conductances (*arrow 2*). The *middle panel* shows the injected (*black*) and re-estimated (*white*) conductances. The *right panel* shows the corresponding V_m distributions (*gray*: experimental; *black solid*: analytic prediction from the re-estimated parameters). There was an excellent agreement between all values. Injected conductances: g_{e0} =2.1 nS, g_{i0} =2.8 nS, σ_e =1.0 nS, σ_i =4.5 nS; re-estimated conductances: g_{e0} =2.2 nS, g_{i0} =2.5 nS, σ_e =0.94 nS, σ_i =4.0 nS. Modified from Rudolph et al. (2004)

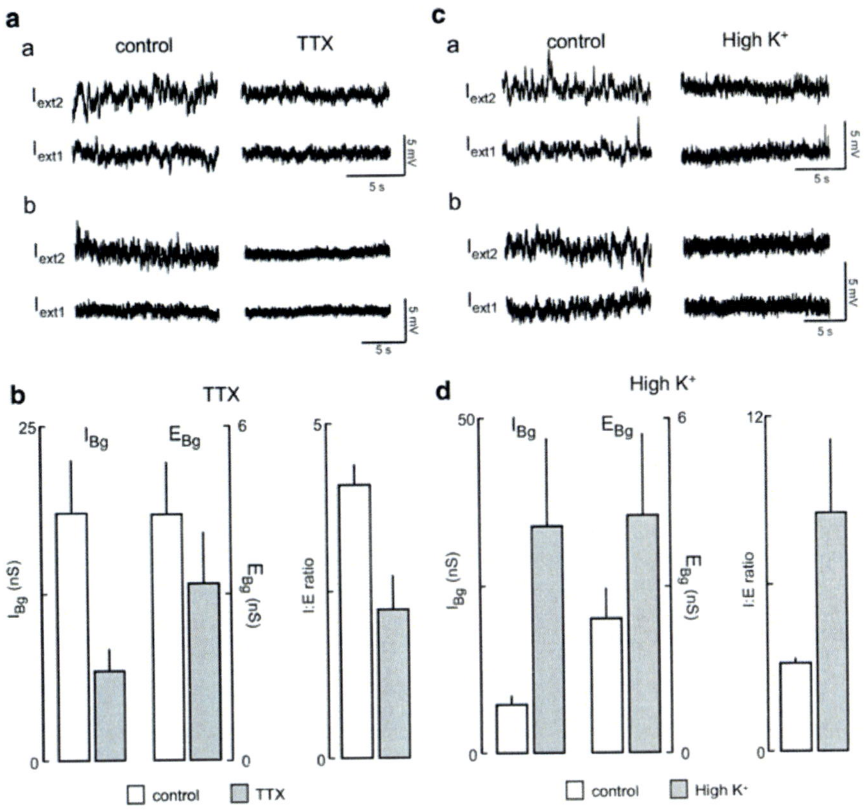

Fig. 8.7 Quantitative evaluation of synaptic noise from intracellular recordings of Layer III neurons in the rat entorhinal cortex in vitro using the VmD method under drug-induced changes in the activity-dependent release. (**a**) Changes in the membrane potential of two cells at two different current levels before and after application of TTX. (**b**) Estimation of excitatory and inhibitory mean conductances and their ratio in both conditions. Inhibition observed a larger reduction than excitatory conductances, with a consequent decrease in the I:E ratio in favor of excitation. (**c**) Examples of voltage recordings before and after increase of the potassium level within the tissue. (**d**) Synaptic noise estimates in both states show a dramatic increase in both excitatory and inhibitory conductances in favor of the former. Modified from Greenhill and Jones (2007)

a reduction in the overall synaptic conductance received my the cells, but also a reduction of the dominance of inhibitory conductances (Fig. 8.7a,b). In contrast, an increase in the overall network excitability by elevating the potassium level in the tissue resulted in an expected increase of both excitatory and inhibitory synaptic conductances, with a clear increase in the dominance of inhibition (Fig. 8.7c,d).

A similar validation of noise estimates obtained via the VmD method was observed when AMPA and $GABA_A$ receptors were directly blocked by application of NBQX and bicuculline, respectively (Greenhill and Jones 2007). In the former case, a more pronounced reduction of excitatory conductances and a general shift of the ratio between inhibition and excitation towards inhibitory dominance could be evidenced. In contrast, blocking GABAergic receptors had the opposite effect, resulting in a clear reduction of inhibitory conductances estimated by utilizing the VmD method. From these experimental results, it was concluded that the proposed VmD method indeed provides a viable and useful approach for the quantitative assessment of synaptic noise and the study of the function of synaptic noise in cortical networks (Greenhill and Jones 2007).

The VmD method was also applied to quantitatively assess the balance between excitatory and inhibitory synaptic conductances in in vitro rodent hippocampal slices exhibiting spontaneous, basal sharp waves (Ho et al. 2009). The latter were split into "sparse" and "synchronous" periods at the intracellular level. In this study, it was found that inhibitory conductances are dominant in both pyramidal cells and putative interneurons, and that the variance of inhibitory conductances dominate during the "synchronous" periods. Furthermore, it was observed that transitions between "sparse" and "synchronous" sharp wave states is accompanied by a several-fold increase in the inhibitory dominance. These results are in general agreement with other results reporting dominance and a decisive functional role of inhibitory synaptic conductances (Rudolph et al. 2007), and show that the VmD method provides a simple way to quantitatively evaluate synaptic noise and even their transient changes from the sole knowledge of intracellular activity.

8.3 The PSD Method: Extracting Conductance Parameters from the Power Spectrum of the V_m

In Sect. 4.4.4, it was shown that, under some assumptions, one can obtain analytic estimates of the PSD of synaptic conductances which could, in principle, be used to analyze experimental data. However, these estimates are only applicable to voltage-clamp experiments. Most experiments (in particular in vivo recordings) are performed in current-clamp mode, in which case the membrane potential (V_m) activity is recorded. One, therefore, needs methods to extract the characteristics of the synaptic inputs under current clamp by analyzing the V_m fluctuations, which is the guiding idea behind the *PSD method*. The theoretical background of this method will be first presented below, followed by a description of its tests using both numerical and dynamic-clamp methods (see details in Destexhe and Rudolph 2004; Piwkowska et al. 2008).

8.3.1 The PSD Method

Assuming a passive isopotential cell, the time evolution of the voltage is given by

$$C_m \frac{dV}{dt} = -g_{leak}(V - E_{leak}) - \sum_j g_j(t)(V - E_{syn}), \qquad (8.9)$$

where V is the membrane potential, $C_\mathrm{m} = 1\,\mu\mathrm{F/cm}^2$ is the specific membrane capacitance, $g_\mathrm{leak} = 0.1\,\mathrm{mS/cm}^2$ and $E_\mathrm{leak} = -70\,\mathrm{mV}$ are the leak conductance and reversal potential, respectively. The membrane is subject to a large number of conductance-based synaptic inputs, $g_j(t)$, with corresponding reversal potential E_syn.

Taking the Fourier transform of the membrane equation yields

$$i\omega C_\mathrm{m} V(\omega) = -g_\mathrm{leak}[V(\omega) - E_\mathrm{leak}\delta(\omega)] - \sum_j g_j(\omega) * [V(\omega) - E_\mathrm{syn}], \quad (8.10)$$

where $*$ is the convolution operator. This equation is, in general, not solvable, precisely because of this convolution being a direct consequence of the multiplicative aspect of conductances.

To mathematically assess this equation, one can make an *effective leak approximation* (e.g., Brunel and Wang 2001; Rudolph and Destexhe 2003b). This approximation is justified by the fact that, in high-conductance states, it was estimated that the total membrane conductance is always about 2 orders of magnitude larger than the (quantal) conductance of single synapses (Destexhe and Paré 1999). In this case, the voltage deflection due to isolated inputs is small compared to the distance to reversal potential, and one can consider the driving force as approximately constant. However, it is still necessary to take into account the high-conductance state of the membrane, which is the temporal average of the sum over all membrane conductances. The membrane equation, then, becomes

$$C_\mathrm{m}\frac{\mathrm{d}V}{\mathrm{d}t} = -g_\mathrm{T}(V - \bar{V}) - \sum_j (g_j(t) - \bar{g}_j)\,(\bar{V} - E_\mathrm{syn}), \quad (8.11)$$

where $\bar{g}_j$ is the average conductance at each synapse,

$$g_\mathrm{T} = g_\mathrm{leak} + \sum_j \bar{g}_j$$

is the total average membrane conductance, and $\bar{V}$ is the average membrane potential. Here, the driving force $(\bar{V} - E_\mathrm{syn})$ is now constant, which is, therefore, equivalent to approximate conductance-based inputs as current-based inputs, but arising on top of a large overall membrane conductance.

Taking the Fourier transform of (8.11), one obtains for $\omega > 0$

$$V(\omega) = \frac{\sum_j g_j(\omega)(E_\mathrm{syn} - \bar{V})}{g_\mathrm{T} + i\omega\, C_\mathrm{m}}. \quad (8.12)$$

The PSD is then given by:

$$P_V(\omega) = |V(\omega)|^2 = \frac{\left|\sum_j g_j(\omega)(E_\mathrm{syn} - \bar{V})\right|^2}{g_\mathrm{T}^2 + \omega^2\, C_\mathrm{m}^2}. \quad (8.13)$$

If all synaptic inputs are based on the same quantal events, then $g_j(\omega) = g(\omega)$, and incorporating the "effective" membrane time constant $\tilde{\tau}_m = C_m/g_T$, one can write

$$P_V(\omega) = \frac{C|g(\omega)|^2}{1+\omega^2\tilde{\tau}_m^2}, \tag{8.14}$$

where $C = \lambda\ (E_{syn} - \bar{V})^2/g_T^2$.

Thus, the PSD of the membrane potential is here expressed as a "filtered" version of the PSDs of synaptic conductances, where the filter is given by the RC circuit of the membrane in the high-conductance state. Taking the example of two-state kinetic synapses (see (4.11) in Chap. 4), the PSD of the membrane potential is given by

$$P_V(\omega) = \frac{C'}{(1+\omega^2\tau_{syn}^2)(1+\omega^2\tilde{\tau}_m^2)}, \tag{8.15}$$

where $\tau_{syn} = 1/\beta$ and $C' = g_{max}^2\alpha^2 C/\beta^2$.

When both excitatory and inhibitory inputs are present (Fig. 8.8d), the theoretical PSD is obtained by a sum of two expressions similar to (8.15):

$$P_V(\omega) = \frac{1}{1+\omega^2\tilde{\tau}_m^2}\left[\frac{A_e\ \tau_e}{1+\omega^2\tau_e^2} + \frac{A_i\ \tau_i}{1+\omega^2\tau_i^2}\right], \tag{8.16}$$

where A_e and A_i are amplitude parameters. This five parameter template is used to provide estimates of the parameters τ_e and τ_i (assuming that $\tilde{\tau}_m$ has been measured). A further simplification consists in assuming that $A_e = A_i$, which can be used for fitting in vivo data, as shown in the next sections.

8.3.2 *Numerical Tests of the PSD Method*

As in the case of the VmD method (Sect. 8.2), a first assessment of the validity of the PSD method is done in the simplest model, namely a single compartment model receiving thousands of Poisson-distributed synaptic conductances. In this case, computing (8.15) numerically for models receiving only excitatory synapses (Fig. 8.8b), or only inhibitory synapses (Fig. 8.8c), shows that the prediction provided by (8.15) (Fig. 8.8b,c, *black solid*) is in excellent agreement with the numerical simulations (*gray*).

Thus, these simulations clearly show that the PSD of the V_m can be well predicted theoretically. These expressions can, therefore, at least in principle, be used to fit the parameters of the kinetic models of synaptic currents from the PSD of the V_m activity. Not all parameters, however, can be estimated. The reason is that several parameters appear combined (such as in the expression of C' and C'' above), in

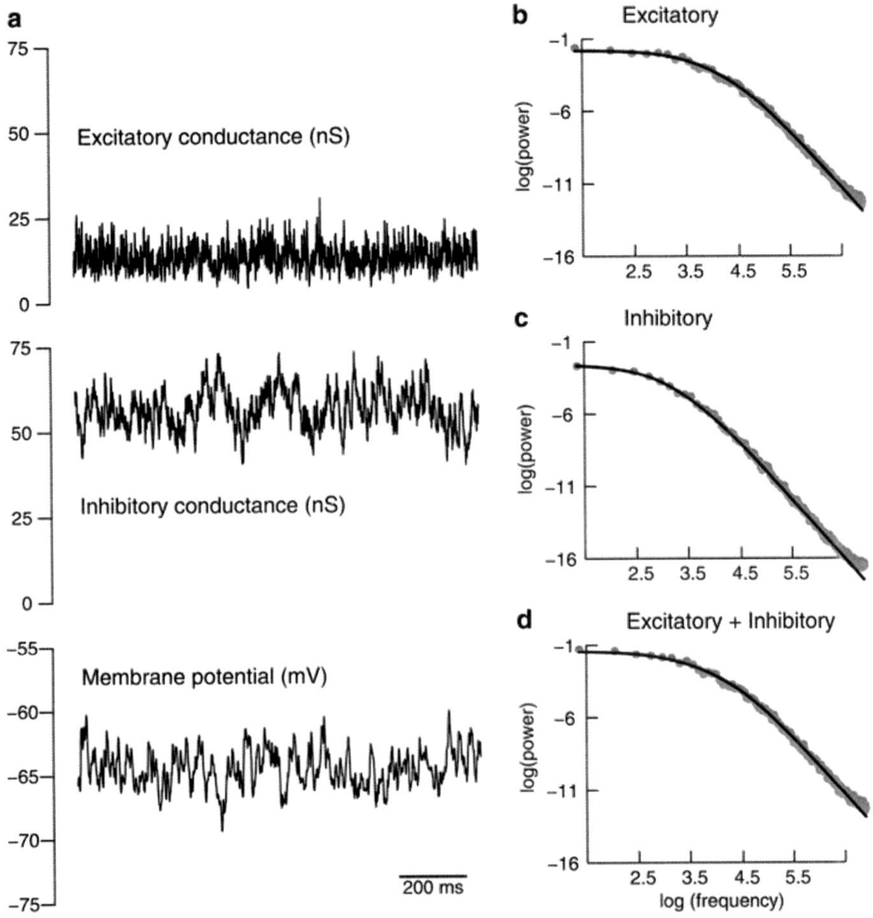

Fig. 8.8 Power spectral estimates of the membrane potential in a model with random synaptic inputs. (**a**) Simulation of a single-compartment neuron receiving a large number of randomly-releasing synapses (4,470 AMPA-mediated and 3,800 $GABA_A$-mediated synapses, releasing according to independent Poisson processes of average rate of 2.2 and 2.4 Hz, respectively). The total excitatory (AMPA) conductance, the total inhibitory ($GABA_A$) conductance and the membrane potential are shown from *top* to *bottom* in the first second of the simulation. (**b**) Power spectral density (PSD) calculated for a model with only excitatory synapses (inhibitory synapses were replaced by a constant equivalent conductance of 56 nS). (**c**) PSD calculated for a model with only inhibitory synapses (excitatory synapses were replaced by a constant equivalent conductance of 13 nS). (**d**) PSD calculated for the model shown in (**a**), in which both excitatory and inhibitory synapses participated to membrane potential fluctuations. In (**b–d**), the continuous curves show the theoretical prediction from (8.15). All synaptic inputs were equal (quantum of 1.2 nS for AMPA and 0.6 nS for $GABA_A$) and were described by two-state kinetic models. Modified from Destexhe and Rudolph (2004)

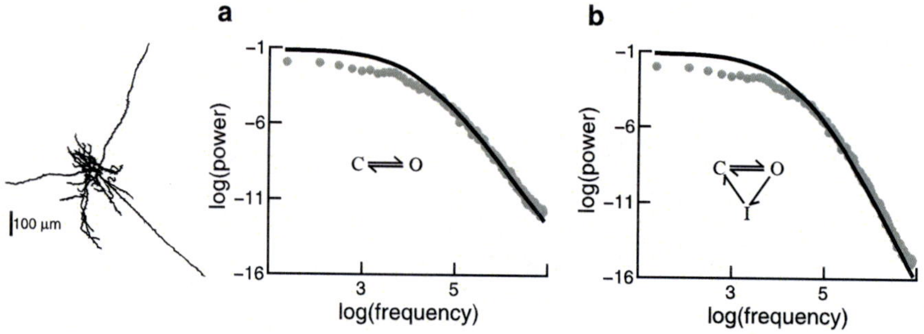

Fig. 8.9 Power spectrum of models with synaptic inputs distributed in dendrites. Simulations of a passive compartmental model of cat layer VI pyramidal neuron (*left*) receiving a large number of randomly releasing synapses (16,563 AMPA-mediated and 3,376 $GABA_A$-mediated synapses, releasing according to independent Poisson processes of average rate of 1 and 5.5 Hz, respectively). This model simulates the release conditions during high-conductance states in vivo (see details in Destexhe and Paré 1999). (**a**) Power spectrum obtained for the somatic V_m in this model when synaptic inputs were simulated by two-state kinetic models (*gray*). The *black curve* shows the theoretical PSD of the V_m obtained in an equivalent single-compartment model. (**b**) Same simulation and procedure, but using three-state (bi-exponential) synapse models. In both cases, the decay of the PSD at high frequencies was little affected by dendritic filtering. Modified from Destexhe and Rudolph (2004)

which case they cannot be distinguished from the PSD alone. Nevertheless, it is possible to fit the different time constants of the system, as well as the asymptotic scaling behavior at high frequencies.

These considerations are valid, however, only for a single-compartment model, and they may not apply to the case of synaptic inputs distributed in dendrites. Because of the strong low-pass filtering properties of dendrites, it is possible that distributed synaptic inputs do affect the scaling behavior of the PSD of the V_m. To investigate this point, a more detailed model of synaptic background activity in vivo (Destexhe and Paré 1999) can be used. The resulting PSD of the V_m in a high-conductance state due to the random release of excitatory and inhibitory synapses distributed in soma and dendrites is shown in Fig. 8.9, *gray*). This PSD can then be compared to the theoretical expressions obtained above for a single-compartment model with equivalent synaptic inputs (Fig. 8.9, *black solid*). Surprisingly, there is little effect of dendritic filtering on the frequency scaling of the PSD of the V_m. In particular, the scaling at large frequencies is only minimally affected (compare *black solid* with *gray* in Fig. 8.9). These simulations, therefore, suggest that the spectral structure of synaptic noise, as seen from the V_m, could indeed provide a reliable method to yield information about the underlying synaptic inputs.

Finally, the theoretical expression for the PSD also matches the V_m fluctuations produced by the point-conductance model (8.1), as shown in Fig. 8.10a. This agreement constitutes a confirmation of the equivalence of the point-conductance model with a model of thousands of Poisson-distributed synaptic conductances, as shown previously (Destexhe and Rudolph 2004; see Sect. 4.4.4).

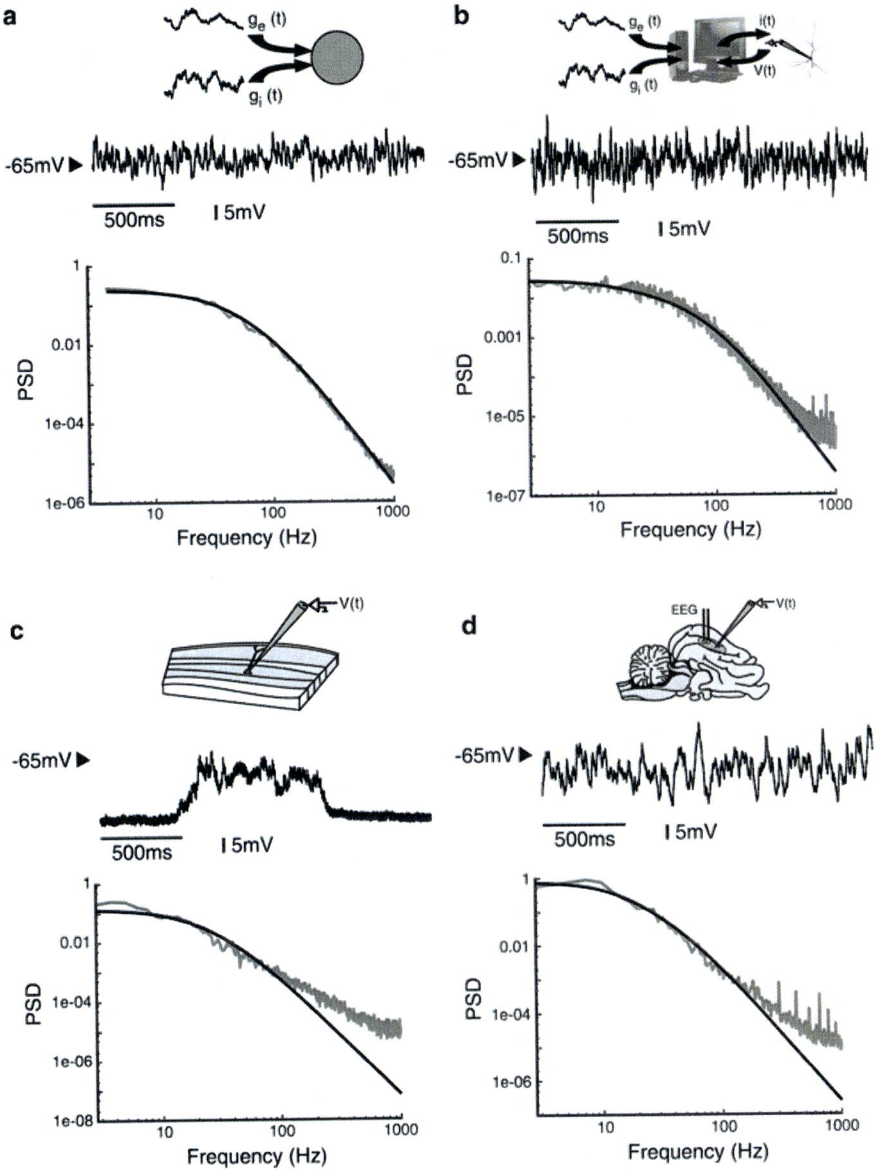

Fig. 8.10 Fit of the synaptic time constants to the power spectrum of the membrane potential. (**a**) Comparison between the analytic prediction (8.15; *black solid*) and the PSD of the V_m for a single-compartment model [(8.1); *gray*] subject to excitatory and inhibitory fluctuating conductances ($\tau_e = 3$ ms and $\tau_i = 10$ ms). (**b**) PSD of the V_m activity in a guinea-pig visual cortex neuron (*gray*), where the same model of fluctuating conductances as in (**a**) was injected using dynamic clamp. The *black curve* shows the analytic prediction using the same parameters as the injected conductances ($\tau_e = 2.7$ ms and $\tau_i = 10.5$ ms). (**c**) PSD of V_m activity obtained in a ferret visual cortex neuron (*gray*) during spontaneously occurring Up-states. The PSD was computed by averaging PSDs calculated for each Up-state. The *black curve* shows the best fit of the analytic expression with $\tau_e = 3$ ms and $\tau_i = 10$ ms. (**d**) PSD of V_m activity recorded in cat association cortex during activated states in vivo. The *black curve* shows the best fit obtained with $\tau_e = 3$ ms and $\tau_i = 10$ ms. Panels (**b**) and (**c**) Modified from Piwkowska et al. (2008); Panel (**d**) modified from Rudolph et al. (2005)

8.3.3 Test of the PSD Method in Dynamic Clamp

The method described above can also be applied to the PSD of V_m fluctuations obtained by controlled, dynamic-clamp fluctuating conductance injection in cortical neurons in vitro utilizing a new, high resolution electrode compensation technique (Brette et al. 2007b, 2008, 2009; see Sect. 6.5). In this case, the scaling of the PSD conforms to the prediction (Fig. 8.10b): the theoretical template (8.16) can provide a very good fit of the experimentally obtained PSD, up to around 400 Hz, where recording noise becomes important. It was found that both templates, (8.15) or (8.16), provide equally good fits. This, thus, shows that the analytic expression for the PSD is consistent not only with models, but also with conductance injection in real neurons in vitro.

Piwkowska and colleagues applied the same procedure as well to the analysis of V_m fluctuations resulting from real synaptic activity (Piwkowska et al. 2008), during Up-states recorded in vitro (Fig. 8.10c) and during sustained network activity in vivo (Fig. 8.10d). In this case, however, it is apparent that the experimental PSDs cannot be fitted with the theoretical template as nicely as for dynamic-clamp data (Fig. 8.10b). The PSD presents a frequency-scaling region at high frequencies, and scales as $1/f^{\alpha}$ with a different exponent α as predicted by the theory (see Fig. 8.10c,d).

The analytic expression (8.15) predicts that the PSD should scale as $1/f^4$ at high frequencies, but performed experiments showed that the exponent α is obviously lower than that value. In a recent study, Bedard and Destexhe investigated reasons for this difference and found that a possible origin is the nonideal aspect of the membrane capacitance, which was the only factor capable of reproducing the results (Bédard and Destexhe 2008). However, incorporation of this findings into the presented approach requires a modification of the cable equations.

This difference, of course, compromises the accuracy of the method to estimate τ_e and τ_i in situations of real synaptic bombardment. Nevertheless, as shown in (Piwkowska et al. 2008), including the values of $\tau_e = 3$ ms and $\tau_i = 10$ ms provided acceptable fits to the low-frequency (<100 Hz) part of the spectrum (Fig. 8.10c,d, *black solid*). However, in this case, small variations (around 20–30%) around these values of τ_e and τ_i yield equally good fits (see also Rudolph et al. 2005). Thus, the method cannot be used to precisely estimate those parameters, but can nevertheless be used to broadly estimate them with an error of the order of 30%.

8.4 The STA Method: Calculating Spike-Triggered Averages of Synaptic Conductances from V_m Activity

In this section, we describe a method to estimate the average conductance variations related to spikes in the neuron under intense synaptic activity. In such a case, one needs to extract the STA conductance patterns prior to the spike. This is a

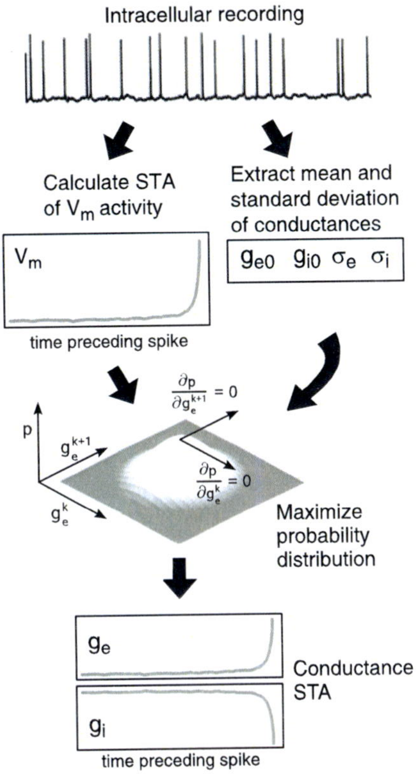

Fig. 8.11 Scheme of the spike-triggered average (STA) method to extract spike-triggered average conductances from membrane potential activity. Starting from an intracellular recording (*top*), the STA membrane potential (V_m) is computed (*left*). From the STA of the V_m, by discretizing the time axis, it is possible to estimate the STA of conductances (*bottom*) by maximizing a probability distribution. This step requires knowledge of the values of the average conductances and their standard deviations (g_{e0}, g_{i0}, σ_e, σ_i, respectively, which must be extracted independently (*right*). Modified from Pospischil et al. (2007)

challenging analysis, because it cannot be done with standard methods, given the fact that spikes are not visible in voltage clamp. This type of analysis must be done in current clamp mode, but in this case the conductance time courses are not easily extracted from the V_m activity. We describe below a procedure to extract STAs from V_m activity, the *STA method*, after which we present tests of this method using numerical simulations and intracellular recordings in dynamic clamp.

8.4.1 The STA Method

The principle of the STA method is to extract spike-related conductances solely based on the knowledge of V_m activity using a maximum likelihood procedure, as illustrated in Fig. 8.11. First, the STA of the V_m is computed from the intracellular recordings. Next, by discretizing the time axis, one estimates the "most likely" conductance time courses that are compatible with the observed STA of V_m. Due to the symmetry of their distribution, the average conductance time courses coincide with the most likely ones, so integration over the entire stimulus space can be replaced by a differentiation and subsequent solution of a system of linear equations. Solving this system provides an estimate of the average conductance time courses.

The first step of the STA method is a discretization of the time axis. With this approach, a probability distribution can be constructed whose maximum gives the most likely conductance path compatible with the STA of the V_m. This maximum is determined by a system of linear equations, which is solvable if the means and variances of conductances are known. To obtain the latter, for example the VmD method (Sect. 8.2) could be employed (Rudolph et al. 2004).

In the mathematical derivation of the STA method, one starts again from the point-conductance model

$$C\frac{\mathrm{d}V(t)}{\mathrm{d}t} = -g_{\mathrm{L}}(V(t)-V_{\mathrm{L}}) - g_{\mathrm{e}}(t)(V(t)-V_{\mathrm{e}}) - g_{\mathrm{i}}(t)(V(t)-V_{\mathrm{i}}) + I_{\mathrm{DC}},$$

$$\frac{\mathrm{d}g_s(t)}{\mathrm{d}t} = -\frac{1}{\tau_s}(g_s(t)-g_{s0}) + \sqrt{\frac{2\sigma_s^2}{\tau_s}}\xi_s(t), \tag{8.17}$$

where g_{L}, $g_{\mathrm{e}}(t)$, and $g_{\mathrm{i}}(t)$ are the conductances of leak, excitatory, and inhibitory currents, V_{L}, V_{e}, V_{i} are their respective reversal potentials, C is the capacitance and I_{DC} a constant current. The subscript s in (8.17) can take the values e, i, which in turn indicate the respective excitatory or inhibitory channel. g_{s0} and σ_s denote the mean and SD of the conductance distributions, $\xi_s(t)$ are Gaussian white noise processes with zero mean and unit SD.

In this model, the voltage STA, defined as an average over an ensemble of event-triggered voltage traces, is considered. Its relation to the conductance STAs is determined by the ensemble average of (8.17). In general, there is a strong correlation (or anti-correlation) between $V(t)$ and $g_s(t)$ in time. However, it is safe to assume that there is no such correlation across the ensemble, since the noise processes $\xi_s(t)$ corresponding to each realization are uncorrelated. Also, the ensemble average is commutative with the time derivative. Thus, one can rewrite (8.17) to obtain

$$\begin{aligned}\frac{\mathrm{d}\langle V(t)\rangle_x}{\mathrm{d}t} &= -\frac{1}{\tau_{\mathrm{L}}}(\langle V(t)\rangle_x - V_{\mathrm{L}})\\ &\quad -\frac{\langle g_{\mathrm{e}}(t)\rangle_x}{C}(\langle V(t)\rangle_x - V_{\mathrm{e}}) - \frac{\langle g_{\mathrm{i}}(t)\rangle_x}{C}(\langle V(t)\rangle_x - V_{\mathrm{i}}) + \frac{I_{\mathrm{DC}}}{C},\\ \frac{\mathrm{d}\langle g_s(t)\rangle_x}{\mathrm{d}t} &= -\frac{1}{\tau_s}(\langle g_s(t)\rangle_x - g_{s0}) + \sqrt{\frac{2\sigma_s^2}{\tau_s}}\langle\xi_s(t)\rangle_x,\end{aligned} \tag{8.18}$$

where $\tau_{\mathrm{L}} = C/g_{\mathrm{L}}$ and $\langle . \rangle_x$ denotes the ensemble average. In other words, the time evolution equations (8.17) also hold in terms of ensemble averages. In the following, the bracket notation is dropped for legibility (ensemble averaged quantities are used, unless otherwise stated.

Discretizing the first equation in (8.18) in time with a step-size Δt and solving it for g_{i}^k yields

$$g_{\mathrm{i}}^k = -\frac{C}{V^k - V_{\mathrm{i}}}\left\{\frac{V^k - V_{\mathrm{L}}}{\tau_{\mathrm{L}}} + \frac{g_{\mathrm{e}}^k(V^k - V_{\mathrm{e}})}{C} + \frac{V^{k+1} - V^k}{\Delta t} - \frac{I_{\mathrm{DC}}}{C}\right\}. \tag{8.19}$$

Since the series V^k for the voltage STA is known, g_{i}^k has become a function of g_{e}^k. In the same way, one can solve the second equation in (8.18) for ξ_s^k, which then become Gaussian distributed random numbers,

$$\xi_s^k = \frac{1}{\sigma_s}\sqrt{\frac{\tau_s}{2\Delta t}}\left(g_s^{k+1} - g_s^k\left(1 - \frac{\Delta t}{\tau_s}\right) - \frac{\Delta t}{\tau_s} g_{s0}\right). \tag{8.20}$$

There is a continuum of combinations $\{g_{\mathrm{e}}^{k+1}, g_{\mathrm{i}}^{k+1}\}$ that can advance the membrane potential from V^{k+1} to V^{k+2}, each pair occurring with a probability

$$\begin{aligned} p^k &:= p(g_{\mathrm{e}}^{k+1}, g_{\mathrm{i}}^{k+1} | g_{\mathrm{e}}^k, g_{\mathrm{i}}^k) = \frac{1}{2\pi}\mathrm{e}^{-\frac{1}{2}(\xi_{\mathrm{e}}^{k2} + \xi_{\mathrm{i}}^{k2})} = \frac{1}{2\pi}\mathrm{e}^{-\frac{1}{4\Delta t}X^k}, \\ X^k &= \frac{\tau_{\mathrm{e}}}{\sigma_{\mathrm{e}}^2}\left(g_{\mathrm{e}}^{k+1} - g_{\mathrm{e}}^k\left(1 - \frac{\Delta t}{\tau_{\mathrm{e}}}\right) - \frac{\Delta t}{\tau_{\mathrm{e}}} g_{\mathrm{e}0}\right)^2 \\ &\quad + \frac{\tau_{\mathrm{i}}}{\sigma_{\mathrm{i}}^2}\left(g_{\mathrm{i}}^{k+1} - g_{\mathrm{i}}^k\left(1 - \frac{\Delta t}{\tau_{\mathrm{i}}}\right) - \frac{\Delta t}{\tau_{\mathrm{i}}} g_{\mathrm{i}0}\right)^2. \end{aligned} \tag{8.21}$$

Here, (8.20) was used. Also, because of (8.19), g_{e}^k and g_{i}^k are not independent and p^k is, thus, a unidimensional distribution only. Given initial conductances g_s^0, one can now write down the probability p for certain series of conductances $\{g_s^j\}_{j=0,\ldots,n}$ to occur that reproduce a given voltage trace $\{V^l\}_{l=1,\ldots,n+1}$:

$$p = \prod_{k=0}^{n-1} p^k. \tag{8.22}$$

Due to the symmetry of the distribution p, the average paths of the conductances coincide with the most likely ones, so the cumbersome task of solving nested Gaussian integrals can be circumvented. Instead, in order to determine the conductance series with extremal likelihood, one solves the n-dimensional system of linear equations

$$\left\{\frac{\partial X}{\partial g_{\mathrm{e}}^k} = 0\right\}_{k=1,\ldots,n}, \tag{8.23}$$

where $X = \sum_{k=0}^{n-1} X^k$, for the vector $\{g_{\mathrm{e}}^k\}$. This is equivalent to solving

$$\left\{\frac{\partial p}{\partial g_{\mathrm{e}}^k} = 0\right\}_{k=1,\ldots,n}$$

and involves the numerical inversion of an $n \times n$-matrix. Since the system of equations is linear, if there is a solution for $\{g_e^k\}$, plausibility arguments suggest that it is the most likely (rather than the least likely) excitatory conductance time course. The series $\{g_i^k\}$ is then obtained from (8.19).

8.4.2 *Test of the STA Method Using Numerical Simulations*

To test this method, Pospischil et al. (2007) first considered numerical simulations of the IF model in four different situations. To that end, high-conductance states, where the total conductance is dominated by inhibition, were distinguished from low-conductance states, where both synaptic conductances are of comparable magnitude. Also the SDs of the conductances was varied, such that for both high- and low-conductance states both cases $\sigma_i > \sigma_e$ as well as $\sigma_e > \sigma_i$ were considered. The results are summarized in Fig. 8.12, where the STA traces of excitatory and inhibitory conductances recorded from simulations are compared to the most likely (equivalent to the average) conductance traces obtained from solving (8.23). In general, the plots demonstrate a very good agreement.

To quantify these results, Pospischil and colleagues investigated the effect of the statistics as well as of the broadness of the conductance distributions on the quality of the estimation. The latter is crucial, because the derivation of the most likely conductance time course allows for negative conductances, whereas in the simulations negative conductances lead to numerical instabilities, and conductances are bound to positive values. One, thus, expects an increasing error with increasing ratio between SD and mean of the conductance distributions. Estimating the root-mean-square (RMS) of the difference between the recorded and the estimated conductance STAs (summarized in Fig. 8.13) yields expected results. Increasing the number of spikes enhances the match between theory and simulation (Fig. 8.13a shows the RMS deviation for excitation, Fig. 8.13b for inhibition) up to the point where the effect of negative conductances becomes dominant. In the shown example, where the ratio SD/mean was fixed at 0.1, the RMS deviation enters a plateau at about 7,000 spikes. The plateau values can also be recovered from the neighboring plots (i.e., the RMS deviations at SD/mean = 0.1 in Fig. 8.13c,d correspond to the plateau values in (a) and (b)). On the other hand, a broadening of the conductance distribution yields a higher deviation between simulation and estimation. However, at SD/mean = 0.5, the RMS deviation is still as low as $\sim$2% of the mean conductance for excitation and $\sim$4% for inhibition.

To assess the effect of dendritic filtering on the reliability of the method, Pospischil et al. (2007) used a two-compartment model based on that of Pinsky and Rinzel (1994), from which all active channels were removed and replaced by an integrate-and-fire mechanism at the soma. Then, the same 100 s sample of fluctuating excitatory and inhibitory conductances in the dendritic compartment were repeatedly injected and two different recording protocols at the soma performed (Fig. 8.14a). In this study, first recordings in current clamp were preformed in order to obtain the V_m time course as well as the spike times. In this case, the leak

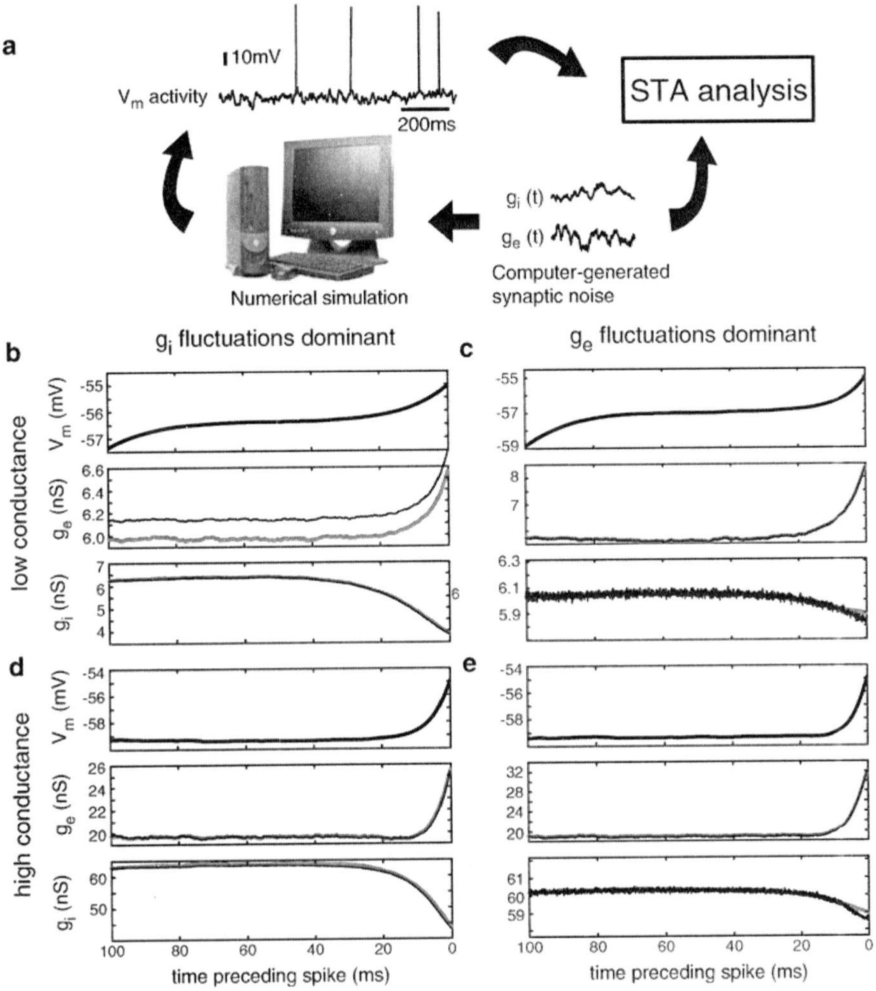

Fig. 8.12 Test of the STA analysis method using an IF neuron model subject to colored conductance noise. (**a**) Scheme of the procedure used. An IF model with synaptic noise was simulated numerically (*bottom*) and the procedure to estimate STA was applied to the V_m activity (*top*). The estimated conductance STAs from V_m were then compared to the actual conductance STA in this model. *Bottom panels*: STA analysis for different conditions, low-conductance states (**b**,**c**), high-conductance states (**d**,**e**), with fluctuations dominated by inhibition (**b**,**d**) or by excitation (**c**,**e**). For each panel, the *upper graph* shows the voltage STA, the *middle graph* the STA of excitatory conductance, and the *lower graph* the STA of inhibitory conductance. *Solid gray lines* show the average conductance recorded from the simulation, while the *black line* represents the conductance estimated from the V_m. Parameters in (**b**) g_{e0} =6 nS, g_{i0} =6 nS, σ_e =0.5 nS, σ_i =1.5 nS; (**c**) g_{e0} =6 nS, g_{i0} =6 nS, σ_e =1.5 nS, σ_i =0.5 nS; (**d**) g_{e0} =20 nS, g_{i0} =60 nS, σ_e =4 nS, σ_i =12 nS; (**e**) g_{e0} =20 nS, g_{i0} =60 nS, σ_e =6 nS, σ_i =3 nS. Modified from Pospischil et al. (2007)

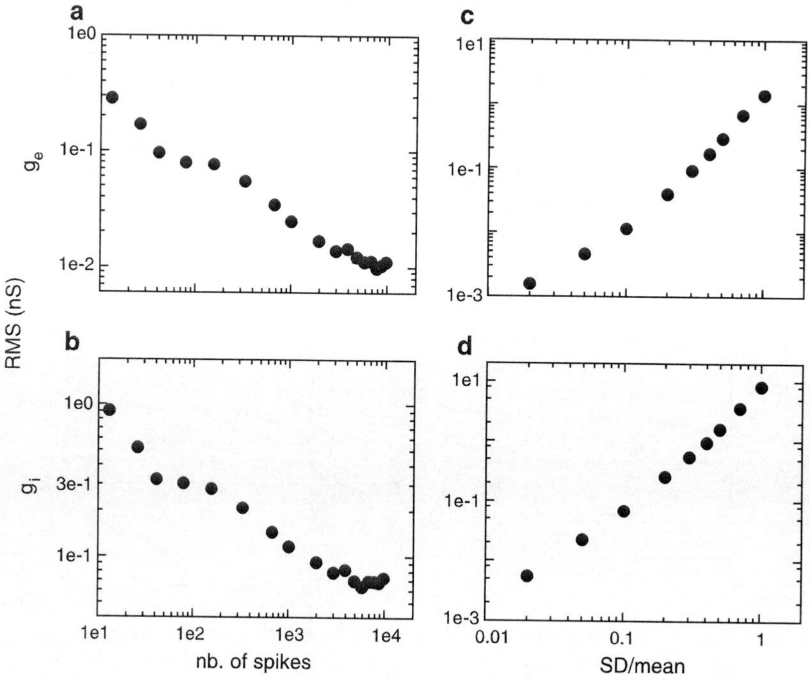

Fig. 8.13 The root-mean-square (RMS) of the deviation of the estimated from the recorded STAs. (**a**) RMS deviation as a function of the number of spikes for the STA of excitatory conductance, where the SD of the conductance distribution was 10% of its mean. The RMS deviation first decreases with the number of spikes, but saturates at ~7,000 spikes. This is due to the effect of negative conductances, which are excluded in the simulation (cf. **c**). (**b**) Same as (**a**) for inhibition. (**c**) RMS deviation for excitation as a function of the ratio SD/mean of the conductance distribution. The higher the probability of negative conductances, the higher the discrepancy between theory and simulation. However, at SD/mean = 0.5, the mean deviation is as low as ~2% of the mean conductance for excitation and ~4% for inhibition. (**d**) Same as (**c**) for inhibition. Modified from Pospischil et al. (2007)

conductance g_L^{so} and the capacitance C^{so} were obtained from current pulse injection at rest. Second, Pospischil and colleagues simulated an "ideal" voltage clamp (no series resistance) at the soma using two different holding potentials (for that, the reversal potentials of excitation and inhibition, respectively, were chosen). Then, from the currents I_{V_e} and I_{V_i}, one can calculate the conductance time courses as

$$g_{e,i}^{so}(t) = \frac{I_{V_{i,e}}(t) - g_L(V_{i,e} - V_L)}{V_{i,e} - V_{e,i}}, \tag{8.24}$$

where the superscript *so* indicates that these are the conductances seen at the soma (somatic conductances). From these, the parameters g_{e0}^{so}, g_{i0}^{so}, σ_e^{so} and σ_i^{so}, the conductance means and SDs were determined.

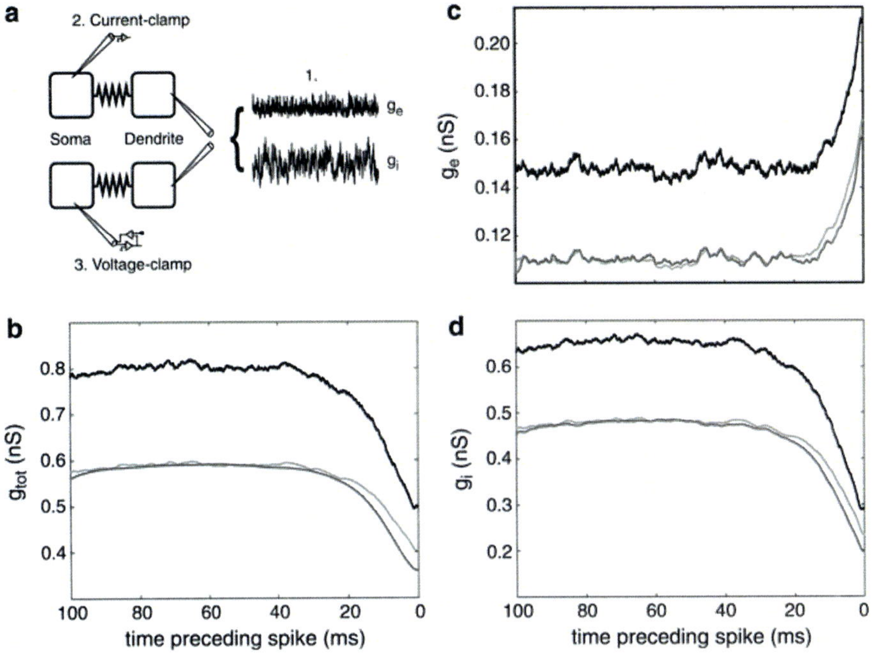

Fig. 8.14 Test of the method using dendritic conductances. (**a**) Simulation scheme: A 100 s sample of excitatory and inhibitory (frozen) conductance noise was injected into the dendrite of a two-compartment model (1). Then, two different recording protocols were performed at the soma. First, the V_m time course was recorded in current clamp (2), second, the currents corresponding to two different holding potentials were recorded in voltage clamp (3). From the latter, the excitatory and inhibitory conductance time courses were extracted using (8.24). (**b**) STA of total conductance inserted at the dendrite (*black*), compared with the estimate obtained in voltage clamp (*light gray*) and with that obtained from somatic V_m activity using the method (*dark gray*). Due to dendritic attenuation, the total conductance values measured are lower than the inserted ones, but the variations of conductances preceding the spike are conserved. (**c**) Same as (**b**), for excitatory conductance. (**d**) Same as (**b**), for inhibitory conductance. Parameters: g_{e0} =0.15 nS, g_{i0} =0.6 nS, σ_e =0.05 nS, σ_i =0.2 nS, g_{e0}^{so} =0.113 nS, g_{i0}^{so} =0.45 nS, σ_e^{so} =0.034 nS, σ_i^{so} =0.12 nS, where the superscript *so* denotes quantities as seen at the soma. Modified from Pospischil et al. (2007)

In contrast to $g_e(t)$ and $g_i(t)$, the distributions of $g_e^{so}(t)$ and $g_i^{so}(t)$ were found to be not Gaussian, and to exhibit lower means and variances. By comparing the STA of the injected (dendritic) conductance, the STA obtained from the somatic V_m using the STA method and the STA obtained using a somatic "ideal" voltage clamp (see Fig. 8.14b–d), the following points could be demonstrated (Pospischil et al. 2007): first, as expected, due to dendritic attenuation, all somatic estimates were attenuated compared to the actual conductances injected in dendrites (compare *light* and *dark gray* curves, soma, with *black curve*, dendrite, in Fig. 8.14b–d); second, the estimate obtained by applying the present method to the somatic V_m (*dark gray curves* in Fig. 8.14b–d) was very similar to that obtained using an "ideal" voltage clamp at the soma (*light gray* curves). The difference close to the spike may be due to the

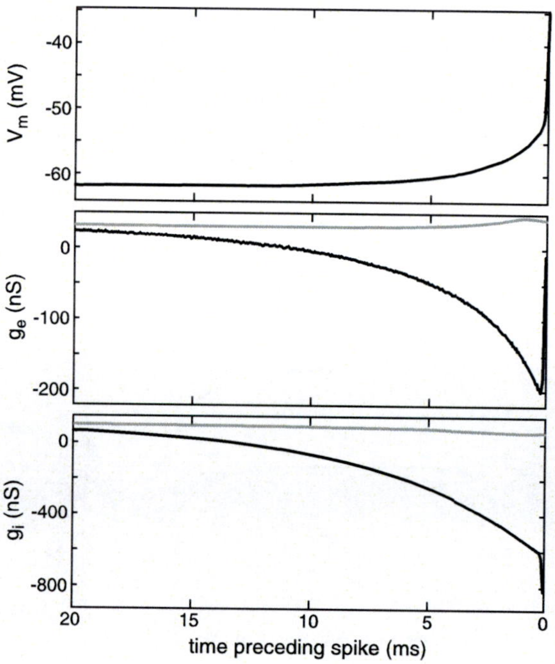

Fig. 8.15 The effect of the presence of additional voltage-dependent conductances on the estimation of the synaptic conductances. *Gray solid lines* indicate recorded conductances; *black, dotted lines* indicate estimated conductances. In this case, the estimation fails. The sharp rise of the voltage in the last ms before the spike requires very fast changes in the synaptic conductances, which introduces a considerable error in the analysis. Parameters used: $g_{e0} = 32$ nS, $g_{i0} = 96$ nS, $\sigma_e = 8$ nS, $\sigma_i = 24$ nS. Modified from Pospischil et al. (2007)

non-Gaussian shape of the somatic conductance distributions, whose tails then become important; third, despite attenuation, the qualitative shape of the conductance STA was preserved. With this, one can conclude that the STA estimate from V_m activity captures rather well the conductances as seen by the spiking mechanism.

8.4.3 Test of the STA Method in Dynamic Clamp

Pospischil et al. (2007) also tested the method on voltage STAs obtained from dynamic-clamp recordings of guinea-pig cortical neurons in slices. In real neurons, a problem is the strong influence of spike-related voltage-dependent (presumably sodium) conductances on the voltage time course. Since in the STA method the *global* probability of $g_e(t)$ and $g_i(t)$ is maximized, the voltage in the vicinity of the spike has an influence on the estimated conductances at all times. As a consequence, without removing the effect of sodium, the estimation fails (see Fig. 8.15). Fortunately, it is rather simple to correct for this effect by excluding the last 1 to 2 ms before the spike from the analysis. The corrected comparison between the recorded and the estimated conductance traces is shown in Fig. 8.16.

Finally, one can check for the applicability of this method to in vivo recordings. To that end, Pospischil et al. (2007) assessed the sensitivity of the estimates with respect to the different parameters by varying the values describing passive

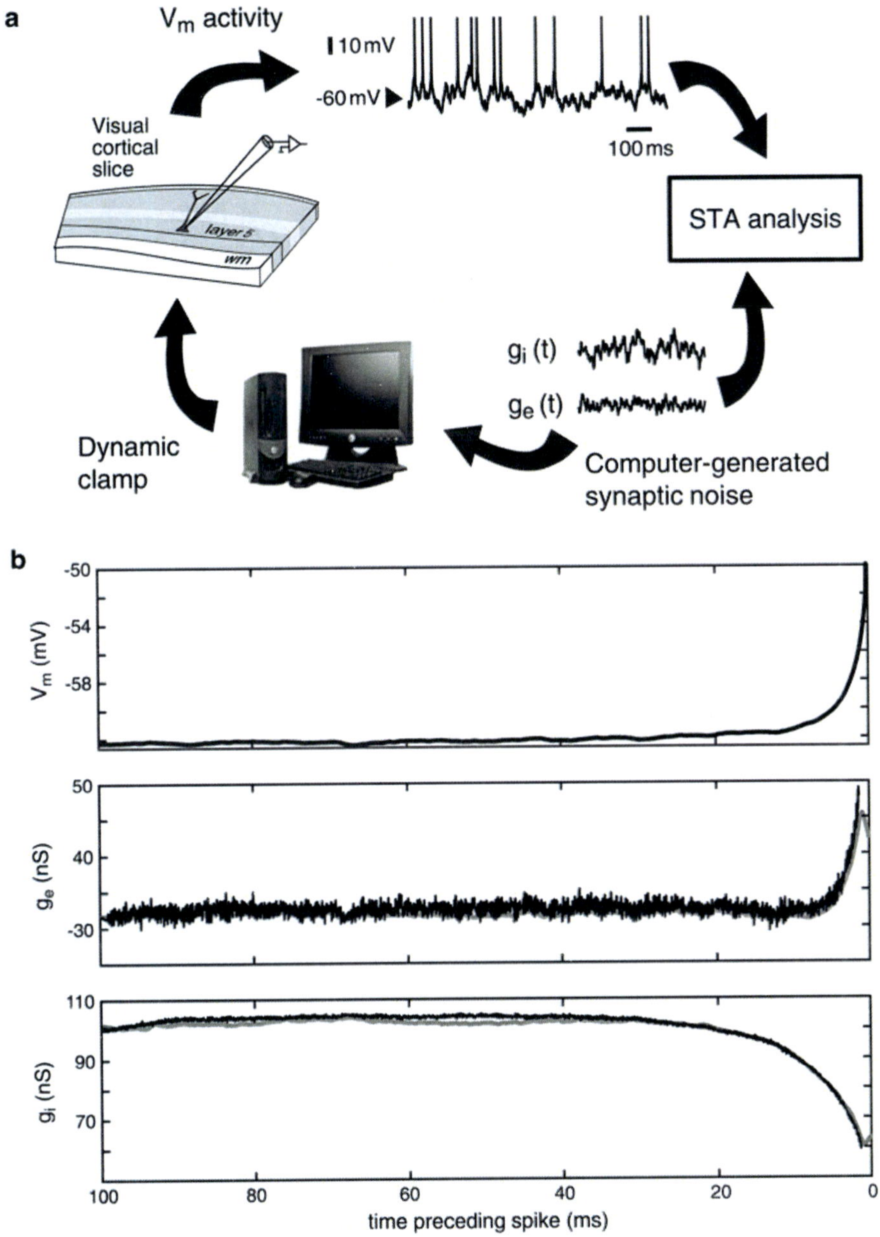

Fig. 8.16 Test of the method in real neurons using dynamic clamp in guinea-pig visual cortical slices. (**a**) Scheme of the procedure. Computer-generated synaptic noise was injected in the recorded neuron under dynamic clamp (*bottom*). The V_m activity obtained (*top*) was then used to extract the STA of conductances, which was compared to the STA directly obtained from the injected conductances. (**b**) Results of this analysis in a representative neuron. *Black lines* show the estimated STA of conductances from V_m activity, *gray lines* show the STA of conductances that were actually injected into the neuron. The analysis was made by excluding the data from the 1.2 ms before the spike to avoid contamination by voltage-dependent conductances. Parameters for conductance noise were as in Fig. 8.15. Modified from Pospischil et al. (2007)

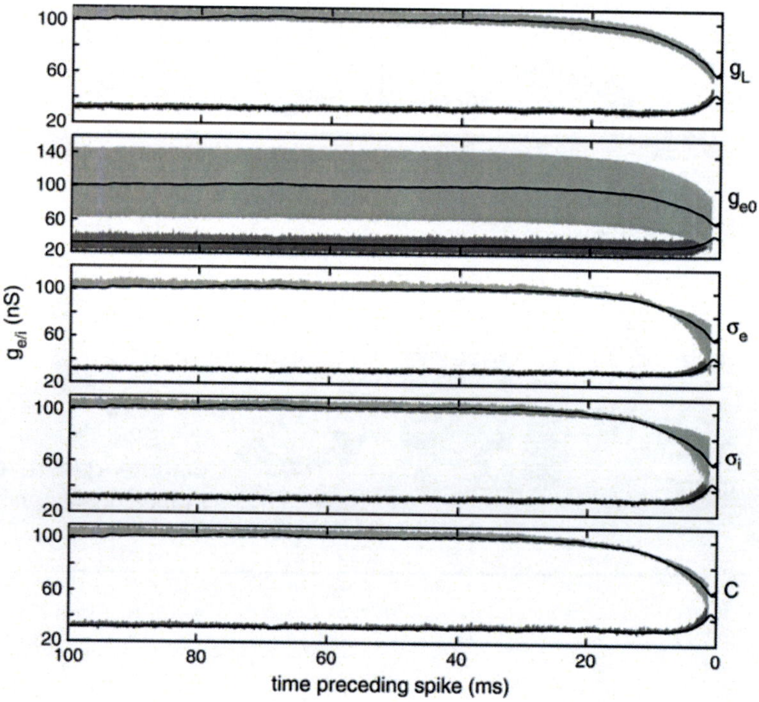

Fig. 8.17 Deviation in the estimated conductance STAs in real neurons using dynamic clamp due to variations in the parameters. The *black lines* represent the conductance STA estimates using the correct parameters, the *gray areas* are bound by the estimates that result from variation of a single parameter (indicated on the *right*) by $\pm$ 50%. *Light gray areas* represent inhibition, *dark gray areas* represent excitation. The total conductance (leak plus synaptic conductances) was assumed to be fixed. A variation in the mean values of the conductances evokes mostly a shift in the estimate, while a variation in the SDs influences the curvature just before the spike. Modified from Pospischil et al. (2007)

properties and synaptic activity. Here, the assumptions were made that the total conductance can be constrained by input resistance measurements, and that time constants of the synaptic currents can be estimated by power spectral analyses (Destexhe and Rudolph 2004). This leaves g_L, C, g_{e0}, σ_e, and σ_i as the main parameters. The impact of these parameters on STA conductance estimates is shown in Fig. 8.17. Varying these parameters within $\pm$50% of their nominal value leads to various degrees of error in the STA estimates. The dominant effect of a variation in the mean conductances is a shift in the estimated STAs, whereas a variation in the SDs changes the curvature just before the spike.

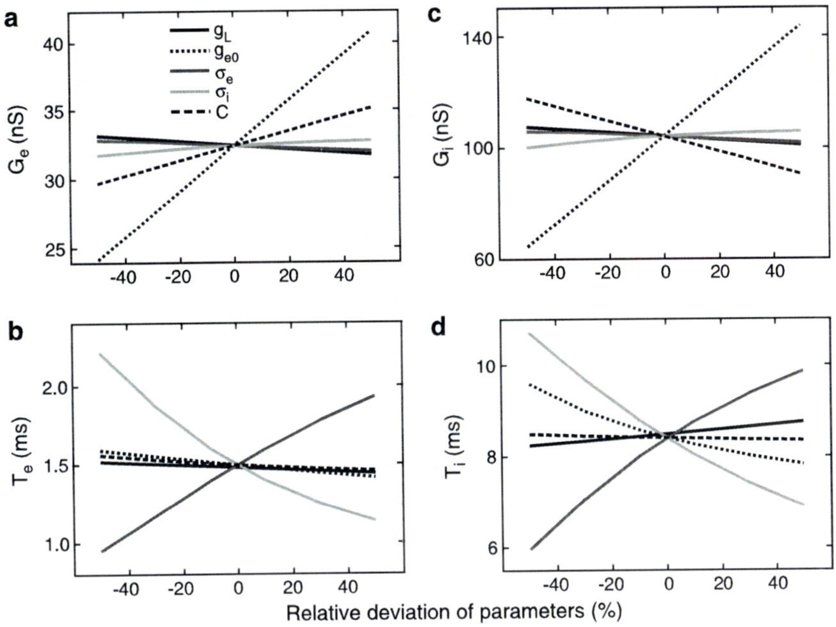

Fig. 8.18 Detailed evaluation of the sensitivity to parameters. The conductance STAs were fitted with an exponential function $f_s(t) = G_s(1 + K_s \exp((t - t_0)/T_s))$, $s = e, i$. t_0 is chosen to be the time at which the analysis stops. Each plot shows the estimated value of G_e, G_i, T_e or T_i from this experiment, each curve represents the variation of a single parameter (see legend). Modified from Pospischil et al. (2007)

To address this point further, Pospischil et al. (2007) fitted the estimated conductance STAs with an exponential function

$$f_s(t) = G_s\left(1 + K_s \mathrm{e}^{\frac{t-t_0}{T_s}}\right). \tag{8.25}$$

Here, t_0 is chosen to be the time at which the analysis stops. Figure 8.18 gives an overview of the dependence of the fitting parameters G_e, G_i, T_e and T_i on the relative change of g_L, g_{e0}, σ_e, σ_i and C. For example, a variation of g_{e0} has a strong effect on G_e and G_i, but affects to a lesser extent T_e and T_i, while the opposite is seen when varying σ_e and σ_i.

8.4.4 STA Method with Correlation

During response to sensory stimuli, there can be a substantial degree of correlation between excitatory and inhibitory synaptic input (Wehr and Zador 2003; Monier et al. 2003; Wilent and Contreras 2005b), and it was recently shown that this could also occur during spontaneous activity (Okun and Lampl 2008). The STA method proposed in Pospischil et al. (2007), was extended in Piwkowska et al. (2008) to include noise correlations. For that, the discretized versions of (8.1) (second equation) are reformulated as follows:

$$\frac{g_e^{k+1} - g_e^k}{\Delta t} = -\frac{g_e^k - g_{e0}}{\tau_e} + \sigma_e \sqrt{\frac{2\Delta t}{\tau_e(1+c)}}(\xi_1^k + \sqrt{c}\,\xi_2^k),$$

$$\frac{g_i^{k+1} - g_i^k}{\Delta t} = -\frac{g_i^k - g_{i0}}{\tau_i} + \sigma_i \sqrt{\frac{2\Delta t}{\tau_i(1+c)}}(\xi_2^{k-d} + \sqrt{c}\,\xi_1^{k-d}). \tag{8.26}$$

Here, instead of having one "private" white noise source feeding each conductance channel, the same two noise sources ξ_1 and ξ_2 contribute to both inhibition and excitation. The amount of correlation is tuned by the parameter c. Also, since there is evidence that the peak of the g_e–g_i–crosscorrelation is not always centered at zero during stimulus-evoked responses ("delayed inhibition"; see Wehr and Zador 2003; Wilent and Contreras 2005b), a nonzero delay d is allowed: for a positive parameter d, the inhibitory channel receives the input that the excitatory channel received d time steps before. Equation (8.26) can be solved for ξ_1^k and ξ_2^k. It is then possible to proceed as in the uncorrelated case, where now, due to the delay, the matrix describing (8.23) has additional subdiagonal entries.

However, the application of this extended method requires the estimation of the usual leak parameters, of conductance distribution parameters, for which the VmD method (Sect. 8.2) cannot be directly used in its current form since it is based on uncorrelated noise sources, as well as knowledge of the parameters c and d. These correlation parameters could, perhaps, be obtained from extracellularly recorded spike trains, provided that simultaneously recorded single units could be classified as excitatory or inhibitory. They can also be derived from the analysis of paired recordings from closely situated neurons receiving mostly shared inputs, as shown recently (Okun and Lampl 2008). Alternatively, different plausible c and d values could be scanned to examine how they could potentially influence the conductance STAs extracted from a given V_m STA. Figure 8.19 shows a numerical test of this extension of the method. The value of the correlation had an influence on the shape of the STA of conductances, and this shape was well resolved using the extended STA method.

8.5 The VmT Method: Extracting Conductance Statistics from Single V_m Traces

In this section, we present a method to estimate synaptic conductances from single V_m traces. The total synaptic conductances (excitatory and inhibitory) are usually estimated from V_m recordings by using at least two V_m levels and therefore cannot

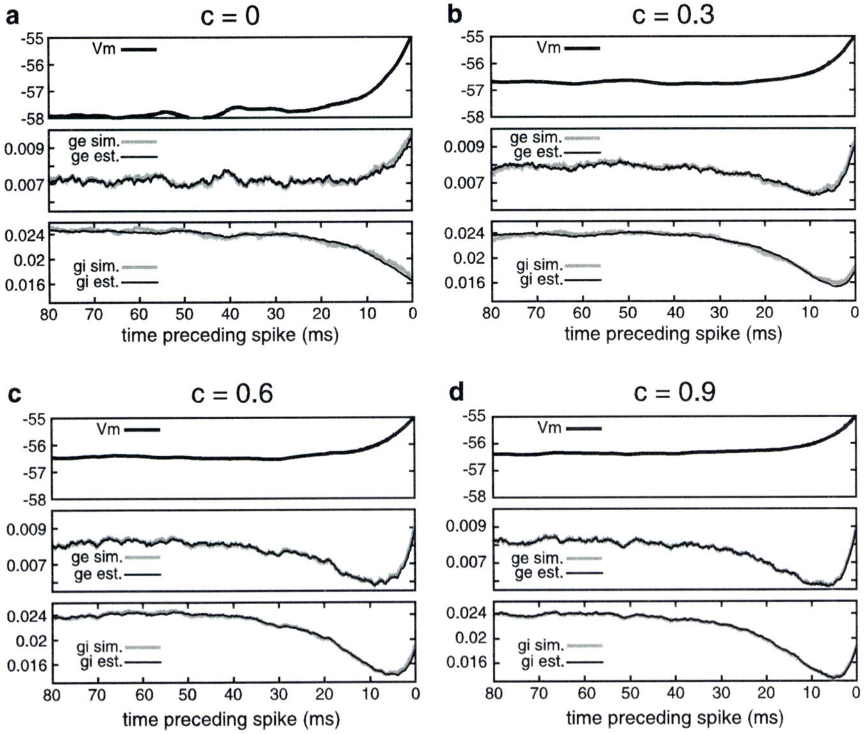

Fig. 8.19 Numerical test of the STA method with different levels of correlation between excitation and inhibition. The point-conductance model was simulated with different levels of correlations between excitation and inhibition, and the V_m activity was used to extract the STA of the conductances. Different levels of correlations were used, $c = 0$ (**a**), $c = 0.3$ (**b**), $c = 0.6$ (**c**), $c = 0.9$ (**d**), all with $d = 0$. In each case, the V_m STA is shown on *top*, while the two *bottom panels* show the excitatory STA and inhibitory STA, respectively (*gray*), together with the estimates using the method (*black*). Modified from Pospischil et al. (2009)

be applied to single V_m traces. Pospischil and colleagues propose a method that can potentially palliate to this problem (Pospischil et al. 2009). This *VmT method* is similar in spirit to the VmD method (see Sect. 8.2), but estimates conductance parameters using maximum likelihood criteria, thus also similar to the STA method presented in the previous section. We start by explaining the VmT method, then test it using models and on guinea-pig visual cortex neurons in vitro using dynamic-clamp experiments.

8.5.1 The VmT Method

The starting point of the method is to search for the "most likely" conductances parameters ($g_{\mathrm{e}0}$, $g_{\mathrm{i}0}$, σ_{e} and σ_{i}) that are compatible with an experimentally observed V_{m} trace. To that end, the point-conductance model (8.1) is discretized in time with a step-size Δt. Equation (8.1) can then be solved for g_{i}^k, which gives:

$$g_{\mathrm{i}}^k = -\frac{C}{V^k - E_{\mathrm{i}}}\left\{\frac{V^k - E_{\mathrm{L}}}{\tau_{\mathrm{L}}} + \frac{g_{\mathrm{e}}^k(V^k - E_{\mathrm{e}})}{C} + \frac{V^{k+1} - V^k}{\Delta t} - \frac{I_{\mathrm{ext}}}{C}\right\}, \tag{8.27}$$

where $\tau_{\mathrm{L}} = C/G_{\mathrm{L}}$. Since the series V^k for the voltage trace is known, g_{i}^k has become a function of g_{e}^k. In the same way, one can solve the second equations in (8.1) for ξ_s^k, which, in turn, become Gaussian-distributed random numbers,

$$\xi_s^k = \frac{1}{\sigma_s}\sqrt{\frac{\tau_s}{2\Delta t}}\left(g_s^{k+1} - g_s^k\left(1 - \frac{\Delta t}{\tau_s}\right) - \frac{\Delta t}{\tau_s} g_{s0}\right), \tag{8.28}$$

where s stands for e, i.

There is a continuum of combinations $\{g_{\mathrm{e}}^{k+1}, g_{\mathrm{i}}^{k+1}\}$ that can advance the membrane potential from V^{k+1} to V^{k+2}, each pair occurring with a probability

$$p^k := p(g_{\mathrm{e}}^{k+1}, g_{\mathrm{i}}^{k+1} | g_{\mathrm{e}}^k, g_{\mathrm{i}}^k) = \frac{1}{2\pi}\mathrm{e}^{-\frac{1}{2}(\xi_{\mathrm{e}}^{k2} + \xi_{\mathrm{i}}^{k2})} = \frac{1}{2\pi}\mathrm{e}^{-\frac{1}{4\Delta t}X^k},$$

$$X^k = \frac{\tau_{\mathrm{e}}}{\sigma_{\mathrm{e}}^2}\left(g_{\mathrm{e}}^{k+1} - g_{\mathrm{e}}^k\left(1 - \frac{\Delta t}{\tau_{\mathrm{e}}}\right) - \frac{\Delta t}{\tau_{\mathrm{e}}} g_{\mathrm{e}0}\right)^2 \tag{8.29}$$

$$+ \frac{\tau_{\mathrm{i}}}{\sigma_{\mathrm{i}}^2}\left(g_{\mathrm{i}}^{k+1} - g_{\mathrm{i}}^k\left(1 - \frac{\Delta t}{\tau_{\mathrm{i}}}\right) - \frac{\Delta t}{\tau_{\mathrm{i}}} g_{\mathrm{i}0}\right)^2. \tag{8.30}$$

These expressions are identical to those derived previously for calculating STAs (Pospischil et al. 2007; see Sect. 8.4.1), except that no implicit average is assumed here.

Thus, to go one step further in time, a continuum of pairs $(g_{\mathrm{e}}^{k+1}, g_{\mathrm{i}}^{k+1})$ is possible in order to reach the (known) voltage V^{k+2}. The quantity p^k assigns to all such pairs a probability of occurrence, depending on the previous pair, and the voltage history. Ultimately, it is the probability of occurrence of the appropriate random numbers ξ_{e}^k and ξ_{i}^k that relate the respective conductances at subsequent time steps. It is then straightforward to write down the probability p for certain conductance series to occur, that reproduce the voltage time course. This is just the probability for successive conductance steps to occur, namely the product of the probabilities p^k:

$$p = \prod_{k=0}^{n-1} p^k, \tag{8.31}$$

given initial conductances g_e^0, g_i^0. However, again, there is a continuum of conductance series

$$\{g_e^l, g_i^l\}_{l=1,\ldots,n+1},$$

that are all compatible with the observed voltage trace. Defining a likelihood function $f(V^k,\theta)$, $\theta = (g_{e0}, g_{i0}, \sigma_e, \sigma_i)$, that takes into account all of these traces with appropriate weight, integrating (8.31) over the unconstrained conductances g_e^k and normalizing by the volume of configuration space, yields

$$f(V^k,\theta) = \frac{\int \prod_{k=0}^{n-1} \mathrm{d}g_e^k \, p}{\int \prod_{k=0}^{n-1} \mathrm{d}g_e^k \, \mathrm{d}g_i^k \, p}, \tag{8.32}$$

where only in the nominator g_i^k has been replaced by (8.27). This expression reflects the likelihood that a specific voltage series $\{V^k\}$ occurs, normalized by the probability, that *any* trace occurs. The most likely parameters θ giving rise to $\{V^k\}$ are obtained by maximizing (or minimizing the negative of) $f(V^k,\theta)$ using standard optimization schemes (Press et al. 2007).

8.5.2 Test of the VmT Method Using Model Data

The VmT method introduced in the last section can be tested in detail with respect to its applicability to voltage traces that are created using the same model (leaky integrator model; see Pospischil et al. 2009). To this end, simulations scanning the (g_{e0}, g_{i0})–plane are performed, followed by attempts to re-estimate the conductance parameters used in the simulations solely from the V_m activity. In Pospischil et al. (2009), for each parameter set $(g_{e0}, g_{i0}, \sigma_e, \sigma_i)$ the method was applied to ten samples of 5,000 data points (corresponding to 250 ms each) and the average was taken subsequently. Moreover, in this study the conductance SDs were chosen to be one third of the respective mean values, while other parameters were assumed to be known during re-estimation ($C = 0.4$ nF, $g_L = 13.44$ nS, $E_L = -80$ mV, $E_e = 0$ mV, $E_i = -75$ mV, $\tau_e = 2.728$ ms, $\tau_i = 10.49$ ms), the time step was $dt = 0.05$ ms. Also, there it was assumed that the total conductance g_{tot} (i.e. the inverse of the apparent input resistance) is known. This assumption is not mandatory, but it was shown that the estimation become more stable. The likelihood function given by (8.32) was, thus, only maximized with respect to g_{e0}, σ_e, and σ_i. Figure 8.20 summarizes the results obtained.

The mean conductances are well reproduced over the entire scan region. An exception is the estimation of g_{i0} in the case where the mean excitation exceeds inhibition by several-fold, a situation which is rarely found in real neurons. The situation for the SDs is different. While the excitatory SD is reproduced very well in the whole area under consideration, this is not necessarily the case for inhibition. Here, the estimation is good for most parts of the scanned region, but shows a considerable deviation along the left and lower boundaries. These are

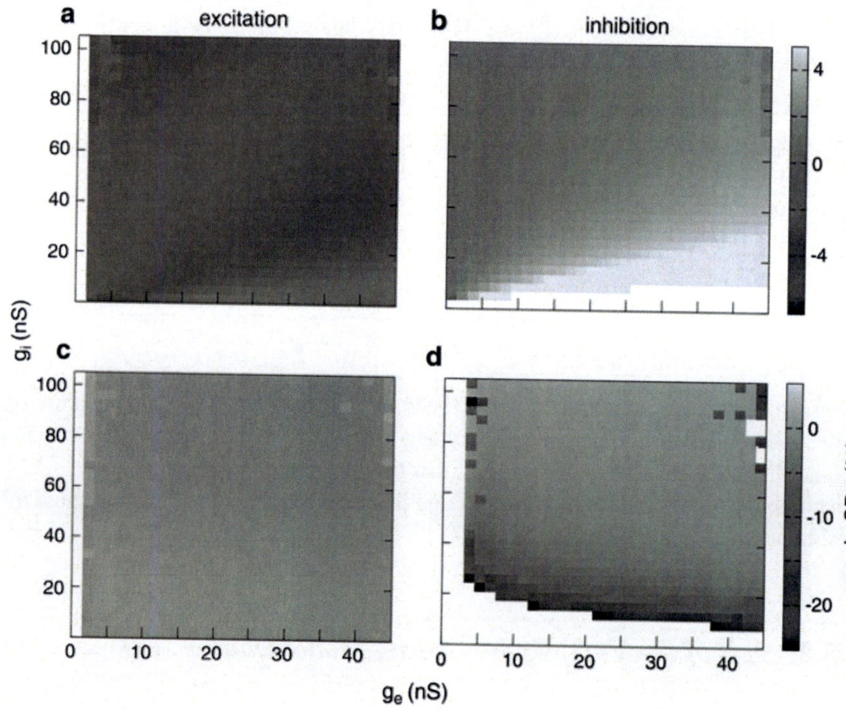

Fig. 8.20 Test of the single-trace VmT method using a leaky integrator model. Each panel presents a scan in the (g_{e0}, g_{i0})–plane. Color codes the relative deviation between model parameters and their estimates using the method (note the different scales for means/SDs). The *white areas* indicate regions where the mismatch was larger, >5% for the means and $< -25\%$ for SDs. **(a)** Deviation in the mean of excitatory conductance (g_{e0}). **(b)** Same as **(a)**, but for inhibition. **(c)** Deviation in the SD of excitatory conductance. **(d)** Same as **(c)**, but for inhibition. In general the method works well, except for a small band for the inhibitory SD. Modified from Pospischil et al. (2009)

regions where the transmembrane current due to inhibition is weak, either because the inhibitory conductance is weak (lower boundary) or because it is strong and excitation is weak (left boundary), such that the mean voltage is close to the inhibitory reversal potential and the driving force is small. In these conditions, it seems that the effect of inhibition on the membrane voltage cannot be distinguished from that of the leak conductance.

This point is illustrated in Fig. 8.21. The relative deviation between σ_i in the model and its re-estimation depends on the ratio of the transmembrane current due to inhibitory (I_i) and leak (I_L) conductance. The estimation fails when the inhibitory current is smaller or comparable to the leak current, but it becomes very reliable as soon as the ratio I_i/I_L becomes larger than 1.5–2. Some points, however, have large errors although with dominant inhibition. These points have strong excitatory conductances (see gray scale) and correspond to the upper right corner of Fig. 8.20d. The error is due to aberrant estimates for which the predicted variance is zero; in

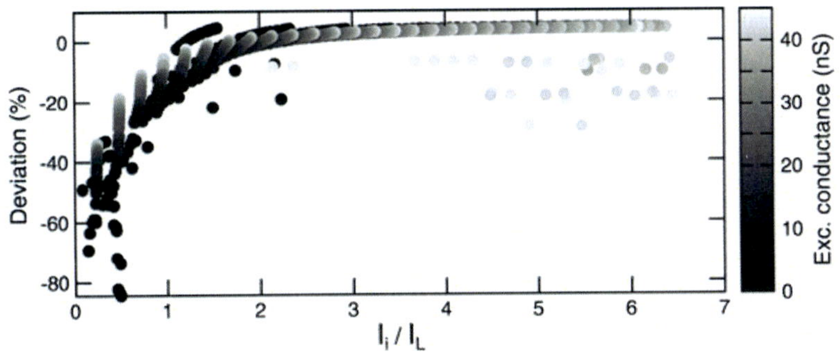

Fig. 8.21 The estimation error depends on the ratio of inhibitory and leak conductances. The relative deviation between the parameter σ_i in the simulations and its re-estimated value is shown as a function of the ratio of the currents due to inhibitory and leak conductances. The estimation fails when the inhibitory component becomes too small. The same data as in Fig. 8.20 are plotted: different dots correspond to different pairs of excitatory and inhibitory conductances (g_{e0}, g_{i0}), and the dots are colored according to the excitatory conductance (see scale). Modified from Pospischil et al. (2009)

principle such estimates could be detected and discarded, but no such detection was attempted in Pospischil et al. (2009). Besides these particular combinations, the majority of parameters with strong inhibitory conductances gives acceptable errors. This suggests that the estimation of the conductance variances will be most accurate in high-conductance states, where inhibitory conductances are strong and larger than the leak conductance.

The unavoidable presence of recording noise may present a problem in the application of the method to recordings from real neurons. Figure 8.22 (left) shows how low-amplitude white noise added to the voltage trace of a leaky integrator model impairs the reliability of the method. To that end, a Gaussian-distributed white noise is added to the voltage trace at every time step, scaled by the amplitude given in the abscissa. In Fig. 8.22, different curves correspond to different pairs (g_{e0}, g_{i0}) colored as a function of the total conductance. The noise has an opposite effect on the estimation of the conductance mean values. While the estimate of excitation exceeds the real parameter value, for inhibition the situation is inverted. However, one has to keep in mind that both parameters are not estimated independently, but their sum is kept fixed. In contrast, the estimates for the conductance SDs always exceed the real values, and they can deviate by almost 500% for a noise amplitude of 10 μV. Here, the largest errors generally correspond to lowest conductance states. Clearly, in order to apply the method to recordings from real neurons, one needs to restrain this noise sensitivity.

Fortunately, this noise sensitivity can be reduced by standard noise reduction techniques. For example, preprocessing and smoothing the data using a Gaussian filter greatly diminishes the amplitude of the noise, and consequently improves the estimates according to the new noise amplitude (see Fig. 8.22, right panels). Too much smoothing, however, may result in altering the signal itself, and may

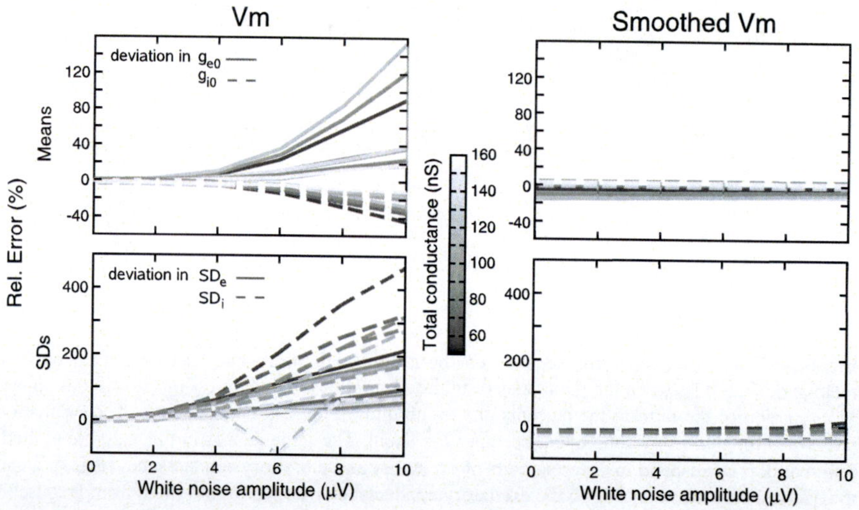

Fig. 8.22 Error of the VmT estimates following addition of white noise to the voltage trace. Gaussian white noise was added to the voltage trace of the model, and the VmT method was applied to the V_m trace obtained with noise, to yield estimates of conductances and variances. *Left*: relative error obtained in the estimation of g_{e0} and g_{i0} (*upper panel*), as well as σ_e and σ_i (*lower panels*). *Right*: same estimation, but the V_m was smoothed prior to the VmT estimate (Gaussian filter with SD of one data point). In both cases, the relative error is shown as a function of the white noise amplitude. Different curves correspond to different pairs (g_{e0}, g_{i0}). The errors on the estimates for both mean conductance and SD increase with the noise. The coloring of the curves as a function of the total conductance (see scale) shows that the largest errors generally occur for the low-conductance regimes. The error was greatly diminished by smoothing (*right panels*). Modified from Pospischil et al. (2009)

introduce errors (Pospischil et al. 2009). It is, therefore, preferable to use smoothing at very short timescales (SD of 1 to 4 data points, depending on the sampling rate). In the next sections, results are shown for which the experimental voltage traces are preprocessed with a Gaussian filter with a SD of three data points.

8.5.3 Testing the VmT Method Using Dynamic Clamp

Finally, the VmT method can be tested on in vitro recordings using dynamic-clamp experiments (Fig. 8.23). As in the model, the stimulus consist of two channels of fluctuating conductances representing excitation and inhibition. The conductance injection spans values from low-conductance (of the order of 5–10 nS) to high-conductance states (50–160 nS). It is apparent from Fig. 8.23 that the mean values of the conductances (g_{e0}, g_{i0}) are well estimated, as expected, because the total conductance is known in this case.

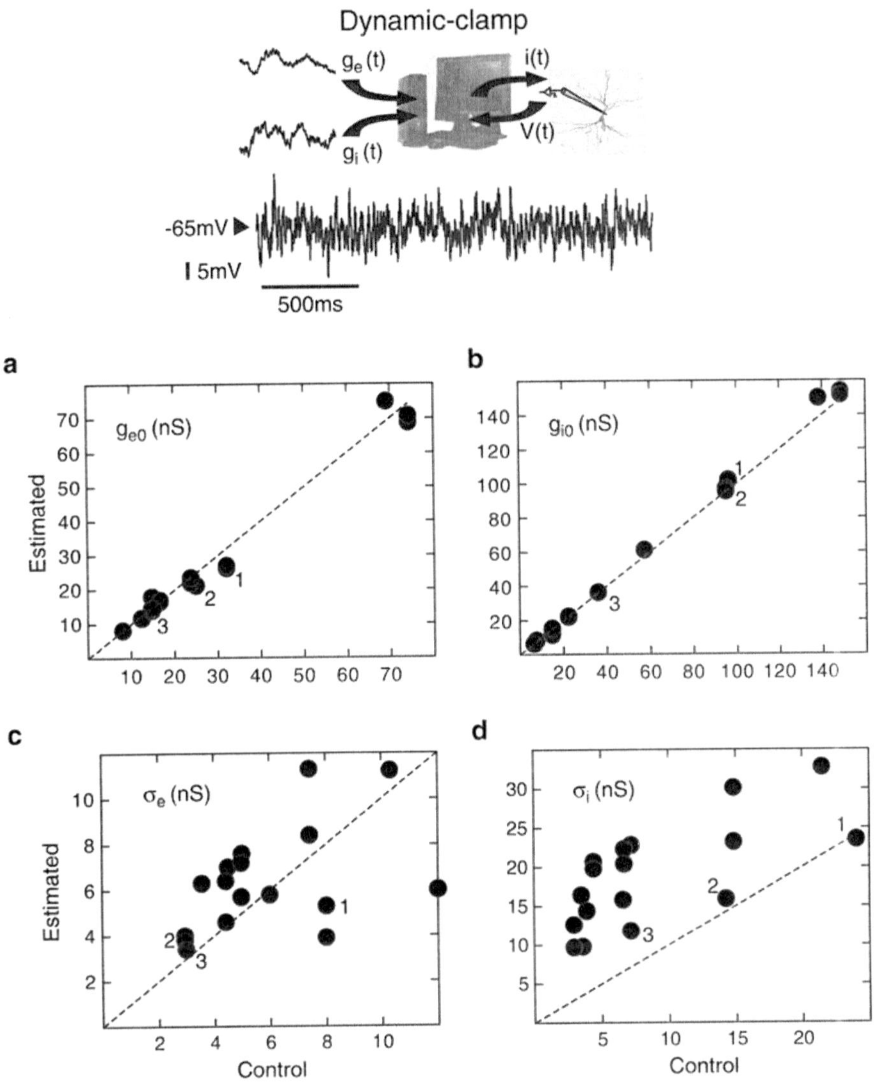

Fig. 8.23 Dynamic-clamp test of the VmT method to extract conductances from guinea-pig visual cortical neurons in vitro. Fluctuating conductances of known parameters were injected in different neurons using dynamic clamp, and the V_m activity produced was analyzed using the single-trace VmT method. Each plot represents the different conductance parameters extracted from the V_m activity: g_{e0} (**a**), g_{i0} (**b**), σ_e (**c**) and σ_i (**d**). The extracted parameter (*Estimated*) is compared to the value used in the conductance injection (*Control*). While in general the mean conductances are matched very well, the estimated SDs show a large spread around the target values. Nevertheless, during states dominated by inhibition (see indexed symbols), the estimation was acceptable. Modified from Pospischil et al. (2009)

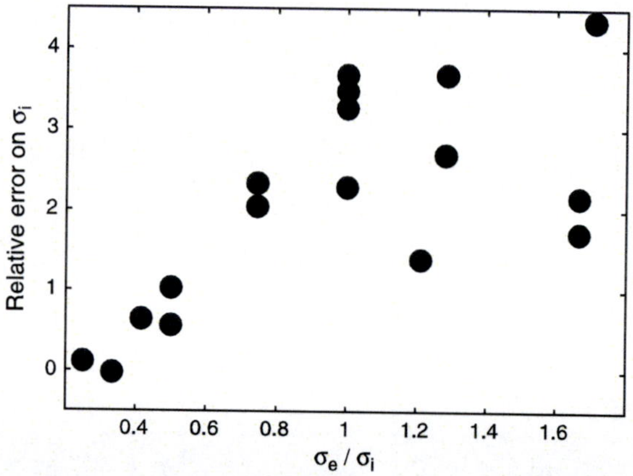

Fig. 8.24 Relative error on inhibitory variance is high only when excitatory fluctuations dominate. The relative mean-square error on σ_i is represented as a function of the σ_e/σ_i ratio. The error is approximately proportional to the ratio of variances. The same data as in Fig. 8.23 were used. Modified from Pospischil et al. (2009)

However, the estimation is subject to larger errors for the SDs of the conductances (σ_e, σ_i). In addition, the error on estimating variances is also linked to the accuracy of the estimates of synaptic time constants τ_e and τ_i, similar to the VmD method (see discussion in Piwkowska et al. 2008). Interestingly, for some cases, the estimation works quite well (see indexed symbols in Fig. 8.23). In the pool of injections studied in Pospischil et al. (2009), there are three cases that represent a cell in a high-conductance state, i.e., the mean inhibitory conductances are roughly three times greater than the excitatory ones, and the SDs obey a similar ratio. For these trials, the estimate comes close to the values used during the experiment. Indeed, Pospischil and colleagues found that the relative error on σ_i is roughly proportional to the ratio σ_e/σ_i for ratios smaller than 1, and tends to saturate for larger ratios (Fig. 8.24). In other words, the estimation has the lowest errors when inhibitory fluctuations dominate excitatory fluctuations. A recent estimate of conductance variances in cortical neurons of awake cats reported that σ_i is larger than σ_e for the vast majority of cells analyzed (Rudolph et al. 2007). The same was also true for anesthetized states (Rudolph et al. 2005), suggesting that the VmT method should give acceptable errors in practical situations in vivo.

8.6 Summary

As shown in Chap. 7, a direct consequence of the simplicity of the point-conductance model is that it enables mathematical approaches. These approaches gave rise to a series of new analysis methods, which were overviewed in the present chapter.

A first method is based on the analytic expression of the steady-state voltage distribution of neurons subject to conductance-based synaptic noise (see Sect. 7.4.6). Fitting such an expression to experimental data yields estimates of conductances and other parameters of background activity. This idea was formulated for the first time less than 10 years ago (Rudolph and Destexhe 2003d) and subsequently gave rise to a method called the *VmD method* (Rudolph et al. 2004), which is detailed in Sect. 8.2. This method was subsequently tested in dynamic-clamp experiments Rudolph et al. (2004).

Similarly, the PSD of the V_m can be approximated by analytic expressions. Fitting these expressions to experiments can yield estimates of other parameters, such as the time constants of the synaptic conductances, as reviewed in Sect. 8.3. This *PSD method* was also tested using dynamic-clamp experiments.

A third method was shown in Sect. 8.4 and is a direct consequence of the ability to estimate conductance fluctuations by the VmD method. Such estimates open the route to experimentally characterize the influence of fluctuations on AP generation. This can be done by estimating the STA conductance patterns from V_m recordings using a maximum likelihood method (Pospischil et al. 2007). This *STA method* is also based on the point-conductance model, and requires the prior knowledge of the mean and variance of excitatory and inhibitory conductances, which can be provided by the VmD method. Similar to the VmD method, the STA method was also tested using dynamic-clamp experiments and was shown to provide accurate estimates (Pospischil et al. 2007; Piwkowska et al. 2008).

Finally, another method was recently proposed and consists of estimating excitatory and inhibitory conductances from single V_m trials. This *VmT method*, outlined in Sect. 8.5, is similar in spirit to the VmD method but uses a maximum likelihood procedure to estimate the mean and variance of excitatory and inhibitory conductances. Like other methods, this method was tested in dynamic-clamp experiments and was shown to yield excellent estimates of synaptic conductances (Pospischil et al. 2009).

While the present chapter was devoted to explaining the methods and testing them, the application of these methods to in vivo recordings are presented in the next chapter.

Chapter 9
Case Studies

In this chapter, we will consider case studies that illustrate many of the concepts overviewed in the preceding chapters. We will review the characterization of synaptic noise from intracellular recordings during artificially activated states in vivo, as well as during Up- and Down-states. In a second example, we will present results from a study which aimed at characterizing synaptic noise from intracellular recordings in awake and naturally sleeping animals, thus providing the first estimates of synaptic conductances and fluctuations in the aroused brain.

9.1 Introduction

As we have reviewed in Chap. 3, intracellular recordings of cortical neurons in awake, conscious cats and monkeys show a depolarized membrane potential (V_m), sustained firing and intense subthreshold synaptic activity (Matsumura et al. 1988; Baranyi et al. 1993; Steriade et al. 2001; Timofeev et al. 2001). It is presently, however, unclear how neurons process information in such active and irregular states. An important step to investigate this problem is to obtain a precise characterization of the conductance variations during activated electroencephalogram (EEG) states. Input resistance measurements indicate that during such activated states, cortical neurons can experience periods of high conductance, which may have significant consequences for their integrative properties (Destexhe and Paré 1999; Kuhn et al. 2004; reviewed in Destexhe et al. 2003a; see Chap. 5).

In anesthetized animals, several studies have provided measurements of the excitatory and inhibitory conductance contributions to the state of the membrane, using various paradigms (e.g., see Borg-Graham et al. 1998; Hirsch et al. 1998; Paré et al. 1998b; Anderson et al. 2000; Wehr and Zador 2003; Priebe and Ferster 2005; Haider et al. 2006). However, no such conductance measurements have been made so far in activated states with desynchronized EEG, such as in awake animals.

A. Destexhe and M. Rudolph-Lilith, *Neuronal Noise*, Springer Series in Computational Neuroscience 8, DOI 10.1007/978-0-387-79020-6_9,

In this chapter, we present two of such studies. The first study (Rudolph et al. 2005) investigated the synaptic background activity in activated states obtained in anesthetized cats following stimulation of the ascending arousal system. The second study (Rudolph et al. 2007) characterized the synaptic background activity from awake and naturally sleeping cats. We finish by overviewing how to estimate time-dependent variations of conductances, and illustrate this estimation from intracellular recordings during Up- and Down-states.

9.2 Characterization of Synaptic Noise from Artificially Activated States

In Chap. 3, Sect. 3.2.4, we have described the intracellular correlates of desynchronized EEG states obtained during anesthesia after artificial stimulation of the PPT nucleus (see Fig. 3.13). In the present section, we describe the conductance analysis of the V_m activity during this artificial EEG activation, as well as computational models based on these measurements (see details in Rudolph et al. 2005).

9.2.1 Estimation of Synaptic Conductances During Artificial EEG Activated States

To estimate the respective contribution of excitatory and inhibitory conductances, one can first use the standard *Ohmic method*, which consists of taking the temporal average of the membrane equation of the point-conductance model (4.2). Assuming that the average activity of the membrane potential remains constant (steady-state), the membrane equation reduces to:

$$\overline{V} = \frac{E_L + r_e E_e + r_i E_i}{1 + r_e + r_i}, \tag{9.1}$$

where $\overline{V}$ denotes the average membrane potential and $r_{\{e,i\}} = \overline{g}_{\{e,i\}}/G_L$ defines the ratio between the average excitatory (inhibitory) and leak conductance. Denoting the ratio between the total membrane conductance (inverse of input resistance R_{in}) in activated states and in states without network activity (induced by application of TTX) with r_{in}, one obtains

$$r_{\{e,i\}} = \frac{r_{in}\overline{V} - E_L + E_{\{i,e\}}(1 - r_{in})}{E_{\{e,i\}} - E_{\{i,e\}}}. \tag{9.2}$$

This relation allows to estimate the average relative contribution of inhibitory excitatory synaptic inputs in activated states. The value of r_{in} for Up-states under

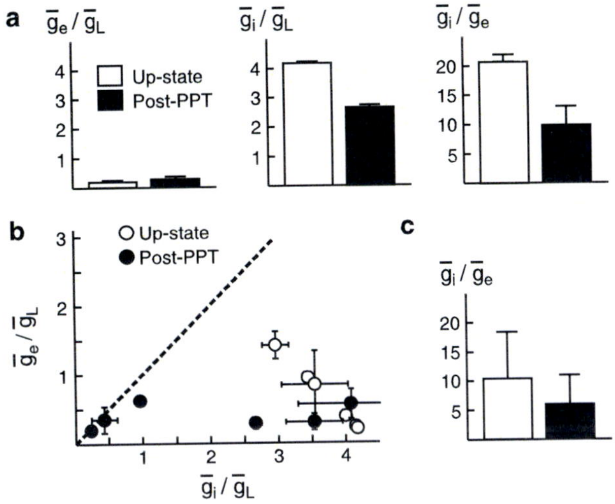

Fig. 9.1 Contribution of excitatory and inhibitory conductances during activated states as estimated by application of the standard method. (**a**) Representative example for estimates of the ratio between mean excitatory and leak conductance (*left*) as well as mean inhibitory and leak conductance (*middle*) in Up-states (*white*) and post-PPT states (*black*). These estimates were obtained by incorporating measurements of the average V_m into the passive membrane equation (estimated values: $r_i = 4.18 \pm 0.01$ and $r_e = 0.20 \pm 0.01$ for Up-state, $r_i = 2.65 \pm 0.09$ and $r_e = 0.28 \pm 0.09$ for post-PPT state), and yield a several-fold larger mean for inhibition than excitation (*left*; $\bar{g}_i/\bar{g}_e = 20.68 \pm 1.24$ for Up-state, $\bar{g}_i/\bar{g}_e = 9.81 \pm 3.23$ for post-PPT state). (**b**) Pooled results for six cells. In all cases, a larger inhibitory contribution was found (the *dashed line* indicates equal contribution). (**c**) The average ratio between inhibitory and excitatory synaptic conductances was about two times larger for Up-states. Modified from Rudolph et al. (2005)

ketamine–xylazine anesthesia is here fixed to $r_{in} = 5.38$, corresponding to a relative change of the input resistance in this states of 81.4% compared to states under TTX (no network activity; see Destexhe and Paré 1999). In Rudolph et al. (2005), the r_{in} for post-PPT states was estimated for each individual cell recorded, based on the ratio between the measured input resistance in Up-states and post-PPT states as well as this assumption of a ratio of 5.38 between the input resistance under TTX and in Up-states.

Rudolph and colleagues used this classical method (9.2) to obtain the respective ratios of the mean inhibitory and excitatory synaptic conductances to the leak conductance. Such estimates were performed for each current level in the linear I-V regime and for each cell recorded. An example is shown in Fig. 9.1a. The pooled results for all available cells indicate that the relative contribution of inhibition is several-fold larger than that of excitation (Fig. 9.1b). This holds for both Up-states and post-PPT states, although inhibition appears to be less pronounced for post-PPT states (paired T-test, $p < 0.015$; Fig. 9.1b,c). Average values are $r_i = 3.71 \pm 0.48$, $r_e = 0.67 \pm 0.48$ for Up-states, and $r_i = 1.98 \pm 1.65$, $r_e = 0.38 \pm 0.17$ for post-PPT states. From this, the ratio between the mean inhibitory and excitatory synaptic conductances can be estimated, yielding 10.35 ± 7.99 for Up-states and 5.91 ± 5.01 for post-PPT states (Fig. 9.1c).

To check for consistency, one can use the above conductance values in the passive equation to predict the average V_m using (9.1) in conditions of reversed inhibition (pipettes filled with 3 M KCl; measured E_i of -55 mV). With the values given above, this predicted $\overline{V}$ equals -51.9 mV, which is remarkably close to the measured value of $\overline{V} = -51$ mV (Paré et al. 1998b; Destexhe and Paré 1999; see Chap. 3). This analysis, therefore, shows that for all experimental conditions (ketamine–xylazine anesthesia, PPT-induced activated states, and reversed inhibition experiments), inhibitory conductances are several-fold larger than excitatory conductances. This conclusion is also in agreement with the dominant inhibitory conductances seen in the cortex of awake cats during spontaneous activity as well as during natural sleep, which we will describe in the next section.

To determine the absolute values of conductances and their variance, the VmD method (Rudolph et al. 2004; see Sect. 8.2) can be used. This analysis makes use of an analytic expression of the steady-state V_m distribution, given as a function of effective synaptic conductance parameters, which can be fit to experimentally obtained V_m distributions. Figure 9.2a illustrates this method for a specific example of Up-state and post-PPT state. Restricting to a linear regime of the I-V-relation (see insets in Fig. 9.2a), by fitting the V_m distributions $\rho(V)$ obtained at different current levels with Gaussians (Fig. 9.2b, left panels), the mean and variance of excitatory and inhibitory synaptic conductances can be deduced (Fig. 9.2b, right panels). Because the VmD method requires two different current levels, in Rudolph et al. (2005) the available experimental data for 3 (or 4) current levels allowed 3 (or 6) possible pairings. In this study, for each investigated cell, the values obtained from all pairings were averaged.

In a first application of the VmD method, Rudolph et al. (2005) estimated the synaptic conductances with reference to the estimated leak conductance in the presence of TTX. This analysis yielded the following absolute values for the mean and variance of inhibitory and excitatory synaptic conductances (see Fig. 9.3a,b): $g_{i0} = 70.67 \pm 45.23$ nS, $g_{e0} = 22.02 \pm 37.41$ nS, $\sigma_i = 27.83 \pm 32.76$ nS, $\sigma_e = 7.85 \pm 10.05$ nS for Up-states and $g_{i0} = 37.80 \pm 23.11$ nS, $g_{e0} = 6.41 \pm 4.03$ nS, $\sigma_i = 8.85 \pm 6.43$ nS, $\sigma_e = 3.10 \pm 1.95$ nS for post-PPT states. In agreement with the results obtained with the standard method (see above), these values show a much larger contribution of inhibitory conductances, albeit less pronounced in post-PPT states (paired T-test, $p < 0.07$ for ratio of inhibitory and excitatory mean, $p < 0.05$ for ratio of inhibitory and excitatory SD in both states). The ratio between inhibitory and excitatory mean conductances was found to be 14.05 ± 12.36 for Up-states and 9.94 ± 10.1 for post-PPT states (Fig. 9.3b, right). Moreover, in this study, inhibitory conductances displayed the largest variance (the SD of the inhibitory synaptic conductance σ_i was 4.47 ± 2.97 times larger than σ_e for Up-states and 3.16 ± 2.07 times for post-PPT states; Fig. 9.2b, right) and have, thus, a determinant influence on V_m fluctuations.

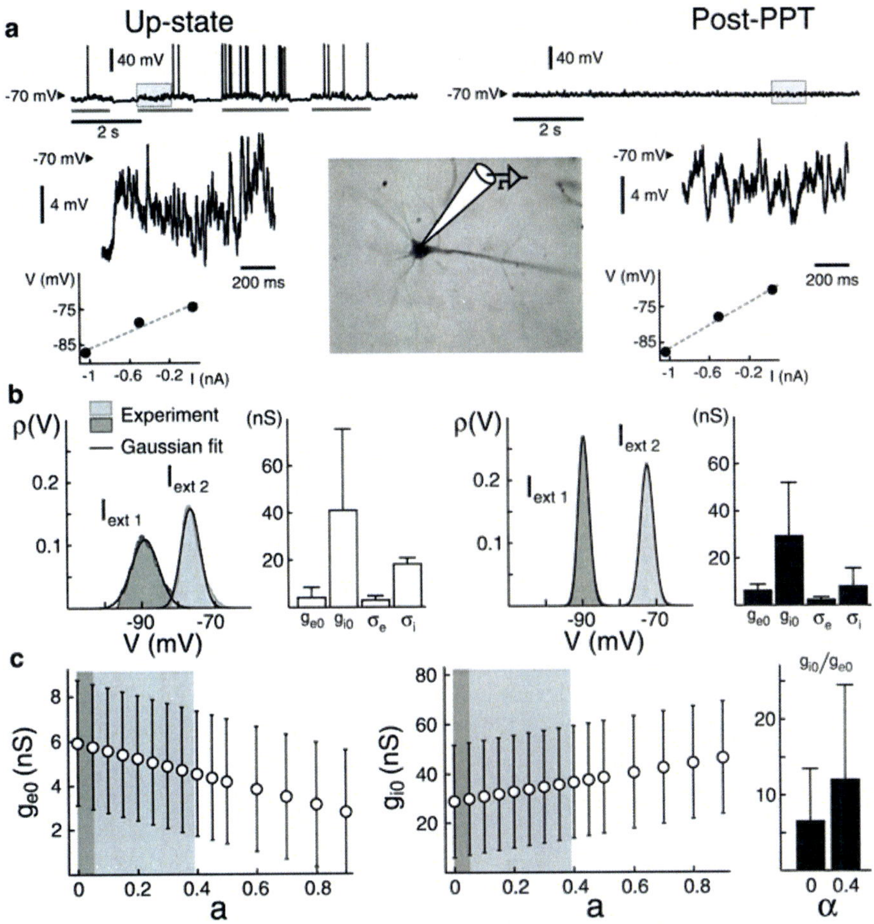

Fig. 9.2 Estimation of synaptic conductances during activated states using the VmD method. (**a**) Examples of intracellular activity during Up-states (*gray bars*) and post-PPT states. Insets show the recorded cell, enlarged intracellular traces (*gray boxes*) and the I-V-curves in the corresponding states. (**b**) Membrane potential distributions $\rho(V)$ (*gray*) and their Gaussian fits (*black*) in both states at two different injected currents ($I_{\mathrm{ext}1} = -1.04\,\mathrm{nA}$, $I_{\mathrm{ext}2} = 0.04\,\mathrm{nA}$). *Right panels* show estimations of the mean (g_{e0}, g_{i0}) and SD (σ_e, σ_i) of excitatory and inhibitory synaptic conductances in both states ($g_{e0} = 4.13 \pm 4.29\,\mathrm{nS}$, $g_{i0} = 41.08 \pm 34.37\,\mathrm{nS}$, $\sigma_e = 2.88 \pm 1.86\,\mathrm{nS}$, $\sigma_i = 18.04 \pm 2.72\,\mathrm{nS}$, $g_{i0}/g_{e0} = 21.13 \pm 16.57$, $\sigma_i/\sigma_e = 7.51 \pm 3.9$ for Up-state; $g_{e0} = 5.94 \pm 2.80\,\mathrm{nS}$, $g_{i0} = 29.05 \pm 22.89\,\mathrm{nS}$, $\sigma_e = 2.11 \pm 1.15\,\mathrm{nS}$, $\sigma_i = 7.66 \pm 7.93\,\mathrm{nS}$, $g_{i0}/g_{e0} = 6.53 \pm 5.52$, $\sigma_i/\sigma_e = 3.06 \pm 2.09$ for post-PPT state). These estimates show an about 20 times larger contribution of inhibition over excitation for the Up-state, and about 10 times larger for the post-PPT state. (**c**) Impact of neuromodulators on the mean excitatory (*left*) and inhibitory (*middle*) conductance estimates. α labels the neuromodulation-sensitive fraction of the leak conductance (9.3). *Light gray* areas indicates the experimentally evidenced parameter regime of a contribution of Down-regulated potassium conductances (Krnjević et al. 1971) which, however, is expected to be small at hyperpolarized levels (McCormick and Prince 1986), thus rendering accompanying changes in conductance estimates negligible (*dark gray*). The *right panel* shows the impact of neuromodulators on the ratio between excitatory and inhibitory mean conductances for the lower and upper limit of the experimentally evidenced parameter regime. Modified from Rudolph et al. (2005)

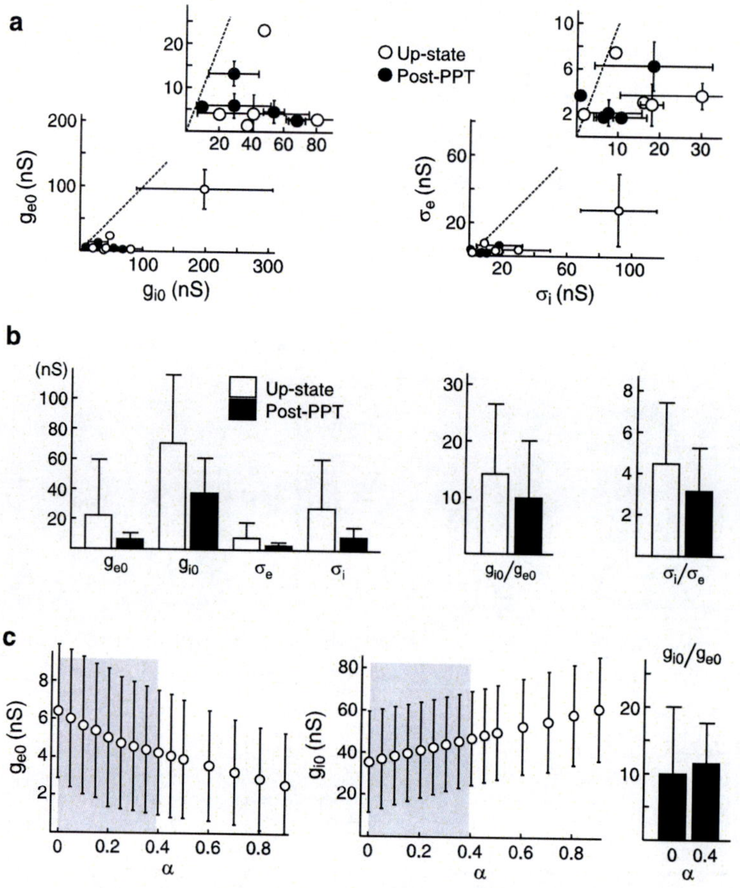

Fig. 9.3 Characterization of synaptic conductances during activated states using the VmD method. (**a**) Pooled result of conductance estimates (mean g_{e0}, g_{i0} and standard deviation σ_e, σ_i for excitatory and inhibitory conductances, respectively) for all cells (*white*: Up-states; *black*: post-PPT states). The insets show data for smaller conductances. (**b**) Mean and standard deviation of synaptic conductances (*left*) as well as their ratios (*middle* and *left*) averaged over whole population of available cells (for estimated values see text). (**c**) Pooled result for the impact of neuromodulator-sensitive potassium conductance on estimates of the mean excitatory (*left*) and inhibitory (*middle*) conductance as a function of the parameter α [see (9.3)]. The *light gray* area indicates the experimentally evidence parameter regime with a contribution of downregulated potassium conductances (Krnjević et al. 1971). The *right panel* shows the only minor impact of neuromodulators on the ratio between g_{i0} and g_{e0} for the lower and upper limit of the experimentally evidenced parameter regime ($g_{i0}/g_{e0} = 9.94 \pm 10.1$ for $\alpha = 0$, and 11.61 ± 6.06 for $\alpha = 0.4$). Modified from Rudolph et al. (2005)

9.2.2 Contribution of Downregulated K$^+$ Conductances

It is known that PPT-induced EEG-activated states are suppressed by systemic administration of muscarinic antagonists (Steriade et al. 1993a). Thus, following PPT stimulation, cortical neurons are in a different neuromodulatory state, likely due to the release of acetylcholine. Given that muscarinic receptor stimulation blocks various K$^+$ conductances in cortical neurons (McCormick 1992), thus leading to a general increase in $R_{\rm in}$ and depolarization (McCormick 1989), Rudolph et al. (2005) assessed the contribution of K$^+$ channels on the above conductance measurements. To this end, the leak conductance $G_{\rm L}$ was decomposed into a permanent (neuromodulation-insensitive) leak conductance $G_{\rm L0}$, and a leak potassium conductance sensitive to neuromodulators $G_{\rm KL}$:

$$G_{\rm L} = G_{\rm L0} + G_{\rm KL} \,. \tag{9.3}$$

Moreover, denoting with $E_{\rm K}$ the potassium reversal potential, the passive leak reversal potential in the presence of $G_{\rm KL}$ takes the form

$$E_{\rm L} = \frac{G_{\rm L0}\,E_{\rm L0} + G_{\rm KL}\,E_{\rm K}}{G_{\rm L0} + G_{\rm KL}} \,, \tag{9.4}$$

where $E_{\rm L0}$ denotes the reversal for the $G_{\rm L0}$ conductance. Introducing the scaling parameter α ($0 \leq \alpha \leq 1$) by rewriting $G_{\rm KL} = \alpha G_{\rm L}$, the impact of the neuromodulator-sensitive leak conductance can be tested. Here, $\alpha = 0$ denotes the condition were the effect of neuromodulators on leak conductance is negligible, whereas for $\alpha = 1$, the totality of the leak is suppressed by neuromodulators.

Experiments indicate that the change of $R_{\rm in}$ of cortical neurons induced by ACh is less than 40% at a depolarized $V_{\rm m}$ between -55 and -45 mV (39% in Krnjević et al. 1971; 26.4 $\pm$ 12.9% in McCormick and Prince 1986), and drops to about 5% at hyperpolarized levels between -85 and -65 mV (4.6 $\pm$ 3.8% in McCormick and Prince 1986). The range of $V_{\rm m}$ in the experiments of Rudolph and colleagues corresponded in all cases to the latter values. One, thus, would expect α to be small around 0.05 to 0.1 (i.e., 5 to 10% $R_{\rm in}$ change). Although the conductance analysis for different values of α shows that there can be up to twofold changes in the values of $g_{\rm e0}$ and $g_{\rm i0}$ (see Fig. 9.2c for specific example and Fig. 9.3c for population result), for α between 0.05 to 0.1 these changes are minimal. Moreover, the finding that synaptic noise is mainly inhibitory in nature is not affected by incorporating the effect of ACh on $R_{\rm in}$ and $E_{\rm L}$ (Figs. 9.2c and 9.3c, right).

9.2.3 Biophysical Models of EEG-Activated States

One of the neurons recorded in the study of Rudolph et al. (2005) was reconstructed by using a computerized tracing system. This reconstructed 3-dimensional pyramidal morphology, shown in Fig. 9.4a, was then integrated into the NEURON simulation environment (Hines and Carnevale 1997; Carnevale and Hines 2006),

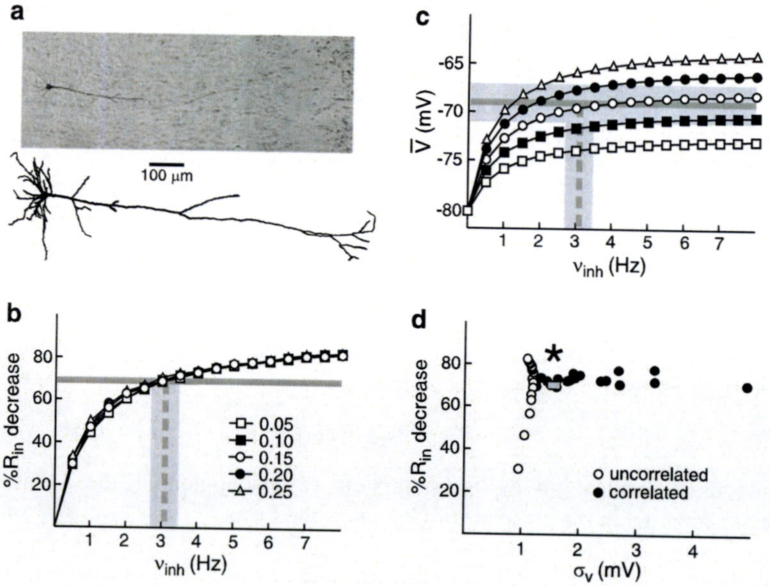

Fig. 9.4 Estimation of synaptic activity in EEG-activated periods elicited by PPT stimulation using biophysically detailed models. (**a**) Morphologically reconstructed layer V neocortical pyramidal neuron of cat parietal cortex incorporated in the modeling studies. (**b**) Dependence of the input resistance R_{in} change on the inhibitory release rate ν_{inh} and the ratio between ν_{inh} and ν_{exc} (ratios of $\nu_{\mathrm{inh}}/\nu_{\mathrm{exc}}$ from 0.05 to 0.25 were considered). (**c**) Dependence of the average membrane potential $\overline{V}$ on ν_{inh} and the ratio between ν_{inh} and ν_{exc} [see legend in (**b**)]. (**d**) Dependence of changes in R_{in} and σ_V on the level of temporal correlation c in the synaptic activity. For uncorrelated synaptic activity, changes in the release rates primarily impact on R_{in} while leaving the fluctuation amplitude nearly unaffected (*white dots*), whereas changes in the correlation for fixed release rates do not change R_{in} but markedly affect σ_V (*black dots*). In all cases the observed experimental and estimated values are indicated by *gray horizontal* and *vertical bars* (mean ± SD), respectively (for values see text). Modified from Rudolph et al. (2005)

and the so constructed model endowed with a realistic density of excitatory and inhibitory synapses, as well as quantal conductances adjusted according to available estimates (see Sect. 4.2). Rudolph et al. (2005) compared this computational model with the corresponding intracellular recordings obtained in the same cell. Specifically, the parameters of synaptic background activity were varied until the model matched these recordings, by utilizing a previously proposed search strategy which is based on matching of experimental constraints (Destexhe and Paré 1999), such as the average V_{m} ($\overline{V}$), its variance (σ_V), and the R_{in} (Fig. 9.4). This method allows to estimate the activity at excitatory and inhibitory synaptic terminals, such as the average release rate and temporal correlation (for an application of this method to Up-states under ketamine–xylazine anesthesia, see Destexhe and Paré 1999; see also Fig. 4.2 in Chap. 4).

In the particular neuron shown in Fig. 9.4a, the post-PPT state is characterized by a R_{in} which is about 3.25 times smaller compared to that estimated in a quiescent network state (corresponding to a R_{in} decrease of about 69%; Fig. 9.4b, *gray solid*). Moreover, at rest the average V_{m} is $\overline{V} = -69 \pm 2\,\text{mV}$ (Fig. 9.4c, *gray solid*) with a SD of $\sigma_{\text{V}} = 1.54 \pm 0.1\,\text{mV}$ (Fig. 9.4d, *gray solid*). The optimal average rates leading to an intracellular behavior matching these measurements are $\nu_{\text{inh}} = 3.08 \pm 0.40\,\text{Hz}$ for GABAergic synapses with a ratio between inhibitory and excitatory release rates of about 0.165, resulting in $\nu_{\text{exc}} = 0.51 \pm 0.10\,\text{Hz}$ (Fig. 9.4b,c, *gray dashed*). In addition, a weak correlation of $c = 0.25$ (see Appendix B for an explanation of this parameter) was found necessary to match the amplitude of the V_{m} fluctuations (Fig. 9.4d, *star*).

To test if the estimated synaptic release rates and correlation are consistent with conductance measurements, one can apply to the computational model both the standard method and the VmD method. This was done in Rudolph et al. (2005) using the results from modeled intracellular activity at nine different current levels ranging from -1 nA to 1 nA (thus yielding 36 current pairings). The averages of the intracellular activity (Fig. 9.5a) as well as the estimates for the mean and standard deviation of synaptic conductances (Fig. 9.5c; estimated values: $g_{\text{e0}} = 5.03 \pm 0.20\,\text{nS}$, $g_{\text{i0}} = 24.57 \pm 0.87\,\text{nS}$, $\sigma_{\text{e}} = 2.12 \pm 0.18\,\text{nS}$, $\sigma_{\text{i}} = 4.74 \pm 0.86\,\text{nS}$) match, indeed, well the corresponding experimental measurements in the post-PPT state (Fig. 9.5c, compare *light* and *dark gray*; estimated values: $g_{\text{e0}} = 5.94 \pm 2.80\,\text{nS}$, $g_{\text{i0}} = 29.05 \pm 22.89\,\text{nS}$, $\sigma_{\text{e}} = 2.11 \pm 1.15\,\text{nS}$, $\sigma_{\text{i}} = 7.66 \pm 7.93\,\text{nS}$). Only in the case of σ_{i} the model yields a slight underestimation of the value deduced from experiments. This mismatch could reflect an incomplete reconstruction, or simply a larger error in the estimation of this parameter. Nevertheless, the ratios between the inhibitory and excitatory means ($g_{\text{i0}}/g_{\text{e0}} = 4.89 \pm 0.15$; model estimate using classical method: $g_{\text{i0}}/g_{\text{e0}} = 4.60 \pm 1.51$), as well as those between the SDs ($\sigma_{\text{i}}/\sigma_{\text{e}} = 2.26 \pm 0.53$), match closely the results obtained by applying the VmD method to the corresponding experimental data (Fig. 9.5c, right; $g_{\text{i0}}/g_{\text{e0}} = 6.526 \pm 5.518$, $\sigma_{\text{i}}/\sigma_{\text{e}} = 3.06 \pm 2.09$; experimental estimation using classical method: $g_{\text{i0}}/g_{\text{e0}} = 9.81 \pm 3.23$), hence cross-validating the different methods utilized.

Finally, to obtain another, independent validation of these results, the conductances underlying synaptic activity, as well as their variances, can be estimated by simulating an "ideal" voltage-clamp (negligible electrode series resistance). For that, Rudolph et al. (2005) run the model at different command voltages (nine levels, ranging from $-50\,\text{mV}$ to $-90\,\text{mV}$) using the same random seed and, hence, the same random activity at each clamped potential. After subtraction of the leak currents, the "effective" global synaptic conductances, $g_{\text{e}}(t)$ and $g_{\text{i}}(t)$, as seen from a somatic electrode, were obtained (Fig. 9.5b middle). The resulting conductance distributions (Fig. 9.5b, right) have a mean ($g_{\text{e0}} = 4.61 \pm 0.01\,\text{nS}$, $g_{\text{i0}} = 28.49 \pm 0.01\,\text{nS}$, $g_{\text{i0}}/g_{\text{e0}} = 6.18 \pm 0.02$) which corresponds quite well with those deduced from the experimental measurements by applying the VmD method (Fig. 9.5c, top). However, the voltage clamp measurements performed in this study yielded, in general, an underestimation of both σ_{e} and σ_{i} (Fig. 9.5b, right, and Fig. 9.5c, bottom; estimated values: $\sigma_{\text{e}} = 1.59 \pm 0.01\,\text{nS}$, $\sigma_{\text{i}} = 4.03 \pm 0.01\,\text{nS}$), whereas the obtained ratio between both standard deviations is in good agreement with that obtained from experimental measurements (Fig. 9.5c, right; $\sigma_{\text{i}}/\sigma_{\text{e}} = 2.54 \pm 0.02$).

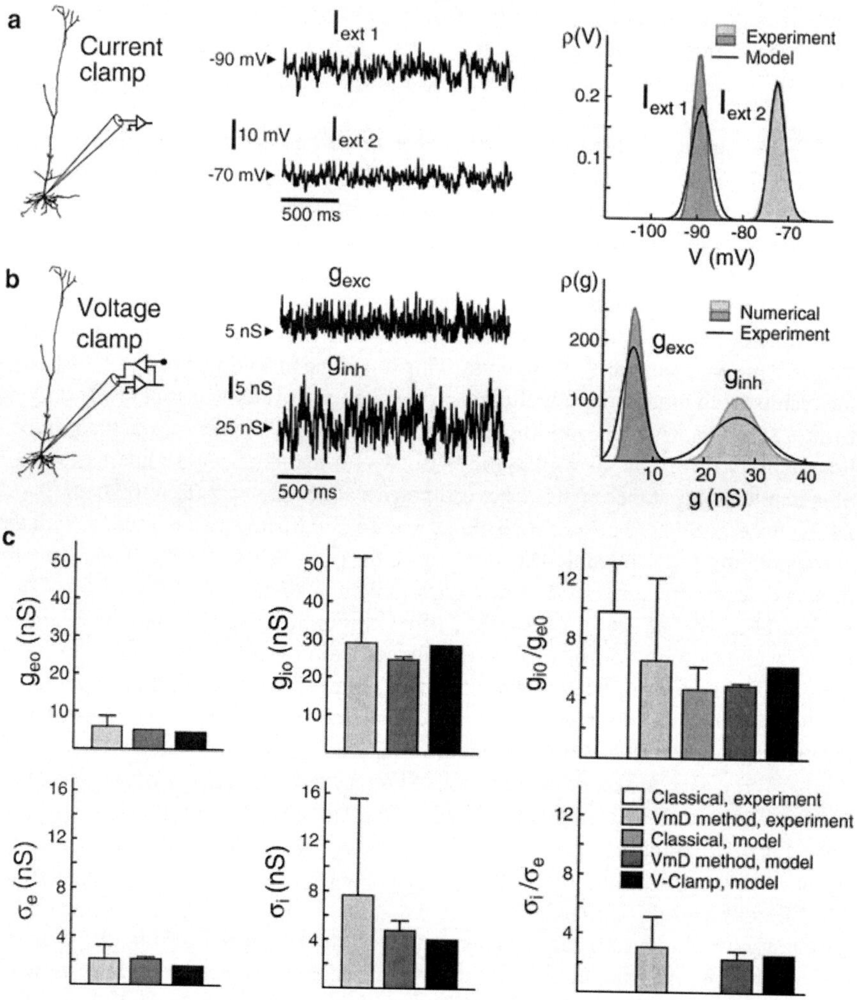

Fig. 9.5 Estimation of synaptic conductances in the detailed biophysical model. (**a**) Estimation of synaptic conductances using the VmD method. Intracellular activity at two different injected constant current levels (I_{ext1} and I_{ext2}, *middle panel*) yields V_m distributions which match well with those seen in the corresponding experiments (*right*). (**b**) An "ideal" voltage clamp (no series electrode resistance) inserted into the soma (*left*) allows to decompose the time course of inhibitory and excitatory synaptic conductances (*middle*) based on pairing of current recordings obtained at two different voltage levels. The conductance histograms (*right*, *gray*), which show the example for one pairing, are compared with Gaussian conductance distributions with mean and standard deviation taken from the VmD analysis of the experimental data. (**c**) Synaptic conductance parameters estimated from various methods applied to experimental results and the corresponding computational model. Whereas the mean conductances are in good agreement across the various methods, the detailed biophysical model shows a slight underestimation of inhibitory standard deviation. Modified from Rudolph et al. (2005)

9.2.4 Robustness of Synaptic Conductance Estimates

In each computational study, the robustness of the obtained results to changes in the parameter space has to be assessed. In the study by Rudolph and colleagues (Rudolph et al. 2005), the robustness and applicability of the employed methods for estimating conductances were tested by their application to more realistic situations with active dendrites capable of generating and conducting spikes. For that, voltage-dependent currents (I_{Na}, I_{Kd} for spike generation, and a slow voltage-dependent K^+ current for spike-frequency adaptation, a hyperpolarization-activated current I_{h}, a low-threshold Ca^{2+} current I_{CaT}, an A-type K^+ current I_{KA}, as well as a voltage-dependent cation nonselective current I_{CAN}) were included in the detailed model using densities typical for cortical neurons (see details in Rudolph et al. 2005).

The presence of voltage-dependent ion currents yields, in general, nonlinear I-V-curves. However, applying the VmD method to the linear regime of these I-V-curves provided synaptic conductance estimates in good agreement with the estimates obtained with the passive model as well as experiments (Rudolph et al. 2005). This suggests that the VmD method constitutes, indeed, a robust way for estimating synaptic contributions to the membrane conductance even in situations where the membrane shows a nonlinear behavior due to the presence of active conductances, but it is critical that only the linear portion of I-V curves is considered.

In models with active conductances, comparing conductance estimates obtained by applying the VmD method with those obtained from "ideal" somatic voltage-clamp simulations show, however, a systematic overestimation of both excitatory and inhibitory mean conductances as well as the SD of excitatory conductance. This finding is in agreement with theoretical as well as experimental results obtained in dynamic-clamp experiments performed in cortical slices (Rudolph et al. 2004). In the particular cell shown in Fig. 9.4a, inserting active conductances for spike generation often led to the presence of a large number of "spikelets" at the soma. The latter result from the arrival of full dendritic spikes, which fail to initiate corresponding somatic spikes. This high probability of spike failure is, very likely, also linked to the incomplete morphological reconstruction of the given cell (see Fig. 9.4a), in particular of its distal dendrites. In the aforementioned study, due to their small and highly variable amplitude, these spikelets could not be detected reliably and hence, were considered as part of the subthreshold dynamics. This, in turn, led to skewed V_{m} distributions, causing the observed deviations in the conductance estimates when the latter are compared to the passive model and ideal voltage-clamp situation, and in particular to an overestimation of excitatory conductances.

To evaluate the impact of a cholinergic modulation other than the K^+ conductance block described above, the voltage-dependent cation nonselective current I_{CAN} (Guérineau et al. 1995; Haj-Dahmane and Andrade 1996) with densities ranging from zero to two times the experimentally reported value of $0.02\,\mathrm{mS/cm^2}$ (Haj-Dahmane and Andrade 1996) were inserted into the model detailed above (see details in Rudolph et al. 2005). In this parameter regime, synaptic conductance estimates performed using the VmD method and an "ideal" somatic voltage clamp are, again, in good agreement with the results obtained from the corresponding experimental recordings and the passive model. Surprisingly, the estimated values for the means as well as SDs of excitation and inhibition are little affected by the I_{CAN} conductance density. This suggests that in the subthreshold regime considered for estimating synaptic conductances, the impact of I_{CAN} is negligible. This relative independence is a direct result of the activation current of I_{CAN}, which takes large values only at strongly depolarized levels, resulting in a nonlinear I-V-relation for the membrane. Moreover, the subthreshold dynamics in this case does not show spikelets and is nearly unaffected by the presence of I_{CAN} in a physiologically relevant regime of conductance densities (see Figures and details in Rudolph et al. 2005).

9.2.5 Simplified Models of EEG-Activated States

From the quantitative characterization of the synaptic conductances in the investigated activated state, it is possible to construct simplified models of cortical neurons in the corresponding high-conductance states (see Sect. 4.4). To that end, Rudolph and colleagues first analyzed the scaling structure of the PSD of the V_{m}. For post-PPT states (Fig. 9.6a), it was found that the PSD $S(\nu)$ followed a frequency-scaling behavior described by

$$S(\nu) = \frac{D\tau^2}{(1+2\pi\tau\nu)^m}, \tag{9.5}$$

where τ denotes an effective time constant, D the total spectral power at zero frequency and m the asymptotic slope for high frequencies ν. The latter is a direct indicator of the kinetics of synaptic currents (Destexhe and Rudolph 2004) and the contribution of active membrane conductances (Manwani and Koch 1999a). Consistent with this, the slope shows little variations as a function of the injected current (Fig. 9.6b, top), and of the membrane potential (Fig. 9.6b, bottom). It was found to be nearly identical for Up-states ($m = -2.44 \pm 0.31\,\mathrm{Hz}^{-1}$) and post-PPT states ($m = -2.44 \pm 0.27\,\mathrm{Hz}^{-1}$). These results indicate that, in these cells, the subthreshold membrane dynamics are mainly determined by synaptic activity, less so by active membrane conductances.

To verify if the values of synaptic time constants obtained previously are consistent with the V_{m} activity obtained experimentally, the PSD method (see Sect. 8.3; Destexhe and Rudolph 2004) can be employed. According to this method, the PSD of the V_{m} should reflect the synaptic time constants, and the simplest expression (assuming two-state kinetic models) for the PSD of the V_{m} in the presence of excitatory and inhibitory synaptic background activity is given by

$$S_V(\nu) = \frac{C_1}{(1+4\pi^2\tau_{\mathrm{e}}^2\nu^2)(1+4\pi^2\tilde{\tau}_{\mathrm{m}}^2\nu^2)} + \frac{C_2}{(1+4\pi^2\tau_{\mathrm{i}}^2\nu^2)(1+4\pi^2\tilde{\tau}_{\mathrm{m}}^2\nu^2)}, \tag{9.6}$$

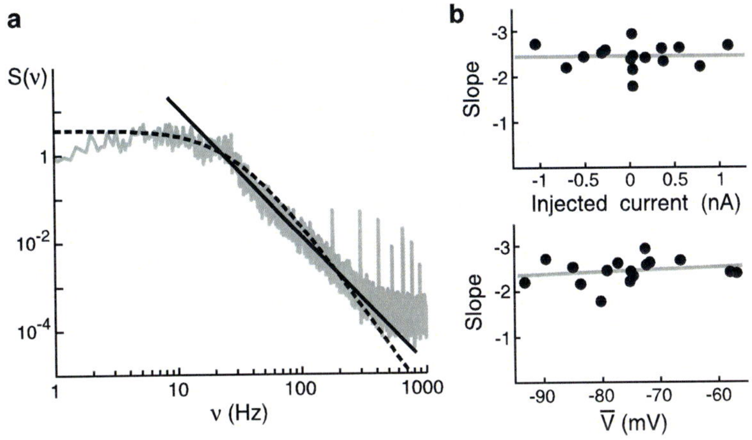

Fig. 9.6 Power spectral densities (PSDs) of membrane potential fluctuations estimated from EEG-activated states. (**a**) Example of the spectral density in the post-PPT state for the cell shown in Fig. 9.2. The *black line* indicates the slope ($m = -2.76$) obtained by fitting the V_m PSD to a Lorentzian $S_\nu = D\tau^2/(1+(2\pi\tau\nu)^m)$ at high frequencies $10\,\text{Hz} < \nu < 500\,\text{Hz}$. The *dashed line* shows the best fit using the analytic form of the V_m PSD [see (9.6)]; fitted parameters: $C_1 = C_2 = 0.183$; $\tau_e = 3$ ms, $\tau_i = 10$ ms, $\tilde{\tau}_m = 6.9$ ms). (**b**) Slope for all investigated cells as a function of the injected current (*top*) and resulting average membrane potential $\overline{V}$ (*bottom*). The slope m was obtained by fitting each V_m PSD to a Lorentzian. Modified from Rudolph et al. (2005)

where C_1 and C_2 are amplitude parameters, τ_e and τ_i are the synaptic time constants and $\tilde{\tau}_m$ denotes the effective membrane time constant in the high-conductance state. Unfortunately, not all those parameters can be extracted from a single experimental PSD. By using the values of $\tau_e = 3$ ms and $\tau_i = 10$ ms estimated previously (Destexhe et al. 2001), Rudolph et al. (2005) obtained PSDs whose behavior over a large frequency range is coherent with that observed for PSDs of post-PPT states (Fig. 9.6a, *black dashed*). However, small variations (~30%) around these values matched equally well, so the exact values of time constants cannot be estimated. The only possible conclusion is that these values of synaptic time constants are consistent with the type of V_m activity recorded experimentally following PPT stimulation.

As detailed in Sects. 4.3 and 4.4, a first type of simplified model can be constructed by reducing the branched dendritic morphology to a single-compartment receiving the same number and type of synaptic inputs as in the detailed biophysical model (Fig. 9.7b). The behavior of this simplified model can then be compared to that of the detailed model (Fig. 9.7a). Rudolph and colleagues found that, in both cases, the generated V_m fluctuations (Fig. 9.7a,b, left) have similar characteristics (Fig. 9.7a,b, middle; $\overline{V} = -70.48 \pm 0.31$ mV and -69.40 ± 0.25 mV, $\sigma_V = 1.76 \pm 0.12$ mV and 1.77 ± 0.07 mV for the detailed and simplified models, respectively; corresponding experimental values: $\overline{V} = -72.46 \pm 0.72$ mV, $\sigma_V = 1.76 \pm 0.26$ mV). Moreover, the PSD of both models

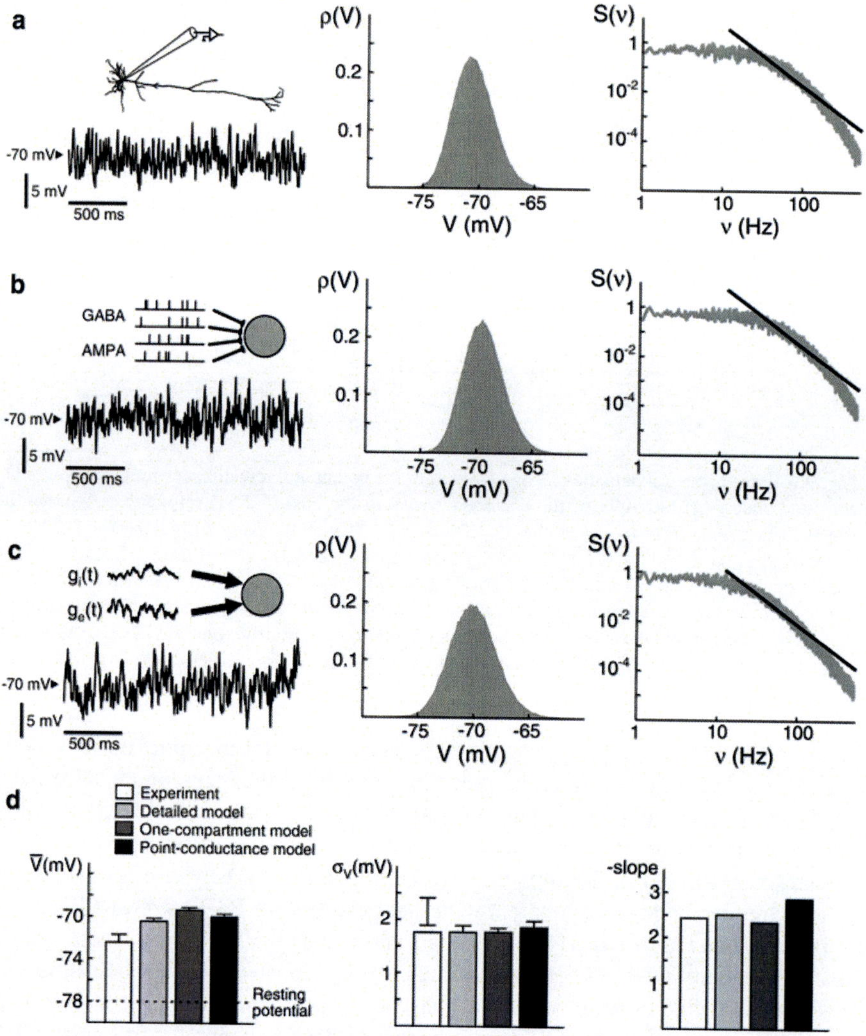

Fig. 9.7 Models of post-PPT states. All models describe the same state seen experimentally after PPT stimulation (Fig. 9.2, *right*). (**a**) Detailed biophysical model of synaptic noise in a reconstructed layer V pyramidal neuron (Fig. 9.4a). Synaptic activity was described by individual synaptic inputs (10,018 AMPA synapses and 2,249 GABAergic synapses, $\nu_{exc} = 0.51$ Hz, $\nu_{inh} = 3.08$ Hz, $c = 0.25$ in both cases) spatially distributed over an extended dendritic structure (area $a = 23{,}877\,\mu m^2$). (**b**) Corresponding single-compartment model with AMPA and GABAergic synapses. (**c**) Point-conductance model with effective excitatory and inhibitory synaptic conductances (parameters: $g_{e0} = 5.9$ nS, $g_{i0} = 29.1$ nS, $\sigma_e = 2.1$ nS, $\sigma_i = 7.6$ nS, $\tau_e = 2.73$ ms, $\tau_i = 10.49$ ms). In all models, comparable membrane potential distributions (*middle panels*) and V_m power spectral densities (*right panels*; the *black line* indicates the high-frequency behavior deduced from corresponding experimental measurements) were obtained. (**d**) Characterization of intracellular activity in models of post-PPT states. Comparison between the experimental data and results obtained with models of various complexity [see (**a**)–(**c**)]. In all cases, the average membrane potential $\overline{V}$, membrane potential fluctuation amplitude σ_V and slope m of the V_m power spectral density showed comparable values. Modified from Rudolph et al. (2005)

displays comparable frequency scaling behavior (Fig. 9.7a,b, right; slope $m = -2.52$ and $m = -2.34$ for the detailed and single-compartment models, respectively; corresponding experimental value: $m = -2.44$). Finally, the V_m fluctuations in the models matches quite well those of the corresponding experiments (Fig. 9.7d), and the power spectra deviates from the experimental spectra only at high frequencies ($\nu > 500$ Hz).

A second type of simplified model is to represent V_m fluctuations by a stochastic process. The OU process (Uhlenbeck and Ornstein 1930) is the closest stochastic process corresponding to the type of noise generated by synapses, using exponential or two-state kinetic models (Destexhe et al. 2001). This type of stochastic process also has the advantage that the estimates of the mean and variance of synaptic conductances provided by the VmD method can directly be used as model parameters. Such an approach (Rudolph et al. 2005) yielded V_m fluctuations with distributions ($\overline{V} = -70.01 \pm 0.30$ mV, $\sigma_V = 1.86 \pm 0.12$ mV) and power spectra (slope $m = -2.86$) similar to the experimental data (Fig. 9.7c; see Fig. 9.7d for a comparison between the different models).

9.2.6 Dendritic Integration in EEG-Activated States

The detailed biophysical model constructed in Sect. 9.2.3 can be used to investigate dendritic integration in post-PPT states. Here, in agreement with previous results (Hô and Destexhe 2000; Rudolph and Destexhe 2003b; see Sect. 5.2), the reduced input resistance in high-conductance states leads to a reduction of the space constant and, hence, stronger passive attenuation compared to states where no synaptic activity is present (Fig. 9.8a, compare *Post-PPT* and *Quiescent*). Thus, distal synapses experience a particularly severe passive filtering, and they must, therefore, rely on different mechanisms to influence the soma significantly.

Previously, it was proposed that during high-conductance states, neurons are in a fast-conducting and stochastic mode of integration (Destexhe et al. 2003a; see Sect. 5.7.2). It was shown that the active properties of dendrites, combined with conductance fluctuations, establish particular dynamics rendering input efficacy less dependent of dendritic location (Rudolph and Destexhe 2003b). Rudolph et al. (2005) investigated whether this scheme also applies to post-PPT states. To this end, sodium and potassium currents for spike generation were inserted into soma, dendrites and axon. To test whether the temporal aspect of dendritic integration was affected by the high-conductance state, subthreshold excitatory synaptic inputs impinging on dendrites (Fig. 9.8b) were stimulated, and the resulting somatic EPSPs with respect to their time-to-peak (Fig. 9.8c, left) and peak height (Fig. 9.8c, right) assessed.

As expected, the lower membrane time constant typical of high-conductance states leads also here to faster rising EPSPs (Fig. 9.8b, compare inset traces for *Quiescent* and *Post-PPT*). Interestingly, both the amplitude and timing of EPSPs are only weakly dependent on the site of the synaptic stimulation, in agreement with

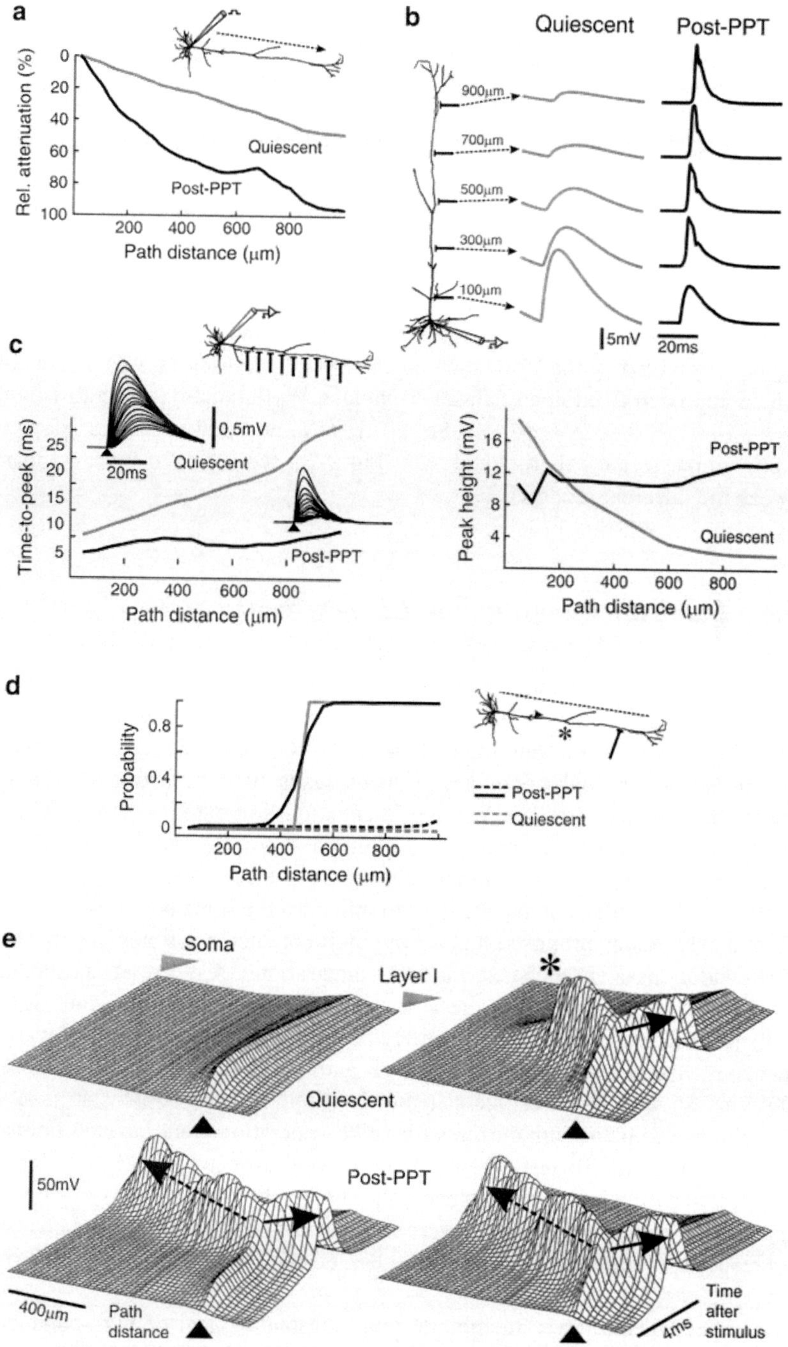
a
Rel. attenuation (%)
Quiescent
Post-PPT
Path distance (μm)
b
Quiescent
Post-PPT
900μm
700μm
500μm
300μm
100μm
5mV
20ms
c
Time-to-peek (ms)
0.5mV
20ms
Quiescent
Post-PPT
Path distance (μm)
Peak height (mV)
Post-PPT
Quiescent
Path distance (μm)
d
Probability
Path distance (μm)
Post-PPT
Quiescent
e
Soma
Layer I
Quiescent
Post-PPT
50mV
400μm
Path distance
4ms
Time after stimulus

previous modeling results (Rudolph and Destexhe 2003b). Moreover, the reported effects were found to be robust against changes in active and passive cellular properties, and to hold for a variety of cellular morphologies (Rudolph and Destexhe 2003b).

The mechanisms underlying this reduced location dependence is linked to the presence of synaptically evoked dendritic spikes (Fig. 9.8d,e). This was first shown in the model constructed in Rudolph et al. (2005) by testing the probability of dendritic spike initiation. In the quiescent state, there is a clear threshold for spike generation in the dendrites (Fig. 9.8d, *gray*). In contrast, during post-PPT states, the probability of dendritic spike generation increases gradually with path distance (Fig. 9.8d, *black*). Moreover, the probability is higher than zero even for stimulation amplitudes which are subthreshold in the quiescent case (Fig. 9.8d, compare *black* and *gray dashed*). Second, the authors found an enhancement of dendritic spike propagation (Rudolph et al. 2005). In post-PPT states, the response to subthreshold (Fig. 9.8e, bottom left) or superthreshold inputs (Fig. 9.8e, bottom right) is facilitated. Local dendritic spikes can also be evoked in quiescent states for large stimulus amplitudes, but these spikes, as reported, typically fail to propagate to the soma (Fig. 9.8e, top). These dynamics seen in this model is very similar to that predicted by a previous model (Rudolph and Destexhe 2003b), although both models correspond to different conductance states.

Fig. 9.8 Models of dendritic integration in post-PPT states. (**a**) Augmented relative passive somatodendritic voltage attenuation at steady state after somatic current injection (+0.3 nA) during activated electroencephalogram (EEG) periods. (**b**) Impact of PPT-induced synaptic activity on the location dependence of EPSP timing probed by synaptic stimuli (24 nS amplitude; stimuli were subthreshold at the soma but evoked dendritic spikes in the distal dendrites in both quiescent and post-PPT conditions) at different locations in the apical dendrite. The somatic EPSPs were attenuated in amplitude (peak height; *bottom panel*) for quiescent conditions (*gray*), but the amplitude varied only weakly with the site of synaptic stimulus in the presence of synaptic activity resembling post-PPT conditions (*black*; average over 1,200 traces). (**c**) The exact timing of EPSPs (represented as time-to-peak between stimulus and somatic EPSP peak) was weakly dependent on the synaptic location only in PPT conditions, suggesting a fast-conducting state during EEG-activated periods. (**d**) Probability of initiating dendritic spikes as a function of path distance. Whereas under quiescent conditions spike initiation occurs in an all-or-none fashion (*gray*), post-PPT conditions display a nonzero probability for evoking spikes at stimulation amplitudes (*dashed*: 1.2 nS, *solid*: 12 nS) and path distances which were subthreshold under quiescent conditions (*black*), suggesting that activated EEG periods favor dendritic spike initiation and propagation. (**e**) Somatodendritic V_m profiles for identical stimuli of amplitude (*left*: 6 nS, *right*: 12 nS) applied in the distal region of the apical dendrite [path distance 800 μm; inset (**d**)] under quiescent (*top*) and post-PPT (*bottom*) conditions. Both forward (*black dashed arrows*) and backward (*black solid arrows*) propagating dendritic spikes were observed, showing that the initiation and active forward-propagation of distal dendritic spikes are favored after PPT stimulation. Modified from Rudolph et al. (2005)

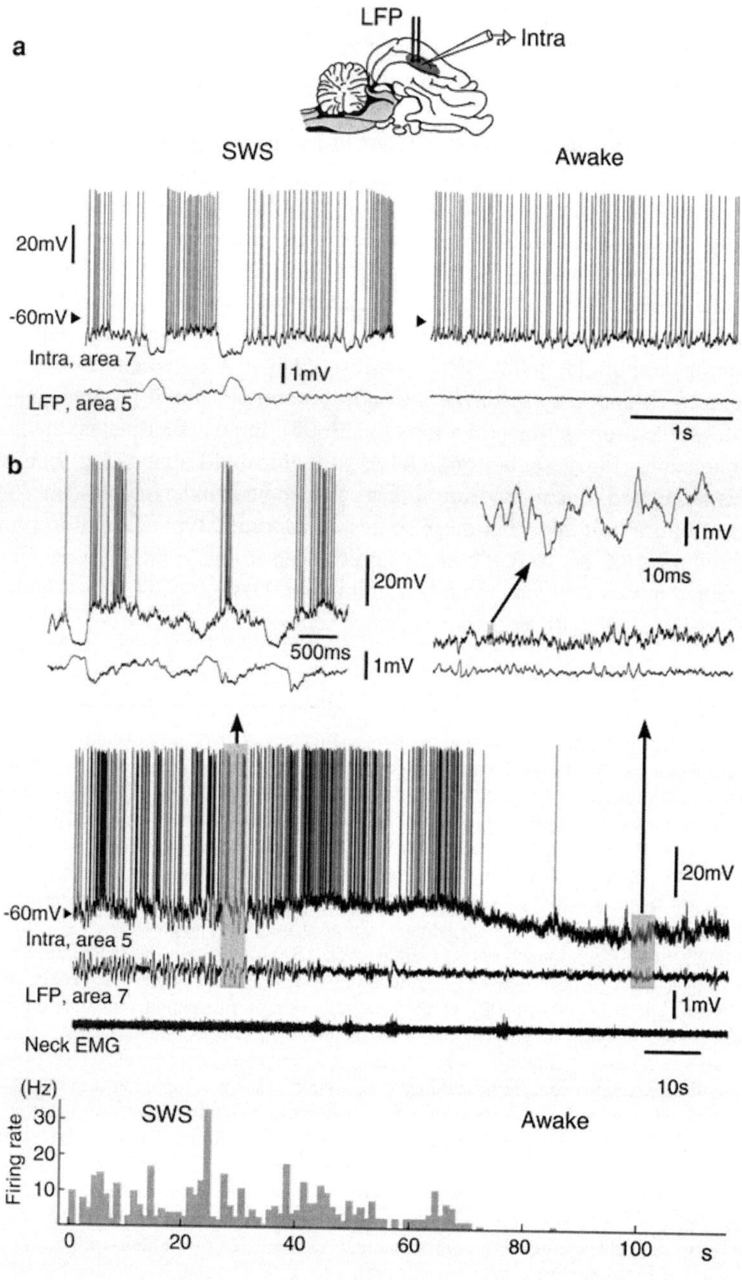
a
LFP
Intra
SWS
Awake
20mV
-60mV
Intra, area 7
1mV
LFP, area 5
1s
b
20mV
500ms
1mV
1mV
10ms
20mV
-60mV
Intra, area 5
LFP, area 7
1mV
Neck EMG
10s
(Hz)
Firing rate
30
20
10
SWS
Awake
0
20
40
60
80
100
s

9.3 Characterization of Synaptic Noise from Intracellular Recordings in Awake and Naturally Sleeping Animals

In Chap. 3, Sect. 3.2.1, we have described intracellular recordings in awake and naturally sleeping cats (Fig. 3.9; recordings obtained by Igor Timofeev and Mircea Steriade). In this section, we present the conductance analysis of such states, as well as their modeling (see details in Rudolph et al. 2007).

9.3.1 Intracellular Recordings in Awake and Naturally Sleeping Animals

As described in Sect. 3.2.1, in a study by Rudolph et al. (2007), intracellular recordings of cortical neurons were performed in parietal cortex of awake and naturally sleeping cats (Steriade et al. 2001). These recordings were done simultaneously with the LFP, electromyogram (EMG) and electrooculogram (EOG) to identify behavioral states. With pipettes filled with K^+-Acetate (KAc), the activities of 96 presumed excitatory neurons during the waking state were recorded and electrophysiologically identified. Of them, 47 neurons revealed a regular-spiking (RS) firing pattern, with significant spike-frequency adaptation in response to depolarizing current pulses, and spike width of 0.69 ± 0.20 ms (range 0.4–1.5 ms). The V_m of RS neurons varied in a range between -56 mV and -76 mV (mean -64.0 ± 5.9 mV). 26 of these RS neurons were *wake-active cells*, in which the firing was sustained all through the wake state, as described previously (Matsumura et al. 1988; Baranyi et al. 1993; Steriade et al. 2001; Timofeev et al. 2001). In these wake-active neurons, the V_m was depolarized (around -65 mV) and showed high-amplitude fluctuations and sustained irregular firing (3.1 Hz on average; range 1 to 11 Hz) during wakefulness (Fig. 9.9a). During SWS, all neurons always showed up and down-states in the V_m activity in phase with the slow waves (Fig. 9.9a, *SWS*), as described in Steriade et al. (2001).

Fig. 9.9 Activity of regular-spiking neurons during slow-wave sleep and wakefulness. (**a**) "Wake active" regular-spiking neuron recorded simultaneously with local field potentials (*LFP*; see scheme) during slow-wave sleep (*SWS*) and wakefulness (*Awake*) condition. (**b**) "Wake-silent" regular-spiking neuron recorded simultaneously with LFPs and electromyogram (EMG) during SWS to wake transition. SWS was characterized by high-amplitude low-frequency field potentials, cyclic hyperpolarizations, and stable muscle tone (expanded in *upper left panel*). Low-amplitude and high-frequency fluctuations of field potentials and muscle tone with periodic contractions characterized the waking state. This neuron was depolarized and fired spikes during initial 30 s of waking, then hyperpolarized spontaneously and stopped firing. A fragment of spontaneous V_m oscillations is expanded in the *upper right panel*. A period with barrages of hyperpolarizing potentials is further expanded as indicated by the *arrow*. Modified from Rudolph et al. (2007)

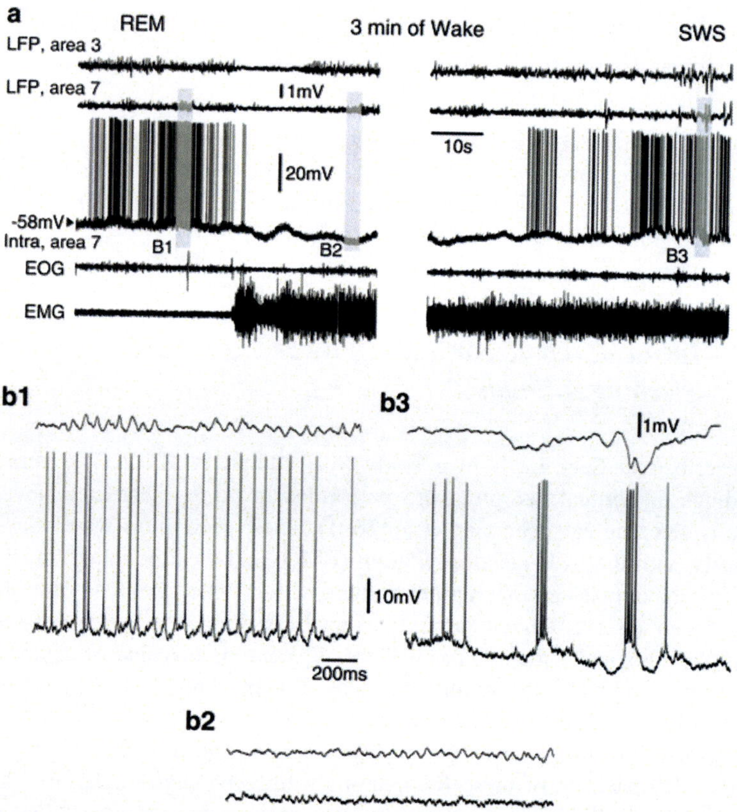

Fig. 9.10 Example of wake-silent neuron recorded through different behavioral states. This neuron ceased firing during the rapid eye movement (REM) to Wake transition (*top left panel*) and restarted firing as the animal drifted towards slow-wave sleep (*top right panel*). The *bottom panels* indicate the membrane potential and LFPs in those different states at higher resolution. Modified from Rudolph et al. (2007)

Almost half of the RS neurons (21 out of 47) recorded in Rudolph et al. (2007) were *wake-silent cells,* which systematically ceased firing during periods of quiet wakefulness (Fig. 9.9b). During the transition from SWS to waking, these wake-silent neurons continued to fire for 10–60 s, and after that period, their V_m hyperpolarized by several mV and they stopped to fire APs as long as the animal remained in the state of quiet wakefulness.

Figure 9.9b illustrates one example of a wake-silent cell which, upon awakening, had a V_m of -53.0 ± 4.9 mV and fired with frequency 10.1 ± 7.9 Hz for about 30 s. Thereafter, the V_m hyperpolarized to -62.5 ± 2.6 mV and the same neuron stopped firing. This observed hyperpolarization during waking state is not due to K^+ load because, on two occasions in the study by Rudolph et al. (2007), it was possible to obtain intracellular recordings from wake-silent neurons during a waking state that was preceded and followed by other states of vigilance (see Fig. 9.10). In this case, the recorded neuron was relatively depolarized and fired action potentials during rapid eye movement (REM) sleep. Upon awakening, this neuron was hyperpolarized by about 10 mV and stopped firing. After 3 min of waking state, the animal went to SWS state and the same neuron was depolarized and started to fire APs.

Moreover, in the mentioned study, on one occasion, extracellularly spikes from two units were recorded. One of the units stopped firing during waking state lasting about 10 min while another unit continued to emit APs. This observation suggests that it is a particular set of neurons and not local networks that stop firing during quiet wakefulness. The mean firing rates for RS neurons were 6.1 ± 6.7 Hz (silent neurons included; 10.1 ± 5.6 Hz with silent neurons excluded). No wake-silent cells were observed for other neuronal classes than RS cells, and all together, wake-silent neurons represented about 25% of the total number of recorded cells in the wake state. This large proportion of wake-silent neurons constitutes a first hint for an important role for inhibitory conductances during waking.

In contrast, in the study by Rudolph et al. (2007), no silent neuron was found for presumed interneurons. During quiet wakefulness, 22 neurons were electrophysiologically identified as fast spiking (FS). They displayed virtually no adaptation and had AP width of 0.27 ± 0.08 ms (range 0.15–0.45 ms). Upon awakening, FS neurons tended to increase firing (Fig. 9.11a,b), and none of them was found to cease firing ($n = 9$). Interestingly, the increase of firing of FS neurons seems to follow the steady hyperpolarization of RS wake-silent neurons (Fig. 9.11a). The mean firing frequency of FS neurons was 28.8 ± 20.4 Hz (Range 1–88 Hz; only two neurons fired with frequency less than 2 Hz), which is significantly higher than that of RS neurons ($p < 0.001$; see Fig. 9.11c). The mean V_m of FS neurons was -61.3 ± 4.5 mV, which is not significantly different from the V_m of RS neurons ($p = 0.059$).

To check for the contribution of K^+ conductances during quiet wakefulness, in Rudolph et al. (2007) the activities of 3 RS neurons with Cs^+-filled pipettes (Fig. 9.12) were recorded as well. The presence of cesium greatly affected the repolarizing phase of APs, demonstrating that Cs^+ was effective in blocking K^+ conductances, but the V_m distribution was only marginally affected by the presence of cesium. The action of intracellular Cs^+ may overlap with the blocking action of neuromodulators on other K^+ conductances (McCormick 1992; Metherate and Ashe 1993), which might explain the absence of effect of Cs^+ on the V_m in the study of Rudolph et al. (2007). This preliminary evidence for a limited effect of cesium during wakefulness indicates that leak and K^+ conductances have no major effect on the V_m distribution, suggesting that it is mainly determined by synaptic conductances.

In the study by Rudolph et al. (2007), the activities of 8 RS and 1 FS neurons with pipettes filled with 1.5 M KCl (see Fig. 9.13) were recorded during quiet wakefulness. The mean V_m was -62.8 ± 4.3 mV ($n = 8$), which is not statistically different from recordings with KAc (-64.0 ± 5.9 mV; $n = 47$). The firing rate of neurons recorded with KCl was 10.7 ± 15.5 Hz, which is significantly larger than neurons recorded with KAc (6.1 ± 6.7 Hz). None of these neurons were classified as wake silent. It is possible that wake-silent cells become wake-active under KCl, but

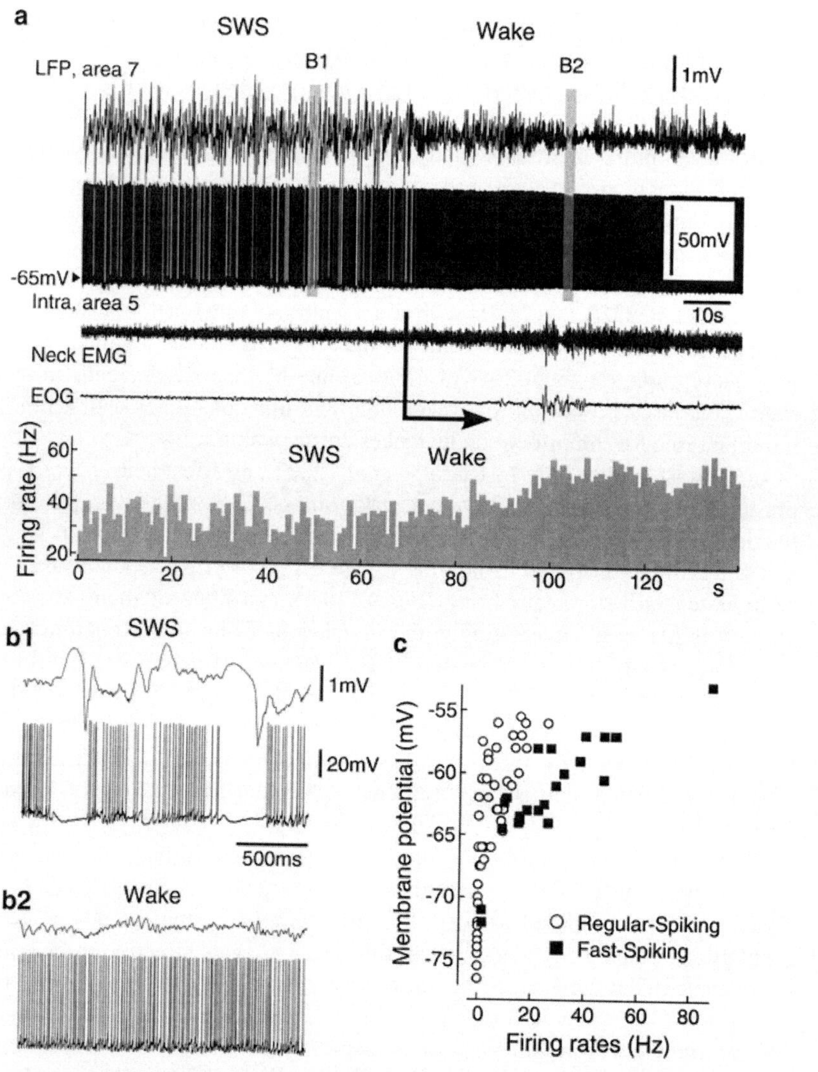

Fig. 9.11 Activity of fast-spiking interneurons upon awakening. (**a**) Intracellular activity of a fast-spiking neuron recorded simultaneously with LFPs, EMG and electrooculogram (EOG) during the transition from slow-wave sleep to wake state. The onset of the waking state is indicated by the *arrow*. Upon awakening, the mean firing rate initially remained the same as during sleep (for about 20 s), then slightly increased (see firing rate histogram at *bottom*). (**b**) Fragments of LFP and neuronal activities during slow-wave sleep and waking states are expanded as indicated in (**a**) by (**b**1) and (**b**2). (**c**) Comparison of firing rates of regular-spiking and fast-spiking neurons in wake states. Pooled results showing the mean firing rate of RS (*open circles*) and FS (*filled squares*), represented against the mean V_m during waking. Modified from Rudolph et al. (2007)

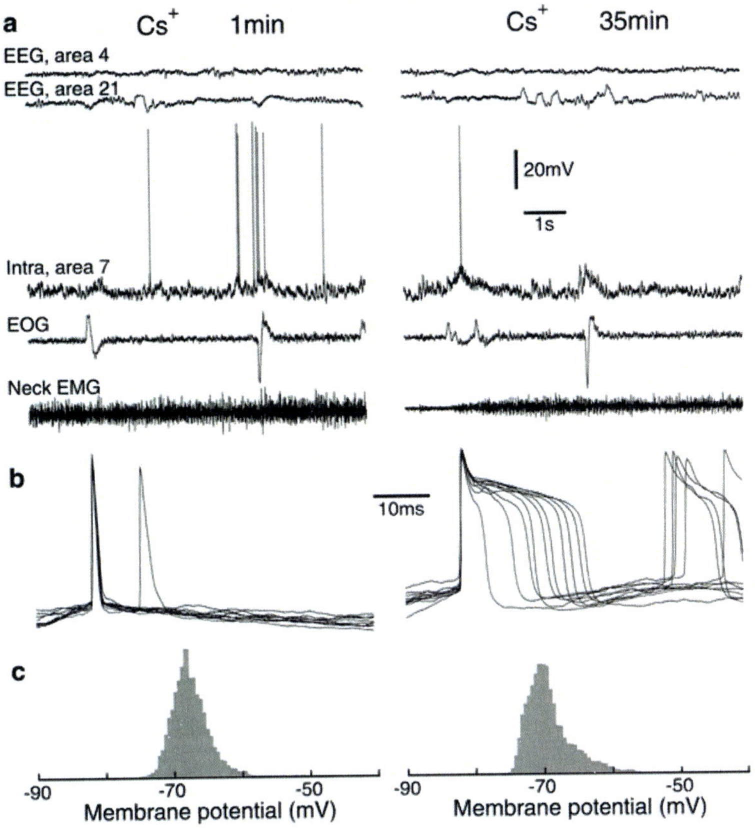

Fig. 9.12 Potassium currents contribute to spike repolarization but not to subthreshold fluctuations during waking. (**a**) Intracellular recording in an awake cat. The waking was defined as fast-frequency and low-amplitude EEG, eye movements and muscle tone. This neuron was recorded with a micropipette filled with 3 M Cs^+-Acetate (*left*: 1 min after impalement, *right*: 35 min later). (**b**) Ten superimposed spikes from early and late periods in (**a**) revealed drastic differences. Just after impalement (*left*), spikes were of about 2 ms width, as in neurons recorded with KAc micropipettes. 35 min after impalement (*right*), the Cs^+ infusion into the cell blocked K^+ currents responsible for spike repolarization, which induced plateau potentials (presumably Ca^{2+}-mediated). (**c**) V_m distributions computed just after impalement (*left*) and after 35 min (*right*). The average V_m and the amount of fluctuations were not statistically different, indicating little contribution of K^+ currents in the subthreshold V_m in the wake state. Modified from Rudolph et al. (2007), and courtesy of Igor Timofeev (Laval University)

in the aforementioned study there is not enough statistics to conclude this point. In individual neurons, Chloride infusion generally depolarized the V_m by a few millivolts (Fig. 9.13). The presumed inhibitory FS neuron fired at a frequency of 51 Hz after Chloride infusion, suggesting a larger effect of chloride infusion in this case. Although there was no control over the effective reversal of Cl^- in those recordings, the presence of hyperpolarizing IPSPs suggests that the Cl^- reversal was still below -60 mV (see expanded panel in Fig. 9.13).

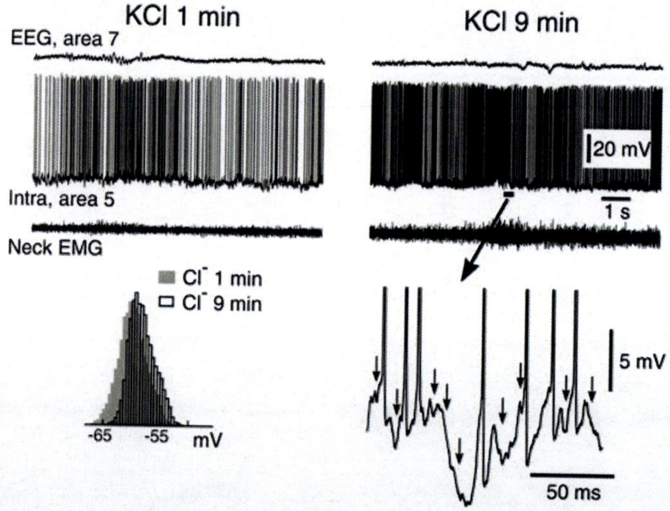

Fig. 9.13 Higher firing and depolarization following chloride infusion during waking. Intracellular recording in an awake cat. The activity of this neuron was recorded with micropipettes filled with 1.5 M of KCl and 1.0 M of K-acetate (*left*: 1 min after impalement, *right*: 9 min after impalement). The firing rate of this neuron immediately after impalement was 20.4 n $\pm$ 6.5 Hz and 9 min later it became 38.1 $\pm$ 5.7 Hz. The increased intracellular levels of Cl^- did not reverse the inhibition, since hyperpolarizing IPSPs were still recorded (indicated by *arrows* in *expanded panel*). The histograms of membrane potential of this neuron show a depolarization of about 3 mV as Cl^- diffused from the pipette. Modified from Rudolph et al. (2007), and courtesy of Igor Timofeev (Laval University)

9.3.2 *Synaptic Conductances in Wakefulness and Natural Sleep*

The primary aim of the study by Rudolph and colleagues (Rudolph et al. 2007) was to determine the relative contribution of excitatory and inhibitory conductances. To that end, the intracellular recordings described above were analyzed using the VmD method (Rudolph et al. 2004; see Sect. 8.2). As input served the V_m distributions, computed for periods of stationary activity during wakefulness and SWS Up-states. Figure 9.14b (*Awake*) shows V_m distributions of two different but representative cells obtained from periods of wakefulness, in which the studied animal (cat) and the LFPs did not show any sign of drowsiness. The obtained V_m distributions are approximately Gaussian, centered around $\bar{V} = -63.1$ mV, and the standard deviation of the V_m (σ_V) is about 3.6 mV. During SWS, the V_m distribution was calculated specifically during Up-states (Fig. 9.14b, *SWS*). It has an approximately similar shape as during wakefulness ($\bar{V} = -62.7$ mV; $\sigma_V = 3.3$ mV). Similar distributions were also observed during REM sleep. In this study, all V_m distributions were computed using several pairs of DC levels, which were selected in the linear portion of the V-I relation (Fig. 9.14a).

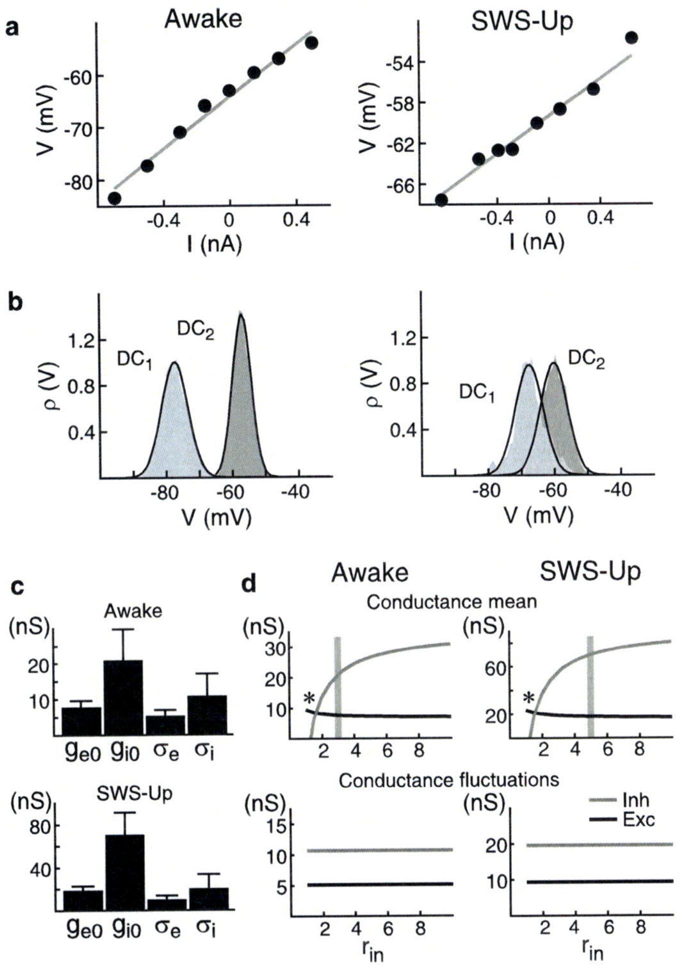

Fig. 9.14 Estimation of conductances from intracellular recordings in awake and naturally sleeping cats. (**a**) Voltage-current (V-I) relations obtained in two different cells during wakefulness (*Awake*) and the Up-states of slow-wave sleep (*SWS-Up*). The average subthreshold voltage (after removing spikes) is plotted against the value of the holding current. (**b**) Examples of V_m distributions $\rho(V)$ obtained in the same neurons as in (**a**). *Solid black lines* show Gaussian fits of the experimental distributions. (**c**) Conductance values (mean and standard deviation) estimated by decomposing synaptic activity into excitatory and inhibitory components using the VmD method (applied to 28 and 26 pairings of V_m recordings at different DC levels for Awake and SWS Up-states, respectively). (**d**) Variations of the value for conductance mean (*top*) and conductance fluctuations (*bottom*) as a function of different choices for the leak conductance. $r_{in} = R_{in}(\text{quiescent})/R_{in}(\text{active})$; * indicates the region with high leak conductances where excitation is larger than inhibition; the *gray area* shows the r_{in} values used for the conductance estimates in (**c**). Modified from Rudolph et al. (2007)

The conductance estimates obtained for several of such pairings are represented in Fig. 9.14c. In the shown example, during both wakefulness and SWS Up-states, the inhibitory conductances are several-fold larger than excitatory conductances. Variations of different parameters, such as the leak conductances (Fig. 9.14d), or the parameters of synaptic conductances, do affect the absolute values of conductance estimates, but always point to the same qualitative effect of dominant inhibition. The sole exception to this behavior is observed when considering high leak conductances, larger than the synaptic activity itself, in which case the excitatory and inhibitory conductance are of comparable magnitude (Fig. 9.14d, *).

The VmD method also provides estimates of the variance of synaptic conductances. Similar to absolute conductance estimates mentioned in the last paragraph, conductance variances were found to be generally larger for inhibition (Fig. 9.14c). However, in contrast to absolute conductance estimates, the estimates of conductance variance do not depend on the particular choice of leak conductances (Fig. 9.14d, bottom panels; Fig. 9.16c for population result). These results suggest that inhibition provides a major participation to the V_m fluctuations.

This pattern was observed in the majority of cells analyzed in Rudolph et al. (2007), although a diversity of conductance combinations are present when considering the different states of vigilance, including periods of REM sleep. In cells for which synaptic conductances were estimated (n = 11 for Awake, wake-active cells only, n = 7 for SWS Up-states, n = 2 for REM), the average V_m and fluctuation amplitude are comparable in all states ($\bar{V} = -54.2 \pm 7.5$ mV, $\sigma_V = 2.4 \pm 0.7$ mV for Awake; $\bar{V} = -58.3 \pm 4.9$ mV, $\sigma_V = 2.7 \pm 0.5$ mV for SWS-Up; $\bar{V} = -67.0 \pm 6.9$ mV, $\sigma_V = 1.9 \pm 0.6$ mV for SWS-Down; $\bar{V} = -58.5 \pm 5.2$ mV, $\sigma_V = 2.1 \pm 0.9$ mV for REM; see Fig. 9.15a). However, the total input resistance shows important variations (16.1 ± 14.5 MΩ for Awake; 12.3 ± 19.6 MΩ for SWS-Up; 22.4 ± 31.7 MΩ for SWS-Down; 8.5 ± 12.1 MΩ for REM), possibly caused by differences in the passive properties and cellular morphologies.

The estimated synaptic conductances spread over a large range of values for both mean (ranging from 5 to 70 nS and 5 to 170 nS for excitation and inhibition; Fig. 9.15b; medians: 21 nS and 55 nS for excitation and inhibition during SWS-Up, 13 nS and 21 nS for excitation and inhibition for Awake; Fig. 9.15c) and SD (ranging from 1.5 to 22 nS and 3.5 to 83 nS for excitation and inhibition; Fig. 9.16a; medians: 7.6 nS and 9.3 nS for excitation and inhibition for SWS-Up, 4.3 nS and 7.7 nS for excitation and inhibition for Awake; Fig. 9.16b). In all states and for reasonable assumptions for the leak conductance (Fig. 9.15d, *gray*), dominant inhibition was found in more than half of the cells analyzed (n = 6 for Awake and n = 7 for SWS-Up had >40% larger mean inhibitory conductance; n = 6 for Awake and n = 4 for SWS-Up had >40% larger inhibitory SD). In the remaining cells studied, inhibitory and excitatory conductance values are of comparable magnitude, with a tendency for a slight dominance of inhibition (except for n = 2 cells in Awake). Moreover, in all cells analyzed, inhibition is more pronounced during the Up-states of SWS (estimated ratios between inhibition and excitation were 2.7 ± 1.4 and 3.0 ± 2.2 for conductance mean and standard deviation; medians: 1.9 and 1.4, respectively) compared to wakefulness (ratios of 1.8 ± 1.1 and 1.9 ± 0.9 for conductance

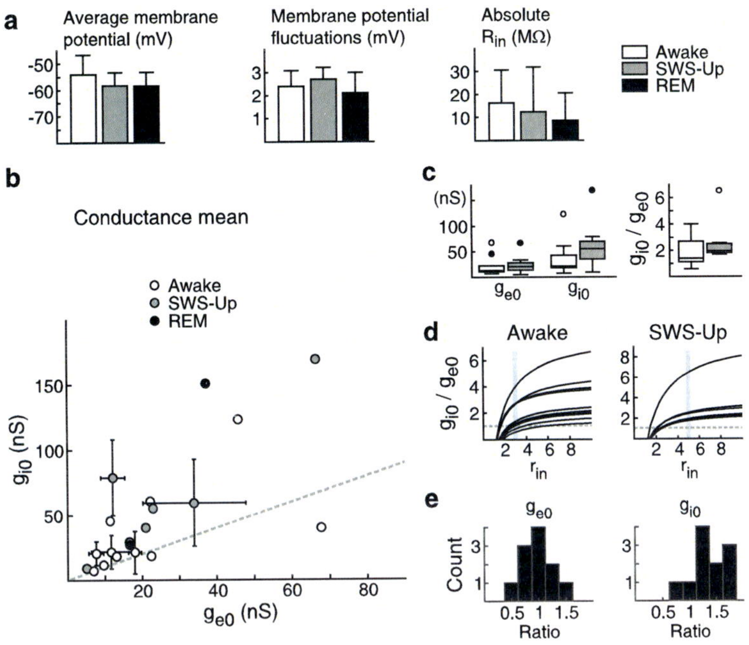

Fig. 9.15 Conductance estimates in cortical neurons during wake and sleep states. (**a**) Average V_m, V_m fluctuation amplitude and absolute input resistance R_{in} during wakefulness (*Awake*), slow-wave sleep Up-states (*SWS-Up*) and REM sleep periods, computed from all cells for which synaptic conductances were estimated. (**b**) Spread of excitatory (g_{e0}) and inhibitory (g_{i0}) conductance mean during wakefulness and slow-wave sleep Up-states. Estimated conductance values show a high variability among the investigated cells, but in almost all states, a dominance of inhibition was observed. (**c**) Box plots of mean excitatory and inhibitory conductance estimates (*left*) and average ratio between inhibitory and excitatory mean (*right*) observed during wakefulness and slow-wave sleep Up-states for the population shown in (**b**). In both states, dominant inhibition was observed, an effect which was more pronounced during SWS-Up. (**d**) Variations of the ratio between inhibitory and excitatory mean conductance values as a function of different choices for the leak conductance. $r_{in} = R_{in}$(quiescent)/R_{in}(active); the *gray area* indicates the values used for conductance estimation plotted in (**b**) and (**c**). (**e**) Histograms of conductance values relative to the leak conductance during the wake state. Modified from Rudolph et al. (2007)

mean and SD; medians: 1.4 and 1.4, respectively; see Fig. 9.15c and Fig. 9.16b, respectively). Renormalizing the conductance values to the leak conductance for each cell in the wake state leads here to values which are more homogeneous (Fig. 9.15e). In this case, the excitatory conductance was found to be of the order of the leak conductance (Fig. 9.15e, left; 0.81 ± 0.26), while inhibition is about 1.5 times larger (Fig. 9.15e, right; 1.26 ± 0.31).

These results obtained in Rudolph et al. (2007) can also be checked using the classic Ohmic conductance analysis (see Sect. 9.2.1). By integrating the V_m measurements in the various active states into the membrane equation [see (9.2)], estimates for the ratio between mean inhibitory (excitatory) conductances and the

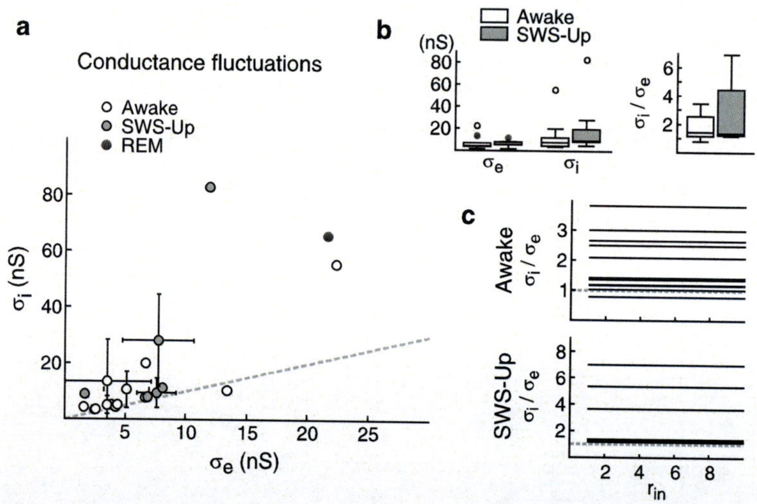

Fig. 9.16 Estimates of conductance fluctuations from cortical neurons during wake and sleep states. (**a**) Spread of excitatory (σ_e) and inhibitory (σ_i) conductance fluctuations during wakefulness and slow-wave sleep Up-states. Estimated conductance values show a high variability among the investigated cells, but in all states, a dominance of inhibition was observed. (**b**) Box plots of excitatory and inhibitory conductance fluctuation amplitude (*left*) and average ratio between inhibitory and excitatory standard deviation (*right*) observed during wakefulness (*Awake*) and slow-wave sleep Up-states (*SWS-Up*) for population shown in (**a**). In all cases, dominant inhibition was observed. (**c**) In the VmD method, estimated values for the ratio between inhibitory and excitatory conductance fluctuations do not depend on different choices for the leak conductance. $r_{in} = R_{in}(\text{quiescent})/R_{in}(\text{active})$. Modified from Rudolph et al. (2007)

leak conductance for each cell (see Fig. 9.17a) are obtained. This and the pooled results for all available cells in the aforementioned study (Fig. 9.17b), also indicate that the relative contribution of inhibition is several-fold larger than that of excitation for both wakefulness and SWS Up-states. Average values are $\overline{g}_i/\overline{g}_e = 3.2 \pm 1.3$ for SWS-Up and $\overline{g}_i/\overline{g}_e = 1.7 \pm 1.1$ during wakefulness. Also here, these values were relatively robust against the choice of the leak conductance (Fig. 9.17c).

Finally, in three of the cells studied by Rudolph and colleagues, the recording was long enough to span across several wake and sleep states, so that SWS and wakefulness could be directly compared. In agreement with the reduction of the average firing rate of RS neurons during the transition from SWS to wakefulness, a reduction of the mean excitatory conductance (values during wakefulness were between 40 and 93% of those during SWS-Up) and its fluctuation amplitude (between 45 and 85% of those observed during SWS-Up) was observed. In contrast to the observed increase of the firing rate of interneurons during sleep–wake transitions, the inhibitory conductances also decreased markedly (values during wakefulness were between 35 and 60% for the mean conductance, and between 10 and 71% for the standard deviation compared to corresponding values during SWS-Up).

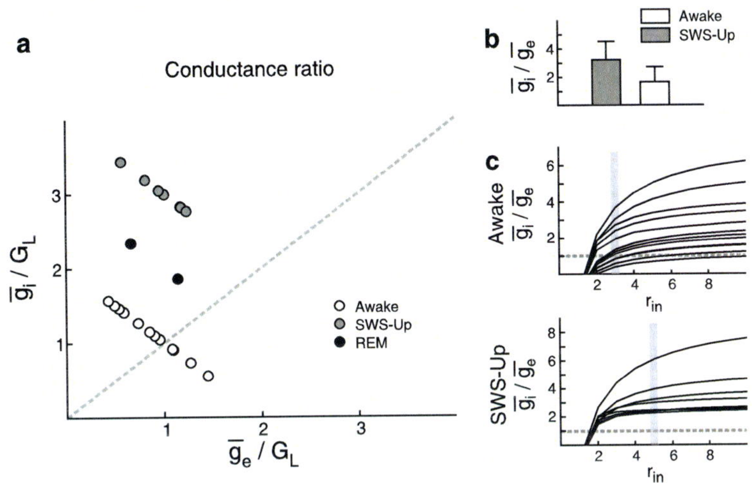

Fig. 9.17 Estimation of relative conductances from intracellular recordings using the Ohmic method. (**a**) Contribution of average excitatory (g_e) and inhibitory (g_i) conductances relative to the leak conductance G_L during wakefulness (*Awake*), slow-wave sleep Up-states (*SWS-Up*) and REM sleep periods (*REM*). Estimates were obtained by incorporating measurements of the average membrane potential (spikes excluded) into the passive membrane equation (Ohmic method, for details see Rudolph et al. 2007, Supplementary Methods). Estimated relative conductance values show a high variability among the investigated cells, but a general dominance of inhibition. (**b**) Average ratio between inhibitory and excitatory mean conductances observed during wakefulness and slow-wave sleep Up-states. Dominant inhibition was observed in both states, and more pronounced during SWS. (**c**) Variations of the ratio between average inhibitory and excitatory conductance values as a function of different choices for the leak conductance. $r_{in}=R_{in(quiescent)}/R_{in(active)}$; the *gray area* indicates the values used for conductance estimation used in (**a**) and (**b**). Modified from Rudolph et al. (2007)

9.3.3 Dynamics of Spike Initiation During Activated States

Another interesting question on the cellular level is, how the excitatory and inhibitory conductance dynamics in such states of wakefulness or natural sleep affects spike initiation. This question can, first, be studied in computational models, constrained by a quantitative characterization of synaptic noise as presented in the last section.

To that end, Rudolph et al. (2007) used a spiking model with stochastic conductances (see (4.2)–(4.4) in Chap. 4), whose parameters are given by the above estimates (see Sect. 9.3.2). Integrating the particular values of conductances shown in Fig. 9.14c led the model to generate V_m activity in excellent agreement with the intracellular recordings (Fig. 9.18a,c, *Awake*). All the present conductance measurements during the waking state were simulated in a similar way and yielded V_m activity consistent with the recordings (two more examples, with clearly dominant excitation or inhibition are shown in Fig. 9.19). Similarly, integrating the

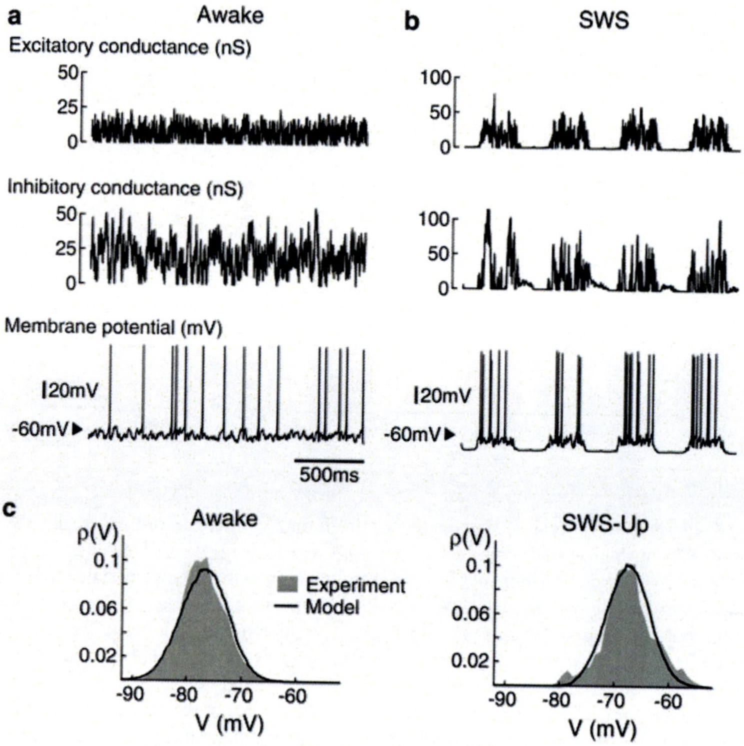

Fig. 9.18 Model of conductance interplay during wakefulness and the Up-states of slow-wave sleep. (**a**) Simulated intracellular activity corresponding to measurements in the wake state (based on conductance values shown in Fig. 9.14c; leak conductance of 13.4 nS). (**b**) Simulated up and down-states transitions (based on the values given in Fig. 9.16b). (**c**) V_m distributions obtained in the model (*black solid*) compared to that of the experiments (*gray*) in the same conditions (DC injection of −0.5 and −0.43 nA, respectively). Modified from Rudolph et al. (2007)

conductance variations, given in Fig. 9.16b, generated V_m activity consistent with the Up–down state transitions seen experimentally (Fig. 9.18b,c, *SWS* and *SWS-Up*). These results show that the conductance estimates obtained above are consistent with the V_m activity recorded experimentally.

Using this simple model, Rudolph et al. (2007) evaluated the optimal conductance changes related to spike initiation in the simulated wake state. Figure 9.20a shows that the STA displays opposite variations for excitatory and inhibitory conductances preceding the spike. As expected, spikes are correlated to an increase of excitation (Fig. 9.20, *Exc*). Less expected is that spikes are also correlated with a decrease of inhibitory conductance (Fig. 9.20, *Inh*), so that the total synaptic conductance decreases before the spike (Fig. 9.20, *Total*). Such a drop of the total conductance was not present in simulated states where inhibition was not dominant (Fig. 9.20b). In Rudolph et al. (2007), these results were also checked using different combinations of parameters, and it was found that a drop of the total conductance

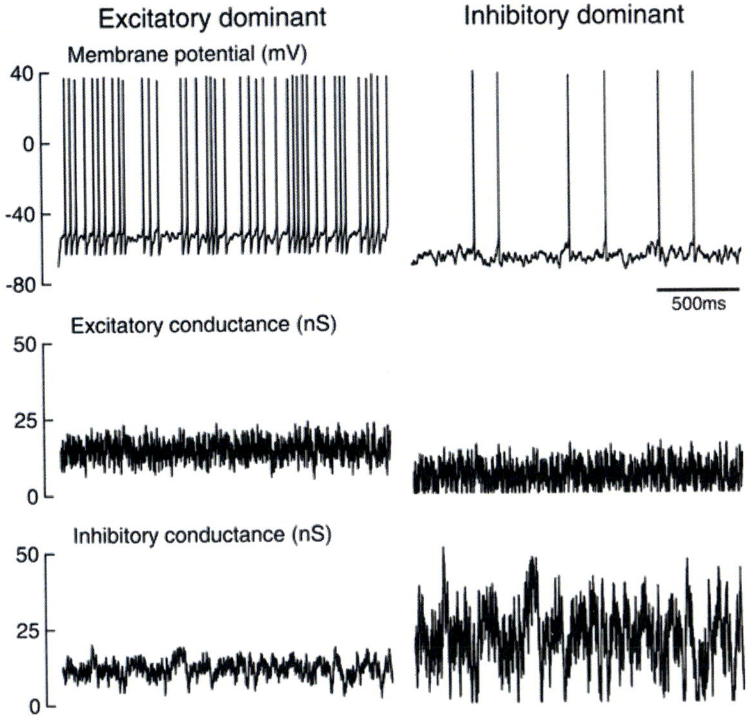

Fig. 9.19 Computational models of two different conductance dynamics in the wake state. Two examples similar to Fig. 9.18a are shown for conductance measurements in two other cells. *Left panel*: neuron where the excitatory conductance was larger than the inhibitory conductance (*Excitatory dominant*). *Right panel*: neuron for which the inhibition was more pronounced (*Inhibitory dominant*; this type of cell represented the majority of cells in the waking state). Same parameters as in Fig. 9.18a, except $g_{e0} = 14.6$ nS, $g_{i0} = 12.1$ nS, $\sigma_e = 2.7$ nS, $\sigma_i = 2.8$ nS (*left panel*); $g_{e0} = 5.7$ nS, $g_{i0} = 22.8$ nS, $\sigma_e = 3.3$ nS, $\sigma_i = 10.0$ nS (*right panel*). Modified from Rudolph et al. (2007)

was always associated with inhibition-dominant states, except when the variance of inhibition was very small. Such a drop of total conductance before the spike, therefore, constitutes a good predictor for inhibition-dominant states, given that conductance fluctuations are roughly proportional to their means.

To test this model prediction from intracellular recording, one can apply the STA method (see Sect. 8.4 in Chap. 8; Pospischil et al. 2007) to evaluate the synaptic conductance patterns related to spikes. From intracellular recordings of electrophysiologically identified RS cells, Rudolph et al. (2007) performed STAs of the V_m during wakefulness and the Up-states of SWS (Fig. 9.21a, *Avg* V_m). The corresponding STA conductances were estimated by discretizing the time axis and solving the membrane equation. This analysis revealed that the STA conductances display a drop of total membrane conductance preceding the spike (Fig. 9.21a, *Total*), which occur on a similar timescale when compared to the model

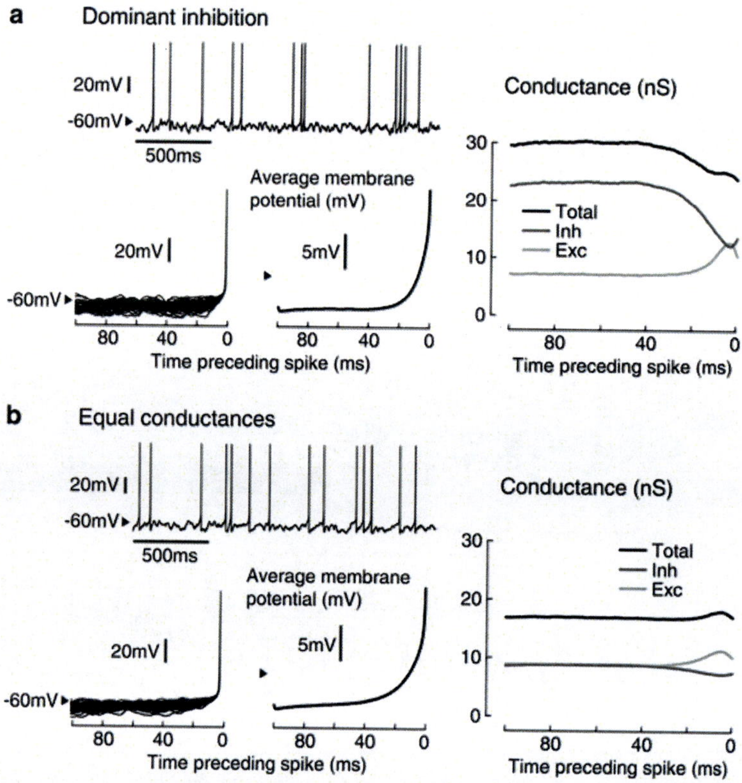

Fig. 9.20 Model prediction of conductance variations preceding spikes. (**a**) Simulated waking state with dominant inhibition as in Fig. 9.18a (*top*). *Left*: selection of 40 spikes; *middle*: spike-triggered average (STA) of the V_m; *Right*: STAs of excitatory, inhibitory and total conductance. Spikes were correlated with a prior increase of excitation, a decrease of inhibition, and a decrease of the total conductance. (**b**) Same STA procedure from a state which displayed comparable V_m fluctuations and spiking activity as in (**a**), but where excitatory and inhibitory conductances had the same mean value. The latter state was of lower overall conductance as compared to (**a**), and spikes were correlated with an increase of membrane conductance. Modified from Rudolph et al. (2007)

(compare with Fig. 9.20a). The decomposition of this conductance into excitatory and inhibitory components shows that the inhibitory conductance drops before the spike, while the excitatory conductance shows a steeper increase just prior to the spike (Fig. 9.21a; the latter increase is probably contaminated by voltage-dependent currents associated with spike generation). Such a pattern was observed in most of the cells tested in Rudolph et al. (2007) (7 out of 10 cells in Awake, 6 out of 6 cells in SWS-Up and 2 out of 2 cells in REM; see Fig. 9.21b,c). An example of a neuron which did not show such a drop of total conductance is given in Fig. 9.22. Most of the cells, however, yielded STAs qualitatively equivalent to that of the model when inhibition is dominant (Fig. 9.20a).

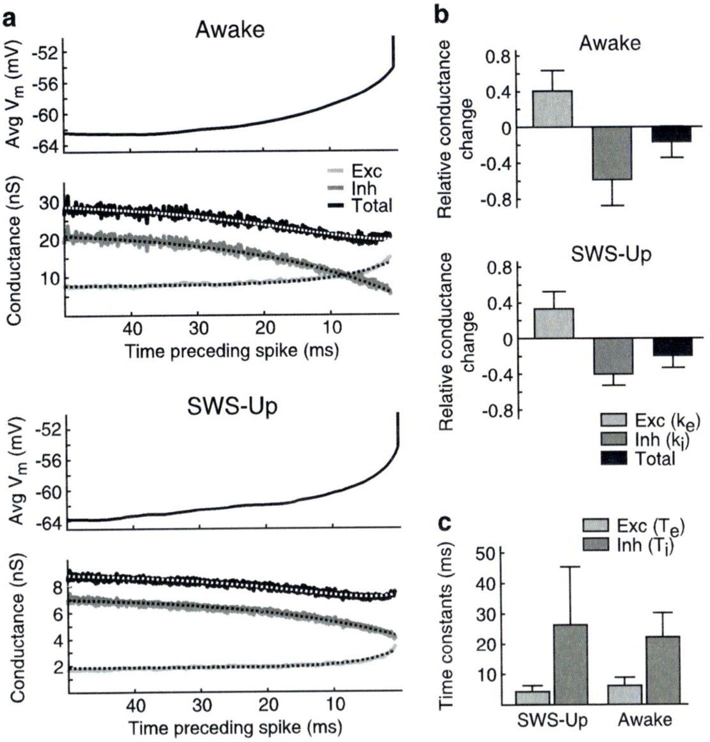

Fig. 9.21 Decrease of membrane conductance preceding spikes in wake and sleep states. (**a**) STA for the membrane potential (*Avg* V_m) as well as excitatory, inhibitory and total conductances obtained from intracellular data of regular-spiking neurons in an awake (*top*) and sleeping (*SWS-Up*, *bottom*) cat. The estimated conductance time courses showed in both cases a drop of the total conductance caused by a marked drop of inhibitory conductance within about 20 ms before the spike. (**b**) Average value for the relative conductance change (k_e and k_i) triggering spikes during wakefulness (*top*) and Up-states during SWS (*bottom*) obtained from exponential fits of the STA conductance time course [using (9.7)], for all investigated cells. A decrease of the total membrane conductance and of the inhibitory conductance is correlated with spike generation, similar to the model (Fig. 9.20a). Estimated values: $k_e = 0.41 \pm 0.23$, $k_i = -0.59 \pm 0.29$, total change: -0.17 ± 0.18 for Awake; $k_e = 0.33 \pm 0.19$, $k_i = -0.40 \pm 0.13$, total change: -0.20 ± 0.13 for SWS-Up. (**c**) Time constants of average excitatory and inhibitory conductance time course ahead of a spike in SWS and wake states. Estimated values: $T_e = 4.3 \pm 2.0$ ms, $T_i = 26.3 \pm 19.0$ ms for SWS; $T_e = 6.2 \pm 2.8$, $T_i = 22.3 \pm 7.9$ ms for Awake. Modified from Rudolph et al. (2007)

To quantify the conductance STA, in Rudolph et al. (2007) the conductance time course was fitted by using the exponential template

$$g_e(t) = g_{e0}\left(1 + k_e \exp\left(\frac{t - t_0}{T_e}\right)\right) \tag{9.7}$$

for excitation, and an equivalent equation for inhibition. Here, t_0 stands for the time of the spike, k_e quantifies the maximal increase/decrease of conductance prior to the

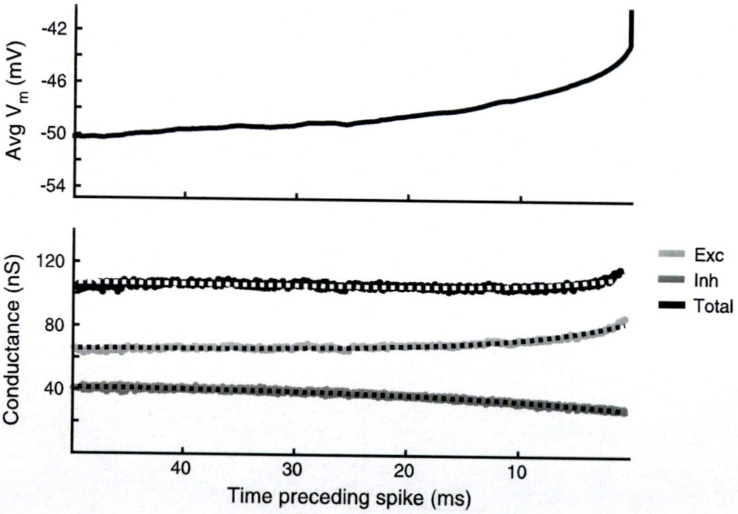

Fig. 9.22 Example of cell showing a global increase of total membrane conductance preceding spikes during the wake state. For this particular neuron recorded during the wake state, the STA showed an increase of total membrane conductance prior to the spike. Same description of panels and curves as in Fig. 9.21a (*Awake*). Modified from Rudolph et al. (2007)

spike, with time constant T_e (and similarly for inhibition). In addition, the *relative conductance change* before the spike was calculated, defined as

$$r_g = \frac{g_{e0}k_e - g_{i0}k_i}{g_{e0} + g_{i0}}. \tag{9.8}$$

Here, the terms $g_{e0}k_e$ and $g_{i0}k_i$ quantify the absolute excitatory and inhibitory conductance change before the spike, respectively. The difference between these two contributions is normalized to the total synaptic conductance. A negative value indicates an overall drop of total membrane conductance before the spike (as in Fig. 9.21a), while a positive value indicates an increase of total conductance (as in Fig. 9.22).

Finally, to relate the STA analysis to the VmD analysis, one can define the *relative excess conductance* by calculating the quantity

$$e_g = \frac{g_{e0} - g_{i0}}{g_{e0} + g_{i0}}. \tag{9.9}$$

Here, a negative value indicates a membrane dominated by inhibitory conductance, while a positive value indicates dominant excitatory conductance. In a similar manner, the *relative excess conductance fluctuations* can be defined by evaluating the quantity

$$s_g = \frac{\sigma_e - \sigma_i}{g_{e0} + g_{i0}}. \tag{9.10}$$

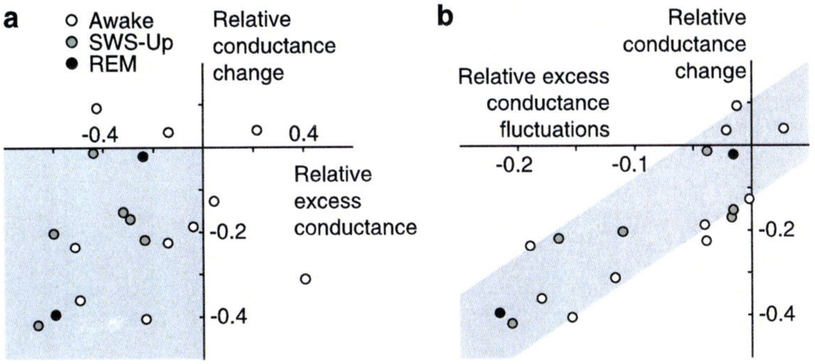

Fig. 9.23 Relation between conductance STA and the estimates of conductance and variances. (**a**) Relation between total membrane conductance change before the spike [*Relative conductance change*, (9.8)] obtained from STA analysis, and the difference of excitatory and inhibitory conductance [*Relative excess conductance*; (9.9)] estimated using the VmD method. Most cells are situated in the lower left quadrant (*gray*), indicating a relation between inhibitory-dominant states and a drop of membrane conductance prior to the spike. (**b**) Relation between relative conductance change before the spike and conductance fluctuations, expressed as the difference between excitatory and inhibitory fluctuations [*Relative excess conductance fluctuations*; (9.10)]. Here, a clear correlation (*gray area*) shows that the magnitude of the conductance change before the spike is related to the amplitude of conductance fluctuations. Symbols: wake = *open circles*, SWS-Up = *gray circles*, REM = *black circles*. Modified from Rudolph et al. (2007)

Rudolph and colleagues investigated whether the dominance of inhibition (as deduced from conductance analysis) and the drop of conductance (from STA analysis) are related, by including all cells for which both analyses could be done (Fig. 9.23). The total conductance change before the spike was clearly related to the difference of excitatory and inhibitory conductance deduced from VmD analysis (*gray area* in Fig. 9.23a), indicating that cells dominated by inhibition generally gave rise to a drop of total conductance prior to the spike. However, there was no quantitative relation between the amplitude of those changes. Such a quantitative relation was obtained for conductance fluctuations (Fig. 9.23b), which indicates that the magnitude and sign of the conductance change prior to the spike is strongly related to the relative amount of excitatory and inhibitory conductance fluctuations. The clear correlation between the results of these two independent analyses, therefore, confirms that most neurons have strong and highly fluctuating inhibitory conductances during wake and sleep states.

Finally, one can also check how the geometrical prediction (see Sect. 6.3.3) relating the sign of total conductance change preceding spikes and the ratio σ_e/σ_i performed for the above data (Fig. 9.24). It was found that the critical value of σ_e/σ_i for which the total conductance change shifts from positive to negative depends on the spike threshold. This parameter was quite variable in the cells recorded in Rudolph et al. (2007), and so a critical σ_e/σ_i value was calculated for each cell. Figure 9.24 shows the lowest and highest critical values obtained (*dashed lines*), and also displays in white the cells which do not conform to the prediction based on

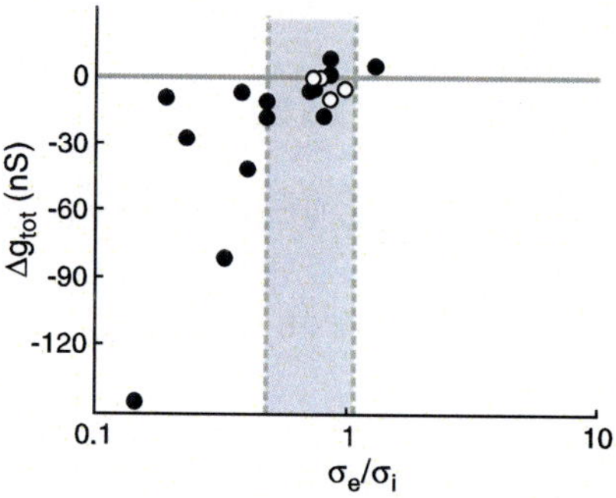

Fig. 9.24 Total conductance change preceding spikes as a function of the ratio σ_e/σ_i from a spike-triggered conductance analysis in vivo. Given the cell-to-cell variability of observed spike thresholds, each cell has a different predicted ratio separating total conductance increase cases from total conductance decrease cases. The two *dashed lines* ($\sigma_e/\sigma_i = 0.48$ and $\sigma_e/\sigma_i = 1.07$) visualize the two extreme predicted ratios. Cells in *white* are the ones not conforming to the prediction (see Piwkowska et al. 2008 for more details). Modified from Piwkowska et al. (2008)

their critical value. This was the case for only 4 out of the 18 investigated cells in the aforementioned study, for three of which the total conductance change is close to zero.

9.4 Other Applications of Conductance Analyses

In this section, we consider different applications of the conductance analyses. We first consider time-dependent conductance analyses, and their connection to stochastic models, and we also consider the estimation of correlation values from conductance measurements.

9.4.1 Method to Estimate Time-Dependent Conductances

In Chap. 8, we have discussed a method to estimate conductances from experimental data, the VmD method (see Sect. 8.2). We will demonstrate the application of this method here, but in a time-dependent framework. Specifically, this variant will yield an estimate, as a function of time, of the means and standard-deviations of the synaptic conductances, g_{e0}, g_{i0}, σ_e, σ_i, in successive time windows, as illustrated in Fig. 9.25.

Using this method, Rudolph and colleagues estimated the time evolution of conductances during Up- and Down-states of SWS (Rudolph et al. 2007). In this case, the V_m distributions were not calculated by accumulating statistics only over time, but also over repeated trials. In this study, several Up-states (one cell, between

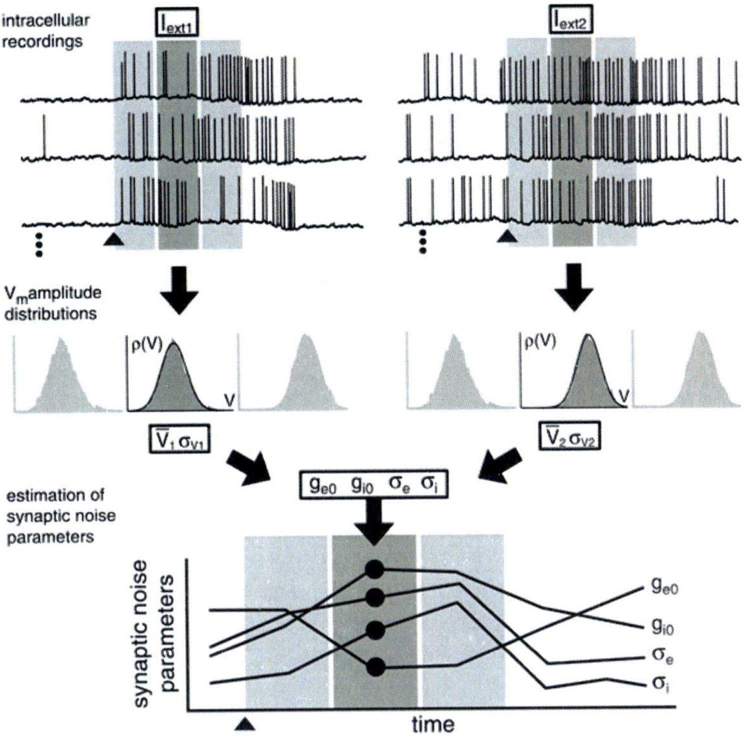

Fig. 9.25 Illustration of conductance time course analysis using the VmD method. *Top*: intracellular data from different trials are accumulated, at two different holding current levels (*left* and *right*), and time is divided into equal bins (*gray*). *Middle*: within each bin, the statistics is accumulated over time and trials to build V_m distributions $\rho(V)$. *Bottom*: from Gaussian fits of the V_m distributions (mean $\bar{V}_i$ and standard deviation σ_{V_i}), the VmD method is used to determine the parameters g_{e0}, g_{i0}, σ_e, σ_i, and the same procedure is repeated for successive time bins

6 and 36 slow-wave oscillation cycles at 8 DC levels) were selected and aligned with respect to the Down-to-Up transition as determined by the sharp LFP negativity (Fig. 9.26, left panels). The V_m distributions were then calculated within small (10 ms) windows before and after the transition. This procedure led to estimates of the time course of the conductances and their variances, as a function of time during Down–Up state transitions, and similarly for Up–down transitions (Fig. 9.26, right panels). Here, conductance changes were estimated relative to the Down-state, and not with respect to rest, as above. This analysis showed that, for the particular cell shown in Fig. 9.26, the onset of the Up-state is driven by excitation, while inhibitory conductances activate with a delay of about 20 ms, after which they tend to dominate over excitation. Also, in this case, inhibition is only slightly larger than excitation, presumably because the reference state is here the Down-state, which does not represent the true resting state. In this cell, also, the end of the Up-state was preceded by a drop of inhibition (Fig. 9.26b, *). The variance of inhibitory conductances was always larger than that of excitatory conductances (see Fig. 9.26b, bottom).

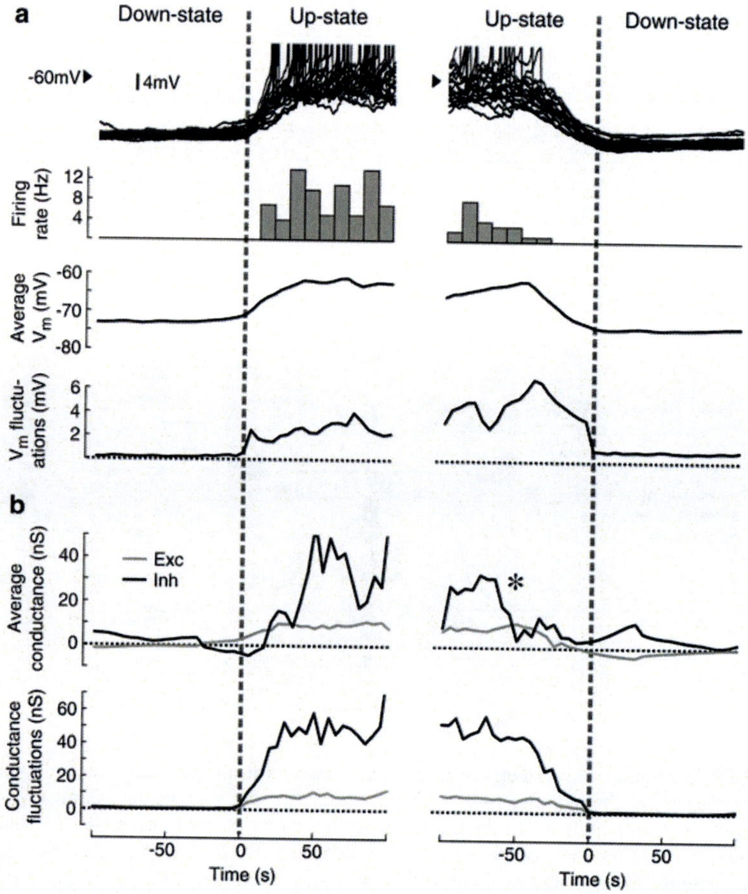

Fig. 9.26 Conductance time course during Up-and Down-states during slow-wave sleep. (**a**) Superimposed intracellular traces during transitions from Down- to Up-states (*left panels*), and Up-to Down-states (*right panels*). (**b**) Time course of global synaptic conductances during Down-up and Up-down transitions. Conductance changes were evaluated relative to the average conductance of the down-state. *Top*: excitatory (g_e, *gray*) and inhibitory (g_i, *black*) conductances; * indicates a drop of inhibitory conductance prior to the Up-down transition. *Bottom*: standard deviation of the conductances for excitation (σ_e, *gray*) and inhibition (σ_i, *black*). Both are shown at the same time reference as for (**a**). Modified from Rudolph et al. (2007)

A similar analysis was also performed from intracellular recordings in the barrel cortex of rats anesthetized with urethane (Zou et al. 2005). In this preparation, cortical neurons display spontaneously occurring slow-wave oscillations, associated with Up- and Down-states. Based on intracellular recordings at two different clamped currents, the same analysis as above can be performed. Here, however, the population activity (*Local EEG*, Fig. 9.27a) was used for the alignment (*dashed lines*) of individual intracellular recordings to the start and end of the Up-states (Fig. 9.27, left and right, respectively). As can be seen, during the Up-state, cells

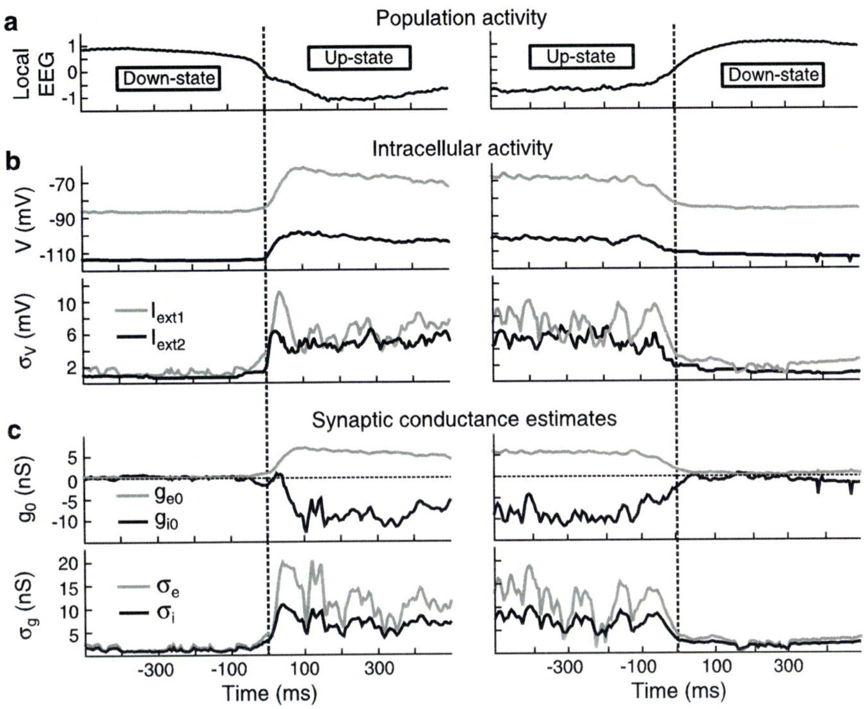

Fig. 9.27 Characterization of intracellular activity during slow-wave oscillations in vivo. (**a**) The population activity (*Local EEG*) is used to precisely align individual intracellular recordings (*dashed lines*) corresponding to the start (*left*) and end (*right*) of the Up-states characterizing slow-wave oscillations. (**b**) Gaussian approximations of the membrane potential distributions yield the mean $\overline{V}$ and standard deviation σ_V as function of time during slow-wave oscillations. Corresponding values for $\overline{V}$ and σ_V at two different currents ($I_{ext1} = 0.014$ nA, $I_{ext2} = -0.652$ nA) are shown. (**c**) With the characterization of the subthreshold membrane potential time course describing intracellular activity during slow waves, changes of the mean g_{e0}, g_{i0} and standard deviation σ_e, σ_i of excitatory and inhibitory conductances relative to corresponding values in the Down-state during slow-wave oscillations can be estimated as a function of time. Modified from Zou et al. (2005)

discharge at higher rate and show a marked depolarization and increase in the V_m variance compared to the Down-state (Fig. 9.27b). Surprisingly, in this preparation, Up- and Down-states are characterized by a similar input conductance for all six cells which were analyzed in this study, as also confirmed by another study in the same preparation (Waters and Helmchen 2006).

In Zou et al. (2005), synaptic conductances were estimated from subthreshold membrane potential fluctuations and revealed that the transition to Up-states is associated with an increase in the mean excitatory and decrease in the mean inhibitory conductance relative to respective synaptic conductances present in the Down-state. The variances of both inhibitory and excitatory conductances increases relative to their values in the Down-state, and show high-frequency fluctuations with periods around 50 ms (Fig. 9.27c, left). The termination of the Up-state shows the opposite pattern of relative synaptic conductance changes (Fig. 9.27c, right). The time of the maximum slope of changes in the conductance mean shows a slight precedence for excitation at the beginning of the Up-state in a slow-wave oscillation. Up-states terminate with a slight precedence of the decrease in the excitatory mean observed in the onset. No temporal precedence is observed in the variance of synaptic conductance changes. In this study, similar results were obtained in all cells recorded, in which local EEG recordings allowed an alignment of the intracellular traces during slow-wave oscillations.

9.4.2 Modeling Time-Dependent Conductance Variations

Two approaches are possible to integrate time-dependent conductance measurements into stochastic models of conductances. The first possibility is to use a model with varying parameters of release rate and correlation. To estimate the variations of these parameters, Zou et al. (2005) performed voltage-clamp simulations on a model with distributed synaptic inputs. By varying the release rates at glutamatergic and GABAergic synapses (ν_{AMPA} and ν_{GABA}, respectively), the mean of the excitatory and inhibitory conductances were changed and adjusted to the experimental estimates (Fig. 9.28a). Similarly, the dependence of the SD of the total synaptic conductances on the temporal correlation in the release activity at synaptic terminals was used to confine c_{AMPA} and c_{GABA} (Fig. 9.28b).

With this, slow-wave oscillations using estimate of conductance parameters were simulated (Zou et al. 2005). Here, it was found that synaptic activity corresponding to the measurements of the Up-states of slow waves lead to a depolarized and highly fluctuating V_{m}, accompanied by an irregular discharge activity. In contrast, during the Down-state the cell rests with low fluctuations at a hyperpolarized value (Fig. 9.29a). A more detailed statistical analysis showed further that the V_{m} distribution obtained for Up-states and Down-states (Fig. 9.29b) matches those found in corresponding experiments. This suggests that a simplified computational model which describes the time course of conductance during a slow-wave oscillation by fast sigmoidal changes is capable to capture the dynamics of slow waves with characteristics consistent with in vivo recordings.

9.4.3 Rate-Based Stochastic Processes

Another possibility for modeling time-dependent conductance patterns is to use stochastic process which have time-dependent parameters. As we have seen in Sect. 4.4.4), it is possible to formally obtain the point-conductance model from a stochastic process consisting of exponentially decaying synaptic events (shot noise). In particular, it is known that the mean and variance of the stochastic process is given by Campbell's theorem (Campbell 1909a,b):

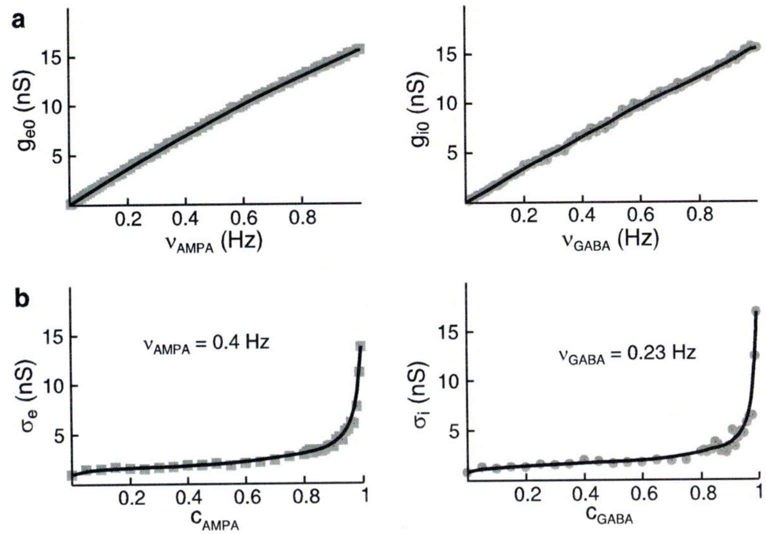

Fig. 9.28 Computing the correspondence between release parameters and global conductance properties. (**a**) Relation between mean conductance (g_{e0} or g_{i0}) and mean rate of release (ν_{AMPA} or ν_{GABA}) at excitatory (*left*) and inhibitory (*right*) synapses. (**b**) Relation between conductance variance (σ_e or σ_i) and the temporal correlation (c_{AMPA} or c_{GABA}) between excitatory (*left*) and inhibitory (*right*) synapses. The mean and variance of conductances were estimated by a somatic voltage clamp (at 0 mV and −80 mV, to estimate inhibition and excitation, respectively). Modified from Zou et al. (2005)

$$x_0 = r\alpha\tau,$$
$$\sigma_x^2 = \frac{r\alpha^2\tau}{2}, \tag{9.11}$$

where r is the rate of the stochastic process, α is the amplitude "jump" of exponential events, and τ is their decay time constant.

This process is well approximated by the following OU model:

$$\frac{dx}{dt} = -\frac{1}{\tau}(x - x_0) + \sqrt{\frac{2\sigma_x^2}{\tau}}\xi(t), \tag{9.12}$$

where $\xi(t)$ is a Gaussian-distributed (zero mean, unit SD) white noise process. Numerical simulations show that such a process can well approximate the synaptic conductances during in vivo–like activity in cortical neurons (Destexhe et al. 2001; Destexhe and Rudolph 2004; see Sect. 4.4.4). Equation (9.12) can be rewritten as

$$\frac{dx}{dt} = -\frac{x}{\tau} + \alpha r + \alpha\sqrt{r}\xi(t), \tag{9.13}$$

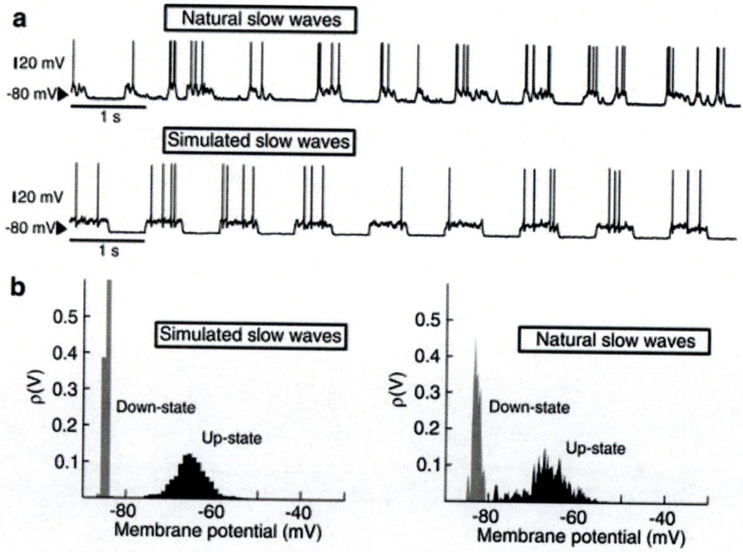

Fig. 9.29 Model of synaptic bombardment during slow-wave oscillations. (**a**) For the Down-state ($R_{in} = 38\,\text{M}\Omega$), none of the excitatory neurons fires, whereas the release rate at GABAergic synapses was $\nu_{\text{GABA}} = 0.67\,\text{Hz}$ ($g_i = 11.85\,\text{nS}$). For upstate, g_i decreased by about 8 nS, corresponding to $\nu_{\text{GABA}} = 0.226\,\text{Hz}$, whereas g_e increases by about 6 nS, corresponding input frequency ν_{AMPA} between 0.3 and 0.4 Hz. σ_i increases by 3 to 4 nS during the transition to the Up-state (corresponding to $c_{\text{GABA}} = 0.8$), whereas σ_e increases by 2 nS (corresponding to $c_{\text{AMPA}} = 0.4$). The periods of slow wave were 1 s. The slopes of changes in the mean and variance of synaptic conductances were fitted to experimental data. (**b**) V_m distribution for up and Down-state. The mean membrane potential was $-64.80\,\text{mV}$ and $-84.90\,\text{mV}$, respectively, in accordance with experimental data. Modified from Zou et al. (2005)

where the mean and variance of x are given by (9.11). In particular, one can see here that in the case the rate r approaches zero, the mean also decreases to zero, and so does the variance.

One can model synaptic background activity by two fluctuating conductances g_e and g_i, each described by such a rate-based OU process, which leads to:

$$C_m \frac{dV}{dt} = -g_L\,(V - E_L) - g_e(V - E_e) - gi(V - E_i),$$
$$\frac{dg_e}{dt} = -\frac{g_e}{\tau_e} + \alpha_e r_e + \alpha_e \sqrt{r_e}\,\xi_e(t),$$
$$\frac{dg_i}{dt} = -\frac{g_i}{\tau_i} + \alpha_i r_i + \alpha_i \sqrt{r_i}\,\xi_i(t), \qquad (9.14)$$

where V is the membrane potential, C_m is the specific membrane capacitance, g_L and E_L are the leak conductance and reversal potential, and E_e and E_i are the reversals of g_e and g_i, respectively. Excitatory conductances are described by the relaxation time τ_e, the unitary conductance α_e, the release rate r_e and an independent Gaussian noise source $\xi_e(t)$ (and similarly for inhibition).

The mean and variance of these stochastic conductances are given by

$$\begin{aligned} g_{e0} &= r_e \alpha_e \tau_e, \\ g_{i0} &= r_i \alpha_i \tau_i, \\ \sigma_e^2 &= \frac{1}{2} r_e \alpha_e^2 \tau_e, \\ \sigma_i^2 &= \frac{1}{2} r_i \alpha_i^2 \tau_i. \end{aligned} \tag{9.15}$$

To obtain these parameters from the time-dependent conductance measurements, one must estimate the "best" time course of the rates that accounts for the measured values of g_{e0}, g_{i0}, σ_e, and σ_i. From the ratio of g_{e0} and σ_e^2 (9.15), and similarly for inhibition, one obtains:

$$\begin{aligned} \alpha_e &= 2\frac{\sigma_e^2}{g_{e0}}, \\ \alpha_i &= 2\frac{\sigma_i^2}{g_{i0}}. \end{aligned} \tag{9.16}$$

These expressions provide a direct way to estimate the values of α_e and α_i from the experimental measurements, for example, by performing an average over all data points in time. For the specific example shown in Fig. 9.26, this procedure gives optimal values of $\alpha_e = 9.7$ nS and $\alpha_i = 68$ nS. These values may seem high for unitary conductances, which probably reflects the fact that the presynaptic activity is synchronized (as also indicated by the high values of σ_e and σ_i compared to the means).

Once the values of α_e and α_i have been estimated, for each data point in time, one calculates the values of r_e and r_i which best satisfy (9.15). This can be done most simply by averaging the two estimates of r_e (9.15), and similarly for inhibition. The result of such a procedure applied to Up–Down-state transitions (Fig. 9.30a) gives a time series of values for r_e and r_i (Fig. 9.30b), which in turn can be used to reconstruct the traces of g_{e0}, g_{i0}, σ_e, and σ_i (Fig. 9.30c). As can be seen, the agreement is quite good in this case.

Thus, the presented method enables to describe the measurements with only one time-varying parameter, the rate, for each conductance. This gives the time course of the mean conductances and of their variances, which reproduce the measurements of means and variances. The stochastic model (9.14) can then be used to simulate individual trials (in dynamic clamp, for example).

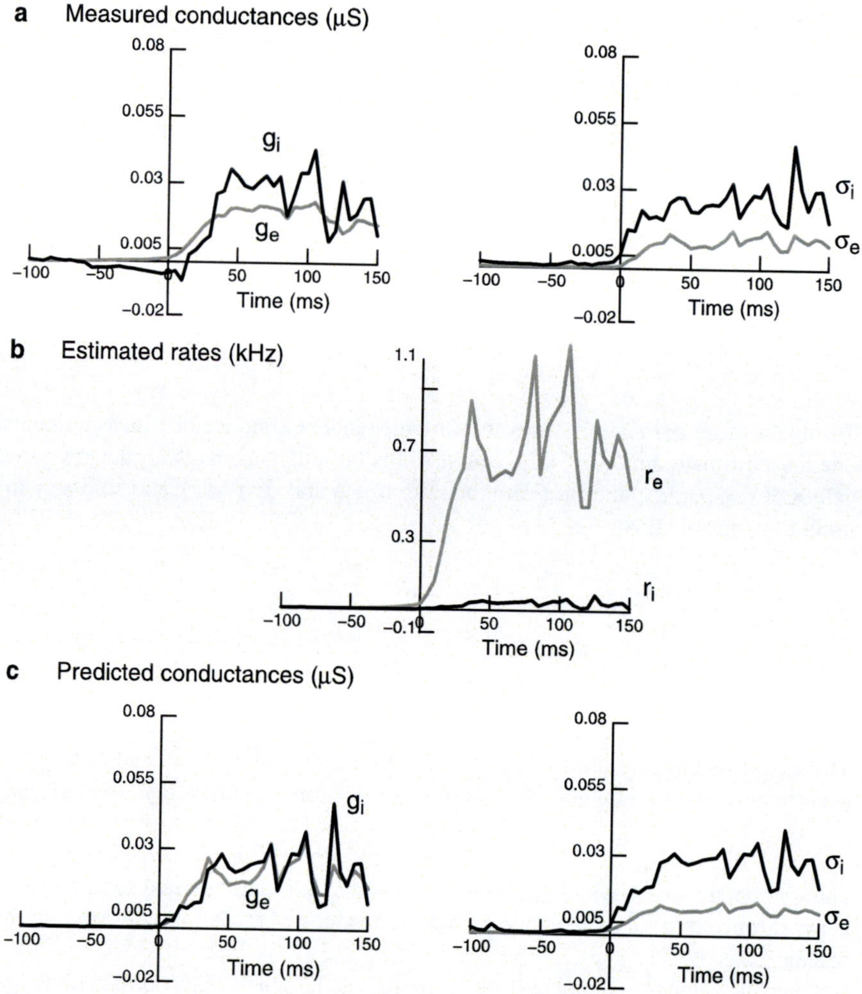

Fig. 9.30 Rate-based stochastic model of Up–Down-state transitions. (**a**) Conductance measurements during the Down-to-Up-state transition (data replotted from Fig. 9.26, *left*). (**b**) Best compromise for the rates r_e and r_i calculated from these data points (with $\alpha_e = 9.7$ nS and $\alpha_i = 68$ nS). (**c**) Predicted means (g_{e0}, g_{i0}) and standard deviations (σ_e, σ_i) of the conductances recalculated from the rate-based model in (**b**) (9.15 were used)

9.4.4 Characterization of Network Activity from Conductance Measurements

In Sect. 4.4.5, we showed that the shot-noise approach provides a powerful method to link statistical properties of the activity at many synaptic terminals, specifically their average release rates λ and correlation c, with the statistical characterization of the resulting effective stochastic conductances. In particular, we showed that the correlation among many synaptic input channels is the primary factor determining the variance of the resulting effective synaptic conductances, whereas the average rate will determine the mean of the resulting effective process.

In Rudolph and Destexhe (2001a), it was demonstrated that even faint (Pearson correlation coefficient of about 0.05) and brief (down to 2 ms) correlation changes led to a detectable change in the cellular behavior. This sensitivity of conductance fluctuations and, thus, the amplitude of resulting V_m fluctuations to temporal correlation among thousands of synaptic inputs, in combination with the monotonic dependence of the mean and variance (4.25) of the total conductance on the channel firing rate λ and temporal correlation c among multiple channels, provides a method for characterizing presynaptic activity in terms of λ and c based on the sole knowledge of $\overline{g}$ and σ_g^2. Mathematically, these relations take the form

$$\lambda = \frac{\overline{g}}{D_1 N},$$
$$c = \frac{(D_2\overline{g}(2N-1) - D_1 N\sigma_g^2)^2}{(N-1)^2(D_2\overline{g} - D_1\sigma_g^2)^2}. \tag{9.17}$$

Experimentally, values for $\overline{g}$ and σ_g^2 can either be obtained by using the voltage-clamp protocol, from which distributions for excitatory and inhibitory conductances can be calculated, or by using the VmD method (Sect. 8.2). Both approaches yield estimates for the mean ($g_{\mathrm{e}0}$ and $g_{\mathrm{i}0}$) and SD (σ_e and σ_i) of excitatory and inhibitory conductances, from which the correlation and individual channel rates for both excitatory and inhibitory subpopulations of synaptic inputs can be estimated with (9.17).

This paradigm was tested by Rudolph and colleagues (unpublished) in numerical simulations of single- and multicompartmental models, in which the temporal correlation and average release rate at single synaptic terminals were changed in a broad parameter regime (Fig 9.31). Synaptic conductances were obtained using the voltage-clamp protocol or the VmD method based on current clamp recordings. For single-compartment models, the estimated values for the average release rate λ matched very precisely the known input values for both excitatory and inhibitory synapses. A good agreement was also obtained for estimates of the correlation c, independent of the protocol used. Similar investigations performed with multicompartment models, however, showed that the suggested method yields an underestimation of λ and c, especially for small c, due to the filtering of synaptic inputs in spatially extended dendrites. Here, a simple linear compensation, which takes into account the conductance actually "seen" at the somatic recording site for individual synaptic inputs (Fig 9.31, top left), can be shown to lead to an estimation of λ and c which closely matches the actual values (Fig 9.31, bottom).

The result of an application of this method to intracellular recordings of a cortical cell during PPT-activated states under ketamine–xylazine anesthesia is shown in Fig. 9.32. Both rate λ and correlation c for AMPA and GABAergic synaptic terminals were estimated from conductance estimates obtained with the VmD

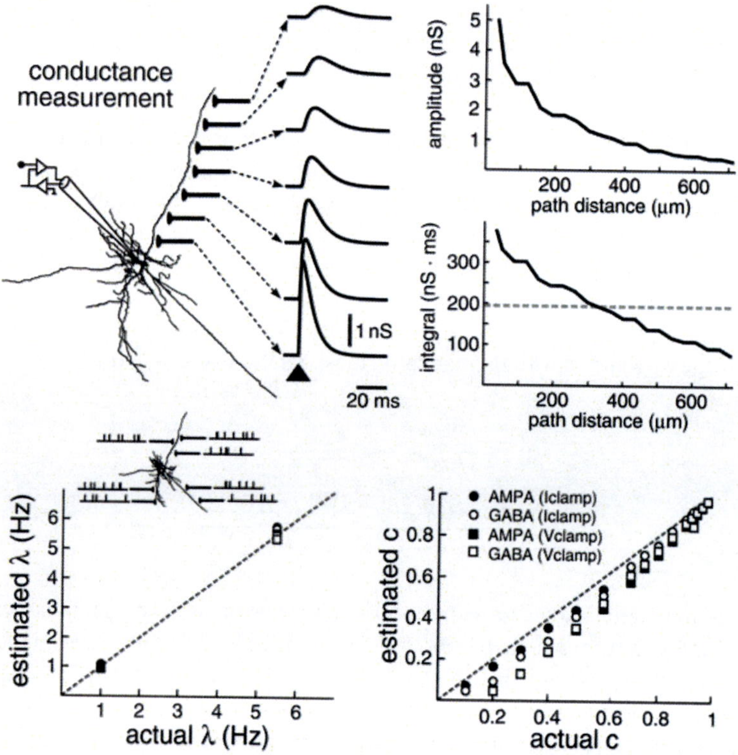

Fig. 9.31 Estimation of the rate λ and correlation c from the characterization of the total conductance distribution for different levels of network activity. Estimated values are shown as functions of the actual values used for the numerical simulations of a detailed multicompartmental model of a cortical neuron. To estimate the attenuation of synaptic inputs in the spatially extended dendritic structure, ideal voltage-clamp simulations were performed to obtain the conductance "seen" at the soma for individual excitatory synaptic (*top left*; stimulation amplitude was 12 nS). The amplitude (*top right, top panel*) and integral (*top right, bottom panel*) of the conductance time course decreased with path distance of the synaptic stimulus. This leads to an correction factor in the equations for estimating $\overline{g}$ and σ_g^2. As a first approximation, the average conductance contribution (*top right, dashed*) was taken, which leads to a correction in the estimation of the rate and correlation (*bottom*). Estimated values for different levels of network activity match remarkably well with the actual values used for the numerical simulations

method (Fig. 9.32 left). The obtained statistical characterization of the network activity was then compared to the results from a detailed biophysical model of the morphologically reconstructed cell from which the recordings were taken. The release rates λ match well with those of the constructed detailed biophysical model (Fig. 9.32 top right). The correlation c for both AMPA and GABAergic release deduced from experimental recordings appear to be overestimated (Fig. 9.32 bottom right). This overestimation was found to be attributable to the lack in the construction of the biophysical model caused by the incomplete reconstruction of the dendritic structure, showing that the utilization of the shot-noise paradigm, if

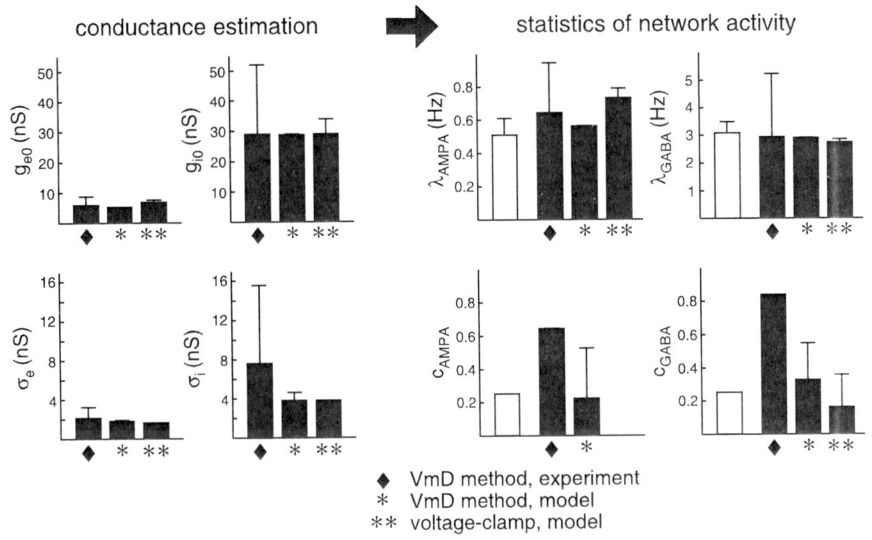

Fig. 9.32 Estimation of the mean (g_{e0} and g_{i0}) and variance (σ_e and σ_i) of excitatory and inhibitory synaptic conductances (*left panels*), as well as statistical properties (rate λ and correlation c for AMPA and GABAergic synaptic terminals) of the network activity (*right panels*) during PPT-activated states under ketamine–xylazine anesthesia. Results for conductance estimates from experimental recordings (using the VmD method), as well as from the constructed biophysical model (using the VmD method and voltage clamp) are shown. The obtained conductance values from the constructed model match well with the estimates from experimental recordings. With these conductance values, the release rate λ and temporal correlation c at synaptic terminals were calculated (*right panels*). The results match well with those of the constructed detailed biophysical model (*white bars*). Only the correlation c for both AMPA and GABAergic release deduced from experimental recordings appears to be overestimated. This deviation is attributable to lack in the construction of the biophysical model caused by the incomplete reconstruction of the dendritic structure as well as peculiarities in the distribution of synapses

applied to experimental recordings, depends on the knowledge of the morphological structure and distribution of synaptic receptors in the dendritic structure. However, it potentially can provide a useful method to characterize statistical properties of network activity from single-neuron activity. Of particular interest are here temporal correlations in the discharge of a large number of neurons, which, although of prime physiological importance, still remains an largely uncharacterized parameter.

9.5 Discussion

In this preceding section, we demonstrated how intracellular recordings in vivo in combination with computational and mathematical models allow to infer synaptic conductances and statistical properties of the network activity. In this final section, we will answer some questions related to the limitations of this methodology.

9.5.1 *How Much Error Is Due to Somatic Recordings?*

In all cases presented here, the excitatory and inhibitory conductances were estimated exclusively from somatic recordings. The values obtained, therefore, reflect the overall conductances as seen from the soma, after dendritic integration, and are necessarily different than the "total" conductance present in the soma and dendrites of the neuron. However, these somatic estimates are close to the conductance interplay underlying spike generation because the spike initiation zone (presumably in the axon; see Stuart et al. 1997a,b) is electrotonically close to the soma.

It is important to note that the present conductance estimates with generally dominant inhibition contrast with the roughly equal conductances measured in voltage clamp during spontaneous Up-states in ferret cortical slices (Shu et al. 2003a) or in vivo (Haider et al. 2006). Although in the study by Rudolph et al. (2007) (see Sect. 9.3) also neurons with roughly equal conductances ($n = 5$ for Wake, none for SWS-Up) were observed, this does not explain the differences. A possible explanation is that those voltage-clamp measurements were performed in the presence of Na^+ and K^+ channel blockers (QX314 and cesium), and these drugs affect somatodendritic attenuation by reducing the resting conductance. Consequently, excitatory events located in dendrites have a more powerful impact on the soma compared to the intact neuron, which may explain the discrepancy. Another possible explanation is that, in voltage-clamp experiments, when the voltage clamp is applied from the soma, the more distal regions of the cell are unlikely to be clamped, which may result in errors in estimating conductances and reversal potentials. Moreover, in this case, the presence of uncompensated electrode series resistance may worsen the estimates or affect the ratio between excitation and inhibition.

In Rudolph et al. (2007), these possible scenarios were tested using simulations of reconstructed pyramidal neurons and a biophysical model of background activity, specifically a Layer VI pyramidal cell with AMPA and $GABA_A$ currents in soma and dendrites. Obtained results are summarized in Table 9.1. The "Control" conditions correspond to a perfect voltage clamp (series resistance $R_s = 0$), which was used to estimate the excitatory and inhibitory conductances visible from the somatic electrode. To simulate electrode impalement, a 10 nS shunt was added in the soma. To simulate recordings in the presence of Cesium, the leak resistance was reduced by 95%, but the shunt was unaffected. As shown in Table 9.1, both the amplitude of the measured conductance, and the ratio of excitation/inhibition, were highly dependent on series resistance. In particular, one can see that a situation where the conductances are dominated by inhibition can be measured as roughly "balanced," principally due to series resistance of the voltage-clamp electrode. In contrast, the presence of a shunt has little effect. This suggests that voltage-clamp measurements introduce a clear bias due to series resistance. This problem should not be present in current clamp (such as the VmD method), because the membrane "naturally computes" the voltage distribution, which is used to deduce the conductances.

Table 9.1 Conductance estimates in voltage clamp using a morphologically reconstructed cortical pyramidal neuron. A model of background activity in a spatially distributed neuron was used (same parameters as given in Destexhe and Paré 1999). The "Control" conditions correspond to a perfect voltage clamp (series resistance $R_s = 0$), which was used to estimate the excitatory and inhibitory conductances visible from the somatic electrode. A 10 nS shunt was added in the soma, and the series resistance was varied. To simulate recordings in the presence of Cesium (Cs^+), the leak resistance was reduced by 95%, but the shunt was unaffected. The last column indicates the ratio between inhibitory and excitatory conductances. Modified from Rudolph et al. (2007)

	g_{e0} (nS)	σ_e (nS)	g_{i0} (nS)	σ_i (nS)	g_{i0}/g_{e0}
Control, $R_s = 0$	13.41	2.68	40.66	2.79	3.03
10 nS shunt, $R_s = 0$	13.44	2.68	40.66	2.78	3.03
10 nS shunt, $R_s = 3$ MΩ	11.3	2.05	30.0	2.02	2.65
10 nS shunt, $R_s = 10$ MΩ	8.31	1.32	15.6	1.24	1.87
10 nS shunt, Cs^+, $R_s = 0$	13.57	2.70	43.9	2.83	3.23
10 nS shunt, Cs^+, $R_s = 3$ MΩ	11.36	2.07	33.97	2.07	2.99
10 nS shunt, Cs^+, $R_s = 10$ MΩ	8.32	1.34	20.3	1.28	2.44
10 nS shunt, Cs^+, $R_s = 15$ MΩ	7.04	1.07	14.6	1.01	2.07
10 nS shunt, Cs^+, $R_s = 25$ MΩ	5.46	0.76	7.46	0.71	1.36

Further conductance measurements should be performed in nonanesthetized animals to address these issues. On the other hand, our results are in agreement with conductance measurements performed in cortical neurons in vivo under anesthesia, which also show evidence for dominant inhibitory conductances (Hirsch et al. 1998; Borg-Graham et al. 1998; Destexhe et al. 2003a; Rudolph et al. 2005).

9.5.2 *How Different Are Different Network States In Vivo?*

Figure 9.33 shows a summary of the conductance measurements presented in this chapter, including measurements during ketamine–xylazine anesthesia following PPT stimulation, as well as awake and naturally sleeping cats. As can be seen, the relative conductances vary between states, but the ratio is always in favor of inhibition (from about twofold in wakefulness, to more than tenfold in the Up-states). There is a tendency to have more inhibition in anesthetized states. Interestingly, the general pattern of conductance is similar in wakefulness compared to the Up-states of SWS (Fig. 9.33a). This similarity also applies to the post-PPT activated state and Up-states under KX anesthesia (Fig. 9.33b), and supports the suggestion that Up-states represent "microwake" episodes, perhaps replaying events during SWS (see Discussion in Destexhe et al. 2007). Nevertheless, there are significant differences in the level, i.e., average values, of excitatory and inhibitory conductances, suggesting that these states are similar but not identical.

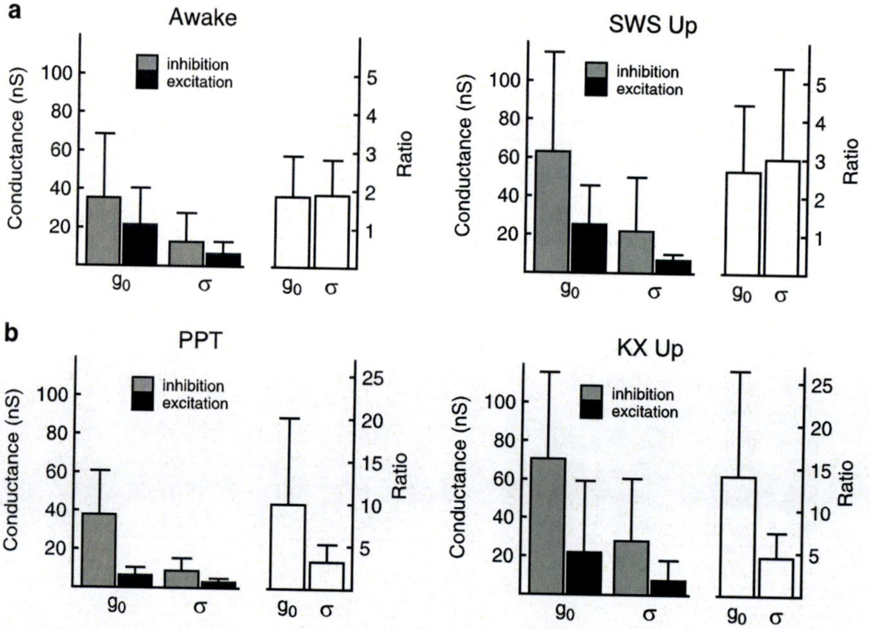

Fig. 9.33 Similar patterns of conductances between Up-states and activated states. Each panel shows on the *left*, the absolute excitatory and inhibitory conductances (g_0), as well as their standard deviation (σ), measured over several cells. On the *right*, the ratio of inhibitory over excitatory mean and σ is shown. The same analysis is compared between wakefulness and the Up-states of slow-wave sleep (**a**), as well as during ketamine–xylazine anesthesia (**b**), comparing activated states following PPT stimulation with the Up-states. (**a**) modified from Rudolph et al. (2007); (**b**) modified from Rudolph et al. (2005)

9.5.3 Are Spikes Evoked by Disinhibition In Vivo?

Not only inhibition seems to provide a major contribution to the conductance state of the membrane but also the conductance variations are larger for inhibition compared to excitation. This suggests that inhibition largely contributes to setting the V_m fluctuations, and, therefore, it presumably has a strong influence on AP firing. This hypothesis can be tested in computational models, which predict that when inhibition is dominant, spikes are correlated with a prior decrease of inhibition, rather than an increase of excitation. This decrease of inhibition should be visible as a membrane conductance decrease prior to the spike, which is indeed what was observed in most neurons analyzed in wake and sleep states (Fig. 9.21). A prominent role for inhibition is also supported by previous intracellular recordings demonstrating a time locking of inhibitory events with APs in awake animals (Timofeev et al. 2001), and the powerful role of inhibitory fluctuations on spiking in anesthetized states (Hasenstaub et al. 2005). Taken together, these results suggest that strong inhibition is not a consequence of anesthesia, but rather represents a property generally seen in awake and natural sleep states, pleading for a powerful role for interneurons in determining neuronal selectivity and information processing.

It is important to note that this pattern is opposite to what is expected from feed-forward inputs. A feedforward drive would predict an increase of excitation closely associated to an increase of inhibition, as seen in many instances of evoked responses during sensory processing (Borg-Graham et al. 1998; Wehr and Zador 2003; Monier et al. 2003; Wilent and Contreras 2005a,b). There is no way to account for a concerted g_e increase and g_i drop without invoking recurrent activity, except if the inputs evoked a strong disinhibition, but this was so far not observed in conductance measurements. Indeed, this pattern with inhibition drop was found in self-generated irregular states in networks of integrate-and-fire neurons (El Boustani et al. 2007). This constitutes direct evidence that most spikes in neocortex in vivo are caused by recurrent (internal) activity, and not by evoked (external) inputs. This argues for a dominant role of the network state in vivo and that inhibition is a key player. These findings are in agreement with recordings in awake ferret visual cortex suggesting that most of the spatial and temporal properties of neuronal responses are driven by network activity, and not by the complex visual stimulus (Fiser et al. 2004). These results support the view that sensory inputs modulate, rather than drive cortical activity (Llinás and Paré 1991).

9.6 Summary

In this final chapter, we have presented a few case studies illustrating the concepts elaborated in other chapters.

First, we have shown the characterization of synaptic noise in various in vivo preparations, such as artificially activated states under anesthesia (Sect. 9.2) or awake and naturally sleeping cats (Sect. 9.3). In all cases, we have applied some of the methods detailed in Chap. 8, such as the VmD method. These measurements showed that the V_m activity during wake and sleep states, Up–Down-states of SWS, or artificially activated states, results from diverse combinations of excitatory and inhibitory conductances, with dominant inhibition in most of the cases.

Such conductance measurements were used to constrain computational models which investigated the properties of dendritic integration. These models showed similar conclusions as outlined in Chap. 5, namely that synaptic noise enhances the responsiveness of cortical neurons, it refines their temporal processing abilities, and it reduces the location dependence of synaptic inputs in dendrites.

Second, we have applied the STA method to determine the optimal conductance patterns triggering action potentials in awake and sleeping cats (Sect. 9.3). This analysis showed that inhibitory conductance fluctuations are generally larger than for excitation and probably determine most of the synaptic noise as seen from the V_m activity. Spike initiation is in most cases correlated with a decrease of inhibition, which appears as a transient drop of membrane conductance prior to the spike. This pattern is typical of recurrent activity and shows that the majority of APs are triggered by recurrent activity in awake and sleeping cat cortex.

Finally, we have illustrated a few additional applications (Sect. 9.4). These include methods to estimate time-dependent variations of conductances or rate-based stochastic processes to model these variations. The expressions derived in Chap. 7 have allowed us to relate the variance of conductances, one of the parameters measured with the VmD method, and the level of correlations in network activity.

These methods show that it is possible to reconstruct properties of network activity from the sole measurement of the synaptic noise in the V_m activity of a single neuron. Thus, in a sense, these allow one to "see" the network activity through the intracellular measurement of a single cell.

In the next concluding chapter, we will briefly elaborate on how networks perform computations in such stochastic states.

Chapter 10
Conclusions and Perspectives

In this book, we have overviewed several recent developments of the exploration of the integrative properties of central neurons in the presence of "noise," with an emphasis on the largest noise source in neurons, synaptic noise. Investigating the properties of neurons in the presence of intense synaptic activity is a popular theme in modeling studies, starting from seminal work (Barrett and Crill 1974; Barrett 1975; Bryant and Segundo 1976; Holmes and Woody 1989), which was followed by compartmental model studies (Bernander et al. 1991; Rapp et al. 1992; De Schutter and Bower 1994). In the last two decades, significant progress was made in several aspects of this problem. The different chapters of this book have overviewed different facets of this exploration. In this final chapter, we first summarize the different facets of synaptic noise, as overviewed in the different chapters. We then speculate on how "noise," and in particular "noisy states," is a central aspect of neuronal computations.

10.1 Neuronal "Noise"

In the course of this book, we explored and reviewed various aspects of synaptic noise, from its experimental discovery and characterization, over the construction of models constrained by these experimental measurements, up to the investigation of neuronal processing in such stochastic states using these models and the development of new methodologies which allow the controlled injection of synaptic noise and its quantification in experiments. In this section, we briefly summarize these different aspects of synaptic noise, as detailed in this book.

A. Destexhe and M. Rudolph-Lilith, *Neuronal Noise*, Springer Series in Computational Neuroscience 8, DOI 10.1007/978-0-387-79020-6_10,

10.1.1 Quantitative Characterization of Synaptic "Noise"

In Chap. 3, we overviewed a first aspect which tremendously progressed in the past years, namely the quantitative measurement of background activity in neurons. The early modeling studies did not use any hard constraint because no measurement of synaptic noise was available at the time. Such a quantitative measurement of synaptic noise was first done for "activated" network states under anesthesia in vivo (Paré et al. 1998b). In this study, the impact of background activity could be directly measured and quantitatively assessed, for the first time, by comparing the same cortical neurons recorded before and after total suppression of network activity. This was done globally for two types of anesthesia, barbiturates and ketamine–xylazine. In a subsequent study (Destexhe and Paré 1999), this analysis was refined by focusing specifically on the "Up-states" of ketamine–xylazine anesthesia, which present very similar "active" network states as in the awake animal, characterized by a locally desynchronized EEG during the Up-state. These analyses evidenced a very strong impact of synaptic background activity on increasing the membrane conductance of the cell into "high-conductance states."

The contribution of excitatory and inhibitory synaptic conductances were later measured in awake and naturally sleeping animals (Rudolph et al. 2007). The availability of such measurements (see details in Chap. 3) can be considered as an important cornerstone, because they allow building precise models and dynamic clamp experiments to evaluate their consequences on the integrative properties of cortical neurons. It is important to note that the availability of such measurements relies on a new generation of stochastic methods, as we have reviewed in Chap. 8. These methods, themselves, were derived utilizing various theoretical and mathematical approaches (Chap. 7), and extensively tested and evaluated by using computational models (Chaps. 4 and 5).

10.1.2 Quantitative Models of Synaptic Noise

The first part of Chap. 4 was devoted to a new class of compartmental models which were directly constrained by the quantitative measurements described in Chap. 3 (Destexhe and Paré 1999). Such models could thus reproduce in vivo–like activity states with an unprecedented level of realism. They extended previous compartmental models of cortical pyramidal neurons (Bernander et al. 1991). Moreover, these models made a number of predictions about the consequences of synaptic noise on integrative properties, as reviewed in Chap. 5.

Chapter 4 also reviewed another aspect that tremendously progressed these last years, namely the formulation of simplified models that replicate the in vivo measurements, as well as important properties such as the typical Lorentzian spectral structure of background activity (see Sect. 4.4). In particular, we have put emphasis on the "point-conductance" model (Destexhe et al. 2001) because this model had many practical consequences: it enabled dynamic-clamp experiments (see Chap. 6), allowed various mathematical approaches (Chap. 7), as well as led to the development of several new analysis methods to characterize synaptic noise in experiments (Chap. 8).

10.1.3 Impact on Integrative Properties

Chapter 5 reviewed the modeling exploration of consequences of synaptic "noise" on the integrative properties of neurons. Consequences on dendritic integration, such as coincidence detection and enhanced temporal processing, were predicted a long time ago (Bernander et al. 1991; Softky 1994). These were confirmed with constrained models (Rudolph and Destexhe 2003b). New consequences were also found, such as enhanced responsiveness (Hô and Destexhe 2000) and location-independent synaptic efficacy (Rudolph and Destexhe 2003b).

Enhanced responsiveness is one of the most spectacular property of neurons in the presence of noise. This property takes the form of a nonzero probability of generating a response (e.g., an action potential) for inputs which are normally much too small to evoke any response. This property of enhanced responsiveness has itself many consequences, such as a modulation of the gain of the neuron in the presence of synaptic noise. This property was investigated (and confirmed) using a dynamic clamp (see Chap. 6). Another consequence of enhanced responsiveness is to markedly affect the properties of dendritic integration in the presence of synaptic noise, leading unexpected properties such as location independence, where the efficacy of synaptic inputs becomes almost independent of their position in the dendritic tree. The latter property still awaits to be investigated experimentally.

10.1.4 Synaptic Noise in Dynamic Clamp

In Chap. 6, we showed that some of these predictions were confirmed by dynamic-clamp experiments. For example, the enhanced responsiveness, initially predicted by models (Hô and Destexhe 2000) was confirmed in real neurons using dynamic clamp (Destexhe et al. 2001; Chance et al. 2002; Fellous et al. 2003; Prescott and De Koninck 2003; Shu et al. 2003a,b; Higgs et al. 2006).

As reviewer in Chap. 4, the formulation of simplified models had important consequences on dynamic-clamp experiments. The point-conductance model (Destexhe et al. 2001) was used in many of the aforementioned dynamic-clamp studies to recreate in vivo–like activity states in neurons maintained in vitro. The main advantage of the point conductance model is to enable the independent control of the mean and the variance of conductances. This allowed investigating their respective role, confirming the model predictions (Hô and Destexhe 2000) about the effect of conductance and the effect of the fluctuations.

In addition to confirm model predictions, dynamic-clamp experiments also took these concepts further and investigated important properties such as gain modulation (Chance et al. 2002; Fellous et al. 2003; Prescott and De Koninck 2003). An inverse form of gain modulation can also be observed (Fellous et al. 2003) and may be explained by potassium conductances (Higgs et al. 2006). It was also found that the intrinsic properties of neurons combine with synaptic noise to yield unique responsiveness properties (Wolfart et al. 2005; see details in Chap. 6).

It is important to note that although the point-conductance model was the first stochastic model of fluctuating synaptic conductances injected in living neurons using dynamic clamp, other models are also possible. For example, models based on the convolution of Poisson processes with exponential synaptic waveforms ("shot noise") have also been used (e.g., see Reyes et al. 1996; Jaeger and Bower 1999; Chance et al. 2002; Prescott and De Koninck 2003). However, it is difficult in such models to independently control the mean and variance of conductances, so the effect of these parameters cannot easily be determined. Such models are also considerably slower to simulate, especially if correlations are included among the synaptic events. It can be shown that these models are, in fact, equivalent at high rates, as the point-conductance model can be obtained as a limit case of a shot-noise process with exponential conductances (Destexhe and Rudolph 2004; Rudolph and Destexhe 2006a). The ability of the point-conductance stochastic model to tease out the respective contributions of the mean and variance of synaptic conductances motivated the development of methods to estimate these parameters from experimental recordings, such as the VmD method (see Chap. 8; see also Chap. 9 for specific applications).

10.1.5 Theoretical Developments

Chapter 7 overviewed another consequence of the availability of simplified models. Their mathematical simplicity enabled mathematical analysis, and in particular the formulation of a number of variants of the Fokker–Planck equation for the membrane potential probability density (Rudolph and Destexhe 2003d; Richardson 2004; Rudolph and Destexhe 2005; Lindner and Longtin 2006; Rudolph and Destexhe 2006b; see details in Chap. 7). One of the main achievement was to obtain excellent analytic approximations of the steady state V_m distribution of neurons in the presence of conductance-based synaptic noise (for a comparison of all the available approximations, see Rudolph and Destexhe 2006b). The practical consequence is that such analytic expressions can be used to analyze real signals, and extract synaptic conductances from the V_m activity, as examined in Chap. 8.

10.1.6 New Analysis Methods

Chapter 8 shows that these theoretical advances have led to several methods to estimate synaptic conductances and their dynamics from V_m recordings (Rudolph et al. 2004; Destexhe and Rudolph 2004; Pospischil et al. 2007, 2009). The *VmD method* is directly derived from the Fokker–Planck analysis and consists of decomposing the V_m fluctuations into excitatory and inhibitory contributions, and estimating their mean and variance (see Sect. 8.2). This method was successfully tested in dynamic-clamp experiments (Rudolph et al. 2004) as well as in voltage-clamp (Greenhill and Jones 2007; see also Ho et al. 2009). The most interesting aspect of the VmD method is that it provides estimates of the variance of conductances or equivalently, conductance fluctuations.

The point-conductance model also enables estimating the kinetics of synaptic conductances from the analysis of the PSD of the V_m. This *PSD method* (Sect. 8.3; Destexhe and Rudolph 2004) was also tested numerically and using dynamic-clamp experiments. It provides an estimate of the decay time constant of excitatory and inhibitory synaptic conductances, as seen from the recording site (the soma in most cases).

Another method, called the *STA method* (Sect. 8.4; Pospischil et al. 2007) was also derived from the same point-conductance model approximation of the V_m activity. In this case, if the mean and variance of the conductances are known (e.g., by applying the VmD method), then one can estimate the optimal conductance patterns leading to spikes in the neuron. This estimate is done using a maximum likelihood estimator. The method was tested numerically, as well as using dynamic-clamp experiments (Pospischil et al. 2007).

Finally, we overviewed a recent method to estimate synaptic conductances from single V_m traces (Sect. 8.5; Pospischil et al. 2009). This *VmT method* is similar in spirit to the VmD method, but estimates conductance parameters using maximum likelihood criteria, thus also similar to the STA method. Like the other methods, it was tested using models and dynamic-clamp experiments. This method enables estimating the mean excitatory and inhibitory conductance from V_m recordings at a single DC current level, which has many possible applications in vivo and is, at the time of the writing of this book, a work in progress.

It is important to note here that the dynamic-clamp experiments found another important application here to formally test methods for conductance estimation. With dynamic clamp, the experimentalist has a complete control over the conductances that are being added to the neuron, which can be compared to the conductance estimated from the V_m activity. Not only this type of testing is an original application of the dynamic clamp, but it is also a powerful way of quantitatively testing such methods (Piwkowska 2007).

10.1.7 Case Studies

These methods were illustrated in the case studies examined in Chap. 9. The VmD analysis was applied to cortical neurons during artificially activated brain states (Rudolph et al. 2005) or in nonanesthetized animals, either awake or sleeping (Rudolph et al. 2007). The latter provided the first quantitative characterization of synaptic conductances and their fluctuations in the intact and nonanesthetized brain.

Chapter 9 also reviewed that this approach can be extended to estimate dynamic properties related to AP initiation. If the information about synaptic conductances and their fluctuations is available (e.g., following VmD estimates), then one can use maximum likelihood methods to evaluate the spike-triggered conductance patterns (see Sect. 8.4 in Chap. 8). This information is very important to deduce which optimal conductance variations determine the "output" of the neuron, which is a fundamental aspect of integrative properties. It was found that in awake and naturally sleeping animals, spikes are statistically related to disinhibition, which plays a permissive role. This type of conductance dynamics is opposite to the conductance patterns evoked by external input, but can be replicated by models displaying self-generated activity. This suggests that most spikes in awake animals are due to internal network activity, in agreement with other studies (Llinás and Paré 1991; Fiser et al. 2004). This dominant role of the network state in vivo, and the particularly strong role of inhibition in this dominance, should be investigated by future studies (for a recent overview, see Destexhe 2011).

10.2 Computing with "Noise"

In the preceding chapters of this book, we outlined experimental, theoretical and computational studies of neuronal noise, specifically synaptic noise, methods of its characterization as well as its impact on the dynamics of single cells, with focus on one of the most "noisy" environments in the mammalian brain, the cerebral cortex. It became clear that the somewhat misused term "noise" does not refer to a detrimental, unwanted entity hindering neurons to perform their duties, but that instead this "noise" transfers advantageous functional properties to neurons.

Single cells, however, are not isolated. The synaptic noise each neuron is exposed to at any time stems from a highly complex spatiotemporal activity pattern in its surrounding recurrent circuitry. At the same time, each neuron's response feeds back into this circuitry, and participates in shaping its spatiotemporal pattern. This leads to a plethora of different and distinguishable states, such as wakefulness or sleep, each of which is endowed with its own functional properties that process and nonlinearly interacts with incoming sensory inputs, outgoing motor commands, internal associative processes and the representation of information. The understanding of this relation between network-state dynamics and information representation and processing is a major challenge that will require developing, in conjunction, specific experimental paradigms and novel theoretical frameworks. In this last section, we will briefly outline some recent developments in this direction.

10.2.1 Responsiveness of Different Network States

The most obvious relation between the spontaneous spatiotemporal activity pattern in the brain on one side, and neuronal and network responsiveness on the other can be observed during the transition between the behavioral states of waking and sleep or during variations in the level of anesthesia (Fig. 10.1). Here, a variety of functional studies in the visual (Livingstone and Hubel 1981; Worgotter et al. 1998; Li et al. 1999; Funke and Eysel 1992; Arieli et al. 1996; Tsodyks et al. 1999), somatosensory (Morrow and Casey 1992), auditory (Edeline et al. 2000; Kisley and Gerstein 1999; Miller and Schreiner 2000), and olfactory (Murakami et al. 2005) systems have shown that slow, high-amplitude activity in the EEG is associated with reduced neuronal responsiveness and neuronal selectivity (see Steriade 2003 for an extensive review).

The cellular correlates of such changes in responsiveness were studied both in vivo and in vitro. It was found that during the transition to SWS or anesthesia, cortical and thalamic cells progressively hyperpolarize, and that this hyperpolarization shifts the membrane potential into the activation range of intrinsic currents underlying burst firing, particularly in thalamic cells. Because of its all-or-none behavior and its long refractory period, thalamic bursting is incompatible with the relay function that characterizes activated states and acts, this way, as the first gate of forebrain deafferentation, i.e., blockade of ascending sensory inputs (Steriade and Deschênes 1984; Llinás and Steriade 2006). Furthermore, synchronized inhibitory inputs during sleep oscillations further hyperpolarize cortical and thalamic neurons and generate large membrane shunting, resulting in a dramatic decrease in responsiveness and a large increase in response variability. Finally, highly synchronized patterns of rhythmic activity (Contreras and Steriade 1997) dominate neuronal membrane behavior and render the network unreliable and less responsive to inputs. Taken together, the above mechanisms result in the functional brain deafferentation that characterizes sleep and anesthesia (Steriade 2000; Steriade and Deschênes 1984).

In contrast to SWS or anesthetized states, the waking state and REM sleep are characterized by a depolarized stable resting membrane potential close to spike threshold. This allows neurons to respond to inputs more reliably and with less response variability. However, despite their striking electrophysiological similarity at the intracellular and EEG levels (Steriade et al. 2001) and the often enhanced evoked potentials during REM (Steriade 1969; Steriade et al. 1969), the understanding of the cellular dynamics is not enough to explain an important paradox posed by these two activated brain states. Waking and REM are diametrically opposite behavioral states (Steriade et al. 1974), with REM sleep being the deepest stage of sleep, hence the stage with the highest threshold for waking up.

In an attempt to explain this paradox, it was shown, using magnetoencephalography in humans, that the main difference in responsiveness during REM sleep and wakefulness is their effect on the ongoing gamma (~40 Hz) oscillations (Llinás and Ribary 1993), i.e., the higher-order dynamical state of the network. Responses

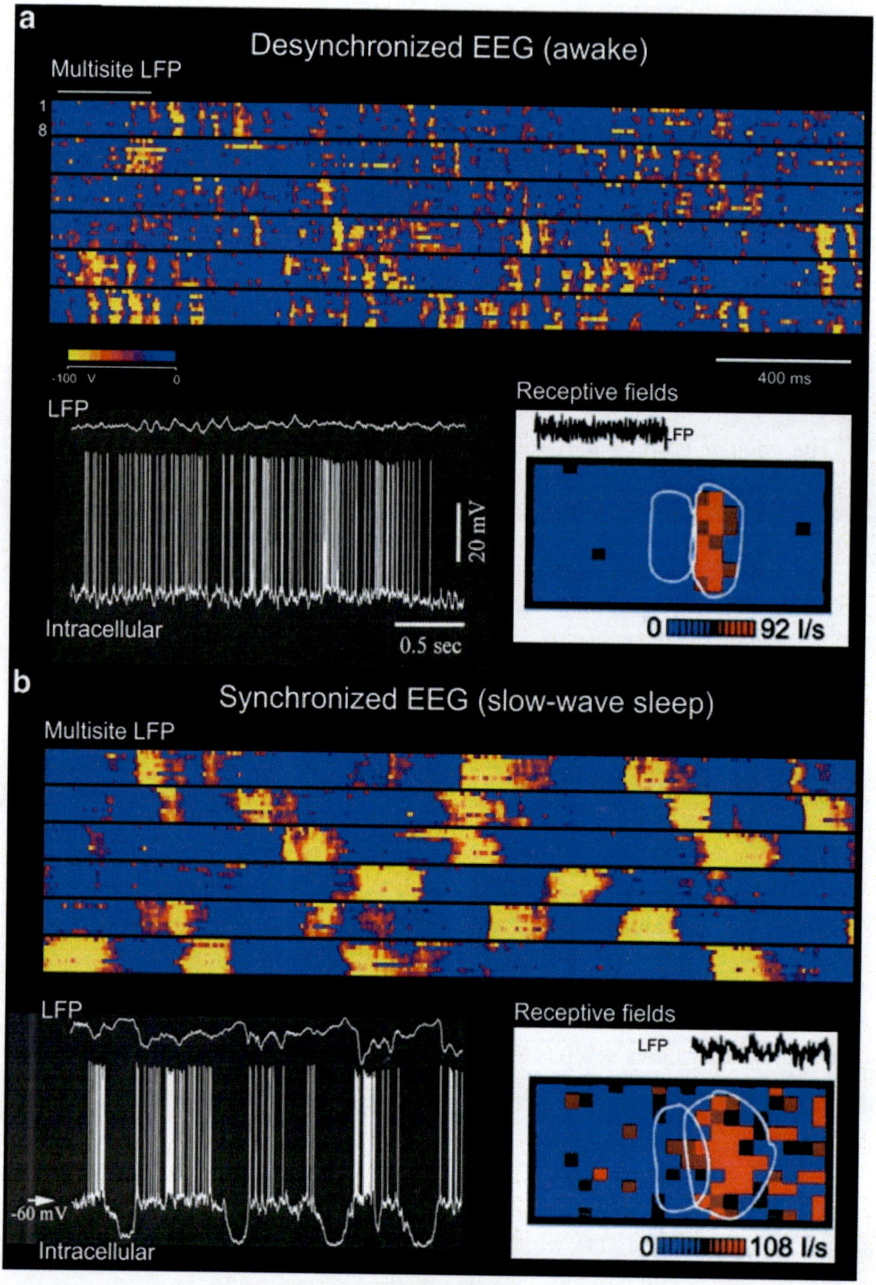

Fig. 10.1 Complex spatiotemporal patterns of ongoing network activity during wake and sleep states in neocortex. (**a**) Spatiotemporal map of activity computed from multiple extracellular local field potential (LFP) recordings in an awake cat. Here, the β frequency-dominated LFPs (15–30 Hz) are weakly synchronized and very irregular both spatially and temporally (modified from Destexhe et al. 1999). Intracellular recordings (*bottom left*) during this state show a sustained

to auditory clicks caused a reset of the ongoing gamma rhythm, whereas during REM, the evoked response did not change the phase of the ongoing oscillation. These findings suggest that, during dream sleep, sensory inputs are not incorporated into the context represented by the ongoing activity (Llinás and Paré 1991). The obvious conclusion is that much smaller changes in network dynamics, or changes at higher statistical order which do not manifest themselves at low order, are critical in determining the processing state of the brain. The failure to detect clear differences in network dynamics that must exist between waking and REM sleep is a clear indication that new approaches are necessary.

10.2.2 Attention and Network State

Another, even more striking example of the role of intrinsic network dynamics in determining neuronal responsiveness is the effect of attention. Even though the parameters of network activity measured with current techniques seem to remain stable, shifts in attentional focus both in space (Connor et al. 1997) and time (Ghose and Maunsell 2002) increase the ability of the network to process stimuli by increasing neuronal sensitivity to stimuli. The neuronal mechanisms underlying such attentional shifts are still unknown. However, the effect of directed attention enhancing neuronal responsiveness and selectivity, as well as behavioral performance (Spitzer et al. 1988), is a clear indication of the critical role played by subtle changes of network dynamics in determining the outcome of network operations.

As mentioned in Chap. 5, the effect of synaptic noise at the level of single cells can be thought as an attentional regulation. This was first proposed by a modeling study (Hô and Destexhe 2000) which noted that modulating the synaptic noise can have effects of enhancing the responsiveness of single neurons and even of networks (see Sect. 5.3.6 in Chap. 5). It was speculated that this effect could play a similar role as attentional modulation (see also Fellous et al. 2003 and Shu et al. 2003b). This link with attention constitutes a promising direction for future work.

Fig. 10.1 (continued) depolarization state with intense fluctuations during wakefulness (courtesy of Igor Timofeev, Laval University) (**b**) Same recording arrangement in a naturally sleeping cat during slow-wave sleep (SWS). The activity consists of highly synchronized slow waves (in the δ frequency range, 1–4 Hz), which are irregular temporally but coherent spatially (modified from Destexhe et al. 1999). The intracellular recordings (*bottom left*) show slow oscillations during this SWS state (courtesy of Igor Timofeev, Laval University). Network state-dependent responsiveness in visual cortex. Cortical receptive fields obtained by reverse correlation in simple cells for ON responses. The procedure was repeated for different cortical states, by varying the depth of the anesthesia (EEG indicated above each color map). (**a**), *bottom right*: desynchronized EEG states (light anesthesia); (**b**), *bottom right*: synchronized EEG states with prominent slow oscillatory components (deeper anesthesia). Receptive fields were always smaller during desynchronized states. Color code for spike rate (see scale). Receptive field maps modified from Worgotter et al. (1998)

10.2.3 Modification of Network State by Sensory Inputs

The reverse problem is of equally critical importance: how much the ongoing network dynamics are modified by sensory inputs. Although cortical and thalamic networks may be strongly activated by specific patterns of stimuli (Miller and Schreiner 2000), such effects are likely due to the engagement of brainstem neuromodulatory systems, which receive dense collaterals from ascending sensory inputs (Steriade 2003). Recordings from visual cortex of awake, freely viewing ferrets (Fiser et al. 2004) revealed that the spatial and temporal correlation between cells while natural scenes were viewed varies little when compared with values obtained during eyes closed. It was also shown that the statistical properties of the V_m fluctuations are identical in spontaneous activity and during viewing of natural images (El Boustani et al. 2009). These subtle variations indicate that most of the spatial and temporal coordination of neuronal firing is driven by internal network activity and not by the complex visual stimulus.

10.2.4 Effect of Additive Noise on Network Models

Experimental studies directed toward identifying the dynamical spatiotemporal patterns which represent and process information at the network level remain up to date very sparse, mostly due to technical challenges and the mathematical complexity in dealing with highly nonlinear dynamical systems. On the other hand, various computational studies attempted to shed light on the relation between network dynamics and neuronal responsiveness, and to identify the minimal set of correlates which give rise to the processing power of brain circuits. One of the simplest type of computational models, namely the modeling of irregular spontaneous network activity as "noise" impinging on individual cells, was the subject of this book. From the results of these studies, it can be concluded that the network activity has a decisive impact on the input–output transformation of single neurons, which, in turn, suggests the mechanisms by which the information-processing capabilities of the network might be altered and shaped.

The obvious continuation of such "noisy" single-cell studies is to embed the latter into networks and consider the effect of noise in neural network models. Although such studies date back many decades to the emergence of the first computers powerful enough to simulate networks of interconnected processing elements (also known as artificial neural networks), only recently the dynamical aspect of information processing in such networks, which constitutes a necessary condition for the emergence of "noise," was recognized. In a number of studies, it was found that noise is not only beneficial in building associative memories by avoiding convergence to spurious states (Amit 1989) but it also enables networks to follow high-frequency stimuli (Knight 1972), boosts the propagation of waves of activity (Jung and Mayer-Kress 1995), enhances input detection abilities (Collins et al.

1995a,b; Stocks and Mannella 2001), and enables populations of neurons to respond more rapidly (Tsodyks and Sejnowski 1995; van Vreeswijk and Sompolinsky 1996; Silberberg et al. 2004). Noisy networks can also sustain a faithful propagation of firing rates (van Rossum 2002; Reyes 2003; but see Litvak et al. 2003) or pulse packets (Diesmann et al. 1999) across successive layers (Fig. 10.2).

The latter results are particularly interesting, because noise allows populations of neurons to relay a signal across successive layers without attenuation (Fig. 10.2c) or prevents a catastrophic invasion of synchronous activity (Fig. 10.2d). The fact that a complex waveform propagates in a noisy network (Fig. 10.2c), but not with low noise levels (Fig. 10.2b), can be understood qualitatively from the response curve of neurons in the presence of noise, for which there is a reliable coding of stimulus amplitude. Indeed, a similar effect is visible in the population response of networks of noisy neurons (Fig. 10.2e). With low noise levels, the nearly all-or-none response acts as a filter, which allows only strong stimuli to propagate and leads to propagation of synfire waves (Fig. 10.2d). With stronger noise levels, comparable to intracellular measurements in vivo, the response curve is progressive, which allows a large range of input amplitudes to be processed (Fig. 10.2c).

10.2.5 Effect of "Internal" Noise in Network Models

The above models considered the effect of additive noise, but in reality the "noise" is provided by the network activity itself. Models of cortical networks have attempted to generate activity comparable to experiments, and several types of models were proposed, ranging from integrate and fire networks (Amit and Brunel 1997; Brunel 2000) up to conductance-based network models (Timofeev et al. 2000; Compte et al. 2003; Alvarez and Destexhe 2004; Vogels and Abbott 2005; El Boustani et al. 2007; Kumar et al. 2008). Although many of such models do not display the correct conductance state in single neurons, it is possible to find network configurations displaying the correct conductance state as well as asynchronous and irregular firing activity consistent with in vivo measurements (Alvarez and Destexhe 2004; El Boustani et al. 2007; Kumar et al. 2008).

Several models explicitly considered the state of the network and its effect on the way information is processed or the responsiveness to external inputs is shaped. Investigating the propagating activity in networks of excitatory and inhibitory neurons that display either silent, oscillatory (periodic), or irregular (chaotic or intermittent) states of activity (Destexhe 1994; Fig. 10.3a), it was found that irregular network states are optimal with respect to information transport. Thus, similar to turbulence in fluids, irregular cortical states may represent a dynamic state that provides an optimal capacity for information transport in neural circuits (Destexhe 1994).

More recent studies have explicitly considered networks endowed with intrinsically generated self-sustained states of irregular activity (Tsodyks and Sejnowski 1995; van Vreeswijk and Sompolinsky 1996; Mehring et al. 2003;

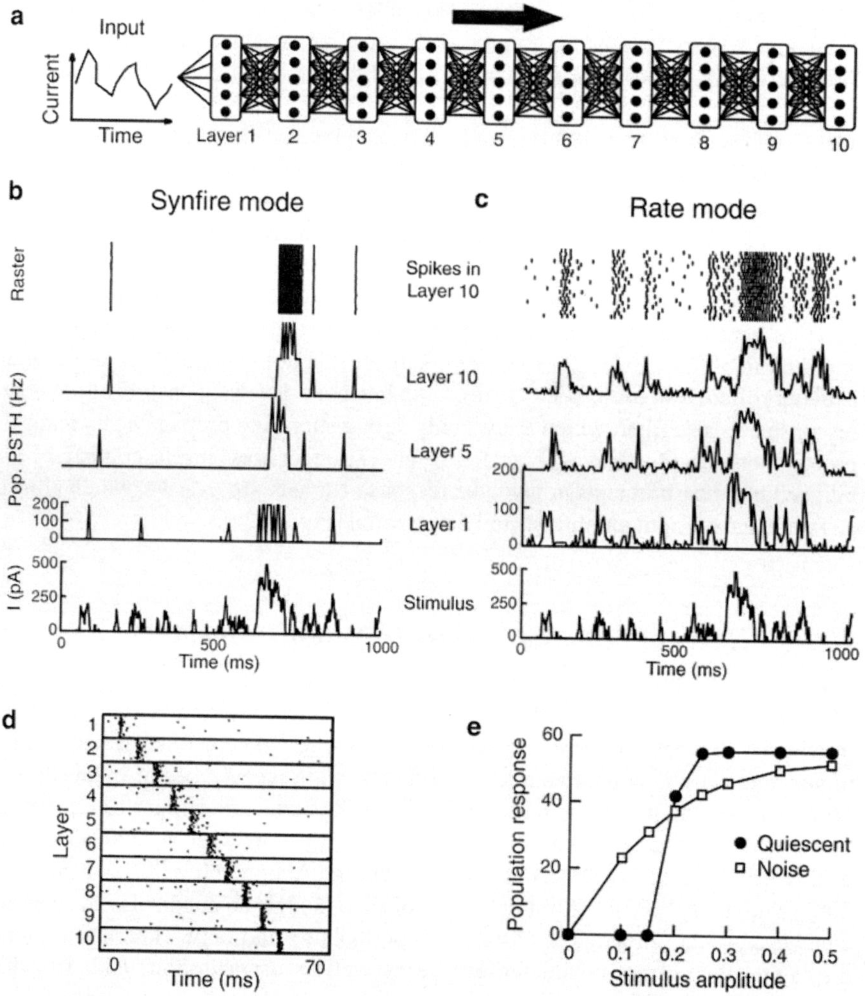

Fig. 10.2 Beneficial effects of noise at the network level. (**a**) Scheme of a multilayered network of integrate-and-fire (IF) neurons where layer 1 received a temporally varying input. (**b**) With low levels of noise (*Synfire mode*), firing was only evoked for the strongest stimuli, and synchronous spike volleys propagated across the network. (**c**) With higher levels of noise (*Rate mode*), the network was able to reliably encode the stimulus and to propagate it across successive layers. (**a**) to (**c**) modified from van Rossum (2002). (**d**) Another example of a network able to sustain the propagation of synchronous volleys of spikes ("synfire chains") only in the presence of noise. Modified from Diesmann et al. (1999). (**e**) Example of population response in a network of noisy neurons (*Noise*), compared with the same network in the absence of noise (*Quiescent*). Network response was close to all-or-none in quiescent conditions, but with noise, the population encoded stimulus amplitude more reliably. Modified from Hô and Destexhe (2000)

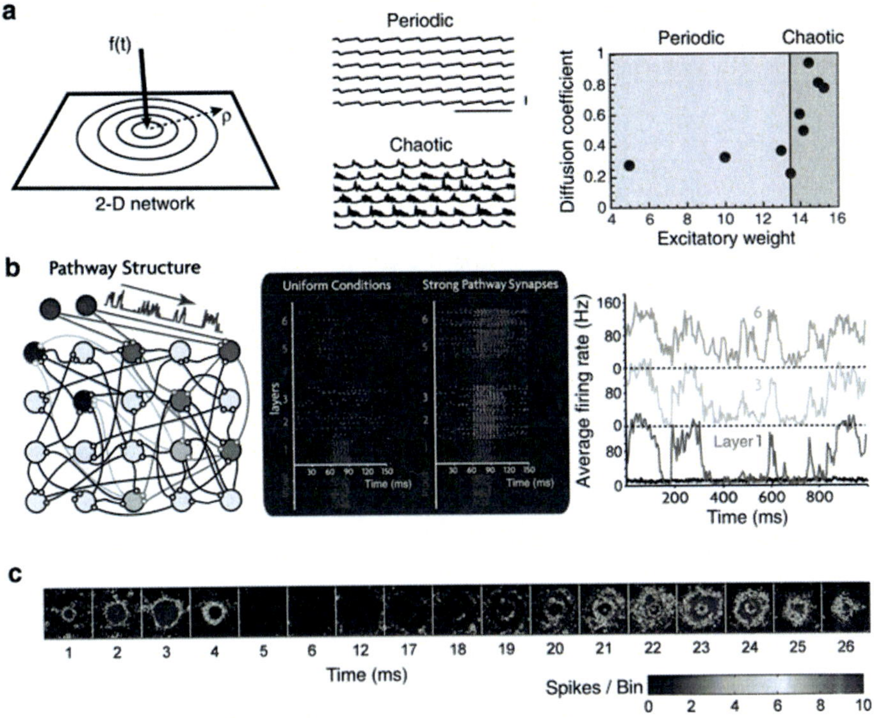

Fig. 10.3 Role of internally generated noise on information propagation in networks. (**a**) *Left*: Stimulation paradigm consisting of injecting a complex waveform ($f(t)$, *left*) and monitoring the spread of activity as a function of distance (r) and state of the network. *Middle*: Example of two self-sustained dynamic states of the network, periodic oscillations (*top*) and irregular activity ("chaotic", *bottom*). *Right*: Diffusion coefficient calculated for Shannon information (method from Vastano and Swinney 1988) as a function of the state of the network. Periodic states (*light gray*) had a relatively low diffusion coefficient, whereas, for irregular or chaotic states (*dark gray*), information transport was enhanced. Modified from Destexhe (1994). (**b**) Propagation of activity in a network of neurons displaying self-sustained irregular states. *Left*: Definition of successive layers and pathways; *middle*: absence of propagation with uniform conditions (*left*) contrasted with propagation when pathway synapses were reinforced (*right*); *right*: propagation of a time-varying stimulus with pathway synapses reinforced. Modified from Vogels and Abbott (2005). (**c**) Propagation of activity in a network with self-sustained irregular dynamics. Successive snapshots illustrate that a stimulus (*leftmost*) led to an "explosion" of activity, followed by silence and echoes. Modified from Mehring et al. (2003)

Vogels and Abbott 2005). However, in contrast to the studies mentioned earlier, propagation was difficult to observe. Firing rates did not propagate unless synapses were reinforced (more than tenfold) along specific feedforward pathways (Vogels and Abbott 2005; Fig. 10.3b), or pulse packets led to explosions of activity ("synfire explosions") in the network (Fig. 10.3c) which could only be avoided by wiring synfire chains into the connectivity to enable stable propagation

(Mehring et al. 2003). Such artificial embedding of feedforward pathways is of course not satisfactory, and the problem of how to obtain reliable propagation in recurrent networks remains an open problem.

10.2.6 Computing with Stochastic Network States

Another type of computational approach considers that inputs and network state are interdependent in the sense that external inputs shape a self-sustained persistent network state. Inputs will necessarily leave a trace in the dynamic spatiotemporal activity pattern of the network in a way that the latter is likely to reflect properties of the inputs and cannot be considered as independent. This type of model is much closer to the in vivo situation, and carries the potential of reflecting more truthfully the immense computational power of biological neural systems. Although first ideas in this direction emerged already in the late 1970s under the term *holographic brain* or *holonomic brain theory* (Pietsch 1981; Pribram 1987), only recently, driven by advancements in computer technology, this type of computational model could be studied in more detail.

In such neural networks, the strict distinction between training and recall, as typical for artificial neural networks, disappears due to self-sustained activity in recurrent networks of spiking neurons. This approach leads to the notion of *any-time* or *real-time computing without stable states, computation with perturbations* or *liquid computing*: information passed into the neural system in form of temporal sequences of spikes is fused with the actual network state, characterized by the overall activity pattern of its constituents. This leads to a new state, which may be viewed as a "perturbation" of the previous one. The recurrent spatial architecture and ongoing activity in the neural network, or "noise," are the basis for a distributed internal representation of information within the temporal dynamics of the neural system. Such high-dimensional complex dynamics can be decoded via low-dimensional "readout networks," as shown in the "liquid state machine." (Fig. 10.4; Maass et al. 2002)

These readout networks are classical perceptrons with fixed weights trained to understand or decode certain aspects in the temporal dynamics without feeding back into the liquid state. It was shown that the computation, which utilizes the temporal dynamics in such networks, is highly parallel, an aspect which can be shown by accessing the network by different readout networks at the same time. Moreover, allowing (unsupervised or self-supervised) plastic changes described by local update rules, such as spike-timing dependent plasticity, not just brings the functional principles of artificial neural networks closer to their biological counterpart, but also provides the basis for spatiotemporal self-organization of the system. The latter can be characterized by the interplay between changes in the spatial connectivity and temporal dynamics, which allows the system to optimize its internal dynamics and representation of neural information. This adds a novel property of real-time

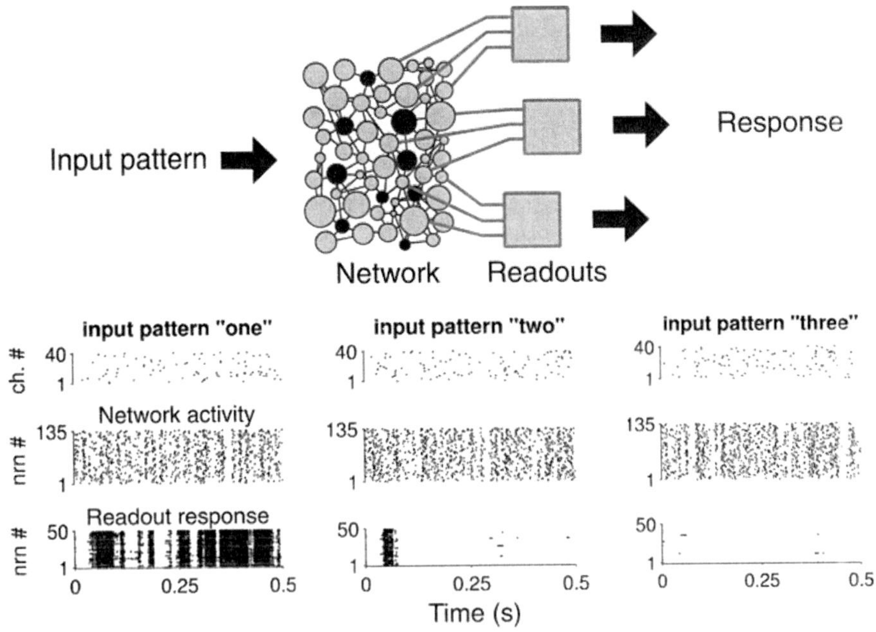

Fig. 10.4 Computing with complex network states. *Top*: Scheme of a computational model that uses a network which displays complex activity states. The activity of a few cells are fed into "readouts" (*black*), which extract the response from the complex dynamics of the network. *Bottom*: Example of computation of different spoken words. The ongoing network activity is apparently random and similar in each case, but it contains information about the input, which can be retrieved by the readout. Modified from Maass et al. (2002)

adaptation and reconfiguration to neural dynamics in artificial networks, and may lead to the emergence of new qualitative behaviors so far unseen in artificial neural networks.

10.2.7 Which Microcircuit for Computing?

What type of neuronal architecture is consistent with such concepts? The most accessible cortical regions are those closely connected to the external world, such as primary sensory cortices or motor cortex. The primary visual cortex (V1) is characterized by the functional specialization of small populations of neurons that respond to selective features of the visual scene. Cellular responses typically form functional maps that are superimposed on the cortical surface. In the vertical axis, V1 cortical neurons seem to obey well-defined rules of connectivity across layers, and make synaptic inputs that are well characterized and typical of each layer. These data suggest a well-constrained wiring diagram across layers, and has motivated the concept of a "cortical microcircuit" (Hubel and Wiesel 1963; Mountcastle 1979; Szentagothai 1983; Douglas and Martin 1991). The cortical microcircuit idea suggests that there is a basic pattern of connectivity, which is canonical and repeated everywhere in cortex. All areas of neocortex would, therefore, perform similar computational operations with their inputs (Barlow 1985). However, even for the primary sensory cortices, there is no clear paradigm in which the distributed activity of neurons, their properties, and connectivity have been characterized in sufficient detail to allow us to the relate structure and function directly, as for the case of oscillations in small invertebrate preparations or in simpler structures such as the thalamus.

An alternative to attempting to explain cortical function on the basis of generic cellular and synaptic properties or stereotyped circuits is to exploit the known wide diversity of cell types and synaptic connections to envision a more complex cortical structure (Fig. 10.5). Cortical neurons display a wide diversity of intrinsic properties (Llinás 1988; Gupta et al. 2000). Likewise, synaptic dynamics are richly variable and show properties from facilitating to depressing synapses (Thomson 2000). Indeed, the essential feature of cortical anatomy may be precisely that there is no canonical pattern of connectivity, consistent with the considerable apparent random component of cortical connectivity templates (Braitenberg and Schüz 1998; Silberberg et al. 2002). Taking these observations together, one may argue that the cortex is a circuit that seems to maximize its complexity, both at the single-cell level and at the level of its connectivity. This view is consistent with models which take advantage of the special information processing capabilities, and memory, of such a complex system. Such large-scale networks can transform temporal codes into spatial codes by self-organization (Buonomano and Merzenich 1995), and as discussed above, computing frameworks were proposed which exploit the capacity of such complex networks to cope with complex input streams (Maass et al. 2002; Bertschinger and Natschläger 2004; Fig. 10.4). In these examples, information is stored in the ongoing activity of the network, in addition to its synaptic weights. This is in agreement with experimental data showing that complex input streams modulate rather than drive network activity (Fiser et al. 2004).

10.2.8 Perspectives: Computing with "Noisy" States

In conclusion, there has been much progress in several paradigms involving various forms of noise (internal, external) to provide computational power to neural networks. To go further in understanding such type of computations, we need progress in two essential directions. First, we need to better understand the different "states" generated by networks. To do this, one needs appropriate simulation techniques which should allow large-scale networks to be simulated in real time for a prolonged time and in a precise manner. Currently available numerical techniques allow to simulate medium-sized networks of up to hundred of thousands of neurons, several orders of magnitude slower than real time. More importantly, most of these

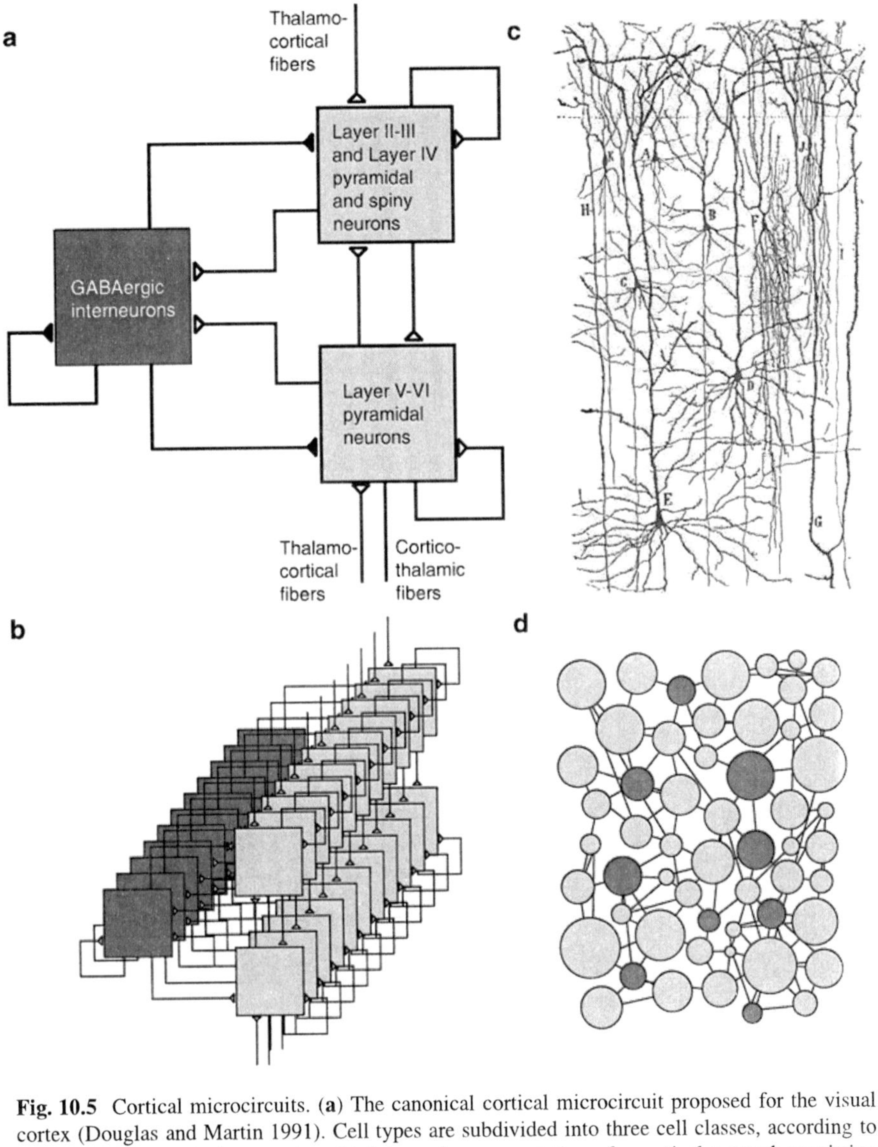

Fig. 10.5 Cortical microcircuits. (**a**) The canonical cortical microcircuit proposed for the visual cortex (Douglas and Martin 1991). Cell types are subdivided into three cell classes, according to layer and physiological properties. (**b**) Schematic representation of a cortical network consisting of the repetition of the canonical microcircuit in (**a**). (**c**) Drawing from Ramón y Cajal (1909) illustrating the diversity of cell types and morphologies in cortex. (**d**) Network of diverse elements as an alternative to the canonical microcircuit. Here, networks are built by explicitly taking into account the diversity of cell types, intrinsic properties, and synaptic dynamics. In contrast to (**b**), this type of network does not consist in the repetition of a motif of connectivity between prototypical cell types, but is rather based on a continuum of cell and synapse properties

techniques restrict their temporal precision in order to solve the huge set of equations governing the dynamics of such networks. However, as it was shown in a number of studies (Hansel et al. 1998; Rudolph and Destexhe 2007), such restrictions can lead to substantial quantitative errors, which may impact on the qualitative interpretation of the result of numerical simulations, especially when considering larger-scale networks with plastic or self-organizing dynamics. Recently, the recognition of these limitations led to the development of novel numerical techniques which are both efficient and precise (Brette et al. 2007c), and there is hope such techniques should be available soon for simulation of neuronal networks of millions of neurons.

Second, we need to understand the computational capabilities of such network "states." This will be possible only through tremendous progress in understanding of the dynamical aspects of nonlinear self-organizing systems. Here also, there is good progress in this field (for an excellent introduction, see Kelso 1995; for a recent review, see Deco et al. 2008), and new mathematical approaches such as graph theory are being explored (Bassett and Bullmore 2006; Reijneveld et al. 2007; Bullmore and Sporns 2009; Guye et al. 2010). We can envision that systematic studies and rigorous mathematical approaches for understanding the functional properties of complex networks should be doable soon. Such approach should naturally explain the "noisy" properties of neurons and networks reviewed here, and how such properties are essential for their computations.

Finally, more work is needed to link this field of biophysical description at the level of conductances and individual neuron responses, with more global models known as *probabilistic models* (Rao et al. 2002). Probabilistic models have been proposed based on the observation that the cortex must infer properties from a highly variable and uncertain environment, and an efficient way to do so is to compute probabilities. Probabilistic or Bayesian models have been very successful to explain psychophysical observations, but their link with biophysics remain elusive. It was suggested that simple spiking network models can perform Bayesian inference, with the probability of firing being interpreted as representing the logarithm of the uncertainty (posterior probability) of stimuli (Rao 2004). This seems a priori consistent with the probabilistic nature of neural responses found the presence of synaptic noise, but other results seems more difficult to reconcile with the Bayesian view. For example, the persistent observation that synaptic noise is beneficial, such as enhanced responsiveness or finer temporal processing, is not easy to link with Bayesian models, where noise is associated to uncertainty. Establishing such links also constitute nice challenges for the future.

Appendix A
Numerical Integration of Stochastic Differential Equations

In this appendix, we will briefly outline methods used in the numerical integration of SDEs. In what follows, we will restrict to the one-dimensional case, but generalization to SDEs of higher orders or systems of (coupled) SDEs is straightforward.

In many physical, and in particular biophysical, stochastic systems, the dynamics of a stochastic variable $x(t)$, the state variable, is governed by the generic SDE

$$\dot{x}(t) = f(x(t)) + g(x(t))\xi(t), \tag{A.1}$$

where $f(x(t))$ is called the *drift term*, and $g(x(t))$ the *diffusion term*. In (A.1), $\xi(t)$ denotes a variable describing a continuous memoryless stochastic process, also called a continuous *Markov process*. Gaussian stochastic processes, such as white noise or OU noise, are just two examples of the latter. In this case, the stochastic variable has a Gaussian probability distribution. For simplicity, we will assume at the moment that $\xi(t)$ has zero mean and unit variance.

In general, SDEs of the form (A.1) are not analytically exact solvable, and numerical approaches remain the only way to obtain approximate solutions. Unfortunately, stochastic calculus itself is ambiguous in the sense that a continuum of parallel notions for the integration of stochastic variables does exist, such as the *Itô* or *Stratonovich calculus*. In contrast to ordinary calculus, these notions yield, in general, different results (for an in depth review of stochastic calculus and various notions, see Gardiner 2002).

Restricting, for the moment, to the Stratonovich calculus (e.g., Mannella 1997), formal integration of (A.1) yields:

$$x(h) - x(0) = \int_0^h \mathrm{d}t \, (f(x(t)) + g(x(t))\xi(t)), \tag{A.2}$$

where h denotes the integration time step. In (A.2) we assume, for simplicity reasons, integration in the interval $[0, h]$ (but a generalization to intervals $[t, t+h]$ is straightforward).

A. Destexhe and M. Rudolph-Lilith, *Neuronal Noise*, Springer Series in Computational Neuroscience 8, DOI 10.1007/978-0-387-79020-6,

Whereas the first term under the integral can be solved utilizing classical calculus, the second term contains the integral over a stochastic variable. The integrated stochastic process

$$Z(h) = \int_0^h \xi(t)\mathrm{d}t \tag{A.3}$$

can shown to be a stochastic process with, in the case considered here, Gaussian probability distribution of zero mean and a SD of $\sqrt{h}$:

$$< Z > = \int_0^h < \xi(t) > \mathrm{d}t = 0$$

$$< Z^2 > = \int_0^h \int_0^h \mathrm{d}t\mathrm{d}s < \xi(t)\xi(s) > = \int_0^h \int_0^h \mathrm{d}t\mathrm{d}s\, \delta(t-s) = h, \tag{A.4}$$

where $< \ldots >$ denotes the statistical average.

A solution of (A.2) can then be obtained by recursion: Taylor expansion of $f(x)$ and $g(x)$ to lowest order in x yields

$$x(h) - x(0) = f_0 h + g_0 Z, \tag{A.5}$$

with $f_0 = f(x(0))$ and $g_0 = g(x(0))$. The next higher order can be calculated by re-inserting the lowest order solution back into (A.2) and collecting the contributions according to power of h. This gives for the value of the stochastic variable x at time increment h and second order in h

$$x(h) = x(0) + g_0 Z(h) + f_0 h + \frac{1}{2} g_0' g_0 Z(h)^2 \tag{A.6}$$

with

$$g_0' = \frac{\partial g(x(t))}{\partial x(t)}\bigg|_{x=x(0)}.$$

Higher orders can be obtained accordingly.

In the last paragraph, the *full integration scheme* with accuracy up to order $O(h^2)$ was developed. Various other integration schemes can be used. In the *Euler scheme*, only the first three terms on the right-hand side of (A.6) are utilized:

$$x(h) = x(0) + g_0 Z(h) + f_0 h, \tag{A.7}$$

yielding accuracy up to order $O(h)$. In the *exact propagator scheme*,

$$\dot{x} = f(x) \tag{A.8}$$

is solved exactly, i.e. analytically or numerically with the desired accuracy. Then, the noise term $Z(h)$ is added. Finally, in the *Heun scheme*, the solution at time increment h is given by

$$x(h) = x(0) + g_0 Z(h) + \frac{h}{2}(f_0 + f(x_1)), \tag{A.9}$$

where

$$x_1 = x(0) + g_0 Z(h) + f_0 h. \tag{A.10}$$

This scheme is accurate up to order $O(h^2)$. Other integration schemes are possible, such as Runge–Kutta (Mannella 1989, 1997).

The above formalism can be extended to more realistic noise models as well. An interesting class which describes noise occurring in many real systems, such as stochastic neuronal membranes, is correlated (linearly filtered white) noise. Considering the simplest case of exponentially correlated Gaussian white noise, the stochastic process $\eta(t)$ is defined by the first-order differential equation:

$$\dot{\eta(t)} = -\frac{1}{\tau}\eta(t) + \frac{\sqrt{2D}}{\tau}\xi(t), \tag{A.11}$$

where τ denotes the time constant and $\xi(t)$ Gaussian white noise of zero mean and unit variance. The mean and correlation of $\eta(t)$ are given by

$$\begin{aligned} < \eta(t) > &= 0 \\ < \eta(t)\eta(s) > &= \frac{D}{\tau}\exp\left(-\frac{|t-s|}{\tau}\right), \end{aligned} \tag{A.12}$$

and its spectral density by a Lorentzian

$$|\hat{\eta}(w)|^2 = \frac{D}{\pi(1+\omega^2\tau^2)}. \tag{A.13}$$

Here, $\hat{\eta}(w)$ denotes the Fourier transform of $\eta(t)$.

Following the same approach as described above for the white noise case, a system or process described by the SDE

$$\dot{x(t)} = f(x(t)) + g(x(t))\eta(t) \tag{A.14}$$

has the following solution up to first order in the integration time step h:

$$x(h) = x(0) + g_0 Z(h) + f_0 h + \frac{1}{2} g_0' g_0 Z(h)^2, \tag{A.15}$$

where

$$\eta(t) = \eta(0) + \frac{\sqrt{2D}}{\tau} w_0$$

$$Z(h) = \int_0^h \mathrm{d}t\, \eta(t) = \tau(1 - \mathrm{e}^{h/\tau})\eta(0) + \frac{\sqrt{2D}}{\tau} w_1\,, \tag{A.16}$$

with

$$w_0 = \int_0^h \mathrm{d}s\, \mathrm{e}^{\frac{s-h}{\tau}} \xi(s)$$

$$w_1 = \int_0^h \int_0^t \mathrm{d}s\mathrm{d}t\, \mathrm{e}^{\frac{s-h}{\tau}} \xi(s)\,. \tag{A.17}$$

Here, w_0 and w_1 are Gaussian variables with zero averages and correlations

$$< w_0^2 > = \frac{\tau}{2}(1 - \mathrm{e}^{-2h/\tau})$$

$$< w_0 w_1 > = \frac{\tau^2}{2}(1 - 2\mathrm{e}^{-h/\tau} + \mathrm{e}^{-2h/\tau})$$

$$< w_1^2 > = \frac{\tau^3}{2}\left(\frac{2h}{\tau} - 3 - \mathrm{e}^{-2h/\tau} + 4\mathrm{e}^{-h/\tau}\right). \tag{A.18}$$

In the remainder of this appendix, we will take a brief look at the general numerical treatment of the stochastic variable itself. Let $\psi(t)$ denote a continuous memoryless stochastic, i.e., Markov, process. Such a process is defined when the following three conditions are met: first, the increment of ψ from time t to some later time $t+\mathrm{d}t$ only depends on the value of ψ at time t and $\mathrm{d}t$, hence a *conditional increment* of $chi(t)$ can be defined as

$$\Psi(t,\mathrm{d}t) = \psi(t+\mathrm{d}t) - \psi(t)\,.$$

Second, the increment $\Psi(t)$ itself is a stochastic variable that depends smoothly on t and $\mathrm{d}t$ only. Finally, $\Psi(t)$ is continuous, i.e., $\Psi(t,\mathrm{d}t) \to 0$ for all t if $\mathrm{d}t \to 0$.

As shown in Gillespie (1996), if these conditions are met, then the conditional increment will take the analytic form

$$\Psi(t,\mathrm{d}t) \equiv \psi(t+\mathrm{d}t) - \psi(t) = A(\psi(t),t)\mathrm{d}t + \sqrt{D(\psi(t),t)}N(t)\sqrt{\mathrm{d}t}\,, \tag{A.19}$$

where $A(\psi(t),t)$ and $D(\psi(t),t)$ are smooth functions of t, and $N(t)$ is a temporally uncorrelated random variable with unit normal distribution and $N(t) \neq N(t')$ if $t \neq t'$.

Equation (A.19) is called the *Langevin equation* for the stochastic process $\psi(t)$ with drift function $A(\psi(t),t)$ and diffusion function $D(\psi(t),t)$.

Despite the fact that $\psi(t)$ defined by (A.19) is continuous, it is generally not differentiable (a hallmark of stochastic continuous Markov processes). However, formally the limit $\mathrm{d}t \to 0$ can be taken, and yields

$$\frac{\mathrm{d}\psi(t)}{\mathrm{d}t} = A(\psi(t),t) + \sqrt{D(\psi(t),t)}\chi t\,. \tag{A.20}$$

Here, $\chi(t)$ denotes a Gaussian white noise process with zero mean and unit SD. Equation (A.20) is also called the (white noise form) Langevin equation, and serves as a definition of the stochastic process $\psi(t)$.

Finally, let us consider a specific example, namely the OU stochastic process we encountered earlier. For this example of a continuous Markovian process, $A(\psi(t),t)$ and $D(\psi(t),t)$ are given by

$$\begin{aligned} A(\psi(t),t) &= -\frac{1}{\tau} \\ D(\psi(t),t) &= c\,, \end{aligned} \tag{A.21}$$

where τ denotes the relaxation time and c the diffusion constant. With this, we obtain for the increment and the defining differential equation of the OU stochastic process:

$$\begin{aligned} \psi(t+\mathrm{d}t) &= \psi(t) - \frac{1}{\tau}\psi(t)\mathrm{d}t + \sqrt{c}N(t)\sqrt{\mathrm{d}t} \\ \frac{\psi(t)}{\mathrm{d}t} &= -\frac{1}{\tau}\psi(t) + \sqrt{c}\chi(t) \end{aligned} \tag{A.22}$$

For an in-depth introduction into the stochastic calculus, we refer to Gardiner (2002). An excellent introduction into Brownian motion and the OU stochastic process can be found in Nelson (1967).

Appendix B
Distributed Generator Algorithm

In numerical simulations of neuron models receiving synaptic inputs through a multitude of individual input channels, such as in simulations of biophysically detailed neurons with spatially extended dendritic structures (see Sect. 4.2), methods are required to shape and distribute the activity among these channels. Whereas the mean release rate of all, subgroups or individual synaptic channels constitutes the lowest-order statistical characterization of synaptic activity, in many models, higher-order statistical parameters, such as the pairwise or average correlation, are needed to more realistically describe the activity at synaptic terminals impinging on a neuron.

In the literature, various methods are described to generate multichannel synaptic input patterns (e.g., Brette 2009). Here, we will briefly describe one of the simplest of these methods, namely the *distributed generator algorithm* (Destexhe and Paré 1999). In this algorithm, correlation among individual synaptic input channels is achieved by selecting the activity pattern randomly from a set of common input channels. This way, a redundancy, or correlation, among individual channels is obtained.

More specifically, consider N_0 independent Poisson-distributed presynaptic spike trains. At each time step, the activity pattern across the spike trains contains N_0 numbers "1" (for release) and "0" (for no release; Fig. B.1, left, *gray boxes*). Each of these numbers is then uniform randomly redistributed among N numbers ($N_0 \leq N$, Fig. B.1, right, *gray boxes*) with a probability p, where

$$p = \frac{N}{N_0} = \frac{N}{N(1-\sqrt{c})+\sqrt{c}}\,. \tag{B.1}$$

Here, c denotes a correlation parameter with $0 \leq c \leq 1$. As this redistribution happens at each time step, an activity pattern at N synaptic input channels is obtained.

From (B.1) it is clear that for $c = 0$ one has $N = N_0$ (no correlation), whereas for $c = 1$ one has $N_0 = 1$, irrespective of N, i.e., all N channels show the same activity

A. Destexhe and M. Rudolph-Lilith, *Neuronal Noise*, Springer Series in Computational Neuroscience 8, DOI 10.1007/978-0-387-79020-6,

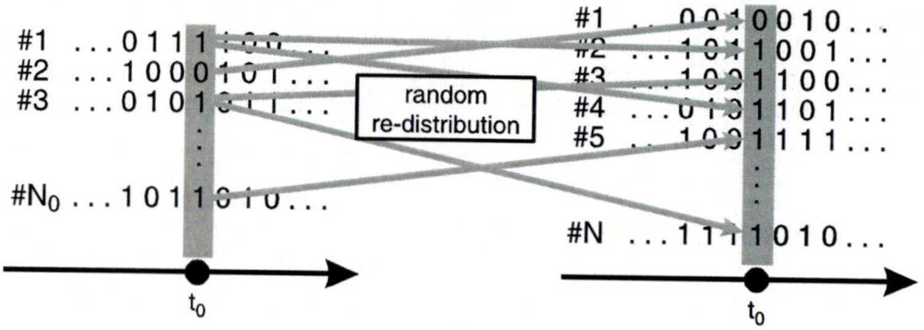

Fig. B.1 Distributed Generator Algorithm. At each time t_0, the activity pattern across N_0 independent Poisson-distributed presynaptic spike trains (*left*) is randomly redistributed among N synaptic channels, thus introducing a redundancy, or correlation, among the N presynaptic spike trains. The number of independent channels N_0 necessary to obtain correlated activity quantified by a correlation parameter c among the N channels is $N_0 = N + \sqrt{c}(1 - N)$

(either release or no release). Vice versa, the number of independent channels N_0 required to achieve a correlation c among N channels is given by

$$N_0 = N + \sqrt{c}(1 - N). \tag{B.2}$$

The advantage of this algorithm is that it allows to control the correlation among synaptic input channels without impairing the statistical signature of each individual channel, such as rate and the Poisson-distributed nature: For $c > 0$, i.e., $N_0 < N$, each synapse still releases randomly according to the same Poisson process, but with a probability for releasing together with other synapses. Moreover, the correlation of the synaptic activity can be changed without affecting the average release frequency at each synapse and, thus, the overall conductance submitted through the cellular membrane.

The distributed generator algorithm is an easy and computationally fast algorithm to control the average correlation independent of the average rate and without impairment of the Poisson characteristics. However, the mathematical link to more commonly used measures, such as pairwise correlation coefficient (Pearson correlation) is hard to draw. In Brette (2009), two methods, the *Cox Method* and *Mixture Method*, were introduced, which extend the algorithm presented above and allow to generate sets of correlated presynaptic spike trains with arbitrary rates and pair-wise cross-correlation functions.

In the Cox method, the activity patterns at each synaptic terminal are described by independent inhomogeneous Poisson process with time-varying rate, called doubly stochastic processes, or Cox processes. It can be shown that the cross-correlation between individual trains can be expressed in terms of the cross-correlation function of the time-dependent rates of these trains. Thus, correlation among synaptic input channels is achieved through Poisson processes with correlated time-dependent rates. The Mixture method describes the generation of correlation among individual channels through selection of activity from a common pool, as described earlier in this appendix, but with generalization to heterogeneous correlation structures.

Appendix C
The Fokker–Planck Formalism

The term *Fokker–Planck equation* originates from the work of the two physicists, A.D. Fokker and M. Planck, who, at the beginning of the last century, arrived at a statistical description of Brownian motion (Fokker 1914; Planck 1917). Shortly after, A.N. Kolmogorov arrived independently at a similar mathematical description, which later became known as the *Kolmorogov forward equation* (Kolmogorov 1931). Despite its original application to describe Brownian motion, the formalism behind the Fokker–Planck equation is very general, and is today widely used to mathematically assess the dynamics and characteristics of stochastic systems (for an throughout modern introduction, see Risken 1984).

In its core, the Fokker–Planck equation describes the time evolution of macroscopic variables, specifically the probability density function of observables, of a stochastic system described by one or more stochastic differential equations. As a complete solution of a high-dimensional macroscopic system based on the knowledge of the dynamics of its microscopic constituents (i.e., the equation of motions for all its microscopic variables) is often complicated or even impossible, the idea one can follow is to introduce macroscopic variables, or observables, which fluctuate around their expectation values. In this sense, the Fokker–Planck equation becomes, then, an equation of motion for the (probability) distribution of the introduced macroscopic fluctuating variables. This equation takes usually the form of one differential equation (or a system of differential equations), first order in time and second order in the observable, for which many methods of solution are readily available.

In its classical form, the Fokker–Planck equation for the time evolution of the probability distribution function $\rho(\{x_i\},t)$ for N (time-dependent) macroscopic variables, or observables, x_i reads

$$\frac{\partial \rho(\{x_i\},t)}{\partial t} = \left[-\sum_{i=1}^{N} \frac{\partial}{\partial x_i} D_1(\{x_i\},t) + \sum_{i,j=1}^{N} \frac{\partial^2}{\partial x_i \partial x_j} D_2(\{x_i\},t) \right] \rho(\{x_i\},t), \quad \text{(C.1)}$$

A. Destexhe and M. Rudolph-Lilith, *Neuronal Noise*, Springer Series in Computational Neuroscience 8, DOI 10.1007/978-0-387-79020-6,

where $D_1(\{x_i\},t)$ denotes the *drift vector* and $D_2(\{x_i\},t)$ the *diffusion tensor*. If only one observable x is considered, (C.1) takes the simpler form

$$\frac{\partial \rho(x,t)}{\partial t} = \left[-\frac{\partial}{\partial x} D_1(x,t) + \frac{\partial^2}{\partial x^2} D_2(x,t) \right] \rho(x,t), \tag{C.2}$$

where the scalars $D_1(x,t)$ and $D_2(x,t)$ are now called *diffusion* and *drift coefficient*, respectively.

Before demonstrating how the Fokker–Planck equation can be deduced from the description of the microscopic system in terms of stochastic differential equations, we will briefly outline how (C.2) arises from the general notion of transition probability in stochastic processes. Let $\rho(x_1,t_1;x_2,t_2)$ denote the joint probability distribution that x takes the value x_1 at time t_1 and x_2 at time t_2, and $\rho(x',t'|x,t)$ the transition probability from value x at time t to x' at t'. For any Markov process, the latter has to obey the following consistency condition

$$\rho(x_3,t_3|x_1,t_1) = \int \mathrm{d}x_2\, \rho(x_3,t_3|x_2,t_2)\rho(x_2,t_2|x_1,t_1), \tag{C.3}$$

which is called *Chapman–Kolmogorov equation.* With this, and the fact that $\rho(x_2,t_2) = \int \mathrm{d}x_1\, \rho(x_2,t_2;x_1,t_1)$, the probability distribution obeys the following identity:

$$\rho(x_2,t_2) = \int \mathrm{d}x_1\, \rho(x_2,t_2|x_1,t_1)\rho(x_1,t_1). \tag{C.4}$$

The Chapman–Kolmogorov equation (C.3) is an integral equation, and it can be shown that it is equivalent to the integro-differential equation

$$\frac{\partial}{\partial t}\rho(x,t|x_0,t_0) = \int \mathrm{d}x' \left[W_t(x|x')\rho(x',t|x_0,t_0) - W_t(x'|x)\rho(x,t|x_0,t_0) \right], \tag{C.5}$$

where $W_t(x_2|x_1)$ is interpreted as the transition probability per unit time from x_1 to x_2 at time t. Equation (C.5) is called *master equation*. With (C.4), the master equation can be written in the form

$$\frac{\partial}{\partial t}\rho(x,t) = \int \mathrm{d}x' \left[W_t(x|x')\rho(x',t) - W_t(x'|x)\rho(x,t) \right], \tag{C.6}$$

from which its heuristic meaning can be deduced: the master equation is a general identity which describes the balance, i.e., the "gain" due to transition from other (continuous) states x' to state x (first term on the right-hand side), and "loss," due to transition from state x into other states x' (second term), for the probability of each state. As we will see below, the Fokker–Planck equation is a specific example of the master equation.

In order to treat the master equation further, one considers "jumps" from one configuration x' to another x. This allows to rewrite (C.6) by means of Taylor expansion with respect to the size of the jumps. This expansion is known as *Kramers–Moyal expansion*, and yields

$$\frac{\partial}{\partial t}\rho(x,t) = \sum_{n=1}^{\infty} \frac{(-1)^n}{n!} \frac{\partial^n}{\partial x^n} \left[a^{(n)}(x,t)\rho(x,t)\right], \tag{C.7}$$

where

$$a^{(n)}(x,t) = \int \mathrm{d}r\, r^n W(x|x')$$

with jump size $r = x - x'$ denotes the jump moments. Mathematically, the Kramers–Moyal expansion (C.7) is identical to the master equation (C.5) and, therefore, remains difficult so solve. However, as we now have a infinite sum due to the Taylor expansion, one can restrict to a finite number of terms and, thus, arrive at an approximation. Restricting to terms up to second order, one obtains

$$\frac{\partial}{\partial t}\rho(x,t) = -\frac{\partial}{\partial x}\left[a^{(1)}(x,t)\rho(x,t)\right] + \frac{1}{2}\frac{\partial^2}{\partial x^2}\left[a^{(2)}(x,t)\rho(x,t)\right], \tag{C.8}$$

which is equivalent to the celebrated *Fokker–Planck equation* (C.2). The deduction for the general multivariable Fokker–Planck equation (C.1) follows a similar approach (for a detailed discussion see, e.g., Risken 1984; van Kampen 1981; Gardiner 2002).

To illustrate how the Fokker–Planck equation is obtained from the description of the underlying microscopic system, we consider the first order stochastic differential equation

$$\frac{\mathrm{d}x(t)}{\mathrm{d}t} = A(x(t),t) + B(x(t),t)\eta(t), \tag{C.9}$$

in which $A(x(t),t)$ and $B(x(t),t)$ denote arbitrary functions in $x(t)$, called drift (transport) and diffusion term, respectively, and $\eta(t)$ a stochastic process. In what follows, we will assume $\eta(t)$ to be a Gaussian white noise process with zero mean and variance $2D$:

$$\begin{aligned} <\eta(t)> &= 0 \\ <\eta(t)\eta(t')> &= 2D\delta(t-t'). \end{aligned} \tag{C.10}$$

Higher-order moments can be deduced from this relations using Novikov's theorem and Wick's formula. Note that this choice of $\eta(t)$ renders the variable $x(t)$ a Markov process, and that for each realization of $\eta(t)$, the time course of $x(t)$ is fully determined if its initial value is given.

In physics, equations of the form (C.9) are typically called *Langevin equation*. However, although the notation used in (C.9) is a common short-hand notation for SDEs, it is mathematically not correct and should always be understood in its differential form

$$\mathrm{d}x(t) = A(x(t),t)\mathrm{d}t + B(x(t),t)\mathrm{d}\eta(t), \tag{C.11}$$

which is meaningfully interpreted only in the context of integration:

$$x(t+\tau)-x(t)=\int_{t}^{t+\tau}\mathrm{d}s A(x(s),s)+\int_{t}^{t+\tau}\mathrm{d}\eta(s)B(x(s),s). \tag{C.12}$$

The first integral on the right-hand side describes an ordinary Lebesgue integral, whereas the second is called *Itô integral* and must be treated in the framework of stochastic calculus.

Since the solution of the Langevin equation is a Markov process, it obeys the master equation (C.7). To obtain the coefficients entering the Kramers–Moyal expansion, we expand $A(x(s),s)$ and $B(x(s),s)$ with resect to x:

$$A(x(s),s)=A(x,s)+\frac{\partial A(x,s)}{\partial x}\Big|_x(x(s)-x)+O((x(s)-x)^2)$$

$$B(x(s),s)=B(x,s)+\frac{\partial B(x,s)}{\partial x}\Big|_x(x(s)-x)+O((x(s)-x)^2).$$

Inserting these into (C.12) and taking the average, one obtains, after utilizing (C.10):

$$\begin{aligned}<x(t+\tau)-x(t)>=&\int_{t}^{t+\tau}\mathrm{d}s A(x(s),s)+\int_{t}^{t+\tau}\mathrm{d}s\,\frac{\partial A(x,s)}{\partial x}\Big|_x(x(s)-x)\int_{t}^{s}\mathrm{d}s' A(x(s'),s')\\&+2D\int_{t}^{t+\tau}\mathrm{d}s\frac{\partial B(x,s)}{\partial x}\Big|_x(x(s)-x)\int_{t}^{s}\mathrm{d}s' B(x(s'),s')\delta(s'-s)+O,\end{aligned} \tag{C.13}$$

where O denotes higher-order contributions.

From the Kramers–Moyal expansion, we have for the first-order jump moment

$$a^{(1)}=\lim_{\tau\to 0}\frac{1}{\tau}<x(t+\tau)-x(t)>|_{x(t)=x} \tag{C.14}$$

from which we obtain

$$a^{(1)}=A(x(t),t)+DB(x(t),t)\frac{\partial B(x,t)}{\partial x}. \tag{C.15}$$

Similarly, with

$$a^{(2)}=\lim_{\tau\to 0}\frac{1}{\tau}<(x(t+\tau)-x(t))^2>|_{x(t)=x} \tag{C.16}$$

one obtains

$$a^{(2)} = 2DB^2(x(t),t) \tag{C.17}$$

for the second-order jump moment. All other coefficients $a^{(n)}, n > 2$ vanish. Thus, the Markov statistic process, given by the Langevin equation (C.9) with Gaussian δ-correlated noise source $\eta(t)$, yields the Fokker–Planck equation

$$\begin{aligned}\frac{\partial \rho(x,t)}{\partial t} &= -\frac{\partial}{\partial x}\left\{\left[A(x(t),t) + DB(x(t),t)\frac{\partial B(x,t)}{\partial x}\right]\rho(x,t)\right\} \\ &+D\frac{\partial^2}{\partial x^2}\left[B^2(x(t),t)\rho(x,t)\right].\end{aligned} \tag{C.18}$$

Here, it is important to note that because all higher order jump moments vanish in the Kramers–Moyal expansion, this Fokker–Planck equation is exact. The term $DB(x(t),t)\frac{\partial B(x,t)}{\partial x}$ which occurs in the drift term (the first term on the right-hand side) together with the deterministic drift $A(x(t),t)$ is called *noise-induced drift*. Moreover, (C.18) allows to directly deduce the Fokker–Planck equation from the microscopic equation of motion given by the differential equation (C.9) by simple inspection.

To illustrate this last point, we take the example of a Wiener process, which is defined by the SDE

$$\mathrm{d}x(t) = \mathrm{d}\tilde{\eta}(t), \tag{C.19}$$

where

$$\begin{aligned} < \tilde{\eta}(t) > &= 0 \\ < \tilde{\eta}(t)\tilde{\eta}(t') > &= \min(t,t').\end{aligned} \tag{C.20}$$

That is, with the above notion, we have

$$\begin{aligned} A(x(t),t) &= 0 \\ B(x(t)t) &= 1 \\ D &= \frac{1}{2}.\end{aligned} \tag{C.21}$$

Inserting this into (C.18), one obtains the corresponding Fokker–Planck equation

$$\frac{\partial \rho(x,t)}{\partial t} = \frac{1}{2}\frac{\partial^2 \rho(x,t)}{\partial x^2}. \tag{C.22}$$

This is the simplest form of a diffusion equation, with an explicit analytic solution

$$\rho(x,t) = \frac{1}{\sqrt{2\pi t}} \mathrm{e}^{-\frac{x^2}{2t}} . \tag{C.23}$$

In general, however, explicit solutions of the Fokker–Planck equation can not be obtained. For an in-depth introduction into the Fokker–Planck formalism and its applications, we refer to Risken (1984).

Appendix D
The RT-NEURON Interface for Dynamic-Clamp

In this appendix, we briefly describe the software interface utilized for most of the dynamic-clamp experiments described in this book, namely a modified version of the well-known NEURON simulator (Hines and Carnevale 1997; Carnevale and Hines 2006). The NEURON simulation environment was also used for most of the computational studies covered in this book, including the models used in the dynamic-clamp experiments. The advantage of this approach is obvious: it allows one to run dynamic-clamp experiments using the same program codes as those used for computational models.

This modification of NEURON was initiated by Gwendal Le Masson and colleagues, and allows interfacing the computational hardware, in real time, with the electrophysiological setup (Le Franc et al. 2001). It was later developed in the laboratory of Thierry Bal at UNIC/CNRS. We describe below this *RT-NEURON* tool (for more details, see Sadoc et al. 2009).

D.1 Real-Time Implementation of NEURON

From the point of view of the NEURON simulation environment (Hines and Carnevale 1997; Carnevale and Hines 2006), the inclusion of a real-time loop is relatively straightforward. In NEURON, the state of the system at a given time is described by a number of variables (V_m, I_m, etc). The value of some of these variables is externally imposed ("input" variables), whereas other variables must be calculated ("output" variables). Their values not only depend on the other variables at present time but also on the values of all variables in the past.

In a conventional NEURON simulation (i.e., not real time), if a fixed-step integration method such as Backward Euler or Crank–Nicholson is chosen the system is first initialized (function `finitialize`), then the state of the system is calculated at each time step, at time $n * \mathrm{d}t$, where n is an integer and $\mathrm{d}t$ is the fixed time step. The function which performs this integration is called `fadvance`

A. Destexhe and M. Rudolph-Lilith, *Neuronal Noise*, Springer Series in Computational Neuroscience 8, DOI 10.1007/978-0-387-79020-6,

in NEURON and calculates the state of the system at time $n * dt$ from the values at preceding times, $(n-1) * dt$, $(n-2) * dt$, ...

To realize a real-time system, one must: (1) associate some input variables of NEURON to the analog inputs of an external device; (2) associate some output variables of NEURON to output variables of this device; (3) make sure that the sequence (read input variables, calculate new state, write output variables) is done at each instant $n * dt$. As a consequence, such a system necessarily requires a fixed-step integration method.

These steps are easily implemented using the recent multifunction data acquisition boards available for PCs. The card must possess analog and digital input and output (I/O) channels, as well as a clock to set the acquisition frequency—these characteristics are relatively standard today. The card must also be able to send an interrupt to the PC at each acquisition, so that computations can be initiated, as well as reading input variables and writing to output variables. This latter requirement is more restrictive and guides the choice of the data acquisition board.

D.1.1 Real-Time Implementation with a DSP Board

RT-NEURON can be set up using a standard data acquisition card with a digital signal processor (DSP) on board. A first and obvious possibility is to use the DSP to program the computational part of the dynamic-clamp system (see also Robinson 2008). However, this solution would require the compilation of the NEURON simulator on the DSP board, which is very difficult technically, and also not flexible because the source code of NEURON would have to be profoundly modified, thus making difficult any upgrade with most recent versions of NEURON.

Instead, an alternative solution is to run NEURON on the operating system, and have a minimal program running on the DSP to handle the timing of I/O events. This minimal program receives simple commands from the PC (such as "Start," "Stop," "Set Clock," etc) and implements a procedure triggered by the DSP clock at each dt. This procedure: (1) reads in inputs and stores the values in a mailbox; (2) triggers an interrupt on the PC; (3) receives the output values in a mailbox and refreshes the analog outputs. In practice, the operation (3) is executed first—the output values of the preceding circle are processed, which introduces a systematic delay equal to the time step dt.

On the PC, in NEURON, the interrupt procedure reads the inputs from the mailbox and stores the values to the variables associated to these entries. The interrupt procedure then calls the procedure `nrn_fixed_step` (which is a low-level version of `fadvance`). It then sends the calculated values of the output variables to the mailbox. Thus, the mechanism consists of having a data acquisition board that sends an interrupt at every time step dt, and this interrupt triggers the execution of a particular I/O procedure linked to NEURON. This procedure allows to run real-time applications even under MS Windows. Other solutions are also possible, such as using a real-time operating system (such as RT-LINUX). Such

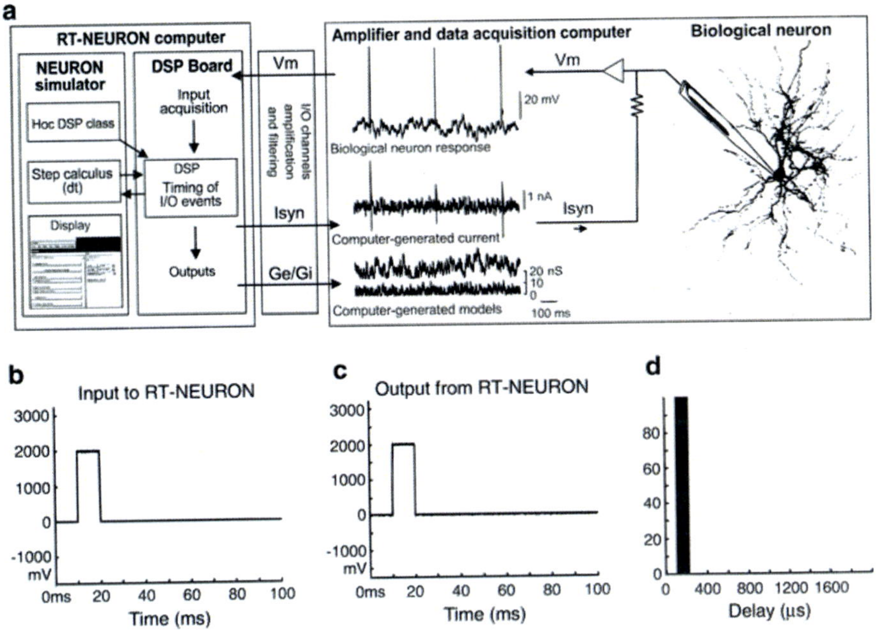

Fig. D.1 The RealTime-NEURON system architecture. (**a**) Software-generated model neuron and conductances run in real time in the RT-NEURON Windows-based computer (*left box*). A DSP board paces the operations of the Pentium processor(s) and controls the input/output data transfer between the model and biological cells. Input variables such as the biological neuron membrane potential (Vm) are sent to the NEURON simulator through the DSP at each dt. In return, output variables such as the command current (Isyn) corresponding to the excitatory and inhibitory conductances (Ge/Gi) in this example, are sent to the amplifier or the acquisition system. Here Isyn is injected in discontinuous current clamp in a thalamocortical neuron through the same pipette that collects the Vm. (**b**) Test of the real time in RT-NEURON-based dynamic clamp. RT-NEURON was used to copy an analog input to an analog output while simultaneously running a simulation. (**b**) Test rectangular signal sent as input to RT-NEURON via the DSP. RT-NEURON was running simultaneously the Ge/Gi stochastic conductance model as in (**a**). (**c**) Signal recorded on the output channel. (**d**) Histogram of delays between (**b**) and (**c**), showing a single peak at the value of 100 ms, which was the value of the cycle dt, thus demonstrating that the system strictly operates in real time. Modified from Sadoc et al. (2009)

a solution is under investigation by M. Hines. The Windows-based RT-NEURON was first implemented in 2001 using version 4.3.1. of NEURON, and running on the Windows NT operating system. This version has now been ported to version 6.0 of NEURON, and runs on all versions of Windows (Fig. D.1).

To manage the DSP in NEURON, a new C++ class directly linked to the tools developed on the board is available. In order to make these functions accessible to the HOC level, NEURON has to be recompiled while registering a corresponding HOC class with methods allowing the direct control and set up of the DSP board. This new HOC class allows to load software on the DSP board, to initialize the internal clock with the chosen integration time step dt, to initialize the gain of the

AD/DA converters, to set the models variables which will be used as input and output, to determine the priority level of the interruption request and finally to start or stop the real-time experiment.

Despite these new tools, the link between NEURON and the acquisition system is not complete. Most of NEURON objects, and more precisely the mechanisms as synapse models or ionic channels, must be linked to an object compartment, which represents a volume of membrane. Thus, in order to calculate the injected current, it is crucial to link the model to a "ghost compartment" which should have a negligible participation to the calculation. This compartment, which acts as an anchoring point for the mechanisms, will be the buffer for the data transfer during the real time, receiving the measured membrane potential of the biological neuron, for example. This "ghost compartment" is a simple cylinder with a total area of 1 cm^2 to avoid any scaling effect on the calculated injected current, as the current is expressed in NEURON by units of Ampere per cm^2. For that, it is possible to add a simple numerical variable attached to the compartment on which the DSP will point to send the measured membrane potential.

In addition to these hoc tools, a graphical interface was developed that allows to access to all these functions and starts the real time sequence (see Sect. D.2.2 below).

D.1.2 MS Windows and Real Time

It is clear that the version presented above is in contradiction with the principles of MS Windows, which was conceived for multitasking, which, by definition, does not respect the timing of interrupts. No process can pretend to have the absolute priority over another process, or over processes triggered by the operating system. Nevertheless, on present PCs and given a few precautions, RT-NEURON works correctly with a dt of 100 ms to 60 ms (10–15 kHz) using the system described above. In this case, the interrupts generated by the DSP board must be given maximum priority. To achieve this, the normal interrupts must be deactivated, such as network cards, USB handling, and all programs that are susceptible to be started at any time. Second, the RT-NEURON software must be given the maximum priority level allowed in MS Windows.

Finally it should be noted that, independently of the type of the dynamic-clamp system used, the time resolution is limited to approximately 3 kHz when using sharp glass pipettes, as it is commonly the case for dynamic clamp in vivo (Brette et al. 2008) or in some cases in vitro (Shu et al. 2003b). Until recently it was necessary to discretize the injection/recording of current using the discontinuous current-clamp (DCC) method. The method of Active Electrode Compensation (AEC; see Sect. 6.5; Brette et al. 2008 suppresses this limitation and was incorporated in the RT-NEURON system.

D.1.3 Testing RT-NEURON

To validate the real-time system,the following test procedure can be used. A program is run on NEURON, and in addition, one asks NEURON at each cycle to copy a supplementary analog input to an analog output. A rectangular periodic signal is sent to the analog input. A second computer, equipped with a data acquisition board (100 kHz), acquires inputs and outputs at high frequency. By comparing these two channels, one can directly evaluate the real time by computing the distribution of measured delays between the input and the output channel. If the real time was perfect, the histogram obtained should have a single peak at the value of the dt (100 ms in this case). This is in general observed, as shown in Fig. D.1b–d.

D.2 RT-NEURON at Work

In this section, we briefly illustrate the use of RT-NEURON, and in particular the new procedures that have been added to run the dynamic-clamp experiments. We will illustrate the steps taken in a typical dynamic-clamp experiment using RT-NEURON.

D.2.1 Specificities of the RT-NEURON Interface

To start a dynamic-clamp experiment, the procedure consists, as in the classical NEURON, of loading the DLL and the HOC files of the chosen models and protocols. A novel *DspControl* box is available with RT-NEURON, which contains the new commands to run the simulations in real time via the control of the DSP (Fig. D.2a).

D.2.2 A Typical Conductance Injection Experiment Combining RT-NEURON and AEC

In order to test a protocol for conductance injection in a biological cell (and also to build the RT-NEURON platform), it is convenient to first simulate the experiment in an electronic RC-circuit (the "model cell" available with some commercial current-clamp amplifiers).

In the illustration of the practical utilization of RT-NEURON (Fig. D.2), a stochastic "point-conductance" model was used to recreate synaptic background activity a in real neuron in the form of two independent excitatory and inhibitory conductances (Ge and Gi) (here, Ge and Gi are equivalent to g_{e0} and g_{i0}, which is

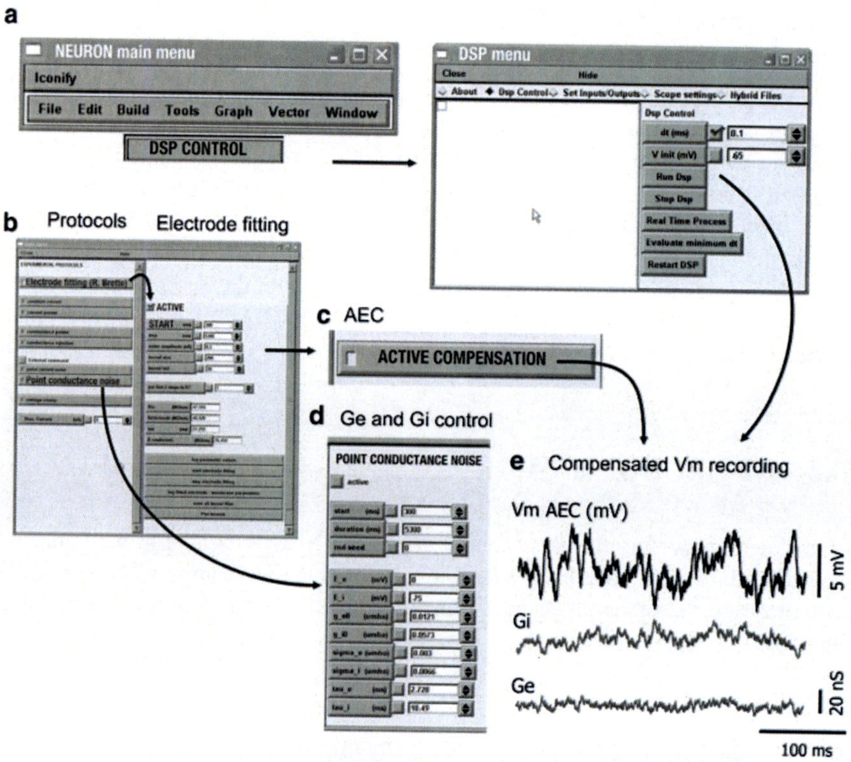

Fig. D.2 Conductance injection using RT-NEURON and AEC. (**a**) Modified "Tools" menu for real-time commands and settings. DSP Control submenu allows to start and stop the DSP and to set the calculation dt. Set inputs/outputs allocates the destination of the input (Membrane voltage, triggers) and output (command current, conductance models, etc.) variables to input and output channels of the DSP, which are connected to various external devices (amplifier, oscilloscope, acquisition system). Sessions containing predefined experimental protocols can be loaded from the menu Hybrid Files. (**b–d**) Examples of programmable display windows for controlling conductance models injection (**b, d**) and Active Electrode Compensation (**c, d**). (**e**) Top trace: error-free membrane potential of a biological cortical neuron recorded in vitro (Vm AEC) in which synaptic conductance noise (lower traces Gi and Ge) is injected via a sharp glass micropipette (modified from Brette et al. 2008, as well as from Sadoc et al. 2009)

the notation used in other chapters) using a point conductance model (Destexhe et al. 2001). In fact, any type of conductance models or current waveforms programmed in the Hoc file can be simultaneously injected in a neuron. The complexity of the model is only limited by the speed of the computer. The On/Off command for injecting the command current corresponding to the modeled ionic currents, and the parameters such as the amplitude of the conductance (Gmax) can be modified online during the recording using the interface (Fig. D.2d).

Mimicking the electrical activity of ionic channels using dynamic clamp relies on the injection of high-frequency currents in the neuron via a glass pipette. However, injection of such high-frequency currents across a sharp microelectrode or a high-impedance patch electrode (such as those used in vivo) is known to produce signal distortions in the recording. These recording artifacts can be traditionally avoided by using the DCC method but with the disadvantage of a limited sampling frequency. They can also be avoided using the AEC method described in Sect. 6.5 (Brette et al. 2008). The dual use of AEC and RT-NEURON allows high temporal resolution dynamic clamp with the sampling frequency only limited by the speed of the computer.

References

Abbott LF and van Vreeswijk C (1993) Asynchronous states in a network of pulse-coupled oscillators. Phys Rev E 48: 1483-1490
Abeles M (1982) Role of the cortical neuron: Integrator or coincidence detector? Isr J Med Sci 18: 83-92
Adrian E and Zotterman Y (1926) The impulses produced by sensory nerve endings. Part 3. Impulses set up by touch and pressure. J Physiol 61: 465-483
Aertsen A, Diesmann M and Gewaltig MO (1996) Propagation of synchronous spiking activity in feedforward neural networks. J Physiology (Paris) 90: 243-247
Aldrich RW (1981) Inactivation of voltage-gated delayed potassium currents in molluscan neurons. Biophys J 36: 519-532
Aldrich RW, Corey DP and Stevens CF (1983) A reinterpretation of mammalian sodium channel gating based on single channel recording. Nature 306: 436-441
Alvarez FP and Destexhe A (2004) Simulating cortical network activity states constrained by intracellular recordings. Neurocomputing 58: 285-290
Amit DJ (1989) Modeling Brain Function. Cambridge University Press, Cambridge
Amit DJ and Brunel N (1997) Model of global spontaneous activity and local structured activity during delay periods in the cerebral cortex. Cereb Cortex 7: 237-252
Amitai Y, Friedman A, Connors BW and Gutnick MJ (1993) Regenerative activity in apical dendrites of pyramidal cells in neocortex. Cerebral Cortex 3: 26-38
Anderson JS, Lampl I, Gillespie DC and Ferster D (2000) The contribution of noise to contrast invariance of orientation tuning in cat visual cortex. Science 290: 1968-1972
Arieli A, Sterkin A, Grinvald, A and Aertsen A (1996) Dynamics of ongoing activity: explanation of the large variability in evoked cortical responses. Science 273: 1868-1871
Armstrong CM (1981) Sodium channels and gating currents. Physiol Rev 62: 644-683
Armstrong CM (1969) Inactivation of the potassium conductance and related phenomena caused by quaternary ammonium ion injection in squid axons. J Gen Physiol 54: 553-575
Azouz R and Gray C (1999) Cellular mechanisms contributing to response variability of cortical neurons in vivo. J Neurosci 19: 2209-2223
Azouz R and Gray CM (2000) Dynamic spike threshold reveals a mechanism for synaptic coincidence detection in cortical neurons in vivo. Proc Natl Acad Sci USA 97: 8110-8115
Badoual M, Rudolph M, Piwkowska Z, Destexhe A and Bal T (2005) High discharge variability in neurons driven by current noise. Neurocomputing 65: 493-498
Bair W and Koch C (1996) Temporal precision of spike trains in extrastriate cortex of the behaving macaque monkey. Neural Computation 15: 1185-1202
Balázsi G, Cornell-Bell A, Neiman AB and Moss F (2001) Synchronization of hyperexcitable systems with phase-repulsive coupling. Phys Rev E 64: 041912

A. Destexhe and M. Rudolph-Lilith, *Neuronal Noise*, Springer Series in Computational Neuroscience 8, DOI 10.1007/978-0-387-79020-6,

Baranyi A, Szente MB and Woody CD (1993) Electrophysiological characterization of different types of neurons recorded in vivo in the motor cortex of the cat. II. Membrane parameters, action potentials, current-induced voltage responses and electrotonic structures. J Neurophysiol 69: 1865-1879

Barlow H (1985) Cerebral cortex as a model builder. In: Models of the Visual Cortex. Rose D and Dobson V (eds). Wiley, Chinchester, pp. 37-46

Barlow H (1995) The neuron doctrine in perception. In: M. S. Gazzaniga (Ed.), The cognitive neurosciences. MIT Press, pp. 415-435

Barrett JN (1975) Motoneuron dendrites: role in synaptic integration. Fed Proc 34: 1398-1407

Barrett JN and Crill WE (1974) Influence of dendritic location and membrane properties on the effectiveness of synapses on cat motoneurones. J Physiol 293: 325-345

Bartol TM and Sejnowski TJ (1993) Model of the quantal activation of NMDA receptors at a hippocampal synaptic spine. Soc Neurosci Abstracts 19: 1515

Bassett DS and Bullmore E (2006) Small-world brain networks. Neuroscientist 12: 512-523

Bédard C and Destexhe A (2008) A modified cable formalism for modeling neuronal membranes at high frequencies. Biophys J 94: 1133-1143

Bédard C and Destexhe A (2009) Macroscopic models of local field potentials the apparent 1/f noise in brain activity. Biophysical J 96: 2589-2603

Bédard C, Kröger H and Destexhe A (2006b) Does the 1/f frequency-scaling of brain signals reflect self-organized critical states? Phys Rev Lett 97: 118102

Bédard C, Rodrigues S, Roy N, Contreras D and Destexhe A (2010) Evidence for frequency-dependent extracellular impedance from the transfer function between extracellular and intracellular potentials. J Comput Neurosci 29: 389-403

Beggs J and Plenz D (2003) Neuronal avalanches in neocortical circuits. J Neurosci 23: 11167-11177

Bell CC, Han VZ, Sugawara Y and Grant K (1997) Synaptic plasticity in a cerebellum-like structure depends on temporal order. Nature 387: 278-281

Bell A, Mainen ZF, Tsodyks M and Sejnowski TJ (1995) "Balancing" of conductances may explain irregular cortical spiking. Tech Report #INC-9502, The Salk Institute

Bernander O, Douglas RJ, Martin KA and Koch C (1991) Synaptic background activity influences spatiotemporal integration in single pyramidal cells. Proc Natl Acad Sci USA 88: 11569-11573

Berthier N and Woody CD (1988) In vivo properties of neurons of the precruciate cortex of cats. Brain Res Bull 21: 385-393

Bertschinger N and Natschläger T (2004) Real-time computation at the edge of chaos in recurrent neural networks. Neural Computation 16: 1413-1436

Bezanilla F (1985) Gating of sodium and potassium channels. J Membr Biol 88: 97-111

Bi GQ and Poo MM (1998) Synaptic modifications in cultured hippocampal neurons: dependence on spike timing, synaptic strength, and postsynaptic cell type. J Neurosci 18: 10464-10472

Bialek W and Rieke F (1992) Reliability and information transmission in spiking neurons. Trends Neurosci 15: 428-434

Bindman LJ, Meyer T and Prince CA (1988) Comparison of the electrical properties of neocortical neurones in slices in vitro and in the anaesthetized rat. Exp Brain Res 69: 489-496

Binzegger T, Douglas RJ and Martin KAC (2004) A quantitative map of the circuit of cat primary visual cortex. J Neurosci 24: 8441-8453

Bliss TV and Collingridge GL (1993) A synaptic model of memory: long-term potentiation in the hippocampus. Nature 361: 31-39

Borg-Graham LJ, Monier C and Frégnac Y (1998) Visual input evokes transient and strong shunting inhibition in visual cortical neurons. Nature 393: 369-373

Braitenberg V and Schüz A (1998) Cortex: statistics and geometry of neuronal connectivity. 2nd edition, Springer-Verlag, Berlin

Braun HA, Wissing H, Schafer H and Hirsch MC (1994) Oscillation and noise determine signal transduction in shark multimodal sensory cells. Nature 367: 270-273

Brennecke R and Lindemann B (1971) A chopped-current clamp for current injection and recording of membrane polarization with single electrodes of changing resistance. TI-TJ Life Sci 1: 53-58

Brennecke R and Lindemann B (1974a) Theory of a membrane-voltage clamp with discontinuous feedback through a pulsed current clamp. Rev Sci Instrum 45: 184-188

Brennecke R and Lindemann B (1974b) Design of a fast voltage clamp for biological membranes, using discontinuous feedback. Rev Sci Instrum 45: 656-661

Brette R (2009) Generation of correlated spike trains. Neural Comput 21: 188-215

Brette R, Piwkowska Z, Rudolph M, Bal T and Destexhe A (2007a) A non-parametric electrode model for intracellular recording. Neurocomputing 70: 1597-1601

Brette R, Piwkowska Z, Rudolph-Lilith M, Bal T and Destexhe A (2007b) High-resolution intracellular recordings using a real-time computational model of the electrode. arXiv preprint: http://arxiv.org/abs/0711.2075

Brette R, Rudolph M, Carnevale T, Hines M, Beeman D, Bower JM, Diesmann M, Morrison A, Goodman PH, Harris Jr FC, Zirpe M, Natschlager T, Pecevski D, Ermentrout B, Djurfeldt M, Lansner A, Rochel O, Vieville T, Muller E, Davison AP, El Boustani S and Destexhe A (2007) Simulation of networks of spiking neurons: A review of tools and strategies. J Computational Neurosci 23: 349-398

Brette R, Piwkowska Z, Monier C, Rudolph-Lilith M, Fournier J, Levy M, Frégnac Y, Bal T and Destexhe A (2008) High-resolution intracellular recordings using a real-time computational model of the electrode. Neuron 59: 379-391

Brette R, Piwkowska Z, Monier C, Gomez Gonzalez JF, Frégnac Y, Bal T and Destexhe A (2009) Dynamic clamp with high resistance electrodes using active electrode compensation in vitro and in vivo. In: Dynamic-clamp: From Principles to Applications, Destexhe A and Bal T (eds). Springer, New York, pp. 347-382

Britten KH, Shadlen MN, Newsome WT and Movshon JA (1993) Response of neurons in macaque MT to stochastic motion signals. Visual Neurosci 10: 1157-1169

Brock LG, Coombs JS and Eccles JC (1952) The recording of potential from monotneurones with an intracellular electrode. J Physiol 117: 431-460

Brown DA (1990) G-proteins and potassium currents in neurons. Ann Rev Physiol 52: 215-242

Brown AM and Birnbaumer L (1990) Ionic channels and their regulation by G protein subunits. Ann Rev Physiol 52: 197-213

Brunel N (2000) Dynamics of sparsely connected networks of excitatory and inhibitory spiking neurons. J Computational Neurosci 8: 183-208

Brunel N and Hakim V (1999) Fast global oscillations in networks of integrate-and-fire neurons with low firing rates. Neural Comput 11: 1621-1671

Brunel N and Sergi S (1998) Firing frequency of leaky integrate-and-fire neurons with synaptic current dynamics. J Theor Biol 195: 87-95

Brunel N and Wang XJ (2001) Effects of neuromodulation in a cortical network model of object working memory dominated by recurrent inhibition. J Computational Neurosci 11: 63-85

Brunel N, Chance FS, Fourcaud N and Abbott LF (2001) Effects of Synaptic Noise and Filtering on the Frequency Response of Spiking Neurons. Phys Rev Lett 86: 2186-2189

Bryant HL and Segundo JP (1976) Spike initiation by transmembrane current: a white-noise analysis. J Physiol 260: 279-314

Bugmann G, Christodoulou C and Taylor JG (1997) Role of temporal integration and fluctuation detection in the highly irregular firing of a leaky integrator neuron model with partial reset. Neural Comp 9: 985-1000

Bullmore E and Sporns O (2009) Complex brain networks: graph theoretical analysis of structural and functional systems. Nature Rev Neurosci 10: 186-198

Buño W Jr, Fuentes J and Segundo JP (1978) Crayfish stretch-receptor organs: effects of length-steps with and without perturbations. Biol Cybern 31: 99-110

Buonomano DV and Merzenich MM (1995) Temporal information transformed into a spatial code by a neural network with realistic properties. Science 267: 1028-1030

Burgard EC and Hablitz JJ (1993) NMDA receptor-mediated components of miniature excitatory synaptic currents in developing rat neocortex. J Neurophysiol 70: 1841-1852

Burns BD and Webb AC (1976) The spontaneous activity of neurons in the cat's visual cortex. Proc. Roy. Soc. London Ser B 194: 211-223.

Busch C and Sakmann B (1990) Synaptic transmission in hippocampal neurons: numerical reconstruction of quantal IPSCs. Cold Spring Harbor Symp. Quant Biol 55: 69-80

Bush P and Sejnowski TJ (1993) Reduced compartmental models of neocortical pyramidal cells. J Neurosci Methods 46: 159-166

Calvin WH and Stevens CF (1968) Synaptic noise and other sources of randomness in motoneuron interspike intervals. J Neurophysiol 31: 574-587

Campbell N (1909a) The study of discontinuous phenomena. Proc Cambr Phil Soc 15: 117-136

Campbell N (1909b) Discontinuities in light emission. Proc Cambr Phil Soc 15: 310-328

Capurro A, Pakdaman K, Nomura T and Sato S (1998) Aperiodic stochastic resonance with correlated noise. Phys Rev E 58: 4820-4827

Carnevale NT and Hines ML (2006) The NEURON book. Cambridge University Press, Cambridge

Celentano JJ and Wong RK (1994) Multiphasic desensitization of the GABAA receptor in outside-out patches. Biophys J 66: 1039-1050

Chance FS, Abbott LF and Reyes AD (2002) Gain modulation from background synaptic input. Neuron 35: 773-782

Chialvo DR and Apkarian AV (1993) Modulated Noisy Biological Dynamics: Three Examples. J Stat Phys 70: 375-391

Chialvo D, Longtin A and Müllller-Gerking J (1997) Stochastic resonance in models of neuronal ensembles. Phys Rev E 55: 1798-1808

Chow CC, Imhoff TT and Collins JJ (1998) Enhancing aperiodic stochastic resonance through noise modulation. Chaos 8: 616-620

Christodoulou C and Bugmann G (2000) Near Poisson-type firing produced by concurrent excitation and inhibition. Biosystems 58: 41-48

Christodoulou C and Bugmann G (2001) Coefficient of variation vs. mean interspike interval curves: What do they tell us about the brain? Neurocomputing 38-40: 1141-1149

Clay JR and Shlesinger MF (1977) United theory of 1/f and conductance noise in nerve membrane. J Theor Biol 66: 763-773

Clements JD and Westbrook GL (1991) Activation kinetics reveal the number of glutamate and glycine binding sites on the N-methyl-D-aspartate receptor. Neuron 258: 605-613

Cole KS and Curtis HJ (1939) Electric impedance of the squid giant axon during activity. J Gen Physiol 22: 649-670

Cole KS and Hodgkin AL (1939) Membrane and protoplasm resistance in the squid giant axon. J Gen Physiol 22: 671-687

Collins JJ, Chow CC and Imhoff TT (1995a) Stochastic resonance without tuning. Nature 376: 236-238

Collins JJ, Chow CC and Imhoff TT (1995b) Aperiodic stochastic resonance in excitable systems. Phys Rev E 52: R3321-R3324

Collins JJ, Imhoff TT and Grigg P (1996) Noise enhanced information transmission in rat SA1 cutaneous mechanoreceptors via aperiodic stochastic resonance. J Neurophysiol 76: 642-645

Colquhoun D and Hawkes AG (1981) On the stochastic properties of single ion channels. Proc Roy Soc Lond Ser B 211: 205-235

Colquhoun D, Jonas P and Sakmann B (1992) Action of brief pulses of glutamate on AMPA/KAINATE receptors in patches from different neurons of rat hippocampal slices. J Physiol 458: 261-287

Compte A, Sanchez-Vives MV, McCormick DA and Wang XJ (2003) Cellular and network mechanisms of slow oscillatory activity (<1 Hz) and wave propagations in a cortical network model. J Neurophysiol 89: 2707-2725

Connor CE, Preddie DC, Gallant JL and Van Essen DC (1997) Spatial attention effects in macaque area V4. J Neurosci 17: 3201-3214

Connors BW, Gutnick MJ and Prince DA (1982) Electrophysiological properties of neocortical neurons in vitro. J Neurophysiol 48: 1302-1320

Constantine-Paton M, Cline HT and Debski E (1990) Patterned activity, synaptic convergence, and the NMDA receptor in developing visual pathways. Ann Rev Neurosci 13: 129-154

Contreras D and Steriade M (1997) State-dependent fluctuations of low-frequency rhythms in corticothalamic networks. Neuroscience 76: 25-38

Contreras D, Timofeev I and Steriade M (1996) Mechanisms of long lasting hyperpolarizations underlying slow sleep oscillations in cat corticothalamic networks. J Physiol 494: 251-264

Contreras D, Destexhe A and Steriade M (1997) Intracellular and computational characterization of the intracortical inhibitory control of synchronized thalamic inputs in vivo. J Neurophysiol 78: 335-350

Cook EP and Johnston D (1997) Active dendrites reduce location-dependent variability of synaptic input trains. J Neurophysiol 78: 2116-2128

Cragg BG (1967) The density of synapses and neurones in the motor and visual areas of the cerebral cortex. J Anat 101: 639-654

Cremer M (1900) Ueber Wellen und Pseudowellen. Zeit fř Biol 40: 393-418

Crochet S and Petersen CC (2006) Correlating whisker behavior with membrane potential in barrel cortex of awake mice. Nature Neurosci 9: 608-610

Crochet S, Fuentealba P, Cisse Y, Timofeev I and Steriade M (2006) Synaptic plasticity in local cortical network in vivo and its modulation by the level of neuronal activity. Cereb Cortex 16: 618-631

Dayan P and Abbott LF (2001) Theoretical Neuroscience. MIT Press, Cambridge, MA

Davies CH, Davies SN and Collingridge GL (1990) Paired-pulse depression of monosynaptic GABA-mediated inhibitory postsynaptic responses in rat hippocampus. J Physiol 424: 513-531

Davis J Jr and Lorente de No R (1947) Studies from the Rockefeller Institute for Medical Research 131: 442-496

Dean A (1981) The variability of discharge of simple cells in the cat striate cortex. Exp Brain Res 44: 437-440

deCharms RC and Zador A (2000) Neural representation and the cortical code. Annu Rev Neurosci 23: 613-647

Deco G, Jirsa VK, Robinson PA, Breakspear M and Friston K (2008) The dynamic brain: from spiking neurons to neural masses and cortical fields. PLoS Comput Biol 4: e1000092

De Felice LJ (1981) Introduction to Membrane Noise. Plenum Press, New York

DeFelipe J and Fariñas I (1992) The pyramidal neuron of the cerebral cortex: morphological and chemical characteristics of the synaptic inputs. Progress Neurobiol 39: 563-607

De Koninck Y and Mody I (1994) Noise analysis of miniature IPSCs in adult rat brain slices: properties and modulation of synaptic GABAA receptor channels. J Neurophysiol 71: 1318-1335

de Polavieja GG, Harsch A, Kleppe I, Robinson HP and Juusola M (2005) Stimulus history reliably shapes action potential waveforms of cortical neurons. J Neurosci 25: 5657-5665

Derksen HE (1965) Axon membrane voltage fluctuations. Acta Physiol. Pharmacol. Neerl. 13: 373-466

de Ruyter van Steveninck RR, Strong SP, Koberle R and Bialek W (1997) Reproducibility and variability in neural spike trains. Science 275: 1805-1808

Desai NS and Walcott EC (2006) Synaptic bombardment modulates muscarinic effects in forelimb motor cortex. J Neurosci 26: 2215-2226

Deschênes M (1981) Dendritic spikes induced in fast pyramidal track neurons by thalamic stimulation. Exp Brain Res 43: 304-308

De Schutter E and Bower JM (1994) Simulated responses of cerebellar Purkinje cells are independent of the dendritic location of granule cell synaptic inputs. Proc Natl Acad Sci USA 91: 4736-4740

Destexhe A (1994) Oscillations, complex spatiotemporal behavior and information transport in networks of excitatory and inhibitory neurons. Phys Rev E 50: 1594-1606

Destexhe A (2001) Simplified models of neocortical pyramidal cells preserving somatodendritic voltage attenuation. Neurocomputing 38: 167-173

Destexhe A (2007) High-conductance state. Scholarpedia 2(11): 1341, http:www.scholarpedia.org/article/High-conductance_state

Destexhe A (2010) Inhibitory "noise". Frontiers Cell Neurosci. 4: 9, 2010

Destexhe A (2011) Intracellular and computational evidence for a dominant role of internal network activity states in cortical computations. Curr Opin Neurobiol 21: in press

Destexhe A and Bal T, eds (2009) The Dynamic-clamp: From Principles to Applications. Springer, New York

Destexhe A and Paré D (1999) Impact of network activity on the integrative properties of neocortical pyramidal neurons in vivo. J Neurophysiol 81: 1531-1547

Destexhe A and Paré D (2000) A combined computational and intracellular study of correlated synaptic bombardment in neocortical pyramidal neurons in vivo. Neurocomputing 32-33: 113-119

Destexhe A and Rudolph M (2004) Extracting information from the power spectrum of synaptic noise. J Computational Neurosci 17: 327-345

Destexhe A and Sejnowski TJ (1995) G-protein activation kinetics and spill-over of GABA may account for differences between inhibitory responses in the hippocampus and thalamus. Proc Natl Acad Sci USA 92: 9515-9519

Destexhe A, Mainen ZF and Sejnowski TJ (1994a) An efficient method for computing synaptic conductances based on a kinetic model of receptor binding. Neural Computation 6: 14-18

Destexhe A, Mainen ZF and Sejnowski TJ (1994b) Synthesis of models for excitable membranes, synaptic transmission and neuromodulation using a common kinetic formalism. J Computational Neurosci 1: 195-230

Destexhe A, Contreras D and Steriade M (1998a) Mechanisms underlying the synchronizing action of corticothalamic feedback through inhibition of thalamic relay cells. J Neurophysiol 79: 999-1016

Destexhe A, Mainen ZF and Sejnowski TJ (1998b) Kinetic models of synaptic transmission. In: Methods in Neuronal Modeling. 2nd edition, Koch C and Segev I (eds). MIT Press, Cambridge, MA, pp. 1-26

Destexhe A, Contreras D and Steriade M (1999) Spatiotemporal analysis of local field potentials and unit discharges in cat cerebral cortex during natural wake and sleep states. J Neurosci 19: 4595-4608

Destexhe A, Rudolph M, Fellous JM and Sejnowski TJ (2001) Fluctuating conductances recreate in-vivo-like activity in neocortical neurons. Neuroscience 107: 13-24

Destexhe A, Rudolph M and Paré D (2003) The high-conductance state of neocortical neurons in vivo. Nature Reviews Neurosci 4: 739-751

Destexhe A, Badoual M, Piwkowska Z, Bal T, Hasenstaub A, Shu Y, McCormick DA, Pelletier J, Paré D and Rudolph M (2003b) In vivo, in vitro and computational evidence for balanced or inhibition-dominated network states, and their respective impact on the firing mode of neocortical neurons. Society for Neuroscience Abstracts 29: 921.14

Destexhe A, Hughes S, Rudolph M and Crunelli V (2007) Are corticothalamic 'up' states fragments of wakefulness? Trends Neurosci 30: 334-342

Diba K, Lester HA and Koch C (2004) Intrinsic noise in cultured hippocampal neurons: experiment and modeling. J Neurosci 24: 9723-9733

Diesmann M, Gewaltig MO and Aertsen A (1999) Stable propagation of synchronous spiking in cortical neural networks. Nature 402: 529-533

Doiron B, Longtin A, Berman N and Maler L (2000) Subtractive and divisible inhibition: effect of voltage-dependent inhibitory conductances and noise. Neural Comp 13: 227-248

Dorval AD and White JA (2006) Synaptic input statistics tune the variability and reproducibility of neuronal responses. Chaos 16: 026105

Douglas RJ and Martin KA (1991) A functional microcircuit for cat visual cortex. J Physiol 440: 735-769

Douglas RJ, Martin KA and Whitteridge D (1991) An intracellular analysis of the visual responses of neurones in cat visual cortex. J Physiol 440: 659-696

Douglass JK, Wilkens L, Pantazelou E and Moss F (1993) Noise enhancement of information transfer in crayfish mechanoreceptors by stochastic resonance. Nature 365: 337-340

Dutar P and Nicoll RA (1988) A physiological role for $GABA_B$ receptors in the central nervous system. Nature 332: 156-158

Economo MN, Fernandez FR and White JA (2010) Dynamic clamp: alteration of response properties and creation of virtual realities in neurophysiology. J Neurosci 30: 2407-2413

Edeline JM, Manunta Y and Hennevin E (2000) Auditory thalamus neurons during sleep: changes in frequency selectivity, threshold, and receptive field size. J Neurophysiol 84: 934-952

El Boustani S, Pospischil M, Rudolph-Lilith M and Destexhe A (2007) Activated cortical states: experiments, analyses and models. J Physiol (Paris) 101: 99-109

El Boustani S, Marre O, Behuret S, Baudot P, Yger P, Bal T, Destexhe A and Frégnac Y (2009) Network-state modulation of power-law frequency-scaling in visual cortical neurons. PLoS Computational Biol 5: e1000519

Engel AK, König P, Kreiter AK, Schillen TB and Singer W (1992) Temporal coding in the visual cortex: new vistas on integration in the nervous system. Trends Neurosci 15: 218-26

Erisir A, VanHorn SC, Sherman SM (1997a) Relative numbers of cortical and brainstem inputs to the lateral geniculate nucleus. Proc Natl Acad Sci USA 94: 1517-1520

Erisir A, VanHorn SC, Bickford ME, Sherman SM (1997b) Immunocytochemistry and distribution of parabrachial terminals in the lateral geniculate nucleus of the cat: a comparison with corticogeniculate terminals. J Comp Neurol 377: 535-549

Ermentrout GB and Kopell NK (1984) Frequency plateaus in a chain of weakly coupled oscillators I. SIAM J Math Analysis 15: 215-237

Ermentrout GB and Kopell NK (1986) Parabolic bursting in an excitable system coupled with a slow oscillation. SIAM J Appl Math 46: 233-253

Evarts EV (1964) Temporal patterns of discharge of pyramidal tract neurons during sleep and waking in the monkey. J Neurophysiol 27: 152-171

Fariñas I and DeFelipe J (1991a) Patterns of synaptic input on corticocortical and corticothalamic cells in the visual cortex. I. The cell body. J Comp Neurol 304: 53-69

Fariñas I and DeFelipe J (1991b) Patterns of synaptic input on corticocortical and corticothalamic cells in the visual cortex. II. The axon initial segment. J Comp Neurol 304: 70-77

Fatt P (1957) Sequence of events in synaptic activation of a motoneurone. J Neurophysiol 20: 61-80

Fatt P and Katz B (1950) Some observations on biological noise. Nature 166: 597-598

Fatt P and Katz B (1952) Spontaneous subthreshold activity at motor nerve endings. J Physiol 117: 109-128

Fellous JM and Sejnowski TJ (2003) Regulation of persistent activity by background inhibition in an in vitro model of a cortical microcircuit. Cereb Cortex 13: 1232-1241

Fellous JM, Rudolph M, Destexhe A and Sejnowski TJ (2003) Synaptic background noise controls the input/output characteristics of single cells in an in vitro model of in vivo activity. Neuroscience 122: 811-829

Feng J and Brown D (1998) Impact of temporal variation and the balance between excitation and inhibition on the output of the perfect integrate-and-fire model. Biol Cybern 78: 369-376

Feng J and Brown D (1999) Coefficient of variation of interspike intervals greater than 0.5. How and when? Biol Cybern 80: 291-297

Feng J and Brown D (2000) Impact of correlated inputs on the output of the integrate-and-fire model. Neural Comput 12: 671-692

Finkel AS and Redman S (1984) Theory and operation of a single microelectrode voltage clamp. J Neurosci Methods 11: 101-127

Fiser J, Chiu C and Weliky M (2004) Small modulation of ongoing cortical dynamics by sensory input during natural vision. Nature 431: 573-578

Fishman HM (1973) Relaxation spectra ofpotassiumchannel noisefromsquidaxon membranes. Proc Natl Acad Sci USA 70: 876-879

Fitzhugh RA (1955) Mathematical models of threshold phenomena in the nerve membrane. Bull Math Biophysics 17: 257-278

Fitzhugh RA (1961) Impulses and physiological states in theoretical models of nerve membrane. Biophysical J 1: 445-466

Fitzhugh RA (1965) A kinetic model of the conductance changes in nerve membrane. J Cell Comp Physiol 66: 111-118

Fitzhugh RA (1969) Mathematical models of excitation and propagation in nerve. In: H.P. Schwan, ed. Biological Engineering. McGraw-Hill, New York

Fokker AD (1914) Die mittlere Energie rotierender elektrischer Strahlungsfelder. Annalen der Physik 43: 810-820

Fourcaud N and Brunel N (2002) Dynamics of the firing probability of noisy integrate-and-fire neurons. Neural Comput 14: 2057-2110

Frehland E (1982) Stochastic transport processes in discrete biological systems. Springer-Verlag, Berlin

Frehland E and Faulhaber KH (1980) Nonequilibrium ion transport through pores. The influence of barrier structures on current fluctuations, transient phenomena and admittance. Biophys Struct Mech 7: 1-16

French CR, Sah P, Buckett KJ and Gage PW (1990) A Voltage–dependent Persistent Sodium Current in Mammalian Hippocampal Neurons. J Gen Physiol 95: 1139-1157

Funke K and Eysel UT (1992) EEG-dependent modulation of response dynamics of cat dLGN relay cells and the contribution of corticogeniculate feedback. Brain Res 573: 217-227

Gammaitoni L, Hänggi P, Jung P and Marchesoni F (1998) Stochastic resonance. Rev Mod Phys 70: 223-287

Gardiner CW (2002) Handbook of Stochastic Methods. Springer Verlag, Berlin and Heidelberg

Gauck V and Jaeger D (2000) The control of rate and timing of spikes in the deep cerebellar nuclei by inhibition. J Neurosci 20: 3006-3016

Gauck V and Jaeger D (2003) The contribution of NMDA and AMPA conductances to the control of spiking in neurons of the deep cerebellar nuclei. J Neurosci 23: 8109-8118

Genovese W and Muñoz MA (1999) Recent results on multiplicative noise. Phys Rev E 60: 69-78

Gerstner W and Kistler W (2002) Spiking Neuron Models. Cambridge University Press, Cambridge, UK

Ghose GM and Maunsell JH (2002) Attentional modulation in visual cortex depends on task timing. Nature 419: 616-620

Gillespie DT (1996) The mathematics of Brownian motion and Johnson noise. Am J Phys 64: 225-240

Gluckman BJ, Netoff TI, Neel EJ, Ditto WL, Spano ML and Schiff SJ (1996) Stochastic resonance in a neuronal network from mammalian brain. Phys Rev Lett 77: 4098-4101

Goaillard JM and Marder E (2006) Dynamic clamp analyses of cardiac, endocrine, and neural function. Physiology (Bethesda) 21: 197-207

Goldberg JM and Brown PB (1969) Response of binaural neurons of dog superior olivary complex to dichotic tonal stimuli: some physiological mechanisms of sound localization. J Neurophysiol 32: 613-636

Golding NL and Spruston N (1998) Dendritic sodium spikes are variable triggers of axonal action potentials in hippocampal CA1 pyramidal neurons. Neuron 21: 1189-1200

Gray CM (1994) Synchronous oscillations in neuronal systems: mechanisms and functions. J Comput Neurosci 1: 11-38

Greenhill SD and Jones RS (2007) Simultaneous estimation of global background synaptic inhibition and excitation from membrane potential fluctuations in layer III neurons of the rat entorhinal cortex in vitro. Neuroscience 147: 884-892

Greenwood PE, Ward LM, Russell D, Neiman A and Moss F (2000) Stochastic resonance enhances the electrosensory information available to paddlefish for prey capture. Phys Rev Lett 84: 4773-4776

Gruner JE, Hirsch JC and Sotelo C (1974) Ultrastructural features of the isolated suprasylvian gyrus. J Comp Neurol 154: 1-27

Guérineau NC, Bossu JL, Gähwiler BH and Gerber U (1995) Activation of a nonselective cationic conductance by metabotropic glutamatergic and muscarinic agonists in CA3 pyramidal neurons of the rat hippocampus. J Neurosci 15: 4395-4407

Gupta A, Wang Y and Markram H (2000) Organizing principles for a diversity of GABAergic interneurons and synapses in the neocortex. Science 287: 273-278

Gutkin BS and Ermentrout GB (1998) Dynamics of membrane excitability determine interspike interval variability: a link between spike generation mechanisms and cortical spike train statistics. Neural Comp 10: 1047-1065

Gutkin B, Ermentrout B and Rudolph M (2003) Spike generating dynamics and the conditions for spike-time precision in cortical neurons. J Computational Neurosci 15: 91-103

Guye M, Bettus G, Bartolomei F and Cozzone PJ (2010) Graph theoretical analysis of structural and functional connectivity MRI in normal and pathological brain networks. MAGMA 23: 409-421

Hagan PS, Doering CR and Levermore CD (1989) Mean exit times for particles driven by weakly colored noise. SIAM J Appl Math 49: 1480-1513

Haider B, Duque A, Hasenstaub AR and McCormick DA (2006) Neocortical network activity in vivo is generated through a dynamic balance of excitation and inhibition. J Neurosci 26: 4535-4545

Haider B, Duque A, Hasenstaub AR, Yu Y and McCormick DA (2007) Enhancement of visual responsiveness by spontaneous local network activity in vivo. J Neurophysiol 97: 4186-4202

Haj-Dahmane S and Andrade R (1996) Muscarinic activation of a voltage-dependent cation nonselective current in rat association cortex. J Neurosci 16: 3848-3861

Han VZ, Grant K and Bell CC (2000) Reversible associative depression and nonassociative potentiation at a parallel fiber synapse. Neuron 27: 611-622

Hänggi P and Jung P (1994) Colored noise in dynamical systems. Adv in Chem Phys 89: 239-329

Hansel D and Sompolinsky H (1996) Chaos and Synchrony in a Model of a Hypercolumn in Visual Cortex. J Comp Neurosci 3: 7-34

Hansel D, Mato G, Meunier C and Neltner L (1998) On numerical simulations of integrate-and-fire neural networks. Neural Computation 10: 467-483

Hanson JE and Jaeger D (2002) Short-term plasticity shapes the response to simulated normal and parkinsonian input patterns in the globus pallidus. J Neurosci 22: 5164-5172

Harsch A and Robinson HP (2000) Postsynaptic variability of firing in rat cortical neurons: the roles of input synchronization and synaptic NMDA receptor conductance. J Neurosci 20: 6181-6192

Hasegawa H (2008) Dynamics of the Langevin model subjected to colored noise: Functional-integral method. Physica A 387: 2697-2718

Hasenstaub A, Shu Y, Haider B, Kraushaar U, Duque A and McCormick DA (2005) Inhibitory postsynaptic potentials carry synchronized frequency information in active cortical networks. Neuron 47: 423-435

Hay E, Hill S, Schürmann F, Markram H and Segev I (2011) Models of neocortical Layer 5b pyramidal cells capturing a wide range of dendritic and perisomatic active properties. PLoS Comput Biol 7: e1002107

Hermann L (1874) Experimentelles und kritisches über electrotonus. Archiv für die gesammte Physiologie des Menschen und der Thiere 8: 258-275

Hermann L (1879) Allgemeine Nervenphysiologie. Handbuch der Physiologie. Ister Theil. 1-196. FCW Vogel, Leipzig

Hermann L (1899) Zur Theorie der Erregungsleitung und der elektrischen Erregung. Pflügers Arch ges Physiol 75: 574-590

Hermann L (1905) Beiträge zur Physiologie und Physik des Nerven. Arch ges Physiol 109: 95-144

Hessler NA, Shirke AM and Malinow R (1993) The probability of transmitter release at a mammalian central synapse. Nature 366: 569-572

Hestrin S (1992) Activation and desensitization of glutamate-activated channels mediating fast excitatory synaptic currents in the visual cortex. Neuron 9: 991-999

Hestrin S (1993) Different glutamate receptor channels mediate fast excitatory synaptic currents in inhibitory and excitatory cortical neurons. Neuron 11: 1083-1091

Hestrin S, Sah P and Nicoll RA (1990) Mechanisms generating the time course of dual component excitatory synaptic currents recorded in hippocampal slices. Neuron 5: 247-253

Higgs MH, Slee SJ and Spain WJ (2006) Diversity of gain modulation by noise in neocortical neurons: regulation by the slow after-hyperpolarization conductance. J Neurosci 26: 8787-8799

Higley MJ and Contreras D (2007) Cellular mechanisms of suppressive interactions between somatosensory responses in vivo. J Neurophysiol 97: 647-658

Hille B (2001) Ionic Channels of Excitable Membranes (3rd edition). Sinauer Associates, Sunderland, MA

Hines ML and Carnevale NT (1997) The NEURON simulation environment. Neural Computation 9: 1179-1209

Hirsch JC, Fourment A and Marc ME (1983) Sleep-related variations of membrane potential in the lateral geniculate body relay neurons of the cat. Brain Res 259: 308-312

Hirsch JA, Alonso JM, Clay Reid R and Martinez LM (1998) Synaptic integration in striate cortical simple cells. J Neurosci 18: 9517-9528

Hô N and Destexhe A (2000) Synaptic background activity enhances the responsiveness of neocortical pyramidal neurons. J Neurophysiol 84: 1488-1496

Ho EC, Zhang L and Skinner FK (2009) Inhibition dominates in shaping spontaneous CA3 hippocampal network activities in vitro. Hippocampus 19: 152-165; erratum in 19: 411

Hobson JA and McCarley RW (1971) Cortical unit activity in sleep and waking. Electroencephalogr. Clin Neurophysiol 30: 97-112

Hodgkin AL (1936) The electrical basis of nervous conduction. Fellowship dissertation. Library of Trinity College, Cambridge

Hodgkin AL (1937a) Evidence for electrical transmission in nerve. Part I. J Physiol 90: 183-210

Hodgkin AL (1937b) Evidence for electrical transmission in nerve. Part II. J Physiol 90: 211-232

Hodgkin AL (1939) The relation between conduction velocity and the electrical resistance outside a nerve fibre. J Physiol 94: 560-570

Hodgkin AL and Huxley AF (1952a) Currents Carried By Sodium And Potassium Ions Through The Membrane Of The Giant Axon Of Loligo. J Physiol 116: 449-472

Hodgkin AL and Huxley AF (1952b) The Components Of Membrane Conductance In The Giant Axon Of Loligo. J Physiol 116: 473-496

Hodgkin AL and Huxley AF (1952c) The Dual Effect Of Membrane Potential On Sodium Conductance In The Giant Axon Of Loligo. J Physiol 116: 497-506

Hodgkin AL and Huxley AF (1952d) A quantitative description of membrane current and its application to conduction and excitation in nerve. J Physiol 117: 500-544

Hodgkin AL and Rushton WAH (1946) The electrical constants of a crustacean nerve fibre. Proc R Soc London B133: 444-479

Hodgkin AL, Huxley AF and Katz B (1952) Measurement Of Current-Voltage Relations In The Membrane Of The Giant Axon Of Loligo. J Physiol 116: 424-448

Hoffman DA, Magee JC, Colbert CM and Johnston DA (1997) K^+ channel regulation of signal propagation in dendrites of hippocampal pyramidal neurons. Nature 387: 869-875

Holmes WR and Woody CD (1989) Effects of uniform and non-uniform synaptic "activation-distributions" on the cable properties of modeled cortical pyramidal neurons. Brain Res 505: 12-22

Holt GR, Softky WR, Koch C and Douglas RJ (1996) Comparison of discharge variability in vitro and in vivo in cat visual cortex neurons. J Neurophysiol 75: 1806-1814

Hoorweg JL (1898) Ueber die elektrischen Eigenschaften der Nerven. Pflügers Arch ges Physiol 71: 128-157

Hoppensteadt F and Izhikevich E (1997) Weakly Connected Neural Nets. Springer-Verlag, Berlin.

Howeling AR, Modi RH, Granter P, Fellous J-M and Sejnowski TJ (2001) Models of frequency preferences of prefrontal cortical neurons. Neurocomputing 38: 231-238

Hubel D (1959) Single-unit activity in striate cortex of unrestrained cats. J Physiol 147: 226-238

Hubel DH and Wiesel TN (1963) Shape and arrangement of columns in cat's striate cortex. J Physiol 165: 559-568

Huguenard JR and McCormick DA (1992) Simulation of the currents involved in rhythmic oscillations in thalamic relay neurons. J Neurophysiol 68: 1373-1383

Huguenard JR and Prince DA (1994) Clonazepam suppresses GABA(B)-mediated inhibition in thalamic relay neurons through effects in nucleus reticularis. J Neurophysiol 71: 2576-2581

Huguenard JR, Hamill OP and Prince DA (1988) Developmental changes in Na+ conductances in rat neocortical neurons: appearance of a slowly inactivating component. J Neurophysiol 59: 778-795

Hunt BJ (1997) Doing science in a global empire: Cable telegraphy and electrical physics, In: Victorian Britain, Victorian Science in Context, Lightman B (ed)

Hunter JD, Milton JG, Thomas PJ and Cowan JD (1998) Resonance Effect for Neural Spike Time Reliability. J Neurophysiol 80: 1427-1438

Ibañes M, García-Ojalvo J, Toral R and Sancho JM (1999) Noise-induced phase separation: mean-field results. Phys Rev E 60: 3597-3605

Isaacson JS, Solis JM and Nicoll RA (1993) Local and diffuse synaptic actions of GABA in the hippocampus. Neuron 10: 165-175

Ivey C, Apkarian AV and Chivalvo DR (1998) Noise-Induced Tuning Curve Changes in Mechanoreceptors. J Neurophysiol 79: 1879-1890

Jack JJB, Noble D and Tsien RW (1975) Electric current flow in excitable cells. Clarendon Press

Jacobson GA, Diba K, Yaron-Jakoubovitch A, Koch C, Segev I and Yarom Y (2005) Subthreshold voltage noise of rat neocortical pyramidal neurones. J Physiol 564: 145-160

Jaeger D and Bower JM (1999) Synaptic control of spiking in cerebellar Purkinje cells: dynamic current clamp based on model conductances. J Neurosci 19: 6090-6101

Jahnsen H and Llinás R (1984) Electrophysiological properties of guinea-pig thalamic neurones: an in vitro study. J Physiol 349: 205-226

Jahr CE and Stevens CF (1990a) A quantitative description of NMDA receptor-channel kinetic behavior. J Neurosci 10: 1830-1837

Jahr CE and Stevens CF (1990b) Voltage dependence of NMDA-activated macroscopic conductances predicted by single-channel kinetics. J Neurosci 10: 3178-3182

Jaramillo F and Wiesenfeld K (1998) Mechanoelectrical transduction assisted by Brownian motion: a role for noise in the auditory system. Nature Neuroscience 1: 384-388

Johnson JB (1927) Thermal agitation of electricity in conductors. Phys Rev 29: 367-368

Johnston D and Wu SM (1995) Foundations of Cellular Neurophysiology. MIT Press, Cambridge, MA

Johnston D, Magee JC, Colbert CM and Cristie BR (1996) Active properties of neuronal dendrites. Annual Rev Neurosci 19: 165-186

Jones EG and Powell TPS (1969) Morphological variations in the dendritic spines of the neocortex. J Cell Sci 5: 509-529

Jones EG and Powell TPS (1970) Electron microscopy of the somatic sensory cortex of the cat. I. Cell types and synaptic organization. Philos Trans R Soc Lond B 257: 1-11

Jung P and Mayer-Kress G (1995) Spatiotemporal stochastic resonance in excitable media. Phys Rev Lett 74: 2130-2133

Kasanetz F, Riquelme LA and Murer MG (2002) Disruption of the two-state membrane potential of striatal neurones during cortical desynchronisation in anaesthetised rats. J Physiol 543: 577-589

Kelso JAS (1995) Dynamic Patterns: The Self-Organization of Brain and Behavior. MIT Press, Cambridge

Kim U, Sanches-Vives MV and McCormick DA (1997) Functional dynamics of GABAergic inhibition in the thalamus. Science 278: 130-134

Kisley MA and Gerstein GL (1999) Trial-to-trial variability and state-dependent modulation of auditory-evoked responses in cortex. J Neurosci 19: 10451-10460

Knight BW (1972) Dynamics of encoding in a population of neurons. J Gen Physiol 59: 734-766

Koch C (1987) The action of the corticofugal pathway on sensory thalamic nuclei: a hypothesis. Neuroscience 23: 399-406

Koch C (1999) Biophysics of Computation. Oxford University Press, Oxford

Koch C and Laurent G (1999) Complexity and the nervous system. Science 284: 96-98

Koch C, Rapp M and Segev I (2002) A brief history of time (constants). Cereb Cortex 6: 93-101

Koh DS, Geiger JR, Jonas P and Sakmann B (1995) Ca(2+)-permeable AMPA and NMDA receptor channels in basket cells of rat hippocampal dentate gyrus. J Physiol 485: 383-402

Kohn AF (1997) Computer Simulation of Noise Resulting from Random Synaptic Activities. Comput Biol Med 27: 293-308

Kolb HA and Frehland E (1980) Noise-current generated by carrier-mediated ion transport at non-equilibrium. Biophys Chem 12: 21-34

Kolmogorov AN (1931) Über die analytischen Methoden in der Wahrscheinlichkeitsrechnung. Math Ann 104: 415-458

König P, Engel AK, Roelfsema PR and Singer W (1995) How precise is neuronal synchronization? Neural Comp 7: 469-485

König P, Engel AK and Singer W (1996) Integrator or coincidence detector? The role of the cortical neuron revisited. Trends Neurosci 19: 130-137

Kreiner L and Jaeger D (2004) Synaptic shunting by a baseline of synaptic conductances modulates responses to inhibitory input volleys in cerebellar Purkinje cells. Cerebellum 3: 112-125

Kretzberg J, Egelhaaf M and Warzecha A-K (2001) Membrane potential fluctuations determine the precision of spike timing and synchronous activity: A model study. J Comp Neurosci 10: 79-97

Krnjević K, Pumain R and Renaud L (1971) The mechanism of excitation by acetylcholine in the cerebral cortex. J Physiol 215: 247-268

Krüger J and Becker JD (1991) Recognizing the visual stimulus from neuronal discharges. Trends Neurosci 14: 282-286

Kuhn A, Aertsen A and Rotter S (2004) Neuronal integration of synaptic input in the fluctuation-driven regime. J Neurosci 24: 2345-2356

Kumar A, Schrader S, Aertsen A and Rotter S (2008) The high-conductance state of cortical networks. Neural Comput 20: 1-43

Lábos E (2000) Codes, operations, measurements and neural networks. Biosystems 58: 9-18

Lampl I, Reichova I and Ferster D (1999) Synchronous membrane potential fluctuations in neurons of the cat visual cortex. Neuron 22: 361-374

Lánský P and Lánská V (1987) Diffusion approximation of the neuronal model with synaptic reversal potentials. Biol Cybern 56: 19-26

Lánský P and Rodriguez R (1999) Two-compartment stochastic model of a neuron. Physica D 132: 267-286

Lánský P and Rospars JP (1995) Ornstein–Uhlenbeck model neuron revisited. Biol Cybern 72: 397-406

Lapicque L (1907) Recherches quantitatives sur l'excitation électrique des nerfs traitée comme une polarization. J Physiol Pathol Gen 9: 620-635

Larkman AU (1991) Dendritic morphology of pyramidal neurons of the visual cortex of the rat. III. Spine distributions. J Comp Neurol 306: 332-343

Larkum ME, Zhu JJ and Sakmann B (1999) A new cellular mechanism for coupling inputs arriving at different cortical layers. Nature 398: 338-341

Larkum ME, Nevian T, Sandler M, Polsky A and Schiller J (2009) Synaptic integration in tuft dendrites of layer 5 pyramidal neurons: a new unifying principle. Science 325: 756-760

Lee S-G and Kim S (1999) Parameter dependence of stochastic resonance in the stochastic Hodgkin–Huxley neuron. Phys Rev E 60: 826-830

Lee S-G, Neiman A and Kim S (1998) Coherence resonance in a Hodgkin-Huxley neuron. Phys Rev E 57: 3292-3297

Lee AK, Manns ID, Sakmann B and Brecht M (2006) Whole-cell recordings in freely moving rats. Neuron 51: 399-407

Le Franc Y, Foutry B, Nagy F and Le Masson G (2001) Nociceptive signal transfer through the dorsal horn network: hybrid and dynamic-clamp approaches using a real-time implementation of the NEURON simulation environment. Soc Neurosci Abstracts 27: 927.18

Leger J-F, Stern EA, Aertsen A and Heck D (2005) Synaptic integration in rat frontal cortex shaped by network activity. J Neurophysiol 93: 281-293
Le Masson G, Renaud-Le Masson S, Sharp AA, Marder E and Abbott LF (1992) Real-time interaction between a model neuron and the crustacean stomatogastric nervous system In: Society for Neuroscience Abstracts, Meeting 18: 1055
Le Masson G, Le Masson S and Moulins M (1995) From conductances to neural network properties: analysis of simple circuits using the hybrid network method. Prog Biophys molec Biol 64: 201-220
Le Masson G, Renaud-Le Masson S, Debay D and Bal T (2002) Feedback inhibition controls spike transfer in hybrid thalamic circuits. Nature 417: 854-858
Leresche N, Hering J and Lambert RC (2004) Paradoxical potentiation of neuronal T-type Ca2+ current by ATP at resting membrane potential. J Neurosci 24: 5592-5602
Lester RAJ and Jahr CE (1992) NMDA channel behavior depends on agonist affinity. J Neurosci 12: 635-643
Levin JE and Miller JP (1996) Broadband neural coding in the cricket sensory system enhanced by stochastic resonance. Nature 380: 165-168
Li B, Funke K, Worgotter F and Eysel UT (1999) Correlated variations in EEG pattern and visual responsiveness of cat lateral geniculate relay cells. J Physiol 514: 857-874
Lin J, Pawelzik K, Ernst U and Sejnowski TJ (1998) Irregular synchronous activity in stochastically-coupled networks of integrate-and-fire neurons. Network 9: 333-344
Lindner B and Longtin A (2006) Comment on "Characterization of subthreshold voltage fluctuations in neuronal membranes." Neural Comput 28: 1896-1931
Litvak V, Sompolinsky H, Segev I and Abeles M (2003) On the transmission of rate code in long feedforward networks with excitatory-inhibitory balance. J Neurosci 23: 3006-3015
Liu XB, Honda CN and Jones EG (1995) Distribution of four types of synapse on physiologically identified relay neurons in the ventral posterior thalamic nucleus of the cat. J Comp Neurol 352: 69-91
Livingstone MS and Hubel DH (1981) Effects of sleep and arousal on the processing of visual information in the cat. Nature 291: 554-561
Llinás RR (1988) The intrinsic electrophysiological properties of mammalian neurons: a new insight into CNS function. Science 242: 1654-1664
Llinás RR and Jahnsen H (1982) Electrophysiology of thalamic neurones in vitro. Nature 297: 406-408
Llinás RR and Nicholson C (1971) Electrophysiological properties of dendrites and somata in alligator Purkinje cells. J Neurophysiol 34: 532-551
Llinás RR and Paré D (1991) Of dreaming and wakefulness. Neuroscience 44: 521-535
Llinás RR and Ribary U (1993) Coherent 40-Hz oscillation characterizes dream state in humans. Proc Natl Acad Sci USA 90: 2078-2081
Llinás RR and Steriade M (2006) Bursting of thalamic neurons and states of vigilance. J Neurophysiol 95: 3297-3308
London M and Segev I (2001) Synaptic scaling in vitro and in vivo. Nat Neurosci 4: 853-855
Longtin A (1993) Stochastic Resonance in Neuron Models. J Stat Phys 70: 309-327
Longtin A (1997) Autonomous stochastic resonance in bursting neurons. Phys Rev E 55: 868-876
Longtin A (2011) Neuronal noise. Scholarpedia, in press, http:www.scholarpedia.org/article/Neuronal_noise
Longtin A and Chialvo DR (1998) Stochastic and Deterministic Resonance for Excitable Systems. Phys Rev Lett 81: 4012-4015
Lu SM, Guido W and Sherman SM (1992) Effects of membrane voltage on receptive field properties of lateral geniculate neurons in the cat: contributions of the low-threshold Ca2+ conductance. J Neurophysiol 68: 2185-2198
Lundstrom I and McQueen D (1974) A proposed 1/f noise mechanism in nerve cell membranes. J Theor Biol 45: 405-409
Maass W, Natschlager T and Markram H (2002) Real-time computing without stable states: a new framework for neural computation based on perturbations. Neural Computation 14: 2531-2560

MacKay D and McCulloch W (1952) The limiting information capacity of a neuronal link. Bull Math Biophys 14: 127-135

Magee JC and Johnston D (1995a) Characterization of single voltage-gated Na+ and Ca2+ channels in apical dendrites of rat CA1 pyramidal neurons. J Physiol 487: 67-90

Magee JC and Johnston D (1995b) Synaptic activation of voltage-gated channels in the dendrites of hippocampal pyramidal neurons. Science 268: 301-304

Magee JC, Hoffman D, Colbert C and Johnston D (1998) Electrical and calcium signalling in dendrites of hippocampal pyramidal neurons. Annu Rev Physiol 60: 327-346

Mainen ZF and Sejnowski TJ (1995) Reliability of spike timing in neocortical neurons. Science 268: 1503-1506

Mainen ZF, Joerges J, Huguenard JR and Sejnowski TJ (1995) A model of spike initiation in neocortical pyramidal neurons. Neuron 15: 1427-1439

Mannella R (1989) In: Noise in Nonlinear Dynamical Systems, III: Experiments and Simulations. Moss F, McClintock PVE (eds). Cambridge Univiversity Press, Cambridge

Mannella R (1997) In: Supercomputation in nonlinear and disordered systems. Vasquez L, Tirado F and Martin I (eds). World Scientific, 100-129

Manwani A and Koch C (1999a) Detecting and estimating signals in noisy cable structures, I: Neuronal noise sources. Neural Computation 11: 1797-1829

Manwani A and Koch C (1999b) Detecting and estimating signals in noisy cable structures, II: information theoretical analysis. Neural Computation 11: 1831-1873

Mar DJ, Chow CC, Gerstner W, Adams RW and Collins JJ (1999) Noise shaping in populations of coupled model neurons. Proc Natl Acad Sci USA 96: 10450-10455

Margrie TW, Brecht M and Sakmann B (2002) In vivo, low-resistance, whole-cell recordings from neurons in the anaesthetized and awake mammalian brain. Pflügers Arch 444: 491-498

Markram H and Tsodyks M (1996) Redistribution of synaptic efficacy between neocortical pyramidal neurons. Nature 382: 807-810

Markram H, Lübke J, Frotscher M, Rodth A and Sakmann B (1997) Physiology and anatomy of synaptic connections between thick tufted pyramidal neurones in the developing rat neocortex. J Physiol 500: 409-440

Marom S and Abbott LF (1994) Modeling state-dependent inactivation of membrane currents. Biophys J 67: 515-520

Maršálek P, Koch C and Maunsell J (1997) On the relationship between synaptic input and spike output jitter in individual neurons. Proc Natl Acad Sci USA 94: 735-740

Martin AR (1977) Junctional transmission. II. Presynaptic mechanisms. In: Handbook of Physiology. Section I: The Nervous System. Kandel ER (ed). Bethesda: American Physiological Society: pp. 329-355

Mason A, Nicoll A and Stratford K (1991) Synaptic transmission between individual pyramidal neurons of the rat visual cortex in vitro. J Neurosci 11: 72-84

Mato G (1998) Stochastic resonance in neural systems: Effects of temporal correlation in the spike trains. Phys Rev E 58: 876-880

Matsumura M, Cope T and Fetz EE (1988) Sustained excitatory synaptic input to motor cortex neurons in awake animals revealed by intracellular recording of membrane potentials. Exp Brain Res 70: 463-469

McBain C and Dingledine R (1992) Dual-component miniature excitatory synaptic currents in rat hippocampal CA3 pyramidal neurons. J Neurophysiol 68: 16-27

McClurkin JW, Optican LM, Richmond BJ and Gawne TJ (1991) Concurrent processing and complexity of temporally encoded neuronal messages in visual perception. Science 253: 675-677

McCormick DA (1989) Cholinergic and noradrenergic control of thalamocortical processing. Trends Neurosci 12: 215-221

McCormick DA (1992) Neurotransmitter actions in the thalamus and cerebral cortex and their role in neuromodulation of thalamocortical activity. Progress Neurobiol 39: 337-388

McCormick DA and Bal T (1997) Sleep and arousal: thalamocortical mechanisms. Annu Rev Neurosci 20: 185-215

McCormick DA and Feeser HR (1990) Functional implications of burst firing and single spike activity in lateral geniculate relay neurons. Neuroscience 39: 103-113

McCormick DA and Huguenard JR (1992) A model of the electrophysiological properties of thalamocortical relay neurons. J Neurophysiol 68: 1384-1400

McCormick DA and Prince DA (1986) Mechanisms of action of acetylcholine in the guinea-pig cerebral cortex in vitro. J Physiol 375: 169-194

McCormick DA, Shu Y-S, Hasenstaub A, Sanchez-Vives M, Badoual M and Bal T (2003) Persistent cortical activity: mechanisms of generation and effects on neuronal excitability. Cerebral Cortex 13: 1219-1231

Mehring C, Hehl U, Kubo M, Diesmann M and Aertsen A (2003) Activity dynamics and propagation of synchronous spiking in locally connected random networks. Biol Cybern 88: 395-408

Metherate R and Ashe JH (1993) Ionic flux contributions to neocortical slow waves and nucleus basalis-mediated activation: whole-cell recordings in vivo. J Neurosci 13: 5312-5323

Migliore M, Hoffman DA, Magee JC and Johnston D (1999) Role of an A-Type K+ Conductance in the Back-Propagation of Action Potentials in the Dendrites of Hippocampal Pyramidal Neurons. J Comp Neurosci 7: 5-15

Miller LM and Schreiner CE (2000) Stimulus-based state control in the thalamocortical system. J Neurosci 20: 7011-7016

Mitchell SJ and Silver RA (2003) Shunting inhibition modulates neuronal gain during synaptic excitation. Neuron 38: 433-445

Mody I, De Koninck Y, Otis TS and Soltesz I (1994) Bridging the cleft at GABA synapses in the brain. Trends Neurosci 17: 517-525

Mokeichev A, Okun M, Barak O, Katz Y, Ben-Shahar O and Lampl I (2007) Stochastic emergence of repeating cortical motifs in spontaneous membrane potential fluctuations in vivo. Neuron 53: 413-425

Monier C, Chavane F, Baudot P, LJ and Frégnac Y (2003) Orientation and direction selectivity of synaptic inputs in visual cortical neurons: a diversity of combinations produces spike tuning. Neuron 37: 663-680

Monier C, Fournier J and Frégnac Y (2008) In vitro and in vivo measures of evoked excitatory and inhibitory conductance dynamics in sensory cortices. J Neurosci Methods 169: 323-365

Moreno-Bote R and Parga N (2005) Membrane potential and response properties of populations of cortical neurons in the high conductance state. Phys Rev Lett 94: 088103

Moreno-Bote R, de la Rocha J, Renart A and Parga N (2002) Response of spiking neurons to correlated inputs. Phys Rev Lett 89: 288101

Morita K, Kalra R, Aihara K and Robinson HP (2008) Recurrent synaptic input and the timing of gamma-frequency-modulated firing of pyramidal cells during neocortical "UP" states. J Neurosci 28: 1871-1881

Morrow TJ and Casey KL (1992) State-related modulation of thalamic somatosensory responses in the awake monkey. J Neurophysiol 67: 305-317

Mortensen RE (1969) Mathematical problems of modeling stochastic nonlinear dynamical systems. J Stat Phys 1: 271-296

Mountcastle VB (1979) An organizing principle for cerebral function: the unit module and the distributed system. In: The Neurosciences: Fourth Study Program. Schmidt FO and Worden FG (eds). MIT Press, Cambridge, pp. 21-42

Mungai JM (1967) Dendritic patterns in the somatic sensory cortex of the cat. J Anat 101: 403-418

Murakami M, Kashiwadani H, Kirino Y and Mori K (2005) State-dependent sensory gating in olfactory cortex. Neuron 21: 285-296

Murthy VN and Fetz EE (1994) Effects of input synchrony on the firing rate of a three-compartment cortical neuron model. Neural Comp 6: 1111-1126

Nagumo J, Arimoto S and Yoshizawa S (1962) An active pulse transmission line simulating nerve axon. Proc IRE 50: 2061-2070

Needleman DJ, Tisienga PHE and Sejnowski TJ (2001) Collective enhancement of precision in networks of coupled oscillators. Physica D 155: 324-336

Neiman A, Schimansky-Geier L, Moss F, Shulgin B and Collins JJ (1999) Synchronization of noisy systems by stochastic signals. Phys Rev E 60: 284-292

Nelson E (1967) Dynamical Theories of Brownian Motion. Princeton University Press

Neumcke B (1978) 1/f noise in membranes. Biophys Struct Mech 4: 179-199

Noda H and Adey R (1970) Firing variability in cat association cortex during sleep and wakefulness. Brain Res 18: 513-526

Nowak L, Bregestovski P, Ascher P, Herbet A and Prochiantz A (1984) Magnesium gates glutamate-activated channels in mouse central neurones. Nature 307: 462-465

Nowak LG, Sanchez-Vives MV and McCormick DA (1997) Influence of low and high frequency inputs on spike timing in visual cortical neurons. Cereb Cortex 7: 487-501

Nozaki D, Mar DJ, Grigg P and Collins JJ (1999) Effects of colored noise on stochastic resonance in sensory neurons. Phys Rev Lett 82: 2402-2405

Nyquist H (1928) Thermal Agitation of Electric Charge in Conductors. Phys Rev 32: 110-113

Offner F, Weinberg A and Young G (1940) Nerve conduction theory: some mathematical consequence of Bernstein's model. Bull Math Biophys 2: 89-103

Okun M and Lampl I (2008) Instantaneous correlation of excitation and inhibition during ongoing and sensory-evoked activities. Nature Neurosci 11: 535-537

Otis TS and Mody I (1992a) Modulation of decay kinetics and frequency of GABA(A) receptor-mediated spontaneous inhibitory postsynaptic currents in hippocampal neurons. Neuroscience 49: 13-32

Otis TS and Mody I (1992b) Differential activation of GABA(A) and GABA(B) receptors by spontaneous transmitter release. J Neurophysiology 67: 227-35

Otis TS, De Koninck Y and Mody I (1992) Whole-cell recordings of evoked and spontaneous GABAB responses in hippocampal slices. Pharmacol Commun 2: 75-83

Otis TS, De Koninck Y and Mody I (1993) Characterization of synaptically elicited $GABA_B$ responses using patch-clamp recordings in rat hippocampal slices. J Physiol 463: 391-407

Panzeri S and Schultz SR (2001) A unified approach to the study of temporal, correlational, and rate coding. Neural Comp 13: 1311-1349

Panzeri S, Schultz S R, Treves A and Rolls ET (1999) Correlations and the encoding of information in the nervous system. Proc R Soc Lond B Biol Sci 266: 1001-1012

Panzeri S, Petersen RS, Schultz SR, Lebedev M and Diamond ME (2001a) The role of spike timing in the coding of stimulus location in rat somatosensory cortex. Neuron 29: 769-777

Papoulis A (1991) Probability, random variables, and stochastic processes, McGraw-Hill, Boston

Paré D, LeBel E and Lang EJ (1997) Differential impact of miniature synaptic potentials on the soma and dendrites of pyramidal neurons in vivo. J Neurophysiol 78: 1735-1739

Paré D, Lang EJ and Destexhe A (1998a) Inhibitory control of somatic and dendritic sodium spikes in neocortical pyramidal neurons in vivo an intracellular and computational study. Neuroscience 84: 377-402

Paré D, Shink E, Gaudreau H, Destexhe A and Lang EJ (1998b) Impact of spontaneous synaptic activity on the resting properties of cat neocortical neurons in vivo. J Neurophysiol 79: 1450-1460

Paz JT, Chavez M, Saillet S, Deniau JM and Charpier S (2007) Activity of ventral medial thalamic neurons during absence seizures and modulation of cortical paroxysms by the nigrothalamic pathway. J Neurosci 27: 929-941

Pei X, Volgushev M, Vidyasagar TR and Creutzfeldt OD (1991) Whole-cell recording and conductance measurements in cat visual cortex in vivo. NeuroReport 2: 485-488

Pei X, Wilkens AL and Moss F (1996) Light enhances hydrodynamic signaling in the multimodal caudal photoreceptor interneurons of the crayfish. J Neurophysiol 76: 3002-3011

Perez-Reyes E (2003) Molecular physiology of low-voltage-activated t-type calcium channels. Physiol Rev 83: 117-161

Peters A and Kaiserman-Abramof IR (1970) The small pyramidal neuron of the rat cerebral cortex. The perikaryon, dendrites and spines. Am J Anat 127: 321-356

Petersen CC, Hahn TT, Mehta M, Grinvald A, Sakmann B (2003) Interaction of sensory responses with spontaneous depolarization in layer 2/3 barrel cortex. Proc Natl Acad Sci USA 100: 13638-13643

Pfeiffer RR and Kim DO (1975) Cochlear nerve fiber responses: Distribution along the cochlear partition. J Acoust Soc Am 58: 867-869

Pietsch P (1981) Shufflebrain: The Quest for the Hologramic Mind. Houghton-Mifflin

Pinsky PF and Rinzel J (1994) Intrinsic and network rhythmogenesis in a reduced Traub model for CA3 neurons. J Comput Neurosci 1: 39-60

Piwkowska Z (2007) Real-time Interactions between Cortical Neurons and Computational Models: Synaptic Conductance Analysis and Digital Compensation of Electrode Artifacts. PhD Thesis, Université Pierre et Marie Curie, Paris

Piwkowska Z, Pospischil M, Brette R, Sliwa J, Rudolph-Lilith M, Bal T and Destexhe A (2008) Characterizing synaptic conductance fluctuations in cortical neurons and their influence on spike generation. J Neurosci Methods 169: 302-322

Piwkowska Z, Bal T and Destexhe A (2009) An introduction to the dynamic-clamp electrophysiological technique and its applications. In: Dynamic-clamp: From Principles to Applications. Destexhe A and Bal T (eds). Springer, New York, pp. 1-30

Planck M (1917) Über einen Satz der statistischen Dynamik und eine Erweiterung in der Quantentheorie. Sitzungsbericht der Preussischen Akademie der Wissenschaften: 324-341

Pongracz F, Firestein S and Shepherd GM (1991) Electrotonic structure of olfactory sensory neurons analyzed by intracellular and whole cell patch techniques. J Neurophysiol 65: 747-758

Pospischil M, Piwkowska Z, Rudolph M, Bal T and Destexhe A (2007) Calculating event-triggered average synaptic conductances from the membrane potential. J Neurophysiol 97: 2544-2552

Pospischil M, Piwkowska Z, Bal T and Destexhe A (2009) Extracting synaptic conductances from single membrane potential traces. Neuroscience 158: 545-552

Poussart DJM (1971) Membrane current noise in lobster axon under voltage clamp. Biophys J 11: 211-234

Prescott SA and De Koninck Y (2003) Gain control of firing rate by shunting inhibition: roles of synaptic noise and dendritic saturation. Proc Natl Acad Sci USA 100: 2076-2081

Prescott SA, Ratté S, De Koninck Y and Sejnowski TJ (2006) Nonlinear interaction between shunting and adaptation controls a switch between integration and coincidence detection in pyramidal neurons. J Neurosci 26: 9084-9097

Press WH, Teukolsky SA, Vetterling WT and Flannery BP (1993) Numerical recipes in C: The art of scientific computing. 2nd Edition. Cambridge University Press, New York

Press WH, Flannery BP, Teukolsky SA and Vetterling WT (2007) Numerical Recipes in C: The art of scientific computing. Cambridge University Press, Cambridge

Pribram KH (1987) The Implicate Brain. In: Hiley BJ and Peat FD (eds) Quantum Implications: Essays in Honour of David Bohm. Routledge

Priebe NJ and Ferster D (2005) Direction selectivity of excitation and inhibition in simple cells of the cat primary visual cortex. Neuron 45: 133-145

Prinz AA (2004) Neural networks: models and neurons show hybrid vigor in real time. Curr Biol 14: R661-662

Prinz AA, Abbott LF and Marder E (2004) The dynamic clamp comes of age. Trends Neurosci 27: 218-224

Pritchard WS (1992) The brain in fractal time: 1/f-like power spectrum scaling of the human electroencephalogram. Int J Neurosci 66: 119-129

Prut Y, Vaadia E, Bergman H, Haalman I, Slovin H and Abeles M (1998) Spatiotemporal structure of cortical activity: Properties and behavioral relevance. J Neurophysiol 79: 2857-2874

Purves RD (1981) Microelectrode Methods for Intracellular Recording and Ionophoresis. Academic Press, London

Rall W (1959) Branching dendritic trees and motoneuron membrane resistivity. Exp Neurol 1: 491-527

Rall W (1962) Theory of physiological properties of dendrites. Ann NY Acad Sci 96: 1071-1092

Rall W (1964) Theoretical significance of dendritic trees for neuronal input-output relations. In: Neural Theory and Modeling. Reiss RF (ed). Stanford University Press

Rall W (1967) Distinguishing theoretical synaptic potentials computed for different soma-dendritic distributions of synaptic inputs. J Neurophysiol 30: 1138-1168

Rall W (1995) The Theoretical Foundation of Dendritic Function. Segev I, Rinzel J, Shepherd GM (eds). MIT Press, Cambridge

Raman IM, Zhang S and Trussell LO (1994) Pathway-specific variants of AMPA receptors and their contribution to neuronal signaling. J Neurosci 14: 4998-5010

Ramcharan EJ, Gnadt JW and Sherman SM (2000) Burst and tonic firing in thalamic cells of unanesthetized, behaving monkeys. Vis Neurosci 17: 55-62

Ramón y Cajal S (1909) Histologie du Système Nerveux de l'Homme et des Vertébrés. (translated by Azoulay L) Maloine, Paris

Rao RPN (2004) Bayesian computation in recurrent neural circuits. Neural Computation 16: 1-38

Rao RPN, Olshausen B and Lewicki M (eds) (2002) Probabilistic Models of the Brain. Cambridge: MIT Press

Rapp M, Yarom Y and Segev I (1992) The impact of parallel fiber background activity on the cable properties of cerebellar Purkinje cells. Neural Computation 4: 518-533

Redman S (1990) Quantal analysis of excitatory synaptic potentials in neurons of the central nervous system. Physiol Rev 70: 165-198

Regehr WG and Stevens CF (2001) Physiology of synaptic transmission and short-term plasticity. In: Synapses. Cowan WM, Südhof TC and Stevens CF (eds). The John Hopkins University Press, Baltimore: pp. 135-175

Reijneveld JC, Ponten SC, Berendse HW and Stam CJ (2007) The application of graph theoretical analysis to complex networks in the brain. Clin Neurophysiol 118: 2317-2331

Reinagel P and Reid RC (2000) Temporal coding of visual information in the thalamus. J Neurosci 20: 5392-5400

Reyes AD (2003) Synchrony-dependent propagation of firing rate in iteratively constructed networks in vitro. Nature Neurosci 6: 593-599

Reyes AD, Rubel EW and Spain WJ (1996) In vitro analysis of optimal stimuli for phase-locking and time-delayed modulation of firing in avian nucleus laminaris neurons. J Neurosci 16: 993-1007

Rhodes PA and Llinás RR (2001) Apical tuft input efficacy in layer 5 pyramidal cells from rat visual cortex. J Physiol 536: 167-187

Ricciardi LM and Sacerdote L (1979) The Ornstein–Uhlenbeck process as a model for neuronal activity. I. Mean and variance of the firing time. Biol Cybern 35: 1-9

Rice SO (1944) Mathematical analysis of random noise. Bell Syst Tech J 23: 282-332

Rice SO (1945) Mathematical analysis of random noise. Bell Syst Tech J 24: 46-156

Richardson M (2004) The effects of synaptic conductances on the voltage distribution and firing rate of spiking neurons. Phys Rev E 69: 051918

Richardson KA, Imhoff TT, Grigg P and Collins JJ (1998) Using electrical noise to enhance the ability of humans to detect subthreshold mechanical cutaneous stimuli. Chaos 8: 599-603

Risken H (1984) The Fokker Planck Equation: Methods of Solution and Application. Springer-Verlag, Berlin

Robinson HPC (2008) A scriptable DSP-based system for dynamic conductance injection. J Neurosci Meth 169: 271-281

Robinson HP and Kawai N (1993) Injection of digitally synthesized synaptic conductance transients to measure the integrative properties of neurons. J Neurosci Methods 49: 157-165

Roelfsema MR, Steinmeyer R and Hedrich R (2001) Discontinuous single electrode voltage-clamp measurements: assessment of clamp accuracy in Vicia faba guard cells. J Exp Bot 52: 1933-1939

Ropert N, Miles R and Korn H (1990) Characteristics of miniature inhibitory postsynaptic currents in CA1 pyramidal neurones of rat hippocampus. J Physiol 428: 707-722

Rose JE, Brugge JF, Anderson DJ and Hind JE (1967) Phase-locked response to low-frequency tones in single auditory nerve fibers of the squirrel monkey. J Neurophysiol 30: 769-793

Rudolph M and Destexhe A (2001a) Correlation detection and resonance in neural systems with distributed noise sources. Phys Rev Lett 86: 3662-3665
Rudolph M and Destexhe A (2001b) Do neocortical pyramidal neurons display stochastic resonance? J Computational Neurosci 11: 19-42
Rudolph M and Destexhe A (2003a) The discharge variability of neocortical neurons during high-conductance states. Neuroscience 119: 855-873
Rudolph M and Destexhe A (2003b) A fast-conducting, stochastic integrative mode for neocortical neurons in vivo. J Neurosci 23: 2466-2476
Rudolph M and Destexhe A (2003c) Tuning neocortical pyramidal neurons between integrators and coincidence detectors. J Computational Neurosci 14: 239-251
Rudolph M and Destexhe A (2003d) Characterization of subthreshold voltage fluctuations in neuronal membranes. Neural Computation 15: 2577-2618
Rudolph M and Destexhe A (2003e) Gain modulation and frequency locking under conductance noise. Neurocomputing 52: 907-912
Rudolph M and Destexhe A (2004) Inferring network activity from synaptic noise. J Physiol (Paris) 98: 452-466
Rudolph M and Destexhe A (2005) An extended analytic expression for the membrane potential distribution of conductance-based synaptic noise. Neural Computation 17: 2301-2315
Rudolph M and Destexhe A (2006a) A multichannel shot noise approach to describe synaptic background activity in neurons. Eur Physical Journal B 52: 125-132
Rudolph M and Destexhe A (2006b) On the use of analytic expressions for the voltage distribution to analyze intracellular recordings. Neural Computation 18: 2917-2922
Rudolph M and Destexhe A (2006) Integrate-and-fire neurons with high-conductance state dynamics for event-driven simulation strategies. Neural Computation 18: 2146-2210
Rudolph M and Destexhe A (2007) How much can we trust neural simulation strategies? Neurocomputing 70: 1966-1969
Rudolph M, Hô N and Destexhe A (2001) Synaptic background activity affects the dynamics of dendritic integration in model neocortical pyramidal neurons. Neurocomp 38-40: 327-333
Rudolph M, Piwkowska Z, Badoual M, Bal T and Destexhe A (2004) A method to estimate synaptic conductances from membrane potential fluctuations. J Neurophysiol 91: 2884-2896
Rudolph M, Pelletier J-G, Paré D and Destexhe A (2005) Characterization of synaptic conductances and integrative properties during electrically-induced EEG-activated states in neocortical neurons in vivo. J Neurophysiol 94: 2805-2821
Rudolph M, Pospischil M, Timofeev I and Destexhe A (2007) Inhibition determines membrane potential dynamics and controls action potential generation in awake and sleeping cat cortex. J Neurosci 27: 5280-5290
Rushton WAH (1950) A theory of the effects of fibre size in medullated nerve. J Physiol 115: 101-122
Sadoc G, Le Masson G, Foutry B, Le Franc Y, Piwkowska Z, Destexhe A and Bal T (2009) Recreating in vivo–like activity and investigating the signal transfer capabilities of neurons: Dynamic-clamp applications using real-time NEURON. In: Dynamic-clamp: From Principles to Applications, Destexhe A and Bal T (eds). Springer, New York, pp. 287-320
Sakai Y, Funahashi S and Shinomoto S (1999) Temporally correlated inputs to leaky integrate-and-fire models can reproduce spiking statistics of cortical neurons. Neural Networks 12: 1181-1190
Sakmann B and Neher E (eds) (1995) Single-Channel Recording (2nd edition). Plenum Press, New York, NY
Salin PA and Prince DA (1996) Spontaneous $GABA_A$ receptor-mediated inhibitory currents in adult rat somatosensory cortex. J Neurophysiol 75: 1573-1588
Salinas E and Sejnowski TJ (2000) Impact of correlated synaptic input on output firing rate and variability in simple neuronal models. J Neurosci 20: 6193-6209
Salinas E and Sejnowski TJ (2001) Correlated neuronal activity and the flow of neural information. Nature Rev Neurosci 2: 539-550
Sanchez-Vives MV and McCormick DA (2000) Cellular and network mechanisms of rhythmic recurrent activity in neocortex. Nature Neurosci 10: 1027-1034

Scott S (1979) Stimulation Simulations of Young Yet Cultured Beating Hearts. PhD Thesis, Buffalo, NY: State University of New York

Segev I and Rall W (1998) Excitable dendrites and spines: earlier theoretical insights elucidate recent direct observations. Trends Neurosci 21: 453-460

Segev I, Rinzel J and Shepherd GM (1995) The theoretical foundation of dendritic function: Selected papers of Wilfrid Rall with commentaries. Cambridge, MIT Press

Segundo JP (2000) Some thoughts about neural coding and spike trains. Biosystems 58: 3-7

Segundo JP, Perkel DH, Moore GP (1966) Spike probability in neurones: Influence of temporal structure in the train of synaptic events. Kybernetik 3: 67-82

Shadlen MN and Newsome WT (1994) Noise, neural codes and cortical organization. Curr Opin Neurobiol 4: 569-579

Shadlen MN and Newsome WT (1995) Is there a signal in the noise? Curr Opin Neurobiol 5: 248-250

Shadlen MN and Newsome WT (1998) The variable discharge of cortical neurons: implications for connectivity, computation, and information coding. J Neurosci 18: 3870-3896

Sharp AA, O'Neil MB, Abbott LF and Marder E (1993a) The dynamic clamp: artificial conductances in biological neurons. Trends Neurosci 16: 389-394

Sharp AA, O'Neil MB, Abbott LF and Marder E (1993b) Dynamic clamp: computer-generated conductances in real neurons. J Neurophysiol 69: 992-995

Sherman SM and Guillery RW (2002) The role of the thalamus in the flow of information to the cortex. Phil Trans Roy Soc Lond Ser B 357: 1695-708

Shinomoto S, Sakai Y and Funahashi S (1999) The Ornstein–Uhlenbeck process does not reproduce spiking statistics of neurons in prefrontal cortex. Neural Comput 11: 935-951

Siebenga E and Verveen AA (1972) Membrane noise and ion transport in the node of Ranvier. Biomembranes 3: 473-482

Sherman SM and Guillery RW (2001) Exploring the Thalamus. Academic Press, New York

Shimokawa T, Rogel A, Pakdaman K and Sato S (1999) Stochastic resonance and spike-timing precision in an ensemble of leaky integrate and fire neuron models. Phys Rev E 59: 3461-3470

Shu Y, Hasenstaub A and McCormick DA (2003a) Turning on and off recurrent balanced cortical activity. Nature 423: 288-293

Shu Y, Hasenstaub A, Badoual M, Bal T and McCormick DA (2003b) Barrages of synaptic activity control the gain and sensitivity of cortical neurons. J Neurosci 23: 10388-10401

Shu Y, Hasenstaub A, Duque A, Yu Y and McCormick DA (2006) Modulation of intracortical synaptic potentials by presynaptic somatic membrane potential. Nature 441: 761-765

Silberberg G, Gupta A and Markram H (2002) Stereotypy in neocortical microcircuits. Trends Neurosci 25: 227-230

Silberberg G, Bethge M, Markram H, Pawelzik K and Tsodyks M (2004) Dynamics of population rate codes in ensembles of neocortical neurons. J Neurophysiol 91: 704-709

Sillito AM and Jones HE (2002) Corticothalamic interactions in the transfer of visual information. Phil Trans Roy Soc Lond Ser B 357: 1739-1752

Simonotto E, Riani M, Seife C, Roberts M, Twitty J and Moss F (1997) Visual perception of stochastic resonance. Phys Rev Lett 78: 1186-1189

Smith DR and Smith GK (1965) A statistical analysis of the continuous activity of single cortical neurons in the cat unanesthetized isolated forebrain. Biophys J 5: 47-74

Smith C and Wise MN (1989) Energy and Empire: A Biographical Study of Lord Kelvin. Cambridge University Press, Cambridge

Softky WR (1994) Sub-millisecond coincidence detection in active dendritic trees. Neuroscience 58: 13-41

Softky WR (1995) Simple codes versus efficient codes. Curr Opin Neurobiol 5: 239-247

Softky WR and Koch C (1993) The highly irregular firing of cortical cells is inconsistent with temporal integration of random EPSPs. J Neurosci 13: 334-350

Spain WJ, Schwindt PC and Crill WE (1987) Anomalous rectification in neurons from cat sensorimotor cortex in vitro. J Neurophysiol 57: 1555-1576

Spencer WA and Kandel ER (1961) Electrophysiology of hippocampal neurons. IV. Fast prepotentials. J Neurophysiol 24: 272-285

Spitzer H, Desimone R and Moran J (1988) Increased attention enhances both behavioral and neuronal performance. Science 240: 338-340

Spruston N and Johnston D (1992) Perforated patch-clamp analysis of the passive membrane properties of three classes of hippocampal neurons. J Neurophysiol 67: 508-529

Spruston N, Schiller Y, Stuart G and Sakmann B (1995) Activity-dependent action potential invasion and calcium influx into hippocampal CA1 dendrites. Science 268: 297-300

Srebo R and Malladi P (1999) Stochastic resonance of the visually evoked potential. Phys Rev E 59: 2566-2570

Srinivasan R and Chiel HJ (1993) Fast calculation of synaptic conductances. Neural Comp 5: 200-204

Stacey WC and Durand DM (2000) Stochastic Resonance Improves Signal Detection in Hippocampal CA1 Neurons. J Neurophysiol 83: 1394-1402

Stacey WC and Durand DM (2001) Synaptic noise improves detection of subthreshold signals in hippocampal CA1 neurons. J Neurophysiol 86: 1104-1112

Stafstrom CE, Schwindt PC and Crill WE (1982) Negative slope conductance due to a persistent subthreshold sodium current in cat neocortical neurons in vitro. J Brain Res 236: 221-226

Standley C, Ramsey RL and Usherwood PNR (1993) Gating kinetics of the quisqualate-sensitive glutamate receptor of locust muscle studied using agonist concentration jumps and computer simulations. Biophys J 65: 1379-1386

Stein RB (1967) Some Models of Neuronal Variability. Biophys J 7: 37-68

Steriade M (1969) Alteration of motor and somesthetic thalamo-cortical responsiveness during wakefulness and sleep. Electroencephalogr Clin Neurophysiol 26: 334

Steriade M (1974) Interneuronal epileptic discharges related to spike-and-wave cortical seizures in behaving monkeys. Electroencephalogr Clin Neurophysiol 37: 247-263

Steriade M (1978) Cortical long-axoned cells and putative interneurons during the sleep-waking cycle. Behav Brain Sci 3: 465-514

Steriade M (2000) Corticothalamic resonance, states of vigilance and mentation. Neuroscience 101: 243-276

Steriade M (2001) Impact of network activities on neuronal properties in corticothalamic systems. J Neurophysiol 86: 1-39

Steriade M (2003) Neuronal Substrates of Sleep and Epilepsy. Cambridge University Press, Cambridge, UK

Steriade M and Deschênes M (1984) The thalamus as a neuronal oscillator. Brain Res Reviews 8: 1-63

Steriade M, Iosif G and Apostol V (1969) Responsiveness of thalamic and cortical motor relays during arousal and various stages of sleep. J Neurophysiol 32: 251-265

Steriade M, Deschênes M and Oakson G (1974) Inhibitory processes and interneuronal apparatus in motor cortex during sleep and waking. I. Background firing and responsiveness of pyramidal tract neurons and interneurons. J Neurophysiol 37: 1065-1092

Steriade M, Amzica F and Nuñez A (1993a) Cholinergic and noradrenergic modulation of the slow ($\sim$0.3 Hz) oscillation in neocortical cells. J Neurophysiol 70: 1384-1400

Steriade M, Nuñez A and Amzica F (1993b) A novel slow (<1 Hz) oscillation of neocortical neurons in vivo depolarizing and hyperpolarizing components. J Neurosci 13: 3252-3265

Steriade M, Timofeev I and Grenier F (2001) Natural waking and sleep states: a view from inside neocortical neurons. J Neurophysiol 85: 1969-1985

Stern P, Edwards FA and Sakmann B (1992) Fast and slow components of unitary EPSCs on stellate cells elicited by focal stimulation in slices of rat visual cortex. J Physiol 449: 247-278

Stevens CF (1978) Interactions between intrinsic membrane protein and electric field. Biophys J 22: 295-306

Stevens CF and Wang Y (1995) Facilitation and depression at single central synapses. Neuron 14: 795-802

Stevens CF and Zador AM (1998) Input synchrony and the irregular firing of cortical neurons. Nature Neurosci 1: 210-217

Stocks NG and Mannella R (2001) Generic noise-enhanced coding in neuronal arrays. Phys Rev E 64: 030902

Stocks NG, Stein ND and McClintock PVE (1993) Stochastic resonance in monostable systems. J Physiol 26: L385-L390

Stratford K, Mason A, Larkman A, Major G and Jack J (1989) The modeling of pyramidal neurones in the visual cortex. In: The Computing Neuron. Durbin A, Miall C, Mitchison G (eds). Addison-Wesley, Workingham, UK, pp. 296-321

Stricker C, Field AC and Redman SJ (1996) Statistical analysis of amplitude fluctuations in EPSCs evoked in rat CA1 pyramidal neurones in vitro. J Physiol 490: 419-441

Stuart GJ and Sakmann B (1994) Active propagation of somatic action potentials into neocortical pyramidal cell dendrites. Nature 367: 69-72

Stuart G and Spruston N (1998) Determinants of voltage attenuation in neocortical pyramidal neuron dendrites. J Neurosci 18: 3501-3510

Stuart G, Dodt HU and Sakmann B (1993) Patch clamp recording from the soma and dendrites of neurons in brain slices using infrared video microscopy. Pflügers Arch 423: 511-518

Stuart G, Schiller J and Sakmann B (1997a) Action potential initiation and propagation in rat neocortical pyramidal neurons. J Physiol 505: 617-632

Stuart GJ, Spruston N, Sakmann B, Häusser M (1997b) Action potential initiation and backpropagation in neurons of the mammalian CNS. Trends Neurosci 20: 125-131

Suter KJ and Jaeger D (2004). Reliable control of spike rate and spike timing by rapid input transients in cerebellar stellate cells. Neuroscience 124: 305-317

Svirskis G and Rinzel J (2000) Influence of temporal correlation of synaptic input on the rate and variability of firing in neurons. Biophys J 79: 629-637

Svoboda K, Denk W, Kleinfeld D and Tank DW (1997) In vivo dendritic calcium dynamics in neocortical pyramidal neurons. Nature 385: 161-165

Szentagothai J (1965) The use of degeneration in the investigation of short neuronal connections. In: Progress in Brain Research, Vol. 14, Singer M and Shade JP (eds). Elsevier, Amsterdam, pp. 1-32

Szentagothai J (1983) The modular architectonic principle of neural centers. Rev Physiol Biochem Pharmacol 98: 11-61

Tan RC and Joyner RW (1990) Electrotonic influences on action potentials from isolated ventricular cells. Circ Res 67: 1071-1081

Tang A (1997) Effects of cholinergic modulation on responses of neocortical neurons to fluctuating input. Cereb Cortex 7: 502-509

Tasaki I and Matsumoto G (2002) On the Cable Theory of Nerve Conduction. Bull Math Biol 64: 1069-1082

Tass P, Rosenblum MG, Weule J, Kurths J, Pikovsky A, Volkmann J, Schnitzler A and Freund H-J (1998) Detection of n:m phase locking from noisy data: Application to magnetoencephalography. Phys Rev Lett 81: 3291-3294

Tateno T and Robinson HP (2006) Rate coding and spike-time variability in cortical neurons with two types of threshold dynamics. J Neurophysiol 95: 2650-2663

Theunissen F and Miller JP (1995) Temporal encoding in nervous systems: a rigorous definition. J Comput Neurosci 2: 149-162

Thomas MV (1977) Microelectrode amplifier with improved method of input-capacitance neutralization. Med Biol Eng Comput 15: 450-454

Thompson SM (1994) Modulation of inhibitory synaptic transmission in the hippocampus. Progress Neurobiol 42: 575-609

Thompson SM and Gähwiler BH (1992) Effects of the GABA uptake inhibitor tiagabine on inhibitory synaptic potentials in rat hippocampal slice cultures. J Neurophysiol 67: 1698-1701

Thomson AM (2000) Facilitation, augmentation and potentiation at central synapses. Trends Neurosci 23: 305-312

Thomson AM and Destexhe A (1999) Dual intracellular recordings and computational models of slow IPSPs in rat neocortical and hippocampal slices. Neuroscience 92: 1193-1215

Thomson AM and Deuchars J (1994) Temporal and spatial properties of local circuits in neocortex. Trends Neurosci 17: 119-126

Thomson AM and Deuchars J (1997) Synaptic interactions in neocortical local circuits: Dual intracellular recordings in vitro. Cerebral Cortex 6: 510-522

Thorpe S, Fize D and Marlot C (1996) Speed of processing in the human visual system. Nature 381: 520-522

Tiesinga PHE and José JV (1999) Spiking statistics in noisy hippocampal interneurons. Neurocomp 26-27: 299-304

Tiesinga PHE, José JV and Sejnowski TJ (2000) Comparison of current-driven and conductance-driven neocortical model neurons with Hodgkin-Huxley voltage-gated channels. Phys Rev E 62: 8413-8419

Timofeev I, Grenier F, Bazhenov M, Sejnowski TJ and Steriade M (2000) Origin of slow cortical oscillations in deafferented cortical slabs. Cereb Cortex 10: 1185-1199

Timofeev I, Grenier F and Steriade M (2001) Disfacilitation and active inhibition in the neocortex during the natural sleep-wake cycle: an intracellular study. Proc Natl Acad Sci USA 98: 1924-1929

Tolhurst DJ, Movshon JA and Dean AF (1983) The statistical reliability of signals in single neurons in cat and monkey visual cortex. Vision Res 23: 775-785

Tovée MJ, Rolls ET, Treves A and Bellis RP (1993) Information encoding and the responses of single neurons in the primate temporal visual cortex. J Neurophysiol 70: 640-654

Traub RD and Miles R (1991) Neuronal Networks of the Hippocampus. Cambridge University Press, Cambridge

Troy JB and Robson JG (1992) Steady discharges of X and Y retinal ganglion cells of cat under photopic illuminance. Vis Neurosci 9: 535-553

Troyer TW and Miller KD (1997) Physiological gain leads to high ISI variability in a simple model of a cortical regular spiking cell. Neural Comp 9: 971-983

Tsodyks M and Sejnowski TJ (1995) Rapid state switching in balanced cortical network models. Network: Computation and Neural Systems 6: 111-124

Tsodyks M, Pawelzik K and Markram H (1998) Neural networks with dynamic synapses. Neural Computation 10: 821-835

Tsodyks M, Kenet T, Grinvald A and Arieli A (1999) Linking spontaneous activity of single cortical neurons and the underlying functional architecture. Science 286: 1943-1946

Tuckwell HC (1988) Introduction to Theoretical Neurobiology. Cambridge Univiversity Press, Cambridge

Tuckwell HC and Walsh JB (1984) Random currents through nerve membranes. I. Uniform poisson or white noise current in one-dimensional cables. Biol Cybern 49: 99-110

Tuckwell HC, Wan FYM and Wong YS (1984) The interspike interval of a cable model neuron with white noise input. Biol Cybern 49: 155-167

Tuckwell HC, Wan FYM and Rospars J-P (2002) A spatial stochastic neuronal model with Ornstein–Uhlenbeck input current. Biol Cybern 86: 137-145

Turner JP, Leresche N, Guyon A, Soltesz I and Crunelli V (1994) Sensory input and burst firing output of rat and cat thalamocortical cells: the role of NMDA and non-NMDA receptors. J Physiol 480: 281-295

Uhlenbeck GE and Ornstein LS (1930) On the theory of the Brownian motion. Phys Rev 36: 823-841

Usher M, Stemmler M, Koch C and Olami Z (1994) Network amplification of local fluctuations causes high spike rate variability, fractal firing patterns and oscillatory local field potentials. Neural Comp 6: 795-836

Vaadia E, Haalman I, Abeles M, Bergman H, Prut Y, Slovin H and Aertsen A (1995) Dynamics of neuronal interactions in monkey cortex in relation to behavioural events. Nature 373: 515-518

Van den Broeck C, Parrondo JMR, Armero J and Hernández-Machado A (1994) Mean field model for spatially extended systems in the presence of multiplicative noise. Phys Rev E 49: 2639-2643

Van Horn SC, Erisir A and Sherman SM (2000) Relative distribution of synapses in the A-laminae of the lateral geniculate nucleus of the cat. J Comp Neurol 416: 509-520

van Kampen NG (1981) Stochastic Processes in Physics and Chemistry. North Holland, Amsterdam

van Rossum MCW (2001) The transient precision of integrate and fire neurons: Effect of background activity and noise. J Comp Neurosci 10: 303-311

van Rossum MC, Turrigiano GG and Nelson SB (2002) Fast propagation of firing rates through layered networks of noisy neurons. J Neurosci 22: 1956-1966

van Vreeswijk C and Sompolinsky H (1996) Chaos in neuronal networks with balanced excitatory and inhibitory activity. Science 274: 1724-1726

Vastano JA and Swinney HL (1988) Information transport in spatiotemporal systems. Phys Rev Lett 60: 1773- 1776

Verheijck EE, Wilders R, Joyner RW, Golod DA, Kumar R, Jongsma HJ, Bouman LN and van Ginneken AC (1998) Pacemaker synchronization of electrically coupled rabbit sinoatrial node cells. J Gen Physiol 111: 95-112

Verveen AA and De Felice LJ (1974) Membrane Noise. Prog Biophys Mol Biol 28: 189-265

Verveen AA and Derksen HE (1968) Fluctuation phenomena in nerve membrane. Proc IEEE 56: 906-916

Vetter P, Roth A and Häusser M (2001) Propagation of action potentials in dendrites depends on dendritic morphology. J Neurophysiol 85: 926-937

Vogels TP and Abbott LF (2005) Signal propagation and logic gating in networks of integrate-and-fire neurons. J Neurosci 15: 10786-10795

Wang MC and Uhlenbeck GE (1945) On the Theory of Brownian Motion II. Rev Mod Phys 17: 323-342

Wang W, Wang Y and Wang ZD (1998) Firing and signal transduction associated with an intrinsic oscillation in neuronal systems. Phys Rev E 57: R2527-R2530

Waters J and Helmchen F (2006) Background synaptic activity is sparse in neocortex. J Neurosci 26: 8267-8277

Wehr M and Zador AM (2003) Balanced inhibition underlies tuning and sharpens spike timing in auditory cortex. Nature 426: 442-446

Werner MJ and Drummond PD (1997) Robust algorithms for solving stochastic partial differential equations. J Comp Phys 132: 312-326

Weyand TG, Boudreaux M and Guido W (2001) Burst and tonic response modes in thalamic neurons during sleep and wakefulness. J Neurophysiol 85: 1107-1118

White EL (1989) Cortical circuits. Birkhauser, Boston, MA

White JA, Rubinstein JT and Kay AR (2000) Channel noise in neurons. Trends Neurosci 23: 131-137

Wiesenfeld K, Pierson D, Pantazalou E, Dames C and Moss F (1994) Stochastic resonance on a circle. Phys Rev Lett 72: 2125-2129

Wiesenfeld K and Moss F (1995) Stochastic resonance and the benefits of noise: from ice ages to crayfish and SQUIDS. Nature 373: 33-36

Wilders R, Verheijck EE, Kumar R, Goolsby WN, van Ginneken AC, Joyner RW and Jongsma HJ (1996) Model clamp and its application to synchronization of rabbit sinoatrial node cells. Am J Physiol 271: H2168-2182

Wilent, W. B., and Contreras D (2005a) Dynamics of excitation and inhibition underlying stimulus selectivity in rat somatosensory cortex. Nature Neurosci 8: 1364-1370

Wilent, W. B., and Contreras D (2005b) Stimulus-dependent changes in spike threshold enhance feature selectivity in rat barrel cortex neurons. J Neurosci 25: 2983-2991

Wilson JR, Friedlander MJ and Sherman SM (1984) Fine structural morphology of identified X- and Y-cells in the cat's lateral geniculate nucleus. Proc Roy Soc Lond Ser B 221: 411-436

Wong RKS, Prince DA and Basbaum AI (1979) Intradendritic recordings from hippocampal neurons. Proc Natl Acad Sci USA 76: 986-990

Wolfart J, Debay D, Le Masson G, Destexhe A and Bal T (2005) Synaptic background activity controls spike transfer from thalamus to cortex. Nature Neurosci 8: 1760-1767

Woody CD and Gruen E (1978) Characterization of electrophysiological properties of intracellularly recorded neurons in the neocortex of awake cats: a comparison of the response to injected current in spike overshoot and undershoot neurons. Brain Res 158: 343-357

Worgotter F, Suder K, Zhao Y, Kerscher N, Eysel UT and Funke K (1998) State-dependent receptive-field restructuring in the visual cortex. Nature 396: 165-168

Xiang Z, Greenwood AC and Brown T (1992) Measurement and analysis of hippocampal mossy-fiber synapses. Soc Neurosci Abstracts 18: 1350

Yamada WM, Koch C and Adams PR (1989) Multiple channels and calcium dynamics. In: Methods in Neuronal Modeling. Koch C, Segev I (eds). MIT Press. Cambridge, MA

Young ED and Sachs MB (1979) Representation of steady-state vowels in the temporal aspects of the discharge patterns of populations of auditory-nerve fibers. J Acoust Soc Am 66: 1381-1403

Yuste R and Tank DW (1996) Dendritic integration in mammalian neurons, a century after Cajal. Neuron 16: 701-716

Zhang S and Trussell LO (1994) Voltage clamp analysis of excitatory synaptic transmission in the avian nucleus magnocellularis. J Physiol 480: 123-136

Zou Q, Rudolph M, Roy N, Sanchez-Vives M, Contreras D and Destexhe A (2005) Reconstructing synaptic background activity from conductance measurements in vivo. Neurocomputing 65: 673-678

Zsiros V and Hestrin S (2005) Background synaptic conductance and precision of EPSP-spike coupling at pyramidal cells. J Neurophysiol 93: 3248-3256

Index

A. Destexhe and M. Rudolph-Lilith, *Neuronal Noise*, Springer Series in Computational Neuroscience 8, DOI 10.1007/978-0-387-79020-6,